Resource Inventory & Baseline Study Methods for Developing Countries

Edited by
Francis Conant, Peter Rogers, Marion Baumgardner, Cyrus McKell, Raymond Dasmann, and Priscilla Reining

Based on a study conducted by the American Association for the Advancement of Science under the auspices of its Committee on Arid Lands and in cooperation with the U.S. National Park Service and the U.S. Agency for International Development.

American Association for the Advancement of Science

Library of Congress Cataloging in Publication Data
Main entry under title:

Resource inventory and baseline study methods for developing countries.

Bibliography: p.
Includes index.
1. Ecological surveys. 2. Environmental monitoring 3. Natural resources. 4. Ecology. 5. Ecological surveys — Developing countries. 6. Environmental monitoring — Developing countries. 7. Natural resources—Developing countries. 8. Ecology—Developing countries. I. Conant, Francis.
QH541.R455 1983   333.7'072   83-15493
ISBN 0-87168-258-3

The work upon which this publication is based was performed pursuant to contract CX-0001-1-0006 with the National Park Service with funding provided by the U.S. Agency for International Development.

The interpretations and conclusions in this report are those of the authors and do not necessarily represent the view of the Board or the Council of the American Association for the Advancement of Science or the members of the Committee on Arid Lands.

AAAS Publication No. 83-3

Printed in the United States of America

Copyright 1983 by the
American Association for the Advancement of Science
1515 Massachusetts Avenue, NW, Washington, DC 20005

CONTENTS

FIGURES xi

TABLES xiii

OVERVIEW xv

PREFACE xix

PANEL MEMBERS xxi

ACKNOWLEDGMENTS xxiii

I. INTRODUCTION AND SYNTHESIS 3
  Introduction 3
    1.1 The Ecosystem Concept 4
    1.2 Natural Resource Inventories and Baseline Studies . . . 5
    1.3 Overview of Sections 8
    1.4 Local Populations and Subsistence Systems 14
  Ecosystems and Resources in Cultural Contexts 16
    2.1 Ecosystem Structure, Functioning, and Resources . . . 17
    2.2 Cultural Ideas and Resource Use 21
    2.3 The Institutional Context 21
    2.4 Tools and Resource Exploitation 22
    2.5 Methods for Discovering Relevant Cultural Contexts . . 22
  Introduction to Integrated Approaches and Methods 24
    3.1 Integrated Inventories 25
    3.2 Sampling for Identification and Appraisal of Resources . 31
    3.3 Levels, Scales and Data Sources Required for Surveys and
        Mapping 33
    3.4 Remote Sensing 34
    3.5 Geographic Information Systems 42
  Comments on Methods, Concepts, and Domains 44
    4.1 Methods 44
    4.2 Concepts 45
    4.3 Parts, Domains and Components 48
  References Cited 49
  Suggested Readings 53

II. AQUATIC ECOSYSTEMS 57
  Introduction 57
    1.1 Aquatic Ecosystems Concept 57
    1.2 Intervention in Aquatic Ecosystems 59
    1.3 Role of Water in Health in Developing Countries . . . 60
    1.4 Defining Baseline Studies and Resource Inventories . . 63
  Aquatic Ecosystem Components 65
    2.1 Ecosystem Properties: Atmosphere, Hydrologic Cycle, Water
        Balance 65
        2.1.1 Terminology. 66
        2.1.2 Atmospheric moisture. 68
        2.1.3 Atmospheric resources and risks. 70
    2.2 Ecosystem Properties: Geomorphology of Soils and
        Watersheds 73
        2.2.1 Biogeochemical cycles. 74
    2.3 Specific Ecosystem Properties: Physical Properties . . 74
        2.3.1 Climate. 74

	2.3.2	Surface water.	78
	2.3.3	Groundwater.	79
	2.3.4	Tides and wave action.	80
	2.3.5	Salinity.	81
	2.3.6	Density-dependent stratification.	82
	2.3.7	Light and transparency.	84
2.4	Specific Ecosystem Properties--Chemical Properties		84
	2.4.1	Dissolved gases.	84
	2.4.2	Dissolved inorganic solids.	86
	2.4.3	Dissolved organic matter.	86
	2.4.4	Inorganic particulates.	86
	2.4.5	Organic particulates.	87
	2.4.6	pH.	87
	2.4.7	Conductivity.	88
2.5	Specific Ecosystem Properties--Biotic Properties		88
	2.5.1	Benthic biota.	88
	2.5.2	Zooplankton.	89
	2.5.3	Phytoplankton.	89
	2.5.4	Littoral vegetation.	91
	2.5.5	Wetlands vegetation.	91
	2.5.6	Periphyton.	92
	2.5.7	Microbiota.	93
2.6	Specific Ecosystem Properties--Functional Properties		94
	2.6.1	Nutrient cycling.	94
	2.6.2	Primary production.	94
	2.6.3	Secondary productivity.	94
	2.6.4	Eutrophication.	95
	2.6.5	Community and ecosystem descriptions.	95
	2.6.6	Water balance.	96
Sampling Methods and Selection Criteria			96
3.1	Recommendations for Baseline Studies with Resource Inventories		96
3.2	Methods for Physical Properties		100
	3.2.1	Climate.	100
	3.2.2	Surface water.	102
	3.2.3	Groundwater.	107
	3.2.4	Tide measurements, waves, and estimation of intertidal elevation.	107
	3.2.5	Current measurement.	110
	3.2.6	Temperature.	112
	3.2.7	Salinity.	113
	3.2.8	Density and stratification.	115
	3.2.9	Light and transparency.	116
3.3	Methods for Chemical Properties		116
	3.3.1	Dissolved gases.	116
	3.3.2	Dissolved inorganic solids.	118
	3.3.3	Dissolved organics.	123
	3.3.4	Inorganic particulates.	123
	3.3.5	Organic particulates.	124
	3.3.6	pH.	124
	3.3.7	Conductivity.	125
3.4	Methods for Biotic Properties		125
	3.4.1	Benthic community analysis.	125
	3.4.2	Zooplankton sampling.	133
	3.4.3	Phytoplankton.	137
	3.4.4	Fish and fisheries.	138
	3.4.5	Littoral vegetation.	142

   3.4.6 Wetlands vegetation. 143
   3.4.7 Periphyton. 144
   3.4.8 Microbiota. 144
  3.5 Methods for Functional Properties 147
   3.5.1 Nutrient cycling. 147
   3.5.2 Primary production. 148
   3.5.3 Secondary productivity. 149
   3.5.4 Eutrophication. 150
   3.5.5 Ecosystem indices. 151
   3.5.6 Water balance. 157
  3.6 Methods for Water Resource Demands 157
 Special Methods: Current Research and Expected Future Methods 159
  4.1 Introduction 159
  4.2 Remote Sensing 160
  4.3 Weather Modification 160
  4.4 Climate-Economic Modeling 162
  4.5 Aquatic Microbiology 162
 References Cited 162
 Suggested Readings 183

III. SOILS 189
 Introduction 189
  1.1 Soils as the Foundation of Food Production: Resource
   Inventories 189
  1.2 Carrying Capacity 191
  1.3 Soils and Man in the Ecosystem: Baseline Studies . . 194
  1.4 Resource Inventories and Baseline Studies: The Basis of
   Soil Management 199
  1.5 Survey of the World's Soils 200
  1.6 Soil Information Management 200
 Development for Sustainable Production: Needed Soils Information 201
  2.1 Perspective on Agriculture 201
  2.2 Categories of Soils Information 202
  2.3 Determining Soils Information Needs 204
   2.3.1 Agricultural intensification. 205
   2.3.2 Conservation. 214
   2.3.3 Specifying information needs. 215
 Classification and Evaluation of Soils and Land Resources . . 216
  3.1 Introduction: Classification and Evaluation 217
   3.1.1 Soil units for classification. 217
   3.1.2 Describing a soil profile or pedon. 218
  3.2 Soil Classification Systems 220
   3.2.1 National systems of soil classification. . . . 220
   3.2.2 International systems of soil classification. . 222
  3.3 Evaluation of Soil and Land Resources 224
   3.3.1 Introduction. 224
   3.3.2 Approaches to land evaluation. 226
   3.3.3 Selection of methods for land evaluation. . . . 227
   3.3.4 Land capability. 228
   3.3.5 Land suitability. 232
   3.3.6 Analysis of indigenous systems. 239
 Soil Survey and Soil Maps 246
  4.1 Concepts and Definitions 246
  4.2 Intensities of Soil Survey and Map Scales 247
  4.3 Soil Classification Units and Soil Map Units 249
  4.4 Costs and Benefits of Soil Surveys 251
  4.5 Skills and Equipment Required for Soil Survey 253

    4.6 Remote Sensing Methods for Soil Survey 254
          4.6.1 Aerial photography. 255
          4.6.2 Satellite data. 255
Field Methods for Soil Inventory, Classification, and Assessment   262
  5.1 Introduction 262
  5.2 Soil Classification 263
  5.3 Evaluation of Soil Suitability For Agriculture 263
  5.4 Assessment of Soil Productivity 267
        5.4.1 Examination of growing crop. 268
        5.4.2 Field productivity studies. 269
        5.4.3 Conducting specific tests 271
  5.5 Assessment of Reclamation Needs 272
  5.6 Laboratory Support for Field Studies 273
        5.6.1 Determination of chemical properties. . . . 273
        5.6.2 Determining physical properties. 273
        5.6.3 Determining mineralogical properties. . . . 274
        5.6.4 Determining microbiological properties. . . 274
        5.6.5 Skills and equipment required. 275
        5.6.6 Evaluation of laboratory results. 275
Methods for Assessing Degradation 275
  6.1 Methods of Assessing Soil or Land Degradation 276
        6.1.1 Water Erosion 276
        6.1.2 Wind erosion. 278
        6.1.3 Salinization. 279
        6.1.4 Waterlogging. 279
        6.1.5 Loss of soil fertility. 279
        6.1.6 Soil compaction. 280
        6.1.7 Surface soil crusting. 280
  6.2 Methods of Assessing Effects of Degradation on
      Productivity 281
        6.2.1 Experimental data. 281
        6.2.2 Predictive models. 282
Institutions 284
References Cited 284
Suggested Readings 303

IV. PLANTS 309
  Introduction 309
  The Resource Base, Its Utilization and Management: A Summary 312
    2.1 The Resource Base 312
    2.2 Resource Use 313
          2.2.1 Origins of plant resource use and domestication. 313
          2.2.2 Adaptation and modernization: Plant resource
                utilization. 315
          2.2.3 Impacts of modernization on plant resources. . 317
    2.3 Management of Plant Resources 319
          2.3.1 Extensive management. 319
          2.3.2 Intensive management. 320
          2.3.3 Use of energy in management. 321
    2.4 Ecosystems: The Basis for Rational Management . . . 322
          2.4.1 Abiotic components of terrestrial ecosystems. . 323
          2.4.2 Biotic components. 324
          2.4.3 Ecosystems dynamics: Vegetation succession and
                climax communities. 324
          2.4.4 Application of ecosystem dynamics to resource
                inventories. 326
          2.4.5 Carrying capacity: A management concept. . . 326

	2.4.6 Conclusion.	328

Review of Problem Analysis and Levels of Plant Resource Surveys ... 329
  3.1 Problem Analysis and Statement ... 330
  3.2 General Planning Considerations ... 331
    3.2.1 Need for quantitative data. ... 332
    3.2.2 Recognizing intersystem dependences. ... 332
    3.2.3 Endangered species. ... 332
    3.2.4 Opportunities for outside expertise. ... 333
  3.3 Plant Resource Assessment ... 334
    3.3.1 Levels of assessment. ... 334
    3.3.2 Background information on multilevel assessment. ... 343

Remote Sensing: A Tool for Vegetation Inventory and Monitoring ... 348
  4.1 Remote Sensing Systems for Plant Resource Studies ... 348
    4.1.1 Aerial mapping camera photography. ... 349
    4.1.2 Landsat multispectral data. ... 350
    4.1.3 Small-format camera aerial photography systems. ... 351
  4.2 Methods for Maximizing the Utility of Remote Sensing Technology ... 352
    4.2.1 Interpretation. ... 352
    4.2.2 Multi-stage sampling methods. ... 354
    4.2.3 Ground data collection (ground truth). ... 355
  4.3 Conclusions ... 357

Field Methods for Plant Resources Inventories ... 358
  5.1 Introduction ... 358
  5.2 Definitions of Plant Resource Parameters ... 358
  5.3 Magnitude and Diversity of Plant Resources ... 359
    5.3.1 The goal of sampling: Statistical reliability. ... 359
    5.3.2 Randomization of sampling for statistical validity. ... 360
    5.3.3 Accuracy versus precision. ... 360
    5.3.4 Other considerations in statistical sampling. ... 361
  5.4 Beyond Single Parameter Sampling: Analytical Methods ... 363
    5.4.1 Comparative analysis. ... 363
    5.4.2 Cross-sectional analysis. ... 363
    5.4.3 Longitudinal analysis. ... 364
  5.5 Plant Cover Estimation Methods ... 365
    5.5.1 Ocular estimates with two dimensional plots. ... 365
    5.5.2 Line-intercept (linear transect) methods for cover (estimation). ... 366
    5.5.3 Point method for sampling plant cover. ... 367
  5.6 Measurement of Plant Density ... 367
    5.6.1 Quadrat or plot methods for density determination. ... 368
    5.6.2 Plotless methods for determining density. ... 370
  5.7 Methods for Determining Vegetation Biomass ... 370
    5.7.1 Direct harvest methods. ... 371
    5.7.2 Double sampling methods. ... 372
    5.7.3 Indirect or regression methods. ... 372

Methods for Laboratory Analysis of Plant Materials ... 373
  6.1 Summary of Laboratory Techniques Available ... 373
  6.2 Sampling for Laboratory Analysis ... 375
  6.3 Handling and Storage of Samples ... 375
  6.4 Conclusion ... 376

Methods for Determining Human Uses of Plant Resources ... 376
  7.1 Investigating Plant Utilization Strategies ... 378
    7.1.1 General guidelines for conducting interviews. ... 378
    7.1.2 Kinds of information required for plant resource use. ... 379

		7.1.3	Time allocation method.	380
		7.1.4	Economic analysis methods.	381
	7.2	Methods for Determining Humans' Responses to Changes in Their Environment		382
		7.2.1	Qualitative methods for measuring human responses to change.	383
		7.2.2	Quantitative methods for measuring changes.	383
Classification of Plant Resources				385
	8.1	The Importance of Classifications		385
	8.2	Vegetation Classification Terminology		386
	8.3	Types of Classifications		388
	8.4	Sampling for Classification Development		389
	8.5	Key Considerations in the Development of a Vegetation Classification System		390
	8.6	Example of a Hierarchical Resource Classification System		391
Sources of Information				393
	9.1	Local Sources of Information		393
	9.2	Archives		394
	9.3	Computer Libraries		394
	9.4	Bibliographies		395
	9.5	Herbaria		395
References Cited				395
Suggested Readings				400

V. WILDLIFE 411

Introduction 411
  1.1 Scope 411
  1.2 Why Study Wildlife 411
      1.2.1 Wildlife studies: Some background concepts. . . . 412
      1.2.2 Integrating development and conservation. 414
  1.3 Objectives 415
      1.3.1 Protected natural areas. 416
      1.3.2 Wildlife utilization. 417
      1.3.3 Environmental impact statements. 418
      1.3.4 Problem species. 418
  1.4 Where Are You? 418
      1.4.1 Ecological regions. 418
      1.4.2 Classification. 418
      1.4.3 Use intensity gradient. 423
  1.5 Approaches to Wildlife Investigations 428
      1.5.1 Strategy and tactics. 428
      1.5.2 Surveys and inventories. 428
      1.5.3 Baseline studies and monitoring. 429
      1.5.4 Management options. 429

Strategic Assessment 430
  2.1 Description 430
  2.2 Development and Organization of a Data Base 431
      2.2.1 Selection of species. 439
      2.2.2 Physical environment. 440
      2.2.3 Biotic environment. 444
      2.2.4 Cultural activities and institutions. 444
  2.3 Analysis of the Data Base 448
      2.3.1 Species-environment relationships in critical habitat areas. 448
      2.3.2 Species-environment-human resource use interactions. 451
  2.4 Management Options 453

		2.4.1 Assigning relative values.	453
		2.4.2 Management considerations.	454
	2.5	Summary of Strategic Assessment	458

Methods for Determining Existing Status of Wildlife 459

	3.1	Decisions in Method Selections	459
		3.1.1 Choice of species.	459
		3.1.2 Choice of methods.	460
		3.1.3 Matching methods and habitats.	460
		3.1.4 Other considerations.	461
	3.2	Resource Inventories and Habitat Surveys	462
		3.2.1 Habitat mapping.	463
		3.2.2 Food and water: Habitat analysis methods.	464
		3.2.3 Factors limiting species abundance.	467
		3.2.4 Habitat fragmentation.	467
	3.3	Animal Inventories, Censuses, and Population Indices	468
		3.3.1 Aerial censuses.	468
		3.3.2 Aerial inventories with photographic records.	468
		3.3.3 Ground observations.	469
		3.3.4 Using habitat parameters to indicate animal abundance.	471
		3.3.5 Animal signs.	471
		3.3.6 Automatic remote cameras for recording wildlife activities.	473
		3.3.7 Mark and recapture methods.	473
		3.3.8 Census methods based on changes in relative abundance.	475
		3.3.9 Local knowledge.	477
	3.4	Baseline Studies and Monitoring	477
		3.4.1 Population structure and dynamics.	478
		3.4.2 Life tables: Survival-fecundity tables.	479
		3.4.3 Body growth and population productivity.	479
		3.4.4 Behavior and travels of animal populations.	480
		3.4.5 What to monitor.	480
		3.4.6 Carrying capacity.	481
	3.5	Aquatic Habitat and Fisheries	484
		3.5.1 Population status.	484
		3.5.2 Ecosystem characteristics.	484
		3.5.3 Censusing and monitoring.	485
		3.5.4 Habitat management.	485

Methods for Determining Past Status of Wildlife 486

Techniques for Management 491

	5.1	Passive vs. Manipulative Procedures	491
	5.2	Core vs. Buffer Areas	491
	5.3	Controlling Habitat vs. Controlling Species	491
	5.4	Exotic Introductions	492
	5.5	Animal Rescues from Hydroelectric Flooding	492
	5.6	Maintenance and Restoration	493
	5.7	Other Essential Management Policies	493

Cultural Ecological Assessment 493

	6.1	Value Systems and Use	494
		6.1.1 Economic/practical values.	494
		6.1.2 Aesthetic/intellectual values.	494
	6.2	Traditional Systems of Wildlife and Fisheries Resource Management	495
		6.2.1 What are traditional systems?	495
		6.2.2 The need for studying traditional systems.	496

		6.2.3 Traditional resource-using systems: A regional overview.	496
	6.3	The Disruption and Breakdown of Traditional Practices	504
		6.3.1 Foreign cultural contact.	504
		6.3.2 Human population increases and associated agricultural development.	505
		6.3.3 Failure of education, administration and enforcement systems.	506
	6.4	Reconciling Traditional Systems with the Modern World	507
		6.4.1 Indicators of acceptance of extant traditional practices.	507
		6.4.2 The reconciliation process: A Pacific Island example.	509
	6.5	Conclusion	510
Organizations and Directories			511
	7.1	Organizations	511
	7.2	Directories of Organizations	512
References Cited			513
Suggested Readings			520
Index			523

FIGURES

SYNTHESIS

2.1 Ecosystem component parts and definitions of terms 18
3.1 Schematic outline for the FAO agroecological zones methodology. 29
3.2 The electromagnetic spectrum and remote sensor sensitivity 36
3.3 Spectral curves for green vegetation, bare soil, and clear water. 37
3.4 Approximate areal coverage of Landsat receiving stations 40

AQUATIC ECOSYSTEMS

1.1 The hydrologic cycle 58
1.2 Components of baseline studies and natural resource inventories. 64
2.1 General distribution of sea level pressure and wind zones. 69
2.2 Average rainfall distribution for four months. 71
2.3 Transport mechanisms for a generalized nutrient cycle. . 75
2.4 The carbon cycle 76
2.5 The nitrogen cycle in the ecosystem. 77
2.6 Tolerance range concept: critical values of an environmental variable 82
2.7 Salinity and its effects on the biota within an estuary. 83
2.8 Zonation of intertidal communities 90
2.9 Major habitat zones of a typical lake ecosystem. 92
3.1 Temperature, light, and primary production rates 114
3.2 Soft-sediment habitats by size distribution. 130
3.3 Transect and quadrat sampling layouts. 131
3.4 A food web for an estuarine community. 150
3.5 Dendrogram for biotic and abiotic sample groupings along a coral reef transect. 154

SOILS

1.1 The nitrogen and phosphorus cycles and the soil. 197
1.2 General relationships along climate, vegetation, and soils. 196
1.3 Relationship of deforestation and crop failure 198
3.1 Soil individual characterized by various features. ...219
3.2 USDA Land capability classifications and safe intensities of uses. 229
3.3 Mapping symbols used in irrigation suitability classification 238
4.1 Spectral map of soil characteristics, Jasper County, Indiana. 258
4.2 Denuded areas and dune encroachment, Wadi Abu Habl, Western Sudan. 259
4.3 Detailed spectral map of dune area south of Wadi Abu Habl in Western Sudan. 260

	4.4	Spectral map units distinguishing severely eroded soils from other soil and vegetation classes 261
	4.5	Reflectance curve for three dark red surface soils 262
	5.1	Soil pH and nutrient availability. 268

PLANTS

	2.1	The concept of trophic levels. 314
	2.2	Productivity increase by management and cultural manipulation 320

WILDLIFE

	1.1	Schematic map of the major biomes of the world 420
	1.2	Biogeographic provinces of Africa. 421
	1.3	Biogeographic provinces of Latin America 422-423
	1.4	Life zones and classification systems of Holdridge 424
	1.5	Use intensity and impact gradient for terrestrial ecosystems 426
	2.1	The strategic assessment process 432
	2.2	Base map with accounting grid. 437
	2.3	Seasonal distribution map of a coastal fish of commercial value 441
	2.4	Seasonal distribution map of a migratory bird. 442
	2.5	Locations of endangered species. 443
	2.6	Composite map of habitat types 445
	2.7	Land use and human activities map. 446
	2.8	Political and institutional jurisdictions map. 447
	2.9	Overlay of animal species and habitats 450
	2.10	Human activity and natural resource spatial conflicts . .452
	3.1	Estimation of animals in a population, continuous mark-ratio method .476

TABLES

SYNTHESIS

- 3.1 Systems of units in ecological land classification 27
- 3.2 Levels of generalization in a hierarchy of ecosystems... 28
- 3.3 Principal band applications of the Thematic Mapper 39
- 4.1 Collection, availability, and need for physical and biological data................................. 46

AQUATIC ECOSYSTEMS

- 1.1 Diseases related to deficiencies in water supply or sanitation....................................... 62
- 2.1 Changes in density in fresh water as a function of temperature.. 85
- 3.1 Baseline studies for assessing impacts on water bodies... 97
- 3.2 Baseline studies for assessing impacts on watershed.... 98
- 3.3 Baseline studies for assessing impacts on airsheds 99
- 3.4 Phases, problems, and techniques for groundwater supply.. 104-105
- 3.5 Some recommended techniques for analysis of groundwater... 106
- 3.6 Recommendation for sampling and preservation of samples... 120
- 4.1 Status of climate parameter monitoring 161

SOILS

- 1.1 Major categories of land 191
- 1.2 Land use and population............................. 192
- 1.3 Land use and population in developing countries...... 193
- 2.1 Soil and site characteristics, land qualities, and methods of assessment......................... 206-207
- 2.2 Some major soil-related constraints to agricultural development.................................... 210-213
- 3.1 Soil and site information to assess land suitability for irrigation 237
- 3.2 Soil classification for engineering projects 239
- 3.3 Resource use and settlement characteristics......... 241
- 3.4 Size of landholding by ethnic population or ethnolinguistic group........................... 245
- 4.1 Orders and types of soil surveys, characteristics, data sources, and uses 250
- 4.2 Cost elements in consultant contract proposals for soil surveys 252
- 5.1 Field inventory and monitoring of soil resources ... 264-265
- 5.2 Characteristics of good, medium, and poor soil classes 267

PLANTS

- 3.1 Plant resource assessment for inventory or baseline studies.. 335

xiv TABLES

    5.1 Cover estimation methodologies in four vegetation
         categories 366
    8.1 Earth surface features and land use classifications. . . .391

WILDLIFE

    1.1 Conservation values and compatible development 416
    1.2 Taxonomy, data sources, and methods for evaluating
         wildlife 425
    2.1 "Animal species" compendium for the Northern Pintail
         in Alaska. 433-436
    2.2 Alternative categories for managing lands. 456-457
    3.1 Food preference ratings and dietary importance of
         forage species 466
    6.1 Traditional wildlife resource management 497-498
    6.2 Attitudes and practices that might be revived, rein-
         forced, or modified in Oceania 500

OVERVIEW

This book is addressed to planners and managers concerned with economic assistance programs in developing countries. Its purpose is to explain, in a single volume, current methodologies for renewable natural resource inventories and environmental baseline surveys that are appropriate for strategic planning and project assessment. Its scope extends from broad aspects of national development planning to the narrower aspects of project design. It responds to a worldwide critical need for more effective management of renewable resources using integrated approaches to regional planning and project design. It is dedicated to strengthening of the technical and institutional capabilities of developing countries in resource and environmental management and to providing staffs of international development assistance agencies with ready access to practical information. We believe that other international institutions will benefit, as will scientists, students, and citizen conservationists.

The book does not concentrate on how to do integrated planning but rather how to fashion the building blocks which resource planning requires; i.e., how to collect, compile, analyze, interpret, and present renewable resources information by four ecosystem components--soils, water, plants, and wildlife.

While it was not possible to eliminate all specialized technical language in the book, scientific jargon is held to a minimum and unfamiliar words are defined in most cases. It is to be used primarily as a reference work. The contents have been especially organized for ease of access and to encourage its use by a wide variety of interests in addition to the staff of the U.S. Agency for International Development: development planners, engineers, investors, project officers, natural and social science consultants, economists, social scientists, and national and international experts in water resources, agriculture, rangelands, forestry, fisheries, nature preserves, and related fields.

In a sense, there are five books within these covers because a separate panel of scientists worked independently on each disciplinary component, or "part". Inter-panel coordination was maintained to ensure common objectives and full coverage but not to press for uniformity of thought or style of approach. In fact, the several reports were meant to be distinct and to reflect the inherently different approach of each panel's mix of disciplines.

Using the Book

A major aim of this book is to assist development planners in designing and managing resource conservation and environmental aspects of economic assistance programs. Since it describes up-to-date methodologies for data collection, compilation, and analysis, the book should help field officers select appropriate experts, methodologies, and levels of effort for resource inventories and baseline surveys.

For example, if you are a project officer attempting to balance economic benefits of a project against long term resource and environmental effects, you may find it difficult to decide what priority to give environmental studies, what level of effort to support, and what type of expertise to hire for environmental assessments or design of mitigation techniques.

Or if you are working on an integrated regional development plan, you may have to know what data are needed for full integration of cultural and natural resource sectors in a regional context, how to acquire and process those data more efficiently, or how to allocate survey work by task assignments to various experts.

As a development planner working on a natural resources inventory for strategic planning purposes, you may need guidance in structuring the survey to ensure that all components of the ecosystem--water, soils, plants, wildlife--are effectively incorporated and that appropriate, state-of-the-art, methods are used in data collection, analysis, and interpretation.

One example of how this book can help you solve problems like the above can be found in Table 2.1 of part IV (Plants) which presents a system for organizing baseline surveys for integrated planning. In the table each of five levels in a multi-stage survey schedule--region-wide to site-specific--is outlined with details for eight categories of data collection and processing (sources, classification, map scales, etc.). Another example is chapter 2 of part V (Wildlife), which describes the "strategic assessment" approach to resource inventory.

Users of the book will find the Table of Contents to be useful in many ways. For example, if you were responsible for designing a comprehensive resource survey, you might turn to the Soils report (part II) to locate, for review, such topics as: Soil Classification Systems (section 3.2), Cost-Benefit of Soils Survey (section 4.4), or Skills and Equipment Required for Soils Survey (section 4.5). For Wildlife, you might select the following for review: Choice of Methods (section 3.1.2), Habitat Mapping (section 3.2.1), and Carrying Capacity (section 3.4.6).

Each of the five parts has a complete list of all literature sources cited as well as a "Suggested Readings", a selective listing of appropriate technical literature.

The comprehensive Index will be useful in checking such specific details as sampling frequency, map scales, or indicator species.

Integrated Planning

The goal of integrated planning is the preparation of a comprehensive plan in which all development sectors have been assessed for their effects on all the resources in a given geographic area. It implies significant coordination among sectors and flexibility to modify activities to assure resource renewability and long term economic productivity. It reflects the shift in emphasis by USAID and other assistance organizations from project-by-project environmental assessment to early incorporation of natural resource information in the planning process. As stated in a recent report by the National Academy of Sciences (NAS): "Ecological information and environmental considerations must be an integral part of development project planning; they must be given equal weight with agriculture, economic, and engineering factors from the outset of project design".*

In a world of rapid population growth and diminishing natural resources, countries that fail to plan their economic development stategy in concert with resource conservation and environmental management may be unable to sustain progress in health, food, housing, energy, and other critical national needs for more than a few decades. Each developing country must have a realistic plan for accommodating its share of the 100 million people per year being added to the world's population. Such basic resources as firewood, water, fertile land, forage, and fish stocks are already in short supply in many countries and their future prospects are in grave doubt. While the presence of integrated planning alone may not assure conservation of the natural resources base of any country, its absence can lead to depletion. The opportunities for development based on excessive exploitation of natural resources are rapidly fading. The future depends on development keyed to resource conservation.

Development planning and management decisions should be based on accurate assessments of the location and quality of individual resources that are up-dated regularly so that trends can be detected and management plans revised appropriately. In this connection, the NAS report stresses the value of a strong data base: "Planning can be made much more effective by expanding and improving natural resources evaluation. Information on climate, soils, vegetation, wildlife, and water resources is essential to environmentally sound planning and facilitates wise selection among project options."

Definition of Terms and Concepts

A resource inventory is the first step in the evaluation of natural resources and their uses and refers to a catalog of resources present. Environmental baseline survey refers to information about the site, quality, and indicators of a resource condition which should be collected at the same time as the inventory. Resource monitoring refers to subsequent assessments of the resources and must be collected in a standardized manner, consistent with the baseline data to allow analysis of trends resulting from resource use (see part II, section 1.4).

Thus, a natural resource inventory is, most simply, a physical itemization of the natural resources of the area, extending over any period of time. A baseline survey is quite different in two ways: 1) it deals not only with the status of physical resources but also with natural and cultural processes and is, therefore, ecosystem oriented; and 2) it is time constant and therefore all parameters are referred to a "time zero." A "monitoring" program is a repeat measurement of some, but not usually all, of the baseline parameters at some "time(s) zero plus." Thus, monitoring programs measure ecosystem change over time against a "time-zero" baseline and are most appropriately done at preplanned intervals using selected, especially sensitive, "indicator" parameters (e.g. see part V, section 4.5).

---

*Savage, Jay M. *et al.* 1982. Ecological Aspects of Development in the Humid Tropics. Washington, D.C.: National Academy Press. 297 pp.

The Ecosystem Approach

A specifically defined ecosystem, which includes both the biotic and the non-biotic features of the environment, should be the basic functional unit of resource management. The ecosystem concept emphasizes the interrelationships and dependencies between these components. Identification of the ecosystems in a given area can be done by resource specialists during a reconnaissance survey or "strategic assessment". A review of ecological classification may be found in part V, section 1.4 (Wildlife), including explanation of the world's natural "Biogeographic Provinces" and "Life Zones."

While resource inventories and baseline surveys are most effectively designed using ecosystems as the organizing and integrating theme, there are also many valuable side products which can result for ecosystem organized survey activities. For example, runoff characteristics--the amount of water produced by watersheds--can be determined for flood water storage and other purposes. For another example, the "carrying capacity" of an ecosystem can be determined for particular human or natural needs. Ecological carrying capacity quantitatively defines the ability of a unit area of land or water to support biota (pounds of biomass or numbers of animals) under a given set of conditions. This type of information is most valuable to economic planners.

In the design of an integrated development plan, planning boundaries should, to the extent possible, approximate ecosystem boundaries. The assignment of ecosystem boundaries is sufficiently elastic to accommodate a reasonable degree of adjustment where necessary to meet practical realities. As recognized in part IV (section 1.4), "...one must define the working ecosystem concept so that it is compatible with the stated problem...level of information."

However, there are limits to ecosystem boundary flexibility. For example, it would usually be inappropriate to use a river as the boundary of a planning district because a whole catchment, including both slopes of the river valley, is the natural ecological unit, or ecosystem. However, a longitudinal segment of the catchment could be defined as an ecosystem, bounded by its catchment divides, with upstream and downstream effects handled as inputs and outputs across the ecosystem boundaries. In effect, ecosystems can be expanded or reduced concentrically but not sliced "across the grain" for either scientific or management purposes.

Ecosystem boundaries not only encompass the physical structure of the system but the network of dynamic processes that enable the system to function. There is danger in using the ecosystem approach of overemphasizing structure compared to process. Analyzing nature as a dynamic network of interrelationships is more useful than simply cataloguing ecozones or assembling overlay maps. It is process, not structure that functionally characterizes an ecosystem.

Hugh Bell Muller
John R. Clark
Jeffrey B. Tschirley

U.S. National Park Service
Washington, D.C.
June 15, 1983

The Ecosystem Approach

PREFACE

Among its many responsibilities, the United States Agency for International Development (AID) is charged with assisting developing countries in building their institutional and scientific capacity to identify, assess, and solve environmental and natural resource problems. In order to meet this charge, AID entered into the Natural Resources Expanded Information Base Project with the National Park Service (NPS) in 1979. This project was to develop and provide information to planners in development organizations, including AID itself, and to planners and scientists in developing countries. The information generated under this Project is divided into: State-of-the-Art Review Papers, Case Studies, Project Design Aids, Information Dissemination, and Training.

For the preparation of one of the State-of-the Art Review Papers, NPS sought the expertise of the scientific community through the American Association for the Advancement of Science (AAAS) to review and report on methods for inventorying renewable natural resources, gathering environmental baseline data, and for carrying out natural resource monitoring in the context of the human use of those resources.

The "Methods Project," as it became known, was started in 1981 when AAAS formed five panels. The first panel, composed of the chairmen of the other four panels, was responsible for formulating the content, for organizing, supervising, and for writing the Introduction and Synthesis part. The other four panels whose products are "parts" addressed methods related to specific aspects of ecosystems: water (which was retitled "aquatic ecosystems"), soils, plants, and animals (which was retitled "wildlife"). Each panel chairman was responsible for organizing and supervising the writing of his particular part. This book represents a unique blend of the methods appropriate to the social, physical and biological sciences; each panel contained such an interdisciplinary mix.

The contributors to this book have first-hand knowledge of the use of the methodologies described and their application to developing countries. They also know the importance of encouraging host country personnel to carry on this work. Therefore, emphasis has been directed toward methodology transfer and to some of the practical considerations of cost, repair and replacement of equipment, reliablity of equipment under field conditions, training requirements, and the need to expand training in existing developing country institutions.

The extensive effort put forth to produce this document reflects the concern of the scientific community about economic development and its relationship to the long-term sustainability of a country's natural and human resources.

The American Association for the Advancement of Science (AAAS) wishes to extend its deep appreciation to the panel chairmen, panel members, and consultants to the panels. The Association especially wishes to acknowledge the role of the AAAS Committee on Arid Lands which had jurisdiction over this project and its chairman, Cyrus McKell, who coordinated work through his chairmanship of the Plants Panel. In addition to project staff, five AAAS employees gave special assistance: Holly Bishop in adapting graphics, Martha Collins and Mary Dorfman as copy editors,

xx   PREFACE

Kathryn Wolff for editing the index, and Susan Cherry in production. The Association also wishes to acknowledge the work of the several staff officers of the Agency for International Development and the U.S. National Park Service; it also recognizes their preparation of the Overview.

STAFFS

The American Association for the Advancement of Science

William Cummings
Marilyn Jones
Nancy Muckenhirn

Priscilla Reining
Joan Spade

Agency for International Development

William M. Feldman
James C. Hester
Molly M. Kux
Steven Lintner

Robert O. Otto
Michael Q. Philley
Charles K. Paul
Albert C. Printz
Jane E. Stanley

National Park Service-Office of International Affairs

John R. Clark
Robert C. Milne

Hugh Bell Muller
Jeffrey B. Tschirley

PANEL MEMBERS

Synthesis Panel

FRANCIS CONANT, Chair. Anthropology, Hunter College, New York, New York 10021.

Chairs of Aquatic Ecosystems, Soils, Plants, and Wildlife.

Aquatic Ecosystems

PETER ROGERS, Chair. Division of Applied Sciences, Harvard University, Cambridge, Massachusetts 02138

LEO R. BEARD. Espey, Huston & Associates, Inc., Austin, Texas 78758-4497

KENNETH D. FREDERICK, Resources for the Future, Washington, DC 20036.

OWEN THOMAS LIND, Biology, Baylor University, Waco, Texas 76798

KENNETH K. SEBENS, Museum of Comparative Zoology, Harvard University, Cambridge, Massachusetts 02138

Soils Panel

MARION BAUMGARDNER, Chair. Laboratory for Applications of Remote Sensing, Purdue University, West Lafayette, Indiana 47906

PIERRE CROSSON, Resources for the Future, Washington, DC 20036

HAROLD E. DREGNE, Agronomy, Texas Tech University, Lubbock, Texas 79409

MATTHEW DROSDOFF, Agronomy, Cornell University, Ithaca, New York 14853

FREDERICK WESTIN, Remote Sensing Institute, South Dakota State University, Brookings, South Dakota 57007

Plants Panel

CYRUS MCKELL, Chair. Native Plants, Inc. Salt Lake City, Utah 84108

CHARLES D. BONHAM, Range Science, Colorado State University, Fort Collins, Colorado 80523

J.R. GOODIN, Biological Sciences, Texas Tech University, Lubbock, Texas 79409

DANIEL R. GROSS, Anthropology, Hunter College, New York, New York 10021

CHARLES E. POULTON, Plant ecology and soils. Consultant in resource ecology and renewable natural resource management, Santa Clara, California 95051

xxii   PANEL MEMBERS

SAMUEL C. SNEDAKER, Biology and Living Resources, Rosenstiel School of Marine and Atmospheric Science, University of Miami Miami, Florida 33143

Wildlife Panel

RAYMOND DASMANN, Chair. Environmental Studies and Ecology, University of California, Santa Cruz, California 95064

GARY KLEE, Environmental Studies, San Jose State University, San Jose, California 95192

THOMAS LOVEJOY, World Wildlife Fund, Washington, DC 20009

GEORGE PETRIDES, Fisheries and wildlife, Michigan State University, East Lansing, Michigan 48823

CARLETON RAY, Environmental Sciences, University of Virginia, Charlottesville, Virginia 22903

Consultants

RALPH MITCHELL, Dean of Applied Sciences, Harvard University, Cambridge, Massachusetts 02138

DEREK WINSTANLEY, Consortium on Energy Impacts. Boulder, Colorado 80544

KENNETH CARLANDER, Animal Ecology, Iowa State University, Ames, Iowa 50011

B. DEAN TREADWELL, Wildlife and rangeland ecology. Moundsville, West Virginia 26041

SALEEM AHMED, East-West Resource Systems Institute, East-West Center, Honolulu, Hawaii 96848

ACKNOWLEDGMENTS

We wish to acknowledge information reprinted from the following sources:

Introduction and Synthesis

Fig. 2.1: From FUNDAMENTALS OF ECOLOGY xx 3rd Edition by Eugene P. Odum. Copyright (c) 1971 by W.B. Saunders Company. Reprinted by permission of Holt, Rinehart and Winston, CBS College Publishing. Fig. 3.1: Food and Agriculture Organization of the United Nations. 1978. Report to the Agro-ecological Zones Project, Vol. 1, Schematic outline for the FAO agroecological zones methodology. Fig. 3.2: R.G. Best. 1982. Handbook of Remote Sensing in Fish and Wildlife Management. Brookings, South Dakota, Remote Sensing Institute. Fig. 3.3: P.H. Swain and S.M. Davis. 1978. Remote Sensing: The Quantitative Approach. New York: McGraw-Hill International Book Company (reproduced with permission). Fig. 3.4: C. Paul & A.C. Mascarenhas, SCIENCE vol. 214, 9 October, 1981, pp. 139-145. AAAS copyright 1981. Table 3.1 and 3.2: R.G. Bailey. 1980. Integrated approaches to classifying land as ecosystems, In Proceedings of the Workshop of Land Evaluation for Forestry: International Workshop of the IUFRO/ISSS. pp. 95-109. Wageningen, The Netherlands: International Institute for Land Reclamation and Improvement. Table 3.3: Landsat Data Users NOTES. July 1982. Issue No. 23. Sioux Falls, South Dakota, EROS Data Center (NOAA/NESDIS).

Aquatic Ecosystems

Fig. 1.1: T. Dunne and L.B. Leopold. WATER IN ENVIRONMENTAL PLANNING. San Francisco, California: Copyright (c) (1978). W.H. Freeman and Company. Fig. 2.1: H.H. Lamb. 1972. Climate: Present, past and future, Vol. I. Fundamentals and climate now. London: Methuen. Fig. 2.2: J.J. Jackson. 1977. Climate, water and agriculture in the tropics. London: Longman Methuen. Fig. 2.3: P.R. Ehrlich. ECOSCIENCE: POPULATION RESOURCES IN ENVIRONMENT. San Francisco, California: Copyright (c) (1977) W.H. Freeman and Company. Fig. 2.4: L. Smith. 1966. Ecology and field biology. New York: Harper and Row. Fig. 2.5: O.S. Owen. Natural resources conservation, 3rd edition. Copyright (c) (1980) New York: Macmillan. Fig. 2.6: S. Charles Kendeigh, ECOLOGY: with special reference to Animals and Man (c) 1974, p. 13. Reprinted by permission of Prentice-Hall, Inc., Englewood Cliffs, New Jersey. Fig. 2.7A and 2.7B D.F. Boesch, In B.C Coull, ed. and B.C. Coull. ECOLOGY OF MARINE BENTHOS. Copyright (c) 1977 by and reproduced by permission of the University of South Carolina Press. Fig. 2.8A and 2.8B: J. Morton and M. Miller. 1973. The New Zealand Seashore, 2nd ed. Glasgow: Wm. Collins and Sons. Fig. 2.9: From ECOLOGY OF INLAND WATERS AND ESTUARIES by G.K. Reid and R. Wood. Copyright (c) 1969 by Van Nostrand Reinhold Co. Reproduced by permission of the publisher. Fig. 3.2A and 3.2B: J. Morton and M. Miller. 1973. The New Zealand Seashore, 2nd ed. Glasgow: Wm. Collins and Sons. Fig. 3.4: R.S.K. Barnes. 1974. ESTUARINE BIOLOGY. London: Edward Arnold Ltd. Reprinted with permission. Fig. 3.5: P.L. Jokiel and J.E. Maragos. 1978. Reef Corals of Canton Island II. Local Distribution. Atoll Research Bulletin. 221. Table 1.1: Saunders/Warford

VILLAGE WATER SUPPLY Copyright (c) 1976 by the International Bank for Reconstruction and Development. By permission of The Johns Hopkins University Press. Table 2.1: P.S. Welch. 1935. LIMNOLOGY. New York: McGraw-Hill. Table 3.4: S. Mandel and Z. Shiftan. 1981. Groundwater resources: investigation and development. New York: Academic Press. Table 3.5: J.W. Lloyd. 1981. Case studies in groundwater resources evaluation. Oxford: at the Clarendon Press. Table 3.6: Environmental Protection Agency. 1979. Methods for chemical analysis of water and wastes, 1978. EPA-600/4-79-020. Springfield, Virginia: National Technical Information Service. Table 4.1: with permission from E. Barrett and D. Martin. 1981. The use of satellite data in rainfall monitoring. Copyright: Academic Press Inc. (London) Ltd.

Soils

Fig. 1.1: A.D. Bradshaw and M.J. Chadwick. 1980. Restoration of Land. Published by the University of California Press. Fig. 1.2: P.R. Erhlich, A.H. Erhlich and J.P.V. Holdren. 1977. ECOSCIENCE. San Francisco, California: W.H. Freeman. Fig. 1.3: R.J.A. Goodland and H.S. Irwin. 1975. Amazon jungle: green hell or red desert? Amsterdam: Elsevier. Fig. 3.1: S.W. Buol, F.D. Hole and R.J. McCracken. 1973. Soil Genesis and Classification. Ames, Iowa: The Iowa State University Press. Fig. 3.2: THE NATURE AND PROPERTIES OF SOILS, 8th Edition, by Nyle C. Brady. (Copyright (c) 1974 by Macmillan Publishing Company.) Fig. 3.2: Bureau of Reclamation. 1953. Land classification handbook. U.S. Department of Interior. Bureau of Reclamation Publication V, Part 2. Figs. 1.4. and 4.3: M.F. Baumgardner. 1982. Remote sensing for resource management today and to morrow. In Remote sensing for resource management. C.J. Johannsen and J.F. Sanders, ed., Ankeny, Iowa: copyright by Soil Conservation Society of America. Figs. 4.2 and 4.4: M.F. Baumgardner. 1979. Assessment of arable land. In Fertilizer raw material resources, needs and commerce in Asia and the Pacific. R. Sheldon et al., eds. Honolulu, Hawaii: East West Center. Fig. 4.5: E.R. Stoner, M.F. Baumgardner, R.A. Weismiller, L.L. Biehl and B.F. Robinson, SOIL SCIENCE SOCIETY OF AMERICA JOURNAL, Vol. 44, 1980, pages 572-574. By permission of the publisher, Soil Science Society of America. Table 1.1: R. Dudal. 1982. Land Degradation in a World Perspective. Journal of Soil and Water Conservation 37(5): 245-249. Copyright (c) Soil Conservation Society of America. Table 1.2 and 1.3: R. Dudal, G.M. Higgins and A.H. Kassam. 1982. Proceedings 12th International Congress of Soils Science, New Delhi. Table 2.1. and 3.1: S.G. McRae and C.P. Burnham. 1981. Land evaluation. Oxford: at the Clarendon Press. Table 2.2: Figure and table from the book: Dudal, R. 1980. Soil-related constraints to agricultural development in the tropics. In Soil-related contraints to food production in the tropics. Los Banos, Philippines: IRRI. Table 3.2: from Soil Survey Interpretations for Engineering Purposes Soil classification for engineering projects. 1973. Food and Agriculture Organization of the United Nations. Table 3.3 (data source): R. M. Netting. 1968. Hill Farmers of Nigeria: Cultural Ecology of the Kofyar of the Jos Plateau. Seattle: University of Washington Press. Table 3.4: W. Allan. 1965. The African Husbandman. Table 4.2: S. Western. 1978. Soil survey contracts and quality control. Oxford: at the Clarendon Press.

Plants

Fig. 2.1: D.K. Northington and J.R. Goodin. (in press). The botanical world. St. Louis: C.V. Mosby (permission given by author). Fig.2.2: R. Merton Love. 1961. The range - Natural plant communities or modified ecosystems? Journal British Grassland Society. 16(2): 89-99. Permission granted by Blackwell Scientific Publications Limited. Table 8.1: Charles E. Poulton. 1972. A comprehensive remote sensing legend system for the ecological characterization and annotation of natural and altered landscapes. In Proceedings 8th International Symposium on Remote Sensing of the Environment. Permission is granted from the Environmental Research Institute of Michigan for the use of information.

Wildlife

Fig. 1.1: V.C. Finch and G.T. Trewartha. 1949. PHYSICAL ELEMENTS OF GEOGRAPHY. McGraw-Hill, Inc. (c) copyright. Figs. 1.2 and 1.3: M.D.F. Udvardy. 1975. A classification of the biogeographical provinces of the world. International Union for Conservation of Nature. Occasional Paper No. 18. Fig. 1.4: Life zones and classification systems of Holdridge from L.S. Holdridge. 1967. Life zone ecology. San Jose, Costa Rica: Tropical Science Center. Request granted by the author. Table 2.2: K.R. Miller. 1978. Planning national parks for ecodevelopment. Fundacion para la Ecologia y para la proteccion del medio ambiente, vols. 1 and 2, Madrid. Table 3.1: G.A. Petrides. 1975. Principal food versus preferred foods and their relations to stocking rate and range condition. Biological Conservation. Permission granted by Applied Science Publishers, Limited.

Part I -- Introduction and Synthesis

Francis Conant, Chair
Peter Rogers, Aquatic Ecosystems
Marion Baumgardner, Soils
Cyrus McKell, Plants
Raymond Dasmann, Wildlife

I. INTRODUCTION AND SYNTHESIS

Introduction

This book is designed to help planners become familiar with the terminology and methods available for making resource inventories and conducting baseline studies in developing countries. Although it also provides guidance in choosing methods best suited for a particular kind of development project and for different levels of technical capability in the host country, the general approach is not "project oriented." The methods are presented within the conceptual framework of ecology and ecosystem functioning. This generic approach has the advantage of placing the methods of resource inventories and baseline studies clearly in the foreground; in the background, of course, are the many and varied environments in developing countries and the equally varied development projects taking place within them. A biome- or project-oriented approach would have meant guessing not only where but what kinds of projects are likely to be required in the future. Readers are assumed to have an interest but not necessarily a professional competence in any of the disciplines involved in an ecological approach.

More than 30 persons have contributed to this book. Among the disciplines represented are the social as well as the natural sciences and engineering. Regular participants and consultants are listed in the Preface. Agronomy, anthropology, biology, botany, climatology, ecology, economics, engineering, geography, hydrology, pedology, plant physiology, range science, soil science, statistics, wildlife ecology, zoology, and many of their subdisciplines were involved. The professionals in these fields were concerned with the following resources or environmental domains: aquatic ecosystems, soils, plants, and wildlife. It should be emphasized that these are natural domains; thus, for example, the Plants Panel does not focus on crops, nor does the Wildlife Panel focus on livestock. For a working definition of "natural," see the box on page 5. A fifth panel, named the Synthesis Panel, composed of the constituent resource panel chairs, was charged with integrating the work of the other panels and providing an overview of ecosystem functioning in these introductory chapters. The purpose and scope of this Introduction and Synthesis part is both to introduce key concepts and provide an overview of what is common to the separate panel reports. The separate panel reports on methods appropriate to each component part or research domain, however, represent the real down-to-earth value of this volume.

The many different points of view that the panelists represent would be impossible to integrate in a single volume were it not for some shared perceptions. These have to do with (i) recognizing the functional interrelatedness of the biotic and abiotic environments, as in the ecosystem concept, (ii) arriving at an agreement on the meanings of the key phrases "natural resource inventory" and "baseline studies," (iii) emphasizing the *practicality* of the methods reviewed for conducting resource inventories and baseline studies, (iv) understanding "intensification" not only as a process often associated with development projects but also as a scale or gradient along which indigenous subsistence systems can be compared, and (v) recognizing the significance of local populations and their subsistence systems as barometers of ecosystem function and dysfunction. The local populations' perceptions of their environment can be of great value as well, especially in helping place baseline studies in the context of inherent resource variability.

1.1 The Ecosystem Concept

The ecosystem concept is essential for understanding the relations between resources that are to be developed and for identifying functional "services" (such as nutrient cycling) necessary for the very existence of the resources. For a development project to succeed where many others have failed, the interrelatedness of all resource domains must be recognized. Examples of failure are legion: lowered water tables, salinized soils, failure of exotic food crops, endangerment to the point of local extinction of once abundant wildlife. It is more than 10 years since the dismal catalog of failure was drawn up which figured so largely in *The Careless Technology* (Farvar and Milton, 1972). The experience gained in the past decade, starting with *Ecological Principles for Economic Development* (Dasmann et al., 1973), has shown the importance of seeing the ecosystem as the interrelated functioning of the total biotic and abiotic environment. It is now realized that the most fundamental resource is not the forest and its potential timber; not the soil and its extractable riches; not even water and its life-sustaining properties. It is the system's functioning: the hydrologic and nutrient cycles, for example, that weld all life together. Methods for monitoring such cycles (or expressions of them) are presented along with methods for estimating more conventional resources such as arable soil, forestry reserves, or rangeland carrying capacity.

The systemic processes on which all resources depend are fundamental to life itself. Monitoring an ecosystem process or function such as an aspect of the hydrologic cycle is taking a very basic pulse indeed. An analogy is sometimes made between ecosystem functioning as a "service," and the resources such as wildlife, as "goods". Humans depend on both. It is a useful analogy because development projects are never free of economic constraints, and goods and services are familiar accounting categories. But the analogy can be carried too far, since ecosystem functions or services such as nutrient cycling cannot be valued in the usual sense. They are basic to the creation and support of life itself.

Properly, ecosystem functioning involves both the creation and the maintenance of resources recognized as valuable. Whether a resource (one already in place or about to be developed) is sustainable depends largely on ecosystem functioning. Given adequate monitoring, environmental degradation can be detected at early stages, and the enormous expense of resource reclamation can be avoided through conservation and wise development.

It has become virtually a truism that economic development and conservation must go forward together. Development, insofar as it involves land, water, and renewable resources, depends for its continuity on conservation of soil and water resources, vegetation, and animal life (see, for example, Agency for International Development, 1982). Conservation practices, however, are unlikely to be carried out unless associated with readily perceived short- or long-term economic benefits to the countries and peoples involved. In planning for economic development, conservation has too often been brought in as an afterthought or undesirable necessity. This has led to unnecessary degradation of renewable resources and to costs and lost benefits that could have been avoided. In this volume the means for bringing conservation and development together are discussed to avoid the problems that will result if either conservation or development is ignored.

For example, many international development agencies (the Agency for International Development, the World Bank, the U.N. agencies) are required to avoid activities that will bring about the endangerment and, by extension, the extinction of species. However, often the activities that are the most threatening (habitat destruction, overhunting, foraging) are a result of rural poverty. Conservation of species cannot be assured unless economic development alleviates rural poverty. In planning a development project, the agency must ascertain that the project is sited in such a way, and has sufficient built-in safeguards, that it will not cause further species extinction, as by flooding, draining, or logging in critical habitats or by opening up critical habitat areas to human settlement (Blake et al., 1980). If the recommendations and methods presented in this volume are followed, development agencies will be more aware of potential conflicts and better able to avoid future difficulties than they have in the past.

1.2 Natural Resource Inventories and Baseline Studies

With so many specialties and disciplines represented among the contributors to this volume, it was perhaps only natural that at the outset there was some uncertainty about the meaning of each term in the heading above. In the literature and in everyday usage, the terms "inventory" and "baseline study" are sometimes used interchangeably with "survey" and even "reconnaissance." In one volume we cannot standardize what is otherwise variable, but we can make our own usage as unambiguous as possible.

The contractual definitions which guided the project follow (NPS, 1980):

> *Resource inventories* identify natural resources in specific geographic areas and indicate their quantity and variety. Evaluation of methods should include data sources such as maps, floral and faunal distribution lists, weather records, species checklists, and instruments such as orbiting satellites and aerial cameras. *Baseline studies* analyze and assess present characteristics, status and ecosystem processes such as productivity, nutrient cycling, seasonality in stream flow, temperature, faunal habitat requirements, endangered species lists, soil analysis (pH), soil moisture and organic content, chemical properties of water. Baseline data, with subsequent updates of ecosystem functioning, create a monitoring capability. Considerations in conducting both inventories and baseline sutdies should include scale, sampling and relative cost.

Under the terms of the contract, methods for resource inventories and baseline studies were to be evaluated for just these four: soils, water, plants, and animals. Under these circumstances characterizing the contractual and the functional relationships of the several parts of this book has been something of a problem. It has been solved in two ways: (i) "water" was renamed "aquatic ecosystems," and by treating water in the (aquatic) ecosystem as *the* resource the contractual requirement has been met and the functional aspect highlighted. Similarly, "animals" was renamed "wildlife," to emphasize the non-domestic but functional aspect. (ii) Secondly, a good deal of thought went into designating parts of the book as "domains" to indicate the generic aspect of each part. An attractive synonym is component but component carries a specialized meaning of direct relevance for the construction of integrated inventory hierarchies, as indeed does domain in Bailey's terminology (1980). Section 3.1 should be read carefully to learn the specialized meaning of the word which overlaps with the more generalized meaning given it elsewhere. By the same token, other words such as "integrated" carry many meanings including its use when coupled with inventory. "Integrated inventory" is coming to have a more specialized meaning for the field of environmental studies. Where possible we prefer to recognize current usage.

Natural resources in developing countries are the focus of this book. In considering natural resources we recognize the effects of what people do for subsistence which affects the environment--for example, farming, livestock management, burning over brush and fields, cutting firewood, setting up markets, establishing communication networks. We do not concentrate on natural resources that occur exclusively in natural areas (see box). Therefore, "resource inventory" and "environmental baseline" may require an intensification gradient in data acquisition and complexity. A resource inventory is conducted to provide predevelopment data that characterize relevant biotic and abiotic features of an ecosystem proposed for development.

> The word "natural" as used here does not imply pristine, primeval or unused by human beings. A *natural resource* is a material or form of energy that occurs in nature and is useful to human cultures. A *natural area* is one in which native or indigenous species predominate, and in which the landscape and vegetation are not shaped or modified primarily by intensive human actions. This is in contrast to a cultivated or cultural area more directly under human control. In many developing countries, in-place subsistence systems and local populations have existed for centuries and millennia. Nature untouched by humans either simply does not exist or exists in such remote and unusable areas as to make siting a development project most unlikely. Natural resources are ubiquitous. Natural areas are rare.

Intensification implies a trend from less to more, from extensive to intensive management of resources. It is a useful concept for organizing the analysis of data from soil and plant inventories. And it is a tempting concept to use for assessing the impact of a development project. However, the direction of the trend toward more intensive--or in some cases toward more extensive--is not always clear. For example, it is difficult to decide whether an Iowa cornfield or an Asian subsistence farmer's plot is the more intensively farmed. Therefore it was decided to use intensification trends only in some of the substantive discussions. Intensification as a process is too complicated to capture in simple diagrammatic form for all processes; although it is useful for some.

Inventory information on aquatic ecosystems would include total and seasonal precipitation, runoff, evaporation rate, stream flow, and aspects of chemical and biological quality. Some typical parameters of soils that would be described by inventory data are major soil complexes, general levels of productivity, and topographic relations of soil groups. Inventory data on plants would include such information as major vegetation types with their dominant species, estimated biomass production, percent cover, and characteristics that influence utilization by wildlife or domestic animals. Data on wildlife most likely to be included in an inventory would be the principal species and their relative abundance along with information on age and sex ratios, migration patterns, and types of habitats.

Inventory information is by its nature usually obtained from extensive surveys conducted over a short period of time and provides little or no indication of variability over seasons or years. Some ecosystem characteristics such as soil texture and groundwater depth vary little over time, and inventory information can provide a basis for predicting the results of future development. However, most biotic ecosystem parameters vary considerably over seasons and years. The only way to take this into account is to utilize any existing data that can be added to the predevelopment resource inventory. An ecological characterization of the favorability or unfavorability of the season(s), e.g., droughts or floods, during which inventory data were collected may help to place the data in a general perspective--high, low, or intermediate--in relation to the long term average generally perceived by local people.

Environmental baseline studies involve intensive effort to gather data prior to project development. Their purpose is to provide details on resource numbers and ecosystem functions that may be affected by development projects. If carried out over a sufficient period of time, baseline studies may enable adjustments to be made in mean values previously acquired in surveys and resource inventories. Further, a wider range of accumulated data for relevant ecosystem outputs and processes will make it possible to calculate the variance of these parameters.

In summary, survey and resource inventory data are used in project planning and design. Information on resource abundance and the vulnerability of critical support processes dictates the magnitude and limits of a potential project. Clearly, data on how much use can be made of particular ecosystem outputs are valuable. Less obvious is the potential for project failure because ecosystem processes are not understood or are overestimated. In project planning optimum use should be made of resource inventory data.

Too often a commitment for continued ecosystem monitoring is omitted from project planning. Data from a monitoring activity can be very useful

8 I. INTRODUCTION AND SYNTHESIS

in tracking project success and external (but vital) influences; they are also useful in making project adjustments as trends are identified.

1.3 Overview of Sections

This book is divided into five principal panel reports, hereafter call "parts": I. Introduction and Synthesis, II. Aquatic Ecosystems, III. Soils, IV. Plants, and V. Wildlife.

The panel chairmen decided at the outset to encourage the presentation of key methods or issues of special importance in *one* of the panel reports rather having them be repetitively covered. The practical effect of this decision also serves as a set of pointers to thematic content.

I. Introduction and Synthesis: a general overview stating the important considerations which integrate the panel reports; sampling, remote sensing, integrated or multiple resource inventories, and geographic information systems.

II. Aquatic Ecosystems: ecosystem components and functioning; this report, especially, addresses the general reader in the first two chapters and the specialist in the second two chapters.

III. Soils: intensification and conservation; a descriptive presentation on remote sensing with special reference to soils.

IV. Plants: sampling design; integration at each level of assessment.

V. Wildlife: stategic assessment for early decisions on areas of critical importance for wildlife.

The social, cultural and economic aspects of natural resource use are included in each part. These sections are in the Synthesis panel (subsistence systems analysis, work with informants), Soils (indigenous systems of classification and management), Plants (human ecology methods) and Wildlife (methods on indigenous conservation and by continent review of indigenous practice, past and present).

References appear at the end of each part, or panel report, rather than being grouped at the very end of the document. A distinction is also made between References Cited and Suggested Readings for the reader who wishes to know more or needs to know the best general references in a particular discipline. Finally a list of institutions is included.

Summaries of each part (Aquatic Ecosystems, Soils, Plants, and Wildlife) are provided here so that the reader can see the "whole" before going on to the individual parts.

Overview of Aquatic Ecosystems, Part II

Water, one of our most important resources, is used in virtually all human activities. Neither plants nor animals can survive very long without it. The growth in size and wealth of human communities places enormous pressures on aquatic ecosystems, and some of the projects undertaken to meet these needs are among the largest ever undertaken. Because human activities often have unintended consequences on aquatic systems, water has become one of the most abused resources. All development projects in tropical countries will have large impacts on the aquatic environment. Water also plays a fundamental role in public health in tropical regions.

The aquatic ecosystems part of this volume is quite long. It includes: freshwater lakes, rivers, streams, and near-shore saline water in the tropics. In discussing these ecosystems considerations other than just the quantity and quality of the water need to be taken into account. The physical, chemical, biotic, and the functional properties of the systems are also considered.

An issue that underlies the discussion on methodology is whether methods recommended for developing countries differ from those for developed countries. There are no easy answers. As the reader will find in the later chapters of aquatic ecosystems there are few, if any, methods that are exclusively for use in either the developed or the developing coutnry. What is more germane is the use of the techniques appropriate to the size of the study, the skill levels of the available staff, ease of access to the field, and the availability of financial resources. The most complex measuring devices might be recommended for some study in developing countries and simple manual techniques for another type of study in developed countries.

The units of spatial analysis for aquatic systems range from large national and international regions for climatic data to small watershed and river reaches for many of the biotic and chemical data. Most data will be of the "point" rather than the "area" type. Hence, different sampling frameworks are needed than those discussed in chapter 3 of part I. Longitudinal, or time-series, data are important for many of the aquatic ecosystem properties.

The Aquatic Ecosystem part is organized to give the general and descriptive material first and then the details of the methodology. Thus the reader who wishes to gain an overview of natural resource inventories and baseline studies of aquatic ecosystems should read chapters 1 and 2. Chapter 3 could serve as reference material for later use as specific problems arise in the area.

In chapter 3 we list the ecosystem properties that may be necessary to study for a wide variety of development projects. The selection required value judgments from the panel members and may not be applicable in all situations. The list provides a starting point for deciding what studies should be done on a particular project. It should be noted that this list does not give any indication of the levels of impact; such impacts should be delayed until a scientific assessment has been made.

We find that in many developing countries the physical properties of aquatic ecosystems are the most likely to have been studied. This is because there are many standard methods available for measurement, engineering staffs are typically trained in these methods and, in many cases, there has been a need to know the state of the physical system for a long time--the amount of water available, for example, is a critical parameter in most developing countries. The relative ease of measurement of some physical properties does not necessarily mean that the data are adequate.

Many data are required in long time series, but few long-data records are available.

Chemical properties are the next set of properties that are relatively easy to measure and include in planning analyses. The skill levels required are not too high, and the equipment is relatively easily available and is inexpensive. In most countries, however, few of these data have been collected systematically, and a substantial investment may be required to start a data collection program.

Biotic properties of the ecosystem and functional properties, which are the integrated sum of the physical, chemical, and biotic properties, have typically not been studied. There are four main reasons for this: (i) the understanding of the need for those data is typically not perceived by the policy-makers in developing countries; (ii) there are few biologists with field skills or interest, (iii) there are few laboratories available for carrying out the laboratory tests, and (iv) there are few handbooks of standard methods available. This part should make it easier to study biotic properties because it was designed to introduce new, and gather previously scattered, information about aquatic ecosystems. But there may still be a need for expatriate consultants and training in the start up phase of any resource inventory or baseline study involving the biotic and functional properties of aquatic ecosystems.

Overview of Soils, Part III

In comparison with other renewable resources soils have been the longest in making. Rates of soil formation and regeneration are among the slowest ecological processes--yet soils are both the most intensely used and the most easily depleted of resources. As a nutrient source and a matrix for plant growth soils are a form of accumulated ecological capital whose sustained use for food production is essential to the ambitions of developing countries and critical to the long-term success of other forms of development. The soils part of this book looks at soils both as a resource for agricultural development and as a part of the larger ecosystem in which such development occurs. Inventory methods adapted to developing country needs which locate and describe the extent, distribution, and quality of soil resources receive the greatest emphasis. Easy and accurate identification of underutilized soil potential is central to agricultural development strategies and the allocation of investments both to intensity of production on land already in use and the expansion of cultivation. Baseline environmental studies which consider the status or condition of soil resources as dynamic components of ecosystems in interaction with man have an importance that will increase with the escalation of demands upon soils accompanying higher yields and the extension of cropping into marginal lands.

Methods for the assessment of those ecological processes which maintain soil quality and fertility and monitor the consequences of development-induced impacts upon them and the ecosystem as a whole are of special concern. Techniques suitable for the prediction and analyses of soil and land degradation are required and are a critical facet of the transfer of soil science and technology to developing countries.

Chapter 1 of Soils provides general discussion of those issues and the social and environmental context of soil resource development in developing countries. Of particular significance are those features of tropical soils and tropical ecosystems which influence the application of standard soil methods in such settings. Chapter 2 draws on recent research on tropical soils and outlines specific soil information needs for resource management and project development at national, regional, and local levels. Aspects of scale and data intensity are discussed as guides to the selection of appropriate methods and emphasis is given to the identification of those soil-related socioeconomic and environmental constraints which present obstacles to development and provide a focus for management action.

Chapter 3 begins the methods portion of the soils part and describes approaches to soil and land classification appropriate to developing country needs which provide the primary basis for both the inventory and interpretation of soil and land resources. Methods for the acquisition of soils data at various intensities through soil survey and sampling are detailed as some of the requisites of survey design. Steps in the transformation of soils data into maps which are useful to decision makers are considered. The increasing value of remote sensing data sources and methods and their contribution to soil resource survey evaluation is presented, as is the continuing value of indigenous information as a source of important soils information.

Chapters 4 and 5 offer explicit discussion of the array of soil science field and laboratory methods and their uses in survey, classification, analyses and interpretation. Factors particular to their application in tropical soils are noted and methods are evaluated with respect to time, costs, and skills required by their use.

Chapter 6 closes the part with a discussion of soil degradation processes associated with development and present methods and models useful in both the monitoring and prediction of erosion salinization, acidification, waterlogging, compaction and crusting, loss of fertility, and determining their impacts on productivity.

Overview of Plants, Part IV

Plants are fundamental to the existence of life because of their ability to take carbon dioxide from the air and transform it into a simple carbohydrate. This simple carbohydrate is the basis for more complex organic compounds from which living organisms, both plant and animal, are constituted. Without this primary production by various types of plants, life on the earth as it is known would not exist. Some plants are more efficient than others in their productive capacity or ability to produce under optimal or stressful environmental conditions. Thus, one of the major roles of development is to increase primary production either by substituting more effective plants for less effective ones, or by increasing plant production through moderation of environmental stress. The carrying capacity of an ecosystem can thus be increased by improving the efficiency of production and management.

Effective project planning and implementation requires data on the kinds and amounts of biomass produced by plants which provide food, fiber, fuel, feed, medicines, and industrial raw materials. Plant-supported processes such as exchange of carbon dioxide for oxygen, uptake and release of soil nutrients, or covering the soil surface with leaves or litter significantly influence the resistance or stability of an ecosystem and its ability to continue to function at an optimal rate.

A baseline study generates data that can be used to assess plant resource production potential as well as key characteristics of the vegetation with regard to its successional status and some of the environmental influences currently in force. An integrated multidisciplinary ecological assessment of an ecosystem targeted for development can help in evaluating the opportunities and the risks that development would incur.

Ecological monitoring, if designed as an integral part of project management, can provide feedback information regarding the performance of a project and whether it is creating adverse ecological impacts that would decrease project effectiveness or shorten its duration. Ecological changes of far reaching consequences may be expressed by a decrease in the numbers of valuable species or the invasion of an important plant community by undesirable or poisonous plants. A monitoring program should detect such changes which could be remedied if known in sufficient time.

Methods for conducting resource inventories and environmental baseline studies of plant resources are numerous. Some methods are precise but require considerable time whereas others are less precise but because they can be done quickly, a large number of samples can compensate for precision. Each method has its advantages in dealing with problems of variabilty and scope of project and should be applied according to statistically-valid criteria. The intensity of the project generally dictates the intensity of inventory and monitoring activities.

To obtain preliminary planning data on large extensive projects a reconnaisance survey may be conducted to provide a general overview and appraisal of project potential, suitability, and impacts. Methods used would be visual estimates from travel through the area, interpretation of aerial photographs and references to existing data.

To provide a basis for more substantive planning, environmental baseline studies of resources may be needed. Survey methods used here must provide generalized but statistically valid data on plant communities, species, productivity, cover, density, and types of existing use.

Intensive surveys may be needed to provide detailed quantified information on critical areas within the larger project scope of influence. Methods of analysis may be similar to those used in a general survey but applied more intensively in providing information on plant species, community relationships, production parameters and special problems such as disease and insects which if not known and corrected could influence other plant communities.

Detailed surveys and monitoring for changes can provide information on plant responses and relationships occurring under unique field conditions that might occur as a result of project development. Specialized methods of plant, chemical, and biological analyses would be prescribed according to recommendations of specialists in microbiology, plant physiology, ecology, genetics, and agronomy. Methods appropriate to these disciplines would be used on samples collected in the field and submitted for laboratory study.

Sampling concepts and statistical procedures are discussed in order to provide an understanding of the importance of obtaining representative samples in adequate numbers to provide a valid representation of plants and ecosystem processes. When informed of the statistical levels of reliability associated with a set of data, decision-makers can weigh the risks against the chances for success involved in developing certain plant resources and ecosystems.

Interpretation and application of data obtained from baseline inventories and environmental monitoring can make a significant difference in immediate and long-term project success. Because baseline data are seldom, if ever, adequate for a developing country, initial data are needed to justify and to plan a project. However, biological data have considerable yearly and seasonal variability which cannot be overcome by merely gathering a large volume of data in a short period. Some help is available however by comparing data from ecologically similar areas. Another possibility is to make an ecological assessment of the ecosystem(s) targeted for development to determine if their present productivity and functions are presently in a stable condition or if they are in a declining trend.

Clearly, once a project is implemented it is wise to monitor key plant functions to assess whether or not adverse impacts are occurring. The same data base can serve as a guide to modify management practice accordingly.

Overview of Wildlife, Part V.

Wildlife is a renewable natural resource which can be developed for economic gain through suitable conservation efforts and properly managed utilization projects. Wildlife is also an integral part of human life support systems which must be maintained if economic gains in other natural resource sectors are to be sustainable. In addition, wild animals serve as indicators of the overall health of the environment in which they occur. Their numbers and specializations reflect the productivity and diversity of their habitats, and as consumer organisms they may concentrate pollutants in the environment through the food web, thereby giving humans an early warning of hazards to their own health. Since animals are dependent on plants, and these in turn depend on soils and water, deficiencies in any of these areas are quickly reflected in wildlife diversity and abundance. Thus wildlife can be the primary focus for a development project, or an important indicator or pest factor to be considered in the development of other renewable resources.

The historic oversight of the contribution of locally consumed bushmeat and other wildlife products in providing rural populations independence from market economies has led planners who depend on trade-derived economic statistics to greatly overestimate the benefits of some earlier development schemes. The expansion of wildlife programs as a primary focus of development projects is a recent phenomenon that coincides with the increasing appreciation of the fact that the productivity of tropical areas is increased through multiple species utilization. Such development plans imitate the complex ecosystems in tropical environments in which species diversity is generally higher but densities of individual species are lower than in temperate regions. This broadening of focus and scope of development programs also coincides with an increasing appreciation for the genetic diversity of natural systems from which marketable products have been derived historically, the global concern for conserving threatened species, and the development industries such as tourism.

I. INTRODUCTION AND SYNTHESIS

In Wildlife, part V, reasons for conducting wildlife studies are discussed, and some of the methods that can be used are evaluated. Methods vary in both kind and intensity depending on the value of the wildlife likely to be affected through development and goal or purpose of the development project. Taxonomy, data sources and methods for evaluating wildlife are summarized in Table 1.2. Methods examined vary with the kinds of ecosystems in the area to be studied, from closed forest ground surveys to open terrain remote sensing techniques. Emphasis is placed on the least intensive and least expensive methods compatible with the objectives of the project.

The techniques to be used in wildlife assessments are discussed in six chapters. In the first chapter, objectives and approaches are discussed. In the second chapter, a strategic assessment of wildlife within the country, region, or area to be developed is described; that assessment is to identify those areas in which tactical, or more detailed studies should be conducted.

A third chapter on methods considers techniques for determining the existing status of wildlife, looking first at habitat evaluation and secondly at studies of animal abundance and population structure and dynamics. The fifth chapter examines some techniques for management.

In the fourth chapter, methods for determining original, or earlier, status of wildife are examined, since the existing abundance and diversity of wildlife in an area may reflect human activity rather than the potential of the areas as wildlife habitat.

A final chapter looks at wildlife from a variety of cultural viewpoints, and stresses the need for cultural assessments, particularly from the viewpoint of incorporating traditional values and indigenous systems of conservation into the development process.

The emphasis throughout the wildlife section is on terrestrial species, with relatively less attention paid to aquatic species and domestic animals. Aquatic biota, particularly fisheries, is discussed in greater detail in part II, Aquatic Ecosystems. Domestic livestock is considered in relation to its interactions with wild populations. Evaluation of the vegetational component of habitat for wild and domestic animals is discussed in greater detail in part IV, Plants.

1.4 Local Populations and Subsistence Systems

The presence of social scientists on the working panels helped keep in mind the role played by local people in creating and maintaining valued resources. "Subsistence," as used here, refers to individual or group activities leading to a regular supply of food. As systems, these activities for some groups may be characterized as gathering or hunting, for others as swidden farming or shifting cultivation, and for still others as practicing a skill or trade in a market economy and using money to purchase food. Thus, subsistence systems and activities can involve relatively simple exchange relations, or complex relations involving specialization, markets, and money. A "subsistence" level economy does not necessarily imply a "poor" standard of living; it does mean that the marketplace is not central to the economy and that other institutions (such as reciprocity and exchange relations) are used in the regular supply of food.

> "Subsistence" is often confused with "bare subsistence." The latter refers to a meagre level of existence, whereas the former simply indicates a system of regularly acquiring food. A subsistence system may involve production of goods and performance of services in exchange for food.

Some subsistence systems such as shifting cultivation and pastoralism have been in place for millennia. Can 20th-century specialists concerned with resources and ecosystem functioning learn from local populations with very different cultural backgrounds and perceptions? They can; and this has helped to draw attention to common methods for eliciting local information about subsistence resources and environmental factors, especially with respect to plants and animals.

Informants in developing areas commonly have extraordinarily detailed recollections of past events related to subsistence activities. The present season, for example, is not compared only to the last, but to the last ten or more seasons. Such recollections, when cross-checked with those of other informants, can provide a perspective on how ecosystem functioning is perceived within a recognized range of variation. This is an important contribution that local people can make in the planning of a project, and in providing a temporal perspective for baseline studies.

Subsistence activities of local populations are an important resource in themselves. They have value in the sense of a favorable balance of energy yields and energy inputs. And insofar as crops are cultivated that have long histories of traditional use, the current yields for such cultigens (cultivated plants) are a barometer of ecosystem functioning.

In-place subsistence systems commonly involve such techniques as intercropping--growing a variety of plants in the same small field at the same time--and fields that are scattered over several contrasting environments. Maize or sorghum, beans, and cucurbits are a common combination in subsistence farming. Thus it is unlikely that all cultigens in the same field will be equally stressed by a common perturbation; some crops will survive a drought, a blight, or an insect pest better than others. And since the fields are variously placed, it is also unlikely that all the fields will be affected equally by the same constraint. Finally, in traditional subsistence farming several varieties of a crop such as sorghum are maintained, perhaps as much by chance as by design (Culture and Agriculture, 1983). Thus genetic diversity is enhanced--an important consideration. Perhaps in any project design, room should be left for continued cultivation in the traditional manner. This would provide a way of evaluating the success of the project compared to the traditional system, as well as a way of monitoring ecosystem functioning and preserving genetic heterogeneity of cultigens.

Ecosystems and Resources in Cultural Contexts

The inclusion of human populations in considering ecosystem structure and functioning is a necessary but enormously complicating factor, and can be treated here only in summary fashion. In this chapter, our strategy is first to consider some of the formal aspects of ecosystem functioning and then to take up three kinds of cultural contexts that affect resource utilization and ecosystem functioning. These cultural contexts are the ideas that a people have about their environment, the institutions that organize cooperation in exploiting resources, and the technology (processing as well as tools) that is available for extracting from the environment the products a people identify as necessary and desirable. The chapter concludes with a brief review of some of the field methods found helpful for learning how a people's ideas, institutions, and technology affect resource exploitation and ecosystem functioning.

Sometimes the most useful definitions are the simplest. Ecology was defined by Eugene Odum (1966) as the study of the structure and function of nature. Odum adds that it should be "thoroughly understood" that humankind is a part of nature, "since we are using nature to include the living world." From the outset, we have avoided setting up an antithesis between nature and culture. The environments in which development has taken place or is being planned have for the most part already been occupied by human populations, sometimes for centuries and sometimes for millennia.

From the points of view of both ecology and development, it is important to realize that a human presence is almost certainly involved in any area about to undergo development, and no matter how slight an effect may appear to be, indigenous peoples influence both the structure and the functioning of the ecosystem of which they are a part. Although some local populations may appear to be passive receivers of ecosystem resources, they may in fact be surprisingly active as managers of that system.

An ecosystem is a biotic community in interaction with its nonliving environment; the latter, the abiotic sector, is composed of the basic elements and compounds of the environment (Odum, 1966). A prominent feature of ecosystems is sometimes referred to as homeostasis; that is, they contain regulatory mechanisms that tend to maintain equilibrium despite fluctuations in external factors such as temperature or rainfall.

Ecosystem functioning is deeply associated with the idea of "feedback" and self-regulatory mechanisms. Margalef (1968) equates ecosystem functioning with many of the principles underlying cybernetics: "Ecology, I claim, is the study of systems at a level in which individuals or whole organisms may be considered elements of interaction, either among themselves, or with a loosely organized environmental matrix. Systems at this level are named ecosystems, and ecology, of course, is the biology of ecosystems."

A broad example of the functioning of an entire ecosystem is energy transfer: energy is emitted by the sun, transmitted through (but modified or filtered by) the earth's atmosphere, and further transformed through the photosynthetic capabilities of green plants into food convertible to energy by animals. Another example of ecosystem functioning is the cycling of nutrients, as between plants, water, and soils. An extended example of the functioning of the hydrologic cycle is given in the introduction to part II, Aquatic Ecosystems, and of nutrient cycling in part III, Soils.

The values of measures of ecosystem functioning made at a given instant in time are likely to be different at some other instant in time. Thus in ecosystem analysis one looks for trends as they become apparent through time. Among the methods presented in this volume are those useful in ecosystem monitoring to establish the kinds of changes within which homeostatic mechanisms (and hence the entire system) can still function. This helps in estimating the limits of change beyond which the system probably cannot function.

A past temporal trend can sometimes be approximated by looking at ecosystem differences spatially, as in the emergence of a "climax," "steady-state," or "mature" community of organisms in one area, with "immature" or less developed communities found elsewhere. Succession in ecosystems, however, can be (and often is) modified by human activities, such as clearing of woodlands or forests for farming. Recollections of local informants can be an important source of data in placing ecosystem changes within a temporal framework.

Resources, the valued properties of an ecosystem, also must be estimated in terms of spatial and temporal variability. Such estimates can give an idea of the *resilience* of resources--their ability to recover or persist in the context of change. Some resources may take a long time, many human generations, to mature--a stand of hardwood, for example--and it is particularly difficult to estimate the resilience of such resources. Fair inferences can sometimes be drawn from shorter time periods than those required for resource maturation; as, for example, by sampling rates of sapling growth and estimating rates of population replacement.

2.1 Ecosystem Structure, Functioning, and Resources

The structure of the abiotic and biotic sectors of the ecosystem is often perceived as having at least four different levels. The "lowest" level is that of the basic elements and compounds in the abiotic sector. Minerals from which soils are formed, the soils themselves, water (as vapor and liquid), and gases in the atmosphere are all examples of basic or elemental resources. The structure of the biotic sector of the ecosystem comprises the remaining three levels. One is the *autotrophic* (self-feeding) level. Organisms such as green plants directly utilize light or energy from the sun and, in combination with abiotic resources, manufacture most of the materials needed for growth and reproduction. Plants and other autotrophic organisms are sometimes known as producers. A second level of the biotic sectors contains *heterotrophic* organisms, often called consumers. These are micro- or macroorganisms that depend on plants or other animals for their food supply. A third biotic level is that of the *saprophytes* or decomposers. The decomposers break down "the complex compounds of dead protoplasm, absorb some of the decomposition products, and release simple substances used by the producers" (Odum,

18   I. INTRODUCTION AND SYNTHESIS

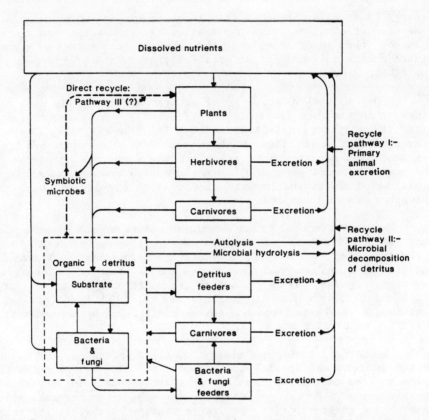

Figure 2.1 Ecosystem component parts and nutrient recycle pathways. Recycle routes I and II correspond to the grazing and detritus food chains. A third more direct recycle pathway (III) has been proposed. The autolysis pathway shown in the diagram could be considered a fourth pathway. [Source: Odum (1971)]

1971). The principal component parts of the ecosystem, and the pathways between them, are illustrated in Fig. 2.1.

The abiotic and biotic sectors and the different levels within them are necessary for the functioning of the entire ecosystem. Within each level, component populations can be identified which are critical for continued system functioning; an example is the role of forests in the maintenance of the carbon cycle. In addition, populations at one trophic level may be essential for the spread and survival of a population at another trophic level. For example, birds and other animals at the heterotrophic level function as vectors for the spread of plant life (including trees) at the autotrophic level.

Finally, there are resources which are needed for the maintenance of *a way of life* but not necessarily for the maintenance of human life. Resources for cultural maintenance are continually being redefined with changes in a people's expectation of what constitutes a good and proper way of life. "Rubber trees" and "iron woods," which at one time were only lightly used as resources, later became important resources for both local and distant populations.

To summarize, there are at least these kinds of resources:

1) Functional resources, as in the interrelations in the carbon cycle on which the continuing existence of the earth's ecosystem is based. Recognition of such relations is by no means complete; discoveries are still being made.

2) Resources or component populations needed for the maintenance of critical functions, as in the functioning of forests to maintain the carbon cycle, and the role of birds and other animals in forest propagation. Here again, knowledge of this subject, especially in the tropics, tends to be fragmentary.

3) Resources necessary for the survival or maintenance of a component population.

4) Resources crucial for a way of life rather than critical for human life itself. As expectations of what should constitute a way of life change, the identifications of these resources will also change.

Recognition of what constitutes the first and second kinds of resources above depends on a sophisticated perception of the environment as a *system* on the one hand and in terms of functional requirements on the other. Recognition of the third and fourth kinds of resources, which are directly and indirectly related to the survival of component (including human) populations, is a cultural matter, will change through time, and does not necessarily require a view of the environment as a system. But for such resources to be sustained, there must be an understanding of how they fit, structurally and functionally, within the larger system. This is a very simple point, but it is often neglected: development of these resources becomes sustainable only with recognition of functional requirements of the larger system.

It is largely because of this realization that the methods presented in the separate panel reports are organized according to "domains" rather than specific resources. Here a domain is considered to be a functioning subdivision of an entire ecosystem; the subdivision is not absolute, but the constituent resources are more like than different. Within each domain there is probably an endless list of resources--for the plant domain yesterday it was rubber trees, today it is thorn thicket to meet local energy needs, and tomorrow it will be some other product not yet identified. Within the domain of plants methods of analysis cross-cut specific resources, and the same is true of soils and wildlife.

Because of its volatility, water requires a somewhat different treatment. The functioning of more than one domain is involved; water exists in the atmosphere and in the ground, and is essential to all parts of the biotic community. This makes presentation of the methods for monitoring water resources particularly elaborate and especially critical from the point of view of total ecosystem functioning.

Just as changes in resources reflect changes in ecosystem functioning, so too are they likely to reflect changes in human cultural systems. To understand a people's interactions with the environment it is not necessary to understand all facets of their culture or of their environment. Netting (1968 and 1974) suggested that both culture and climate may have critical parameters which make possible an abbreviated approach to the study of human ecology. Among these parameters are precipitation, soils, population density, division of labor, rights to land, and the size of land holdings.

Important changes in resources can, of course, result from causes totally beyond human control or origins. Shifts in high-altitude jet streams affect the onset of monsoon rainy seasons, causing a change in the frontier between desert and semidesert and hence changes in the subsistence systems of local populations (Harris, 1980). Ash deposits from volcanic eruptions (Mount St. Helens; El Chichon, Mexico most recently; Krakatao, Indonesia, ca. 1910) cause profound regional changes in plant cover, again with basic effects on local populations and their subsistence activities. Of greater interest here are changes in resources more immediately relatable to human activities--that is, resource changes that take place in some human cultural context.

Examples are everywhere in the present and the past. Extending the availability of water resources is a case in point. Such extensions can range from simple ditches scratched out by neighborhood work groups along an escarpment wall from some permanent source of water to a distant field, to monumental aqueducts designed by specialists and involving the labor of thousands of persons in highly complex societies.

Nutrients in soils constitute another resource affected by a range of cultural practices--for example, seasonal burns of drying or dead vegetation in fields under cultivation or the far more elaborate manufacture and application of chemical fertilizers. The change in the soil resource base takes place within cultural contexts ranging from the very simple to the highly complex. The same is true of plant and wildlife resources. The degree to which plant regrowth approaches a woody state is affected by changes in the length of time that old fields are allowed to lie fallow.

Changes in basic resources such as water, soil, plants, and wildlife do not only take place passively within different cultural contexts. Resource changes may also be *initiated* and *managed* by local populations with highly contrasting cultural configurations: low population densities, few specialists, commodity exchange relations rather than markets, and an exceedingly simple technology. Perhaps because of the stark simplicity of such cultures, it may be difficult to perceive a local population as either initiating or managing changes in its environment. Sometimes such a population may be acknowledged as having an *effect* on resources, but without direct "awareness," "consciousness," or "understanding" of what it is doing. This is a point discussed further below.

By cultural contexts are meant those aspects of human behavior directly related to the use and exploitation of the surrounding environment and its resources. The relevant contexts can be labeled as belonging to (i) the realm of ideology or the ideas and beliefs related to the local resources and their maintenance, (ii) sociology or the institutions that control subsistence activities or access to valued resources, and (iii) technology or the artifacts and tools available for production, distribution, and consumption.

Ideas, institutions, and technology are not completely separable as contexts for analysis of resource change. A change in beliefs may facilitate use of a new kind of tool, with accompanying changes in the nature and organization of the work force. These interactions are almost certain to affect the way in which a resource such as arable land is used. A brief commentary on each of these contexts follows.

2.2 Cultural Ideas and Resource Use

The close relation between resource changes and local ideologies is not always appreciated. Sometimes it is assumed that only change directed from an outside source amounts to "planned" change. Rarely are local peoples credited with conscious manipulation of their environment, whether for "good" or for "bad." But the record accumulated from almost 100 years of fieldwork (in the modern sense) testifies overwhelmingly to the conscious awareness a people have of their environment and, in their belief, appropriate use of its resources. Many of these ideas about resources and their use are embedded in metaphors which translate strangely, if at all, into the language of science and scientists. This does not mean, however, that conscious awareness of environmental resources is lacking.

What are the beliefs, and who knows them? While some informants will know more than others, and few may be identified as having special knowledge (which sometimes can only be expressed in ritual settings), the underlying perceptions of environmental resources belong to the group at large, and hence are widely shared. Commitment to a set of beliefs can be the hallmark of belonging to a wider ethnic grouping. The reaction (positive or negative) of villagers to a local development project may well involve regional and even national attitudes. It is a basic mistake to assume that ideas expressed in a remote settlement or village somehow do not count because of the apparent isolation of the people. Ideas travel, and they count, as much among preliterate populations as among the literate.

2.3 The Institutional Context

The way in which a people organize for subsistence activities affects their manipulation and utilization of local resources. The household is a common unit of production in the developing world, especially where much of the labor is by hand. For certain tasks, however, the household workforce is insufficient; and is supplemented by the communal or group labor force. Group labor is widely used in the tropics and subtropics because of the critical timing of the agricultural cycle, especially at times of seeding and harvesting. Even physically large compounds with complex households composed of several sets of extended families may find it necessary to contribute to and benefit from a group labor force. Examples of communal labor tasks include initial clearing of woody overgrowth, planting, burning over an area used as rangeland, harvesting, and initial processing of the crop for storage and later distribution.

With communal labor available, even a small household may have quite scattered holdings, with fields located in different environments providing insurance against failure in any one ecozone. Family herds are also spread out and dependent on the availability of labor in much the same way. The institution of communal labor facilitates exploitation of distant and scattered resource areas, thereby reducing the intensity with which any one area is exploited and insuring against failure in any one area.

Commonly in the development process, it is the men who leave in search of wage labor, wherever it may be found. Women become, in effect, heads of households which now must feed themselves through their own labor in fields in the immediate area. Subsistence farming for home consumption becomes locally focused, where before it was likely to have been areally extensive. Intensification of land use around the homestead or the settlement is an almost certain result. Fallow periods are shortened and, to avoid exhaustion of soil nutrients, additional inputs must be made in the form of fertilizer and improved seed stock. These inputs require money, in large part earned by men as wage laborers elsewhere. If agricultural loans are made, they go to men with some record of savings or earning capacity; women remain in the household primarily as laborers and caretakers rather than managers of the land. Whereas before development the division of labor by sex was such that women were active--managers of some of the land and part of the herds--with the change to a cash economy, the disappearance of communal labor, and the need to purchase agricultural inputs even to sustain household production, women tend to have far less involvment in land management, resource utilization, and food production for their own households. Food must often be purchased to supplement food grown for home consumption even in rural households. Local resource use has intensified partly as a result of institutional changes involving wage labor.

2.4 Tools and Resource Exploitation

While changes in ideas and institutions indirectly affect resource utilization, changes in technology do so far more directly. The importance of tools in human history is well recognized and needs little further comment here except perhaps in two respects. First, a local technology, even (perhaps especially) the simplest, is likely to be finely tuned to resource management and exploitation. Studying the household tool kit leads one directly into questions of when it is used, for what, and by whom. Talking about tools is often easy and informative. Second, no matter how simply tools and artifacts are made--a digging stick with a fire-hardened tip, for example--some degree of cooperation within or between communities is required. Attention to sources of supply, processes of manufacture, and patterns of use, together with observations on who has been involved is almost certain to inform the investigator about the kind and extent of community connections sustained by even the simplest, the smallest, and most remote local settlements.

2.5 Methods for Discovering Relevant Cultural Contexts

Field methods for discovering cultural contexts that interact with changes in resources (including those brought about by development projects) are spelled out more fully in Plants (part IV). This is fitting because rural horticultural and agricultural peoples are usually most often involved in development projects. For the present review, some of the methods and their advantages and disadvantages are presented in summary form as follows:

Participant observation is probably the best known method of field work. The field researcher learns through daily personal participation in community activities over a period of a year or two. By being there, he or she converses and observes. The observer gains perceptions and insights that are probably not available with any other method. How better to learn what wild plants are of value than to go with the foragers? How better to learn what is hunted than to hunt?

But the disadvantages are important. It is difficult to do because the language must be learned to converse and the researcher's age and sex form constraints. Moreover, the results are difficult to duplicate, that is, time and those informants, cannot be replicated as if they were in a laboratory. After almost a century of fieldwork, there are still almost no rules; perhaps it is more an approach or even a philosophy than a method. The results are highly subjective. Interpretation requires extensive cross-checking with other informants and, if possible, repeated participation in the events of interest. If these events occur in the horticultural or agricultural cycle, from field preparation to crop harvest, considerable amounts of time are involved, and there is likely to be a high degree of variability from one informant to the next, and from one season to the next.

Interviewing key informants, with control over translation accuracy, is probably the quickest and most reliable method of fieldwork. The choice of informants is further discussed in part III. One must be sure to include an informant *not* identified by the local official, to include women, and, whatever kinds of informants are chosen, to talk with more than one in each group. Interviews can be more or less tightly structured. The more open they are the greater the risk that, if translation is involved, the translator takes over and directs the interview while the investigator stands by.

Sampling strategies in the selection of informants for interviews and companions for participant observation are best left open until some familiarity with the local population is gained. Who are the important people and who are unimportant according to your informants? The latter should be included in the sampling strategy. It is likely that the cultural contexts of interest are heterogeneous enough that there is little utility, and indeed some danger, in attempting to work with a randomly assigned set of informants. If there is enough time and familiarity with local society a stratified approach may be useful with informants selected at random from each stratum. Working with one or more old persons may be a necessary precursor of establishing a sampling strategy or even knowing what questions to ask. This is particularly true if the investigator is interested in how the local environment is perceived, what changes have been taking place, and who knows most about farming, herding, marketing, and so on.

A sampling strategy based on a social unit such as the household can be effective, but households may vary in size and composition in any society. There are regularities, of course, in the variability that households display, and once these are understood, then sampling can proceed.

Much the same is true of *settlements*. A single settlement is best looked at as part of a *system*, which is probably to some extent hierarchical. Inhabitants of a single settlement are integrated in many ways with other such settlements, as well as with other kinds of settlements; marketing administration, religion, and education are only some of the special functions a settlement may have. Some settlements may be ecologically specialized in that their inhabitants are committed to the use of a particular resource, such as rangeland or moorland. It is never safe to assume that such remote settlements are in any sense cut off from the larger system. Sometimes it is helpful to perceive the local settlement as a type unit in a larger community including the same and different types of units.

Field assistants play an important role in an investigation into the cultural contexts of resource utilization and change. Many developing countries have secondary school graduates or university students with some training in the social sciences, although not necessarily in fieldwork. However, older, noneducated informants often become exceedingly defensive when interviewed by younger, educated persons. If there are strong gender biases (and there commonly are), both women and men should be on the field staff. In many developing areas the division of labor betwen men and women in subsistence activities is rapidly changing; informants may still verbalize the traditional point of view, but the reality may be quite different.

Finally, local field assistants, may be best able to draw out informants on matters that are of great interest to the development and planning officer (or the human ecologist) but are strongly marked by *tabu*, especially before strangers. The tabu may have little or nothing to do with religion, superstition, or custom. This is the case in demographic matters, particulary with respect to women's reproductive histories. While religious beliefs about spirits may be involved, sometimes pain and the amnesia that blanks it out are behind women's tabu about speaking of the dead (as aborted, stillborn, or dying in infancy or in later childhood). Death rates at different ages may be among the most sensitive indicators of ecological functioning, together with length of intervals between births, length of nursing period, and so on. Much of this information is very difficult to get; women investigators and field assistants improve the chance of getting it.

Introduction to Integrated Approaches and Methods

Integrated inventories, sampling, remote sensing, data requirements at various levels and for various purposes, and georeferenced information systems are common to the methods detailed in the separate panel reports. It is redundant to discuss, in four places, the characteristics of the electromagnetic spectrum. Yet those characteristics are fundamental to the methods that have become available through orbiting satellites and the sensors mounted on them and on other platforms. Because integrated inventories are designed to pull together diverse data sets, we begin this section with a review of some prominent efforts to develop them.

3.1 Integrated inventories

Two general approaches may be used in classifying natural resource: the *integrated* or *holistic* ecosystem approach, and the *component* or *element* approach. An integrated, multiple-resource inventory unites all parts of the landscape--flora, fauna, soil, landforms, climate, and water--to form as complete a description as possible of an ecosystem. The goal is to express the interactive character of the land's parts and compare them to surrounding systems so that their spatial functional relations are also understood (Rowe, 1972; Bailey, 1980; Laban, 1981; Journal of Forestry, 1978).

In the component approach each part of the land--soil, vegetation, landforms, water, and climate--is described to create a classification. Although most resource inventories use the component element approach, they do not consider a single component of the ecosystem in isolation from other components. For instance, soil surveys, besides describing the physical and chemical properties of different soil entities, include information about the landform, vegetation, moisture, and temperature regimes under which the soils were formed.

The past decade has witnessed a worldwide realization of the need to consider resource management problems in an ecological context. As a result, several integrated inventory systems have evolved. Work has been under way since 1976 to *combine* the integrated and component approaches to create a hierarchical system for the classification of natural terrestrial ecosystems in terms of a combination of biotic and abiotic criteria for the United States (Driscoll et al., 1978). Merkel (in preparation) reported in Bailey (1980):

> The framework consists of four ecosystem components--vegetation, soil, landform, and water--which are examined to describe and define taxonomic ecosystems and ecosystem association in relation to their geographic arrangement.

Note, however, that animals are not included in this list of components. Furthermore, the current draft of this classification system has reduced its scope to plants and soils only.

Earlier, Bailey et al. (1978) discussed classification problems and procedures:

> Classifications may be built by aggregation or subdivision. By aggregation we mean beginning with a universe of individual objects and grouping them into classes based on similarities. In contrast, subdivision begins with a whole (e.g. continent) and subdivides into smaller and smaller units. As a general rule, taxonomies are based on aggregation and regionalizations are based on subdivisions.

The major difficulty remaining is that of interrelating the hierarchical classifications of different resources (plants, soils, landforms) at any level. This difficulty apparently exists whether the classification is constructed by aggregation or by subdivision, and it is a barrier to easy use of the results of the several approaches and classification schemes. Nevertheless, the utility of the approaches for planning and management still depends on definition of the problem.

The most feasible approach to integrating systems is by selecting those individual classifications (taxonomic and regionalization) that meet the needs of a particular planning job. The planner chooses, in each system, the level of hierarchy that will provide the best answers to the questions being asked, and then (and only then) integrates the information. (Bailey et al., 1978)

Two aids for orientation in the integrated resource inventory literature are provided by Bailey: (i) an international view of units in ecological land classification and (ii) levels of generalization in a hierarchy of ecosystems. They are reproduced in Tables 3.1 and Table 3.2. The domains and scale levels in Table 3.2 are similar to those in tables prepared for the parts of this book on soils, plants, and wildlife.

Integrated approaches show promise of making more effective use of data on vegetation, wildlife, soils, landforms, climate, and water. Marked changes in techniques of data collection (remote sensing) and analysis and storage (computers) have led to an active period for the research community (Hutchison, 1981). Two other examples of integrated resource surveys are briefly reviewed here.

The inability of the component approach to provide credible estimates of global arable lands and their production potential led FAO to initiate an "integrated approach." A study known as the agroecological zones project was began in 1976 to obtain a first-order estimate of the production potential of the world's land resources. Building on the 1:5,000,000 FAO/UNESCO Soil Map of the World, project scientists superimposed a climatic inventory characterizing temperature and moisture regimes matched to crop requirements. The land-evaluation principle developed over the plast 10 years by FAO and Dutch interdisciplinary land evaluation groups was used (FAO, 1976). The 13 steps and the logic of the methodology are depicted in Fig. 3.1 from Higgins and Kassam (1981). See also FAO (1978a, 1980a, and 1980b).

Since the productivity potential, under a given set of climate and soil conditions is largely dependent on the level of inputs and the technology applied, a land suitability assessment for each crop is made for two different levels of input assumptions: low input and high input. The former approximates a low technological level and involves hand cultivation. The high-input level involves chemical, mechanical, or other investments in cultivation.

Twenty-two alternative land uses are considered in the study. Other input assumptions, appropriate to specific environmental and sociological circumstances, can be used in the model.

The approach has many attractive features, and it may provide the best estimates yet compiled of potential productivity of the global land base. However, the study has some serious limitiations. Since it is being conducted at a very small scale (1:5,000,000), there is insufficient detail for use in management decisions at local and subnational levels. For many countries, there is insufficient data even for national planning. One of the major constraints in the study is the dearth of good data for many countries.

TABLE 3.1 Systems of units in ecological land classification. [Source: Bailey (1980)]

Australian Land Research Approach	British Land Unit Approach	Canadian Ecological Land Classification	Soviet Union Landscape Approach	United States Land Systems/ Ecosystem Approach
			Zone	
	Land Zone			Domain
	Land Region	Ecoregion		Division
	Land District	Ecodistrict	Province	Province
				Section
			Landscape	
Land System	Land System	Ecosection		District
	Land Type	Ecosite	Urochishcha	Landtype Association
Land Unit				
Land Type	Land Phase			Landtype
Site		Ecoelement		Landtype phase
			Facia	Site

A second example of an integrated survey is that sponsored by the Organization of American States (OAS, 1969). In the OAS casebook, the subject of a seminar in 1967, Freeman (1969) recommended the use of reconnaissance scale aerial photographs to survey land use and determine land use capability. However, the practical problems of integrating the several data sets, including those derived from aerial photos, are candidly set forth in the OAS casebook. Such an effort is an example of an aggregation procedure. Freeman cautions against lumping data, as follows:

> Much pertinent data on present land use is related to forestry, agronomy, economics, engineering and sociology-specializations which deal with specific factors influencing land use. The convenience of lumping together information from these different disciplines, and treating them as "land use data", is questionable, particularly for the purposes of development studies. In the first place, the cartographic portrayal of such data on a map of present land use does not seem feasible and, up to now, has defied systematic classification. Secondly, even if such a comprehensive classification were successfully developed and were cartographically possible, its value in development work would be doubtful because it would be contrary to a fundamental principle of geography, which is highly relevant to development studies: the need to isolate factors which are pertinent to an analysis of the study problem. In order to insure this possiblity, raw data should be preserved in such a form that a single factor affecting land use can be easily manipulated independently of other factors. The inclusion of a multiplicity of factors in a single land use classification system works against the flexibility needed for factorial analysis, particularly for the geographic analysis of individual factors. Finally, since it can be assumed that the classification would have to be structured so as to be used in the map legend, considerations relating to cartographic portrayal very likely would further complicate the classification, and would dilute its effectiveness.

Many resource inventory projects fail to include in planning and implementation a provision to monitor significant changes in the landscape. Baseline studies, single component, and integrated inventory approaches lend themselves to monitoring, and provision for monitoring

TABLE 3.2 Levels of generalization in a hierarchy of ecosystems. [Source: Bailey (1980)]

Levels of generalization and common scales of mapping	Current definitions
1. Domain 1:3,000,000 and smaller	Subcontinental areas of broad climatic similarity identified by zonal heat and water balance criteria.
2. Division 1:1,000,000 to 1:3,000,000	A part of a domain identified by macroclimatic criteria generally at the level of the basic climatic types of Köppen (Trewartha 1943).
3. Province 1:500,000 to 1:1,000,000	A part of a division identified by bioclimatic and soil criteria at the level of soil orders and classes of vegetation formations. Highland regions (e.g. mountain systems) with complex climate-vegetation zonation are distinguished at this level.
4. Section 1:250,000 to 1:500,000	A part of a province identified by a single climatic vegetation climax at the level of Küchler's (1964) potential vegetation types.
5. District 1:125,000 to 1:250,000	A part of a section identified by Hammond's (1964) land-surface form types.
6. Landtype association 1:20,000 to 1:125,000	A part of a district determined by isolating areas whose form expresses a climatic-geomorphic process (e.g. fluvial, glacial, etc.).
7. Landtype 1:10,000 to 1:20,000	A part of a landtype association having a fairly uniform combination of soils (e.g. soil series) and chronosequence of vegetation at the level of Daubenmire's (1968) habitat types.
8. Landtype phase 1:2,500 to 1:10,000	A part of a landtype based on variations of soil and landform properties such as soil drainage and slope that affect the productivity of the habitat type.
9. Site 1:2,500 and greater	A part of a landtype phase that is homogeneous in respect to all components, their appearance, potential to produce biomass, limitations to use and response to management. It is the basic geographic cell of the ecological classification.

change should be an integral part of the planning in any of these approaches. Further, the problems of composite mapping have been addressed by Mouat and Treadwell (1978). Composite mapping can be done at more than one level, but it is most successful if those doing it realize from the start that the boundaries of each parameter will not coincide with other boundaries. Twenty-five years of experience of conducting surveys in Africa with a number of variables (vegetation, climate, anthropozoic, edaphic) by Le Houerou (1981) has resulted in succinct recommendations on the need to consider these variables in relation to each other and cost estimates to do so.

In a program designed to inventory and monitor the natural resources of an area, the ideal approach would be to integrate all components of the ecosystem in the process. However, it is seldom, if ever, possible to assemble at one time or even over a short period of time the skills, equipment, and supporting infrastructure needed to accomplish a truly integrated inventory. An attractive alternative is to anticipate integration by using a common data base system--that is, plan and conduct each of the component inventories in such a way that the data can be combined and integrated into a common data base. Useful information may then be extracted for a wide variety of planners, decision-makers, and policy-makers. If integration is intended, some important decisions must be

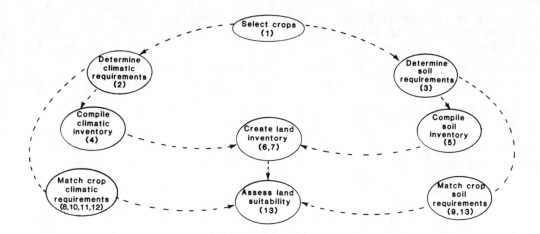

1. Selection and definition of land-utilization types (crop and product, production type, input level).
2. Division of the crops of the study into groups based on differences in their eco-physiological response, temperature and radiation, and compilation of crop climatic adaptability inventory.
3. Assemblage of information on the soil requirements of the crops.
4. Compilation of a quantitative climatic inventory based on major climates (characterizing temperature regimes) and lengths of growing periods (characterizing the time when water and temperature permit crop growth).
5. Computer assemblage of a soil inventory, by countries from the FAO/Unesco *Soil Map of the World*.
6. Overlay of the climatic inventory (from 4) on the soil map (5) and area measurement of resultant climate/soil units.
7. Computation (from 5 and 6) of country extensive soil units (by slope class, texture class and phase) by major climates and lengths of growing period zones (thirty-day intervals).
8. Matching of the climatic inventory (4) with the crop groups (2) and, where the temperature requirements of the crop groups are met, calculation of the biomass constraint-free individual crop yield by growing period zones.
9. Matching of the soil requirements of crops (3) with the soil units, slope classes, texture classes and phases of the soil map, by rating soil limitations to growth and production.
10. Compilation and rating of the various agro-climatic constraints to crop production occurring in the various major climates and growing period zones.
11. Application of the agro-climatic constraints (10) to the constraint-free crop yields (8) to derive anticipated (agro-climatically attainable) crop yields, by growing period zones.
12. Classification of the agro-climatic suitability of each growing period zone according to the anticipated crop yields (11).
13. Application of the soil-limitation ratings to the agro-climatic suitability classification of each growing period zone (12) according to the soil composition of the zone, to arrive at the land suitability classification, that is, extents of land variously suited to the production of the crop.

Figure 3.1 Schematic outline for the FAO agroecological zones methodology. [Source: Higgins and Kassam (1981)]

made during the planning stages of a natural resource inventory program. Since the data load will probably be substantial and interactions among ecosystem components complex, computer-implemented analysis will be essential.

The following considerations should be included in planning for integrated resource inventories:

Substantive Matters	Practical Considerations
Ecosystem components to be included (now or in the future)	Scale, detail of mapping
Compatibility of data sets	Map projection to be used
Intensity of mapping	Extrapolation possibilities
Definition of geographic area	Minimum grid or polygon size
Aggregation of possibilities	Ability to purge and update data
Level of hierarchy	Desired precision and accuracy

Some of the most serious obstacles to integration of existing component resource inventories are differences in scale and map detail, incompatibility of inventory objectives, differences within single component classification systems, and inconsistencies of survey techniques. The lack of integrated resource information systems makes multiple-use programs extremely difficult or impossible.

The potential of current approaches and some of the problems involved are illustrated by the ECOCLASS method developed by the U.S. Forest Service (1973, 1977). This method combines an ecological classification of terrestrial and aquatic ecosystems with landform attributes and integrates existing systems of soil, vegetation, and aquatic distributions. Animals are not included. The ECOCLASS system was designed for the Rocky Mountains of the United States (Butlery, 1978), but is not yet operational. The method was prepared by an interdisciplinary task force, which sought to provide ecological subdivisions of major forest types to be used for improved description and coordinated management of multiple resources. It attempts to offer a unifying framework or hierarchy appropriate for a variety of organizational and management needs.

Of some significance for developing countries is a recent collaborative effort undertaken by the FAO in conjunction with the International Institute for Land Reclamation and Improvement (ILRI). By the 1970's many nations had developed their own systems of integration and evaluation. Many were modeled after of the U.S. Department of Agriculture's Land Capability Classification (Klingebiel and Montgomery, 1961), but a need for standardization and international discussion was recognized (FAO, 1972). A summary of different approaches, strengths, and weaknesses was prepared by FAO (1974); a comprehensive approach to land evaluation was advanced in 1976. Subsequent work led to the *Land Evaluation for Agricultural Development* (Beek, 1978), which was written for Latin America. The FAO/ILRI approaches have found considerable application in developing countries. These and other approaches are considered in greater detail in Soils (part III).

The United Nations has under way or has proposed a series of projects that include monitoring of natural resources. One is the Global Assessment of Forest Resources. Another involves "integrated monitoring" of pollutant transfers from the atmosphere to soils, plants, and animals. Aquatic sectors are not specified. Two other projects are also relevant. One is to establish a unified geographic information system at an estimated cost of $800,000 (U.S.). The second is to determine minimum area requirements for the "sustained productivity of species and ecosystems."

3.2 Sampling for Identification and Appraisal of Resources

The ultimate goal of stratifying natural resources into homogeneous units is finding combinations of vegetation, soils, topography, and so on that will have similar responses to practices of intensive land management. Not all such units will occur in contiguous areas; rather there will be a patchwork or mosaic within the study area. Background resource information for the area may be available in the form of maps, published reports, satellite imagery, and, in some cases, aerial photographs.

Once strata of homogeneous units have been determined, field sampling can be initiated to obtain more detailed data from each stratum. The importance of the stratified approach for sampling purposes may be more readily appreciated when one considers that a strictly random sample of resource measurements for the study area will often overemphasize frequent classes of data and underrepresent less frequent ones. For example, random sampling for vegetation cover will produce estimates that overrepresent species dominant in the study area, while species restricted to certain habitats are undersampled. Stratification of the area according to vegetation cover characteristics will provide vegetation cover values weighted according to the area occupied (see Box 1). Another example of stratification is erosion intensity (see Box 2).

Stratification can be accomplished in several ways. A practical way is to divide and map the study area into homogeneous units by field data variables. These may include vegetation cover, cultivation, grazing, relative relief, and erosion intensity.

For specific discussions of field data needed by domain, refer to the appropriate chapter. Details of stratification and sampling of various resources in large areas are given by Olsson and Stern (1981).

After a region has been mapped according to resource data categories (vegetation, soils, topographic relief) and intensities (for example, erosion) overlays are combined to show areas of homogeneity that may be related to management treatments or development potential. Homogeneous areas are treated as units for statistical purposes, especially sampling. These units are commonly referred to as strata and are then used to obtain further field data. The sample unit should be weighted by size of area, by percent of total area, and by importance for intended project use, in order to assign relative importance in collecting or interpreting the data.

Once strata have been identified, care should be taken in obtaining more detailed data on each resource within each stratum. But it must be remembered that statistical procedures for randomization cannot always dictate how ecological and environmental data are obtained. Random data may end up as "random events" that cannot be interpreted for resource development. Common sense must be used by everyone responsible for the resource.

Box 1

The *vegetation* cover map may include:

      Desert - barren land   Bush grassland
      Grassland   Dry forest

A *cultivation* map may include:

      No cultivation   Mechanized cultivation
      Traditional cultivation   Irrigated cultivation

A map of *grazing* may be divided into:

      Seasonal "open range"   Moderate intensity of
      grazing   grazing at a moderate
                                 distance

      Low level of distant   High intensity of
      grazing   constant grazing near
                                 holding pens

In field sampling, specific methodologies are used to obtain information on soils, plants, animals, and water. For each resource certain criteria may be used to select a measurement technique. For example, the method selected to estimate soil moisture may depend on available laboratories and levels of training experience. However, accuracy and precision are different, and this should be kept in mind when selecting field techniques for resource measurements. A rule of thumb is to consider the criterion of repeatability. Repeatability is related to the precision of the technique, and means that different observers obtain estimates that are fairly close to one another; that is, there is not much spread in data values for the same observations. However, this is not an indication of how accurate the technique is--that is, how close the values obtained are to the actual values. For example, estimates of vegetation cover classes at 10 percent intervals (0-10, 11-20, 21-39, and so on) may yield very precise estimates. That is, several observers using the technique should obtain an average that falls within the same interval (say, 11-20 percent). However, the accuracy of the estimate is unknown; the cover may be 13 percent, for example. If cover estimates need to be within one percent for individual species, then accuracy is not obtained in this example.

Each resource measurement should be considered in light of accuracy and precision. In particular, remotely sensed data should be evaluated to determine whether the accuracy obtained is acceptable. Then the levels of precision desired should be determined, keeping in mind economics of obtaining the needed information. In many cases, mitigating efforts must be made to obtain an optimum trade-off between accuracy and precision for a given measurement technique. Common sense may dictate which procedure should be used. Statistical methods do not always take into consideration the economics and sociology of the planning process for developing resources.

```
Box 2

Erosion intensity subareas:

   No erosion marks
   Low degree of erosion
   Moderate degree of erosion
   High degree of erosion
```

Sampling of all resources cannot be accomplished in a simultaneous fashion. The homogeneous units formed for stratification may not be equally efficient for obtaining data on vegetation, soils, animals, and water characteristics of the study area. Vegetation is the easiest to deal with in stratified approaches, but soils do not always occur in a constant relation with vegetation types, large animals do not graze only in one vegetation type, and water is not correlated with any vegetation other than aquatic forms. So each resource may have to be measured in terms of broad strata, which can be subdivided as needed for detailed measurements of a resource. Remote sensing plays an important role in constructing such strata and partitioning the study area in manageable units for acquiring field data.

3.3 Levels, Scales and Data Sources Required for Surveys and Mapping

Before discussing geographic information systems (section 3.5), it seems appropriate to briefly summarize problems of scale and mapping in relation to data sources, theoretical needs, size of phenomena, and project requirements. Contributors to this volume have paid special attention to the selection of appropriate methods for assessing spatial variability in resources. One problem is that of scale. At one scale, say 1 to 1,000,000, there may be so much loss of detail that an entire area appears homogeneous, while at a scale of 1 to 250,000 sufficient detail can be represented that the area appears heterogeneous.

Tables designed to identify regularities in levels of data, mapping scales, data sources, and taxonomic categories were prepared:

1) Soils (part III), Table 4.1 Soil survey types, scales and remote sensing sources.

2) Plants (part IV), Table 2.1 Levels of plant resource assessment.

3) Wildlife (part V), Table 1.2 Data sources and methods for evaluating wildlife.

It is notable that soil orders run from small areas as the first order to large area as the fifth order, while for plants the usage is reversed: first-order plant surveys are made for preliminary reconnaissance. At the same time, an attempt was made to determine the most suitable data sources for phenomena. Discussion of choice of method in relation to scale is

34   I. INTRODUCTION AND SYNTHESIS

found in parts III-V. Although no table was prepared, a discussion of scale is also found in Aquatic Ecosystems, section 2.1.

3.4 Remote Sensing

Remote sensing means gathering information about objects or conditions without direct contact. Although the eye is the original remote sensing device, the term commonly implies the use of cameras or other sensing devices carried aboard aircraft or spacecraft. The first photoreconnaissance with a balloon was carried out in 1859, and aerial photography has been used for natural resource mapping (especially of soils, rangelands, and forests) and interpretation for at least five decades.

The past two decades have seen dramatic changes in remoting sensing capabilities for surveying and monitoring the earth's surface environment. This includes both data acquisition and data analysis methodologies. Major advances have been made since 1960 in instruments, cameras, other remote sensing devices, equipment, and sampling strategies for observing, characterizing, and monitoring a target or scene. Use of new data acquisition methods in the laboratory, in the field, and from air and space platforms are providing views never available before, from the $1m^2$ field plot to thousands of square kilometers in a single synoptic view of the earth's surface.

In the past, the ability of the observer to acquire data far exceeded the ability to analyze and interpret the data. The computer places at our disposal an unprecedented capability for storing, retrieving, overlaying, analyzing, and interpreting vast quantities of data. Relations between the many complex facets of global resource management--physical, biological, chemical, cultural, economic, political and social--can now be explored. This capability for relating data sets in a digitized format is discussed in greater detail in Geographic Information Systems (section 3.5).

A third area of new technology of importance for the observation and management of earth resources is communication. It is now possible to transmit virtually instantaneously from one place to another sound, images, and large amounts of data. These three technologies are integrated in the Land Satellite (Landsat) program. Four Landsats have been launched by the National Aeronautics and Space Administration during the past 10 years.

Although cameras have been around for over a century, the past five decades have seen vast technological advances in precision, small (35- and 70-mm), medium (5 x 5 inch), and large (9 x 9 inch) format aerial mapping cameras and lenses, photointerpretation and photogrammetric equipment, and other peripherals. New filters and films have enhanced capabilities for recording data at diagnostic wavelengths of light reflected from different surfaces. Every object interacts with incoming solar energy by absorbing reflecting, transmitting, and emitting the radiant energy at different rates and in different portions of the spectrum. The resulting spectral response (pattern of reflected and emitted energy that varies according to wavelength) can be used to characterize the object. Figure

3.2 illustrates the electromagnetic spectrum, showing the various regions such as x-rays, visible light, and microwaves, the wavelengths in each region, and the areas in which different remote sensing systems operate (Best, 1982). Figure 3.3 graphically illustrates spectral responses and shows how they differ for soil, vegetation, and water (Swain and Davis, 1978).

"Spectral signature" is a phrase frequently used in literature on remote sensing. In the early days of Landsat, it was thought that characteristic spectral responses could be uniquely identified for a wide variety of surfaces on the earth. Such unique identifications could be called "signatures". However, attempts to observe the same signature repetitively led to the realization that a spectral signature atlas is not feasible under present circumstances.

In current usage, the preferred phrase is "spectral response." For vegetation, response at the level of plant formation or of plant associations is influenced by (i) the characteristics of the leaf, (ii) the background in which it occurs (soil, slope, aspect, and any other feature of the immediate environment), and (iii) the plant canopy (vertical and horizontal orientation of the leaves). For soils there are similar qualifications, as discussed in Part III of this volume. The reliability of the interpretation at these levels is substantially improved if good multitemporal data are available, i.e., repetitive coverage of the same scene.

A noteworthy advance in film types was the development of camouflage detection film during World War II. This film, used in conjunction with a minus blue filter, records light reflected in the near-infrared region of the spectrum. Since this region is invisible to the human eye, infrared films shift the color-sensitive emulsions up one notch by convention assigning red light (the longest wavelength in the visible part of the spectrum) to the near-infrared. Thus, green vegetation, being highly reflective at these wavelengths, appears red. The term false color is often applied to these pictures. A useful reference on multispectral films is Wenderoth and Yost (1972).

Parallel to the development of spectral capabilities were advances in spatial resolution, or the ability to distinguish between objects. High-resolution films, when used with precision optics, now allow the discrimination of very small features from great distances above the surface of the earth. With stereo aerial photography (normally acquired with a 60 percent overlap between successive frames in the direction of flight), landscapes and vegetation resources can be analyzed in three dimensions. Stereo parallax and height perception can be further enhanced for special applications by use of convergent aerial photography, which employs both forward- and backward-looking cameras to greatly increase the distance between successive frames covering the same scene.

Photographic remote sensing systems, even with the various films and filter combinations used to segregate spectral variations in reflectance, are still limited to recording information at wavelengths between 0.3 and 0.9 micrometer. Perhaps the most significant advance in remote sensing technology for earth resource surveys have been the development of non-photographic systems capable of measuring spectral variations of earth surface features from the ultraviolet to the emissive infrared region of the spectrum. This wide spectral range increases the utility of

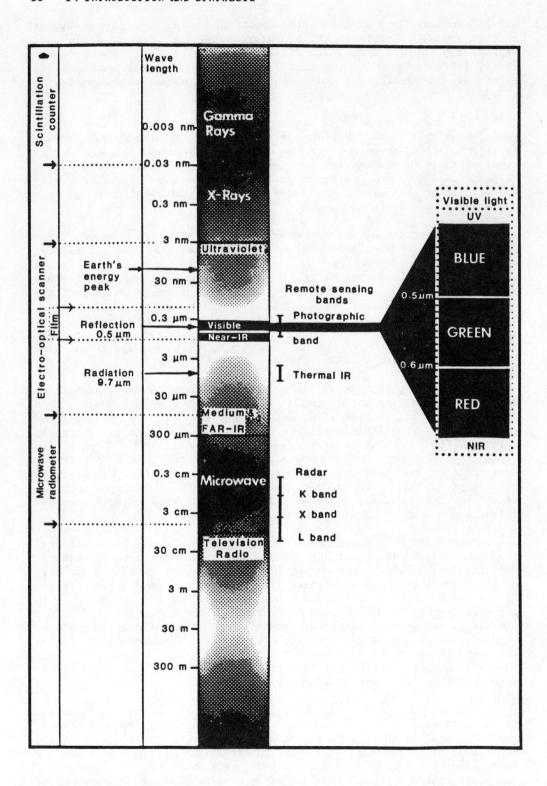

Figure 3.2 The electromagnetic spectrum and remote sensor sensitivity. [Source: Best (1982)]

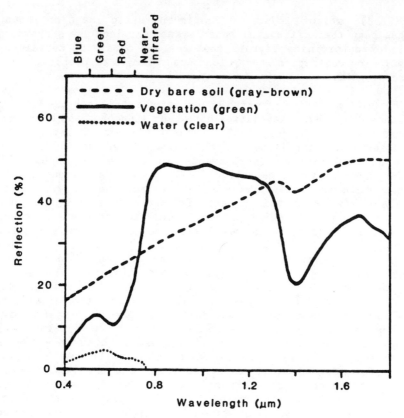

Figure 3.3 Typical spectral curves for green vegetation, bare soil and clear water. [Source: Swain and Davis (1978)]

remotely acquired data for land use mapping. Photographic pictures result from chemical reactions between reflected light and the sensitive emulsions of the films. Non-photographic systems use electronic detectors (roughly analogous to light meters), which respond to the amount of energy reflected or emitted from the object and produce an electronic signal in proportion to the incident (incoming) energy. The non-photographic sensors used in natural resource studies include thermal scanners, certain radar systems, the multispectral scanner (MSS), and with the recent launch of Landsat 4, the thematic mapper (TM). The MSS is by far the most commonly used non-photographic system, but thermal scanners and radar systems have specific applications.

Thermal scanners detect the part of the absorbed solar energy that is reemitted in the far infrared region of the magnetic spectrum. Their primary use in vegetation assessment is for detection of forest fires, although other applications are theoretically possible (Morain, 1976). These sensors can operate at night, but they cannot "see" through clouds.

Radar systems operate in the microwave region of the spectrum. They are unique among sensing instruments used in resource studies because they provide their own energy source. A signal is transmitted, and data are recorded from the part of the signal that is reflected by the surface. Since radar systems do not depend on solar illumination, they can acquire imagery when most other sensors are inoperable. Side-looking airborne radar (SLAR), which produces a continuous strip image on photographic film, has been the most useful radar system for land use surveys, particularly, for determining terrain features and obtaining certain types of information on vegetation where cloud cover limits or prohibits the use of other sensing instruments.

The Landsat series of earth observation satellites has been in operation since 1972. Four satellites have been launched, the most recent one being Landsat 4 on 16 July 1982. Landsat 4 begins a new generation of earth observation satellites, with a redesigned sensor (the thematic mapper) and substantial improvements in resolution, spectral bands, selection orbit control, geometric fidelity, and data transmission capabilities. Data from this system should be routinely available in 1984. The overall success of this program is also indicated by the increasing numbers of data receiving stations, as well as the plans of other countries including France, Japan, and Brazil, to launch their own satellites.

The Landsat satellites are in a sun-synchronous, near-polar orbit which provides repetitive overflights at mid-morning local time. Landsats 1 to 3 could scan the entire land surface of the earth every 18 days, and Landsat 4 can do so every 16 days. However, the distribution of line-of-sight receiving stations, limited on-board data storage capability, and priority image acquisition have all imposed limitations on the frequency of repetitive coverage. Cloud and dust cover and occasional sensor malfunctions have also affected the availability and quality of the imagery. For many areas of the world that have had receiving stations operating for several years (Brazil, Canada, the United States), images may be available for more than 100 different days. For other areas, very few images are available. Nonetheless, repetitive coverage is a powerful tool for monitoring change. If such information will be of use in a project, the availability of the imagery should be investigated. Computer printouts of all available images for a specified geographic area can be acquired from the EROS Data Center, Sioux Falls, South Dakota. As additional receiving stations are constructed, tape recorder problems are overcome, and with the Tracking and Data Relay Satellite (TDRS) put in proper orbit, monitoring change by earth resource observation satellites will become more practicable. Planners of future resource inventory projects should be alert for this development.

The MSS includes four sensors that are sensitive to light in two wavelength bands in the visible spectrum (green and red) and two segments of the near-infrared. Band designations, spectral ranges, and principal applications of the thematic mapper (TM) appear in Table 3.3 (Landsat Data Users NOTES, 1982).

The resolution of these scanners is expressed in terms of picture element (pixel) size. For the MSS, the area of ground surface discretely detected measures 56 by 79 meters, or approximately 0.4 hectare. Within each pixel, the reflectance values for that ground surface area are averaged for each of the four spectral bands. The TM pixel will increase this resolution to represent 30 by 30 meters on the ground, less than 0.1 hectare, for all bands except the thermal band, which will measure 120 by 120 meters or 1.44 hectares.

The electrical signals produced by these sensors are digitized for transmittal, storage and manipulation. This is roughly analogous to producing a television picture. Since the scanners operate continuously,

TABLE 3.3 Principal applications, band designations and spectral ranges of the Thematic Mapper.

Thematic Mapper
The TM operates in seven spectral bands. The selection of bands for this sensor was the subject of considerable study and debate. The band designations, spectral ranges, and principal applications are as follows: • Band 1 (0.45–0.52 μm) Designed for water body penetration, making it useful for coastal water mapping. Also useful for differentiation of soil from vegetation, and deciduous from coniferous flora. • Band 2 (0.52–0.60 μm) Designed to measure visible green reflectance peak of vegetation for vigor assessment. • Band 3 (0.63–0.69 μm) A chlorophyll absorption band important for vegetation discrimination. • Band 4 (0.76–0.90 μm) Useful for determining biomass content and for delineation of water bodies. • Band 5 (1.55–1.75 μm) Indicative of vegetation moisture content and soil moisture. Also useful for differentiation of snow from clouds. • Band 6 (10.40–12.50 μm) A thermal infrared band of use in vegetation stress analysis, soil moisture discrimination, and thermal mapping. • Band 7 (2.08–2.35 μm) A band selected for its potential for discriminating rock types and for hydrothermal mapping. These spectral bands were chosen primarily for vegetation monitoring. The one exception is band 7, which was added primarily for geological applications.

the data have been subdivided into nominal "scenes" or areas of coverage. The coverage for each scene has remained essentially the same during the entire Landsat program and measures 185 by 185 kilometers (34,225 kilometers squared).

There are two different modes in which this digital reflectance information can be displayed. The first mode converts the digital data to black-and-white positive transparencies for each of the four bands. Three of these transparencies are optically registered and sequentially exposed on film through appropriate color filters. This results in a hard copy photolike image called a color composite. Since the infrared band is used, terminology similar to that used for color infrared films is employed and the product is called a false color composite. Because it is possible to vary both the color (by using various filters) and the intensity of the light passed through these transparencies for each spectral band, there is a great capability for enhancing different subjects of interest.

The second mode is to analyze the digital information from computer-compatible tapes (CCT's). This mode allows sophisticated digital enhancement of images (including geometric corrections and registration, haze removal, contrast stretching, edge enhancement, and other transformations). The data can then either be converted to single-band trans-

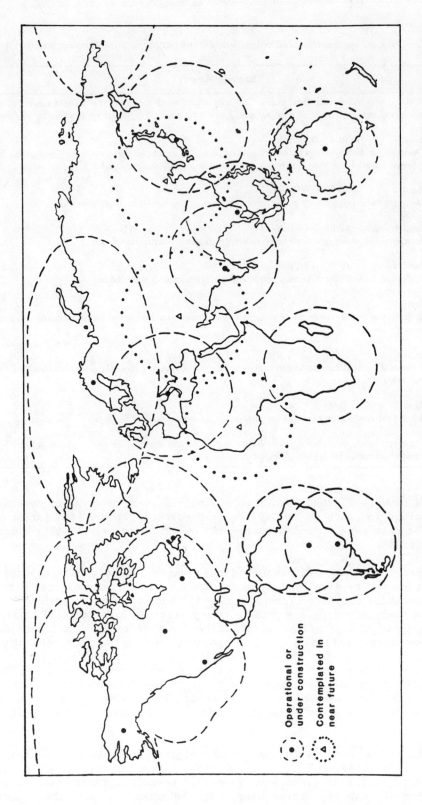

Figure 3.4 Approximate areal coverage of Landsat receiving stations. Source: Paul and Mascarenhas (1981)

parencies and printed as composite images by the photographic process described above, or used to classify subject matter by afor computer-aided pattern recognition based on four-dimensional cluster analysis techniques. The clusters, when statistically validated, represent spectral signature groups, which can often be related to subjects of interest by knowledgeable analysts.

In the past, the ability to acquire data has far exceeded the ability to analyze and interpret data. For instance, each of the four spectral sensors in the MSS records data for 7.5 million pixels per scene. This means that a total of 30,000,000 data points are transmitted for each scene every 23 seconds. The thematic mapper will increase this data flow to over 10 million data points per second.

Data from the Landsat sensors are transmitted to receiving stations in digital form. Receiving stations in operation in December 1982 included the following.

United States:

    Greenbelt, Maryland
    Goldstone, California
    Fairbanks, Alaska

Canada:

    Prince Albert, Saskatchewan
    Shoe Cove, Newfoundland

Brazil: Buiaba

Argentina: Mar Chiquita

Australia: Alice Springs

India: Hyderabad

Indonesia: Djakarta

Italy: Fucino

Japan: Tokyo

Sweden: Kiruna

South Africa: Johannesburg

Thailand: Bangkhok

The growing number of receiving stations helps make Landsat data locally available in short periods of time. Receiving stations are under construction in the People's Republic of China and under consideration in other countries. The common reception radius of ground receiving stations is 2780 kilometers. The area covered by the Landsat receiving stations is shown in Fig. 3.4. At present, few stations have been upgraded to receive thematic mapper data from Landsat 4.

Since the launch of Landsat 1 in 1972, resource scientists in many nations have used satellite data as an aid to inventory and monitor land, water, forest, crop, range, and mineral resources.

The Kenya Rangeland Ecological Monitoring Unit (KREMU) has completed a national forest cover inventory through visual analysis of color composites and some digital analysis done at the Canada Centre for Remote Sensing, in Ottawa. A more detailed study of four forest areas is being done with visual interpretation and digital analysis of Landsat MSS data. Changes in these forests will be assessed over a 20-year period by a combination of aerial photography and Landsat data.

The UNESCO Integrated Project in Arid Lands (IPAL) has been using aerial photography to map the vegetation of Marsabit District in Kenya and some satellite imagery to determine areas of overgrazing in relation to vegetation types. The Remote Sensing Section of KREMU is preparing a land use map of all of Kenya, using Landsat imagery supplemented with sample ground observation from a light aircraft. The land use map of the high potential areas will give some estimate of available wood fuel biomass.

In many resource inventory programs, Landsat data can serve two important functions: to integrate ecosystem components (such as soils, geology, landforms, vegetation, land cover, human settlements, cultivation and crop areas), and to provide a synoptic view for project planning and field sample selection. It can also be done in a timely and cost effective manner (Ackerson and Fish, 1980). Landsat data have served both functions in a project designed to develop a multiple resource georeferenced information management system for Bolivia (Bartolucci and Phillips, 1980, 1981, 1982).

Land resource scientists are also using Landsat to map and monitor land degradation caused by wind erosion, water erosion, alkalinization, and flooding. An important example of denudation and sand dune encroachment is discussed in the remote sensing section of Plants (part IV).

Remotely sensed data from aerial or space sensors are seldom used as the single source of information on which to base management decisions. Instead, they are used with other data sources to analyze and interpret a scene or condition at the surface of the earth. A whole new area of technology is being developed to register to the same scale multiple sources of data related to specific geographic areas. Such geographic information systems provide a framework within which remotely sensed data can be related to other sources of data. Advances in low-cost microprocessors, interactive analytical devices, and geographic information system technology will place these systems within reach of many developing countries.

3.5 Geographic Information Systems

The most common and probably still the most familiar example of a geographic information system is a map. Even a crude sketch map, examined by an experienced observer, can be used to integrate observations on such diverse matters as slope, water resources, soil types, plants, wildlife and a variety of human activities. Such a system is inherently multidisciplinary, as it integrates data collected by researchers in many fields and hence perceived from several different points of view. The models of reality that can be constructed from such an information system incorporate both the diverse points of view involved and the diverse subject matters. Geographic information systems are essential in modeling environmental constraints and opportunities.

A geographic information system can also incorporate historical information--what informants recall of rainfall variability, for example, or when a road was built, and the sequence of pioneer human settlements. Information on land ownership, land rights, and land use classification should also be included.

The problem remains, however, of retrieving all the information that went into making the map. One way of doing this is the overlay approach. Separate maps of vegetation, soils, settlements, rainfall distribution, and so on can be prepared in detail, and then placed over and registered to one another. The overlay method is old, but it is still central to the conceptual development of geographic information systems. See also chapter 2 in part V, Wildlife.

The development of information systems has been greatly affected by satellite digital data, computers, and digitization. Precise comparisons with other data sets are facilitated by digitization, a procedure in use for 20 or more years, and demand for the technique has led to further refinements. Now any information that can be mapped can be digitized. Geometric corrections (for instance, by use of the Universal Transverse Mercator) can be made for each digitized data set so that the data from different sources match precisely. Digitization is an expensive procedure, however. At present, computer-based information systems make it possible to represent at the same scale information on a variety of different subjects, including water resources, plant communities, soil types, wildlife distribution, and human activities. These different kinds of data can be aggregated in three ways: by subject, by spatially registering different areas or regions, and temporally, with repeated sets of data. The combination of spatial and temporal dimensions enhances both detection of changes and elucidation of the feedback relations in ecosystem functioning. In a well-designed system it should be possible for encoded data to be disaggregated by subject, space, or time to serve the purposes of different investigators and projects.

Modeling ecosystem functioning is central to current research in ecology. A properly updated geographic information system can provide information on changes in resource inventories and ecosystem functioning which are essential in formulating development policy. As development projects are initiated, the changes that are effected feed back on the projects themselves; planning must take into account these feedback relations.

Spatial representation of information related to natural resources can yield clues about causes, processes, and results. Aerial photography, satellite data, and other forms of remotely sensed data (radar, thermal infrared) have in common the fact that the information they contain is spatially arranged. But each map, photograph, or image is a static representation of a dynamic set of relations. Repeated mapping, photography, and satellite observations are essential for ecosystem dynamics or the effects of development projects to become evident. These repeated observational sets must be formatted to the same scale and georeferenced to one another.

In addition, the information system must be capable of integrating textual and other kinds of data with the geographic data base. Georeferencing data sets that are not spatially arranged is one of the main functions of a information system. Textual materials in the form of historical observations, government reports, statistical arrays, and the like, when represented as events on the ground, greatly enhance the interpretive value of the photography or the satellite data. In a fully functioning information system, qualitative observations must be encoded as well and referenced to the other data sets (Conant, 1981).

Entry of data into the system should be accomplished so as to facilitate retrieval and the aggregation and disaggregation of data sets. These are not trivial tasks. Even such basic questions as grid cell size in relation to accuracy in resource inventories and maps have not yet been resolved (Wehde, 1982).

It is likely that no one geographic information system will serve all purposes. Although it is possible and worthwhile to plan for a generalized system, (Calkins and Tomlinson, 1977), no universal system has yet emerged (Nagy and Wagle, 1979). In devising an information system for Marinduque Island in the Philippines, Coiner and Bruce (1978) found it necessary to experiment with several components of the system before arriving at a configuration appropriate to the island and the development taking place on it. Despite enthusiasm for the use of satellite data in developing countries (Paul and Mascarenhas, 1981), and the potential value of ways to concatenate different kinds of data, such as area frame analysis (Wigton and Borman, 1978), no generally applicable information system has yet emerged. Attempts are under way, however, such as the UDMS system designed by Robinson and Coiner (1980) for the U.N. Centre for Human Settlements (HABITAT). The UDMS system takes advantage of microcomputer technology for storing, retrieving, and manipulating a variety of kinds of data.

To be widely acceptable, an information system must be sufficiently refined to capture differences in ecosystem functioning in contrasting biomes. Environmental monitoring in dryland areas presents one set of problems (Rapp and Hellden, 1979), and monitoring in the humid tropics presents another (Savage, 1982). Thus functional differences between biomes present one kind of difficulty that has yet to be satisfactorily resolved.

Comments on Methods, Concepts, and Domains

4.1 Methods

Investigations of the functional properties of ecosystems for environmental baseline studies differ systematically in the aquatic ecosystems, soils, plants, and wildlife parts of this volume. In aquatic ecosystems, methods of collecting physical data are comparatively simple (collecting gauges) and have been widely deployed. Data are readily available. Chemical methods are more complex and the data fewer. Biotic methods are still more difficult and very few data are available. The reverse is true for plants: methods for determining the status of biota are well developed and cover a range of levels of accuracy, whereas chemical or physical properties of plants are determined only in special situations where substantial detail is required. Much the same could have been said of wildlife methods until recently. Comparatively new biochemical techniques may already be altering the simple presentation, prepared by the panel, in Table 4.1. Methods for soils are similar to those for aquatic ecosystems, and physical and chemical tests are of first importance. However, microbial processes are receiving increased attention. Point sampling is characteristic of aquatic methods. An area approach, heavily dependent on sampling in both space and time, is characteristic of methods for plants and wildlife.

4.2 Concepts

Some differences in the use of terms in this volume occur, depending on whether economic or ecological meanings are intended. For example, in the discussion of methods for estimating water balance in the aquatic ecosystems report, ecological orientation is clear, while in the following discussion of water demands (especially in the context of development projects) the outlook is more economic. Estimating "balance" and "demand" involves many of the same considerations as estimating "carrying capacity," the term used in the report soils, plants, and wildlife. A prime consideration is whether available water resources are sufficient to sustain rural development projects involving irrigation, or industrial development requiring more water for manufacturing processes or power generation. Increasing population size and density is also a major factor in estimating demands on water resources. Water balance (see part II) integrates hydrometeorology, soil physics, and ground water hydrology into the system. Although it is a simple method finding acceptable data is difficult; it is thus analogous to carrying capacity.

Carrying capacity is given direct ecological expression in the reports on soils, plants, and wildlife. The report on soils, presents the methods used to assess carrying capacity and some of the controversy about the application of the concept to human populations. A major problem is the number of simplifying assumptions that have to be made. These assumptions concern both the environment and the cultural context. For example, assumptions must be made about best and worst cases of rainfall, and this can only be done if records are sufficient to establish the pattern of rainfall variability. Further assumptions must be made about demographic trends, including the likelihood of significant population movements into, within, or out of an area. Development often creates a need for wage labor, and this can set up significant rural migration patterns, which must be considered along with seasonal or permanent migration from the countryside to urban centers. Finally, assumptions must be made about the relative stability or instability of various cultural contexts, including the belief systems, institutional arrangements, and available technology.

TABLE 4.1 Ease of collection, need for, and available data: A simplified summary of methods for functional properties of the ecosystem as reviewed by the Aquatic Ecosystems, Soils, Plants, and Wildlife panels.

		Physical	Chemical	Biotic
Aquatic Ecosystems	Point Sampling	Data available Easy to collect	Some data available Tests somewhat difficult	Few data available Difficult to collect
Soils		Data available Readily collected and essential	Data available Tests routine	Few data available Difficult to obtain (but regarded as significant)
Plants	Area Sampling	Very few data Only collected if precise methods are mandatory	Same as physical	Data available Widely collected with well established methods
Wildlife		New biochemical methods likely to become much more important		Data available Widely collected

Still, if constants can be identified, the attempt to estimate ecosystem carrying capacity for human populations should provide at least a rough guide for the kind and degree of development a particular environment can sustain. If carrying capacity estimates are used in a relative rather than an absolute sense, then the concept may be helpful in evaluating the chances of success of one kind of project rather than another in the same area; or in deciding which of several areas might best sustain a particular kind of project without destroying the functioning of the ecosystem. Even in this relative usage, carrying capacity estimates must include such complicating factors as subsistence specialization in traditional settlements, including patterns of indigenous trade and exchange. Many other uncertainties are involved in our understanding of the relations between climatic factors, soil types, plant cover and variety, and productivity in developing countries in the tropics and subtropics.

In the report on plants the terms intensive, extensive, intensification, and even carrying capacity are characterized by both ecological parameters and styles of management. For example, savanna or grasslands with inherently low biomass production are usually managed with low inputs of labor and existing plant resources. Intensive management, with high labor and/or capital inputs is found in areas of high potential productivity. It commonly involves increased reliance on mechanical sources of energy, chemical fertilizers, reduced numbers of plant species cultivated, and attempts to obtain ever greater yields to meet the costs of the necessary inputs. According to the plants report, the emphasis on maximum return on investments often leads to the ecological or external costs being overlooked.

The carrying capacity concept in the report on plants is similarly referenced to management styles and costs, especially in the rangeland management. The concept involves the idea of sustained production, with herd sizes and frequency of grazing in balance with plant biomass productivity, so that overgrazing and deterioration of the resource base are avoided. To determine what this balance should be it is necessary to know about past variability in plant productivity and the climatic conditions under which it occurred. Estimating carrying capacity is elusive and, in terms of making practical decisions about stock rates and grazing intensities, may not be necessary. It may be preferable to approximate plant production and make management decisions on this basis.

Comments on Methods, Concepts, and Domains 47

In the wildlife report, carrying capacity is related to the maximum number of animals that a specified area can support in a healthy and vigorous condition within a stated time period. Carrying capacity is not static. It depends on a range of climatic and environmental conditions and the dynamic relation between plant cover and the numbers, kinds, and condition of the animals it supports. Methods for estimating habitat functioning and population well-being are presented, as well as techniques for recording the age and sex structures of the wildlife populations. These are important aspects of a baseline study for monitoring through time, allowing determination of reproductive and survival rates, population increases or decreases, and changes in the minimum and maximum ages of fertility and longevity.

Intensity and intensification are presented in the wildlife report in the sense of a gradient of usage of terrestrial environments:

Wilderness

    Low-intensity forestry or grazing

        High-intensity forestry or grazing

            Low-intensity farming

                Intensive agriculture

                    Human settlements

                          Industrial or commercial

There is a similar usage gradient for aquatic environments:

Uninhabited estuarine systems

    Low-intensity subsistence fishing

        Small-scale human settlement with some resultant pollution

            High-intensity commercial fishing and transportation

                Urban industrial seaport

In cultural terms, human settlements may be arranged along an intensification gradient as follows:

Seasonal encampments (least intense management of resources)

    Dispersed households, loosely organized into "village areas"

        Clustered farmsteads

            Villages

                Towns

                    Cities, and urban centers

Commonly, more than one settlement type may be found in the same general area; this is an almost certain indicator, of great importance to the development planner, of subregional contrasts in resource potentials.

To some extent, terrestrial and aquatic usage gradients can help determine strategy for choosing among development options. The least intensively used areas offer the greatest number of options for conversion to more intense use. Areas of high intensity of use--irrigation agriculture, for example--offer fewer options. A limited economic analogy is drawn between wild areas and development capital to be invested with caution.

Several significant points emerge from the foregoing. Apart from simple analogies, such as likening wilderness areas to capital, there is little cross over between ecology and economics. The reasons for this are perhaps best stated in part II on aquatic ecosystems. One reason is that some water uses, such as preservation of riparian plant and animal life, are not vendable; another is that users may not so much "consume" water (that is, prevent its return to surface or underground systems of storage and transport) as alter its value for other users. Water is a "fugitive resource," difficult to value or to place within a market system.

The carrying capacity concept, although attractive, is difficult to apply. Its main value may be to demonstrate that ecosystem functioning is multifactorial and limits exist to the number of organisms that can be sustained, unless inputs to the system are made that increase productivity (reduce risk). A figure of summation (which a carrying capacity estimate tends to be) is or should be more a measure of process than an end result. To be useful, carrying capacity should be calculated between known limits of variability in several dimensions. While this task appears to be susceptible to computer modeling and approximations of the "what if" type, the necessary prior monitoring of critical parameters may well take too long and be far too costly for practical consideration in many developing areas. A static approach to carrying capacity may so oversimplify reality as to provide little guidance in setting development strategy.

Development options and their impact on ecosystem resources may be more clearly seen in terms of an intensification gradient ranging from wilderness areas to those zoned for commercial and industrial use. Methods for placing an area along an intensification gradient are simpler to apply and a good deal less costly than those needed for carrying capacity estimates. But placement along an intensification gradient, while helpful in planning initial development strategy, is little or no help in estimating the impact of a development project on existing resources. Particularly as agricultural development proceeds and relative shifts occur in the distribution of population and economic activity much land formerly in crops can be expected to move "backward" into pasture or forest. For this, the planner is advised to return to the basic ecosystem concept and proceed with the necessary resource inventories and baseline studies.

4.3 Parts, Domains and Components

Each of the domains discussed in the separate reports that follow in parts II to V of this volume is affected by the other domains. Water quality can be affected by the soil erosion: soil stability can be related to plant cover type; plant covers do better in some soils than others and react to the presence or absence of some forms of wildlife or domesticated animals. Wildlife distribution itself is dependent on the quantity and quality of available water and plant cover. And the human presence, of course, depends on the integration of all these domains and affects them all.

References Cited

Ackerson, V.C. and E.B. Fish. 1980. An evaluation of landscape units. *Photogrammetric Engineering and Remote Sensing.* 46:(3): 347-358.

Agency for International Development. 1982. *Proceedings of the U.S. Strategy Conference on Biological Diversity, November 16 - 18 1981.* Department of State Publication 9262. International Organization and Conference Series 300. Washington, DC: U.S. Dept. of State.

Bailey, R.G., R.D. Pfister, and J.A. Henderson. 1978. Nature of land resource classification--a review. *Journal of Forestry* 76(10): 650-655.

Bailey, R.G. 1980. Integrated approaches to classifying land as ecosystems, In *Proceedings of the workshop of land evaluation for forestry: International Workshop of the IUFRO/ISSS*, pp. 95-109. Wageningen, the Netherlands: International Institute for Land Reclamation and Improvement.

Bartolucci, L.A. and T.L. Phillips. 1980. Digital information system for the Oruro Department, Bolivia. ATN/SF-1812 BO Quarterly Progress Report for August 1, 1980 to October 31, 1981. Contract Report number 110180. West Lafayette: Laboratory for Applications of Remote Sensing.

Bartolucci, L.A. and T.L. Phillips. 1981. Digital information system for the Oruro Department, Bolivia. ATN/SF-1812 BO Quarterly Progress Report for November 1, 1980 to January 31, 1981. Contract Report number 020181. West Lafayette: Laboratory for Applications for Remote Sensing.

Bartolucci, L.A. and T.L. Phillips. 1982. Digital information System for the Oruro Department, Boliva. ATN/SF-1812 BO Quarterly Progress Report for November 1, 1981 to January 31, 1982. Contract Report number 020282. West Lafayette: Laboratory for Applications of Remote Sensing.

Beek, K.J. 1978. *Land evaluation for agricultural development.* Publication No. 23. Wageningen, The Netherlands: International Institute for Land Reclamation and Improvement.

Best, R.G. 1982. *Handbook of remote sensing in fish and wildlife management.* Brookings, S.D.: Remote Sensing Institute, South Dakota State University.

Blake, R.O., B.J. Lausche, S.J. Scherr, T. B. Stoel, Jr., G. A. Thomas. 1980. *Aiding the environment.* Washington, DC: Natural Resources Defense Council, Inc.

Butler, R. F. 1978. Modified ecoclass - A forest service method for classifying ecosystems. In *Integrated inventories of renewable natural* resources: Proceedings of the workshop. General Technical Report RM-55. Washington, D.C.: U.S. Dept. of Agriculture.

Calkins, H.W. and R.F. Tomlinson. 1977. *Geographic information systems, methods and equipment for land use planning. Resource and land investigation (RALI).*, Reston, Virginia: United States Department of International Geological Survey.

Coiner, J.C. and R.C. Bruce. 1978. *Use of change detection in assessing development plans: A Philippine example.* Ann Arbor: Environmental Research Institute of Michigan.

Conant, F.P. 1981. *Five contexts for a geographic information system.* Conference on Remote Sensing Education, 1981. NASA Publication 2197. West Lafayette: Purdue University.

Culture and Agriculture. 1983. Issue 18. Tucson: Department of Anthropology, University of Arizona.

Dasmann, R., J.P. Milton, and P.H. Freeman. 1973. *Ecological principles for economic development.* London: John Wiley & Sons.

Driscoll, R.S., J.W. Russell, and M.C. Meier. 1978. *Recommended national land classification system for renewable resource assessments.* Fort Collins, Col.: USDA Forest Service, Rocky Mountain Forest and Range Experiment Station.

FAO. 1972. Background document. Expert consultation on land evaluation for rural purposes. *AGL/LERP.* 72/1 Oct. 1972. Rome: FAO.

FAO. 1974. *Approaches to land classification.* Soils Bulletin 22. Rome: FAO.

FAO. 1977. *A framework for land evaluation.* FAO Soils Bulletin No. 32. Publication No. 22. Wageningen: International Institute for Land Reclamation and Improvement.

FAO. 1978a. *Soil map of the world.* Vols. 1-X. Paris: UNESCO.

FAO, 1978b. Report to the agro-ecological zones project. Vol.1. *Methodology and results for Africa.* World Soil Resources Report No. 48/1. Rome: FAO.

FAO. 1980a. Report on the agro-ecological zones project. Vol. 4. *Results for South-east Asia.* World Soil Resources Report no. 48/4. Rome: FAO.

FAO. 1980b. *Report on the second FAO/UNFPA expert consultation on land resources for populations of the future.* Rome: FAO.

Farver, M.T. and J.P. Milton. 1972. *The careless technology: Ecology and international development.* New York: Natural History Press, Doubleday.

Freeman, P. 1969. The survey and classification of present land use in integrated resource development studies in Latin American tropics. In *Physical resource investigations for economic development, a casebook of*

OAS field experience in Latin America, pp. 255-276. Washington, DC: General Secretariat, Organization of American States.

Harris, D.R., ed. 1980. *Human ecology in savanna environments.* London: Academic Press.

Higgins, G.M. and A.H. Kassam. 1981. Regional assessments of land potential: A follow-up to the FAO/UNESCO *Soil Map of the World. Nature and Resources* 17(4): 11-23.

Hutchison, C.F. 1981. Use of digital Landsat data for integrated survey. In *Arid land resource inventories: Developing cost-efficient methods.* Forest Service General Technical Report WO-28, pp. 640-649. Washington, D.C. U.S. Department of Agriculture.

Journal of Forestry. 1978. Land Classification Series. *Journal of Forestry* 76(10): 644-673.

Klingebiel, A.A. and P.H. Montgomery. 1961. *Land capability classification.* Agricultural Handbook 210, Soil Conservation Service. Washington, D.C.: U.S. Government Printing Office.

Laban, P., ed. 1981. *Proceedings of the workshop on land evaluation for forestry.* Nov. 10-14, 1980. Publication no. 28. Wageningen: International Institute for Land Reclamation and Improvement, Reclamation and Improvement.

Landsat Data Users NOTES. July 1982. Issue No. 23. Sioux Falls, South Dakota: EROS Data Center.

Le Houerou, H.N. 1981. Phytoecological surveys for land use planning and agricultural development: 25 years of experience in the arid zones of Africa. In *Arid land resource inventories: Developing cost-efficient methods.* Forest Service General Technical Report WO-28, pp. 154-158. Washington, DC: US Department of Agriculture.

Margalef, R. 1968. *Perspectives in ecological theory.* Chicago: Univ. of Chicago Press.

Morain, S. 1976. Use of radar for vegetation analysis. In *Remote sensing of electromagnetic spectrum*, ed. A.J. Lewis. Washington, D.C.: Association of American Geographers.

Mouat, D.A. and B.D. Treadwell. 1978. *Resource Inventory of the Pima County and the Papago Indian Reservation, a manual to accompany the map atlas.* Tucson, Arizona: The Univ. of Arizona Office of Arid Land Studies.

Nagy, G. and S. Wagle 1979. Geographic data processing. *Computing Surveys* 2(2): 139-181.

Netting, R.Mc. 1968. *Hill Farmers of Nigeria, cultural ecology of the Kofyar of the Jos Plateau.* Seattle: Univ. of Washington Press.

Netting, R.Mc. 1974. Agrarian Ecology. In *Annual Reviews in Anthropology.* Palo Alto, Calif.: Annual Reviews, Inc.

Odum, E. 1966. *Ecology.* New York: Holt, Rinehart and Winston.

Odum, E. 1971. *Fundamentals of ecology.* 3rd ed. Philadelphia: W.B. Saunders.

Olsson, L. and M. Stern. 1981. *Large area data sampling: for remote sensing applications and statistical analysis of environment.* Rapporter Och Notiser 49. Lund: Universitets Naturgeografiska Insitution.

Organization of American States. 1969. *Physical resource investigations for economic development. A casebook of OAS field experience in Latin America.* Washington, D.C.: Organization of American States, General Secretariat.

Organization of American States. 1978. *Environmental quality and river basin development: A model for integrated analysis and planning.* Washington, D.C.: OAS.

Organization of American States. 1981a. *Activities of technical cooperation, department of regional development, energy and natural resources.* Washington, D.C.: OAS

Organization of American States. 1981b. *Results of technical cooperation, department of regional development, energy, and natural resources.* Washington, D.C.: OAS

Organization of American States. 1982. *OAS/DRD-NPS cooperative agreement, case studies project working document.* Washington, D.C.: OAS.

Paul, C. and A.C. Mascarenhas. 1981. Remote sensing in development. *Science* 214(4157): 139-145.

Rapp, A. and U. Hellden. 1979. *Research on environmental monitoring methods for land-use planning in African drylands.* Lund University Department of Physical Geography Report no. 42. Lund, Sweden: Lund University.

Robinson, V.B. and J.C. Coiner. 1980. Microcomputer geoprocessing as a minimum technology solution for data management in developing countries. In *Proceedings of the 1981 Urban and Regional Information Systems Association*, Chicago. Washington, D.C.: URISA.

Rowe, J.S. 1972. *Forest regions of Canada.* Publication 1300. Ottawa: Canadian Forest Service.

Savage, J. 1982. *Ecological aspects of development in the humid tropics.* Washington, D.C.: National Academy Press.

Swain, P.H. and S.M. Davis. 1978. *Remote sensing: The quantitative approach.* New York: McGraw-Hill.

U.S. Forest Service. 1973. *ECOCLASS - A method for classifying ecosystems - a task force analysis.* Washington, D.C: USDA Forest Service.

U.S. Forest Service. 1977. *Modified ECOCLASS - A method of classifying ecosystems.* Washington, D.C.: USDA Forest Service.

Whittaker, R.H. 1975. *Communities and ecosystems.* New York: Macmillan.

Wehde, M. 1982. Grid cell size in relation to errors in maps and inventories produced by computerized map processing. *Photogrammetric Engineering and Remote Sensing,* 68(8): 1289-1298.

Wenderoth, S. and E. Yost. 1972. *Multispectral photography for Earth resources.* Huntington, New York: West Hills Printing Co.

Wigton, W.H. and P. Borman. 1978. *A guide to area frame and area sampling frame construction utilizing satellite imagery.* New York: United Nations Outer Space Affairs Division.

Suggested Readings

Anderson, R.S., P.R. Brass, E. Levy, and B.M. Morrison, eds. 1982. *Science, politics, and the agricultural revolution in Asia.* AAAS Selected Symposium 70. Boulder, Co.: Westview Press.

Brann, T.B., L.O. House, IV and H.C. Lund, eds. 1981. *In-place resource inventories: Principles and practices. Proceedings of a National Workshop. Orono, Maine.* Washington, D.C.: Society of American Foresters.

Calkins, H.W. and R.F. Tomlinson. 1977. *Geographic information systems, methods and equipment for land use planning. Resource and land investigation (RALI).* Reston, Virginia: U.S. Department of International Geological Survey.

Conant, F.P. 1981. *Five contexts for a geographic information system. Conference on Remote Sensing Education, 1981.* NASA Publication 2197. West Lafayette: Purdue University.

Daly, H.E. and A.F. Umana, eds. 1981. *Energy, economics, and the environment. Conflicting views of an essential interrelationship.* AAAS Selected Symposium 64. Boulder, Co.: Westview Press.

Dasmann, R., J.P. Milton, and P.H. Freeman. 1973. *Ecological principles for economic development.* London: John Wiley & Sons.

FAO. 1977. *A framework for land evaluation.* FAO Soils Bulletin No. 32. Publication No. 22. Wageningen, The Netherlands: International Institute for Land Reclamation and Improvement.

FAO, 1978b. *Report to the agro-ecological zones project.* Vol. 1. *Methodology and results for Africa.* World Soil Resources Report No. 48/1. Rome: FAO.

Journal of Forestry. 1978. Land classification series. *Journal of Forestry* 76(10): 644-673.

Miller, H.H. and R.R. Piekarz, eds. 1982. *Technology, international economics, and public policy.* AAAS Selected Symposium 68. Boulder, Co.: Westview Press.

Olsson, L. and M. Stern. 1981. *Large area data sampling: for remote sensing applications and statistical analysis of environment.* Rapporter Och Notiser 49. Lund: Universitets Naturgeografiska Insitution.

Savage, J. 1982. *Ecological aspects of development in the humid tropics.* Washington, D.C.: National Academy Press.

Wigton, W.H. and P. Borman. 1978. *A guide to area frame sampling and area frame construction utilizing satellite imagery.* New York: United Nations Outer Space Affairs Division.

Williams, N.M. and E.S. Hunn, eds. 1982. *Resource Managers: North American and Australian Hunter-Gatherers.* AAAS Selected Symposium 67. Boulder, Co.: Westview Press.

Part II -- Aquatic Ecosystems

Peter Rogers, Chair
Leo R. Beard
Kenneth D. Frederick
Owen T. Lind
Kenneth P. Sebens

II. AQUATIC ECOSYSTEMS

Introduction

The name for the aquatic ecosystem panel grew out of our concern to ensure broad coverage of the topic. Often the direct examination of "water" itself, its quantity and quality, has not included an evaluation of the complex ecosystem that depends directly on the water. Other concerns deal with the types of scientific disciplines required for adequate study of aquatic ecosystems. This is reflected in the composition of our aquatic ecosystem panel. Persons involved in environmental quality analyses, studies, and control in the United States are engineers. But in developing and evaluating methods for environmental assessment, engineers have a lot to learn from scientists in other disciplines. In our panel, therefore, the two engineers were balanced by two biologists and an economist.

The purpose of this study and report is not to make an environmental assessment, nor to list the environmental consequences of projects; rather we intend to describe the methods available, or soon to be available, for natural resource inventories and environmental baseline studies, with an emphasis on tropical countries.

1.1 Aquatic Ecosystems Concept

In an ecosystem everything is connected to everything else. This probably makes sense from the point of view of describing a system but it makes analysis of the ecosystem particularly difficult. Aquatic ecosystems tend to be more difficult to analyze and understand than terrestrial ecosystems because of the complex system of transport of chemicals and nutrients by water.

Our panel's task has been to look at the aquatic ecosystems of fresh-water lakes, rivers, streams, and near-shore saline waters of the tropics from their physical limits--the topography, geomorphology, and the hydrology of the watersheds. We need to consider climatic and meteorological conditions as well as soils and near-surface geology. For the aquatic ecosystems themselves we will consider their functional properties.

One way to visualize the aquatic ecosystem is to consider the hydrologic and nutrient cycles. The hydrologic cycle (Fig. 1.1) accounts for the passage of water in its various phases through the lithosphere and atmosphere. Aquatic ecosystems exist wherever water concentrates in this cycle--in lakes, rivers, marshes, oceans, in soil moisture, and in the groundwater. Because the hydrologic cycle intersects with many terrestrial ecosystems, it is a useful conceptual model with which to identify the boundaries of the ecosystems.

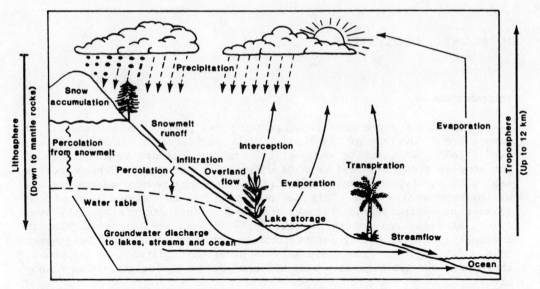

Figure 1.1 Schematic diagram of the hydrologic cycle. [From Dunne and Leopold (1978)]

This concept of the hydrologic cycle is basic to the understanding of aquatic ecosystems (Jackson, 1977a; Riehl, 1954; Critchfield 1974). It is a natural machine, a constantly running distillation and pumping system. The sun applies heat energy, which together with the force of gravity, keeps water moving. Although this water cycle has neither beginning nor end, from our point of view the oceans are the major source, the atmosphere is the delivery agent, and the land is the user. In this system there is no water lost or gained, but the amount of water available to the user may fluctuate because of variations at the source, or more usually, in the delivery agent. In the past, large alterations in the atmosphere and the oceans have produced deserts and ice ages. Even now small local alterations of the patterns of the hydrologic cycle produce floods and droughts.

The three most important stages of the cycle are evaporation, condensation, and precipitation. The atmosphere acquires moisture by evaporation from oceans, lakes, rivers, and damp soil, and by transpiration from plants. These processes are referred to as evapotranspiration. Air currents transport the water vapor over large distances. Condensation, cloud formation, and precipitation may then occur. During precipitation, some evaporation occurs in the air but much of the water reaches the ground, a water surface, or vegetation. Of the precipitation that reaches the vegetation, some is held by the canopy and eventually evaporates again and some will run down, drip from the canopy, or be shaken off by the wind and reach the ground. Of the water that reaches the ground, some will infiltrate into the surface, some will accumulate in surface depressions, and some will begin to move over the surface.

Some of the water that enters the soil may evaporate back into the atmosphere, and some will be taken up by plants. In the upper layers of the soil, subsurface lateral movement of water may be considerable and may reach streams or lakes. Some water may percolate to deeper layers into the groundwater, where it may be held for long periods. But fossil water in some of the aquifers under the Sahara desert is millions of years old and essentially non-renewable.

Man can interfere with virtually any part of the hydrologic cycle by making changes in the vegetation cover, by extracting groundwater or by stimulating precipitation artificially. Interference at one stage has repercussions throughout the whole cycle, and we do not yet know the extent to which the atmosphere can recover from human disruption or at what point changes become irreversible.

In addition to water, nutrient cycles such as those of nitrogen, phosphorus, and sulphur and other geochemical cycles such as oxygen, hydrogen, and carbon are extremely important in the aquatic ecosystems. Whenever possible in the report of our panel we will indicate the potential intersection of development projects with these cycles.

1.2 Intervention in Aquatic Ecosystems

Water is used in virtually everything we do; it is used more than any other material by industry, it is used both directly and indirectly to produce energy, it provides the basis of much of our outdoor recreation as well as an important transportation medium, and it serves as a vehicle for disposing of society's wastes. Some examples of human demands placed on aquatic ecosystems are for activities such as those given in Dunne and Leopold (1978) and Kazenelson et al. (1976): (i) domestic supply, (ii) industrial and commercial supply, (iii) irrigation, (iv) transportation, (v) generating power, (vi) cooling purposes, (vii) fishing and aquaculture, (viii) water-based recreation, and (ix) disposal of wastes.

Indeed demands for water have generated some of the largest construction projects ever undertaken as well as efforts at weather modification. Such activities produce drastic changes in the underlying ecosystems, and many produce unintended impacts on aquatic systems. Water has become one of our most abused resources. Management of water projects for which inadequate allowance is made for potentially adverse impacts on the ecosystems and human populations have led to some very unfortunate results. For example, some water projects have caused the spread of schistosomiasis, a debilitating infection by a parasitic worm which affects an estimated 200 million people (Rosenfield, 1979). Even activities not classified as water projects may alter aquatic ecosystems in detrimental ways. For instance, agriculture and construction activities lead to soil and nutrient losses, and the nutrients are transferred to lakes and streams. Failure to allow and perhaps compensate for such impacts can be costly indeed. Human use and abuse of water supplies can convert once valued streams and lakes into unproductive bodies of water that are sources of disease and poisons. If a society is to benefit from its resources and natural environment, it is essential to understand how alternative activities affect the ecosystem. This knowledge would enable us to reduce the range and severity of unintended negative impacts and to design and select activities with greater net benefits.

Since in almost every case aquatic ecosystems are an important part of the geochemical recycling of materials through the lithosphere, it is impossible to list all types of interventions in aquatic ecosystems.

Instead, we will discuss the methods to analyze classes of interventions and not relate them to specific project choices. Some examples of specific projects and their impacts are given in White (1978), El-Ashry (1980), and Tillman (1981).

All development projects in tropical countries will affect the aquatic environment to some degree. In this report, ways of monitoring these impacts are presented. The reader should not forget that most, if not all, of the impacts upon aquatic ecosystems can be engineered out of a particular project at some cost. Legal and institutional frameworks influence the use and management of a region's water resources and are critical to the pressures placed on the resources and, therefore, the need for these large water projects. For example, water is often treated as a free good or made available to users at prices well below real costs. Such policies increase the pressures placed on aquatic ecosystems and may result in socially inefficient allocations of scarce water resources. Thus, development of a strategy for meeting a society's water demands should include analysis of legal, institutional, social, and engineering alternatives for meeting the demands (see Whyte, 1977).

1.3 Role of Water in Health in Developing Countries

Some understanding of the relation between water use and health has been around since at least the time of Frontinus, the Water Commissioner of Rome in A.D. 97 (Babcock and Matera, 1973), but the exact relation is not good even now. A World Bank report (1976) said, "...other things being equal, a safe and adequate water supply is generally associated with a healthier population." Briscoe (1977), who is quoted extensively below, discussed the empirical studies of the relation between water supply and health, and problems with the analyses and findings.

The large literature on the empirical relationships between water supply and health consists of both cross-sectional and longitudinal studies. The cross-sectional studies, such as those of the World Health Organization (WHO) Diarrheal Diseases Advisory Team (see van Zijl, 1966), have been plagued by the existence of high multicolinearity in "independent" variables (for instance income and nutritional status are usually highly correlated with quality of water supply) and by the existence of simultaneity (a correlation between water supply and quality and health may imply that the communities or individuals with better health status are more healthy because of the quality of their water supply or that a more healthy community has taken steps to improve the quality of its water supply). It is somewhat surprising and unfortunate, given the frequency with which the multicolinearity question has arisen, that no analysis of cross-sectional data with multivariate techniques has been attempted. The longitudinal studies are generally prospective studies in which water supply improvements are made in an "experimental" community, while the health of this and a "similar" control community are monitored. The assumption that the two communities are similar in all important respects has proved to be a major problem. Frequently adequate pre-intervention monitoring has not taken place and often there have been differential changes in the "experiment" and "control" communities. Particularly serious is the fact that the communities have often been exposed to quite different probabilities of infection due to the occurrence of an epidemic on one of the communities only. One of the most carefully conducted longitudinal studies, the INCAP study in the 1960's (see Scrimshaw, 1970), failed to yield any definitive conclusions since the interventions failed, for a variety of reasons, to substantially alter either water use or defecation patterns.

The most widely used classification of water-related disease is based on the work of Bradley (cited in Feachem et al., 1977), as presented by Saunders and Warford (1976) in Table 1.1. In this scheme, water and disease relationships are divided into five major categories:

1) Waterborne diseases--water acts only as a passive vehicle for the infecting agent. All of these diseases also depend on poor sanitation.

2) Water-washed diseases--a necessary part of the life cycle of the infecting agent takes place in an aquatic animal. Some are also affected by waste disposal. Infections spread other than by contact with or ingestion of water have been excluded.

3) Water-based diseases--in these worm infection diseases water is an important part of the host-parasite relation.

4) Water-related insect vectors--infections are spread by insects that breed in water or near it. Adequate piped-in supplies may remove people from the breeding areas or enable them to dispense with water storage jars where the insects breed. These vectors are not affected by waste disposal.

5) Diseases related to fecal disposal and very little affected by water more directly--these are one extreme of a spectrum of diseases, mostly water washed, together with a group of water-based type of infections likely to be acquired only by eating uncooked fish or other large aquatic organisms.

Many of the water-related diseases can be spread or introduced into areas as a result of development of water resources. The classical case of this was the spread of schistosomiasis in Africa and Asia after the introduction of irrigation projects. Schistosomiasis infects more than 200 million people. Eggs passed in urine or feces produce a ciliated larva, called a miracidium, which hatches on contact with water. The miracidium must contact a suitable snail (family Planorbidae) host within 24 hours or perish. Within the intermediate host the schistosomes move to the digestive glands and change into larval stages. After about one month the larvae leave the snail and seek to penetrate the skin of a mammalian host. In humans, the schistosomes enter blood vessels and eventually reach the bladder or intestines where they achieve sexual maturity. The worms live for several years and are capable of laying several hundred eggs per day. The disease tends to be debilitating rather than fatal in most cases (see Tillman, 1981). Similar health problems related to irrigation include malaria, filariasis, yellow fever, onchocerciasis, and sleeping sickness.

The objective set for United Nations International Drinking Water Supply and Sanitation Decade (1980-1990) is to have "clean drinking water

TABLE 1.1 Diseases related to deficiencies in water supply or sanitation. [Source: Saunders/Warford VILLAGE WATER SUPPLY, p. 32. Copyright © 1976 by the International Bank for Reconstruction and Development. By permission of The Johns Hopkins University Press]

Group	Diseases	Route leaving man[a]	Route entering man[a]
Waterborne diseases	Cholera	F	O
	Typhoid	F, U	O
	Leptospirosis	U, F	P, O
	Giardiasis	F	O
	Amoebiasis[b]	F	O
	Infectious hepatitis[b]	F	O
Water-washed diseases	Scabies	C	C
	Skin sepsis	C	C
	Yaws	C	C
	Leprosy	N (?)	?
	Lice and typhus	B	B
	Trachoma	C	C
	Conjunctivitis	C	C
	Bacillary dysentery	F	O
	Salmonellosis	F	O
	Enterovirus diarrheas	F	O
	Paratyphoid fever	F	O
	Ascariasis	F	O
	Trichiuriasis	F	O
	Whipworm *(Enterobius)*	F	O
	Hookworm *(Ankylostoma)*	F	O, P
Water-based diseases	Urinary schistosomiasis	U	P
	Rectal schistosomiasis	F	P
	Dracunculosis (guinea worm)	C	O
Water-related vectors	Yellow fever	B	B mosquito
	Dengue plus dengue hemorrhagic fever	B	B mosquito
	West-Nile and Rift Valley fever	B	B mosquito
	Arbovirus encephalitides	B	B mosquito
	Bancroftion filariasis	B	B mosquito
	Malaria[c]	B	B mosquito
	Onchocerciasis[c]	B	B *Simulium* fly
	Sleeping sickness[c]	B	B tsetse
Fecal disposal diseases	Hookworm *(Necator)*	F	P
	Clonorchiasis	F	Fish
	Diphyllobothriasis	F	Fish
	Fasciolopsiasis	F	Edible plant
	Paragonimiasis	F, S	Crayfish

[a] F = feces; O = oral; U = urine; P = percutaneous; C = cutaneous; B = bite; N = nose; S = sputum.
[b] Though sometimes waterborne, more often water washed.
[c] Unusual for domestic water to affect these much.

and sanitation for all by 1990." This is certainly an ambitious goal, especially since Feachem et al. (1977) estimate that 86 percent of the world's rural population lack an adequate water supply and 92 percent lack adequate facilities for waste disposal. Saunders and Warford (1976) estimate that for the developing countries 71 percent of the populations live without adequate water supply and 75 percent without adequate sewage disposal.

One goal that has much more likelihood of success, is eradication of guinea worm (Bourne, 1982). Guinea worm disease (dracunculiasis) which afflicts 20 to 40 million people worldwide, is particularly vulnerable to a concerted eradication program because it is transmitted exclusively by water. Other water-borne diseases such as typhoid, infant diarrhea, and schistosomiasis can only be reduced by 60 to 70 percent at best because other modes of transmission are not affected by making water supplied potable.

Many other health related problems have been introduced by new agricultural practices. For example, many chemicals used in agriculture, particularly the chlorinated hydrocarbons, accumulate in animal tissues and concentrate in the food chain. High concentrations of fertilizer in water supplies have been known to cause blood diseases.

1.4 Defining Baseline Studies and Resource Inventories

In this volume, the appropriate methods for taking natural resource inventories and making environmental baseline studies in tropical countries are reviewed. Unfortunately, the terms "resource inventories" and "baseline studies" may have different meanings for different readers. For the purposes of this study we have chosen to use the definition presented by the U.S. National Park Service and used by the chairmen of the various panels (see part I).

In our opinion the baseline studies are of paramount concern since they require careful monitoring of everything including the resources. In Fig. 1.2, we, the panel, have attempted to illustrate some further distinctions between resource inventories and baseline studies. A distinction is made between exploited resources and unexploited resources. In order for something to be a resource there has to be some use for it--either it is demanded by humans or by other parts of the ecosystem to meet human demands. There are many things currently not viewed as resources that in the right circumstances could become important resources, perhaps as a new source for drugs or food.

After a baseline is established and some intervention takes place in the environment, there may be perturbations in the ecosystem that can be considered as environmental impacts. The ecosystem is then divided into areas that are not affected by the intervention, and can thus be used as controls, and other areas that are affected and may show different parameter values.

The dynamic nature of the process is illustrated in Fig. 1.2. A baseline study should come before a project intervention; resource inven-

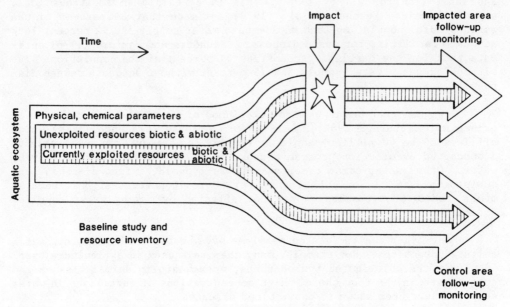

Figure 1.2 The components of baseline studies and natural resource inventories.

tories can proceed at any time but are also best conducted *before* exploitation begins and environmental impacts emerge.

Examples of the baseline study and resource inventory for a project that uses water directly and one that uses water only indirectly are described below.

1) An example of a project with direct use of water as a resource is a hydroelectric dam. Ideally a baseline study should be conducted for several years (or more for some data) to measure the important physical, chemical, and biotic properties of the river, both upstream and downstream from the proposed dam site. Since the resource to be exploited in this case is the moving water, the resource inventory should include extensive hydrologic and hydraulic studies. These will require extensive data gathering both in the stream and in the contributing watershed. Rainfall, runoff to streams, and sediment yield time series data are required, in the form of time-series data over many years. The physical properties of the river, of the bed and its subsurface geology typically only need to be studied close to the time of project design. In addition to baseline data acquired for several years both upstream and downstream of the proposed project, similar data should be collected for the area that will be submerged by the reservoir when the dam is built. These data need to be collected seasonally for at least one full year. These data are necessary to evaluate the resources that are likely to be lost as a result of the project.

2) An example of a project that uses water indirectly is a highway. Such construction projects often have major impacts on the aquatic environment, not only during the building phase but also during their useful life. The construction typically involves large

amounts of earth moving and clearing, which leads directly to changes in the sediment and nutrient loads entering nearby surface waters. Often construction of embankments, bridges, and culverts involves the ponding or other hindering of water flow.

We do not have space to consider all types of projects in this report. We have, therefore, characterized development projects by the types of effects they are likely to have on aquatic ecosystems and where the initial effects occur--directly on water, on the watershed, or on the atmosphere. Development projects with a direct effect on the water include impoundments, withdrawal and return water uses, dredging, introduction of chemicals, and biotic resource harvesting.

Projects with watershed effects can be categorized generically as involving land modification, land reclamation, and introduction of chemicals.

For those that affect the atmosphere, the most pertinent effect is the introduction of chemicals.

One project can cause two or three different types of generic effects. For example, irrigation projects may require the impoundment and withdrawal of water. At the same time irrigation may require clearing and the application of chemicals. These generic effects are related to the components of baseline studies described later in section 3.1.

Aquatic Ecosystem Components

2.1 Ecosystem Properties: Atmosphere, Hydrologic Cycle, Water Balance

Because the atmosphere is a key influence on in all ecological systems, knowledge of its composition, structure, and behavior as well as of prevailing climatic conditions is of fundamental importance. Of particular significance is knowledge of the behavior of moisture in the atmosphere, its transformation from one state to another, its movement, and its geographic and temporal variations.

Climatic conditions not only influence life on earth and shape the physical and biological environment, but these in turn influence the state and composition of the atmosphere; that is, as is characteristic of ecosystems, there are strong interactions and feedback loops between the various components. Human activities are influenced greatly by weather and climatic conditions and, in turn, man can both purposefully and inadvertently modify weather, climate, and the moisture cycle.

In this section some basic facts about the composition, structure, and behavior of the atmosphere are presented (Lamb, 1972; Barry and Chorley, 1978; and Stringer, 1972). There is also some discussion of the content and scope of the sciences of meteorology and climatology, on the various geographic and temporal scales of climatic variability that are of relevance in resource inventories, baseline studies, and impact assessment, and on the use of weather and climate data.

In atmospheric science, as in ecology, systems exist on various scales, and these interact to a greater or lesser degree making the precise delineation of boundaries difficult. More frequently there are gradients of climatic conditions.

The scale chosen for environmental studies will depend on the nature of the particular project or program. Food systems, for example, range in size and complexity from relatively simple village-level subsistence to the highly complex world food system, which involves interactions between nations. To analyze weather and climate in a particular location or region, atmospheric systems and processes must be studied on local scales (several kilometers) and on up to the global scale. The approach we take is to start with a discussion of some general global principles of atmospheric behavior on a global scale and work down in scale; the intention is to provide information that will help planners appreciate the different geographic and temporal scales that are important in understanding weather and climate and thus allow them to deduce those that are of greatest relevance to specific projects. First, users must become familiar with meteorological and climatological concepts and terminology.

2.1.1 Terminology.

Atmospheric scientists use a vocabulary that has fairly precise definitions, and it is important to clarify this terminology.

The term *weather* is defined as a set of atmospheric conditions at a specified time and thus refers to events or episodes. *Climate* is the sum total of weather experienced at a place or throughout a region in the course of the year and over the years. The statistical characteristics of climate at a place over some past period are used to describe the climate at that place and can be used to derive estimates of what climatic conditions can be expected in the future.

The term *climatic fluctuation* or variation applies to persistent departures from "normal" over relatively short periods. One might speak of a climatic fluctuation lasting a few years, or a few decades, but not several centuries. Variations in average climatic conditions over periods of decades, centuries, and millennia fall into the category of climatic change.

Climatology, a branch of atmospheric science, is concerned with the types and causes of atmospheric conditions and with general laws or principles governing the behavior of atmospheric conditions: it may also focus on the relations between atmospheric conditions and other components of ecological systems, including human activities. *Meteorology* is the study of the composition, behavior, and structure of the earth's atmosphere. There is no rigid delineation between the two sciences in subject matter, methods of measurement, data collection, storage, or analysis. Both deal with improving forecasting--meteorology on a time scales of hours to several months and climatology on longer time scales.

Composition of the atmosphere. Air is a mechanical mixture of gases, not a compound, consisting mainly of nitrogen (about 78 percent) and oxygen (about 21 percent). There are much smaller concentrations of other gases, including carbon dioxide and ozone. In addition, there is a variable admixture of water vapor (Barry and Chorley, 1978; Stringer, 1972; Miller and Thompson, 1970).

These gases are highly compressible, so that the lower layers of the atmosphere are much denser than the upper layers. More than three-quarters of the atmosphere's mass is contained within the troposphere, the lowest 12 kilometers, where the weather is produced. In addition to the gases, the atmosphere contains various amounts of suspended matter such as condensed water and miscellaneous impurities. Many impurities result from natural processes, but with industrialization and urbanization, the proportion contributed by human activities is increasing. Water vapor accounts for up to 4 percent of the atmosphere by volume near the surface and is supplied to the atmosphere by evaporation or by transpiration.

The water vapor content of the atmosphere is directly related to air temperature--greatest in summer and in low latitudes. There are exceptions to this generalization, such as the deserts of the tropics. There is little rainfall over these deserts because of the persistence of mechanisms that hinder the ascent of air and, hence, the precipitation process--such a mechanism is the slow descent of air in high pressure systems.

Atmospheric energy and motion. The source of the energy that heats the air, the ground and the seas and drives the winds and the ocean currents is the sun. By comparison, the flow of heat from the earth's interior is quite negligible. Since the earth's main heat supply comes by radiation from the sun, radiation exchanges are the first stage in the production of the climates we observe. The circulation of the winds in the atmosphere and of the water currents in the oceans are important in transporting and redistributing heat and available water vapor.

The atmosphere acts like gigantic heat engine in which the temperature differences maintained between the poles and the equator provide the energy supply needed to drive atmospheric circulation. The conversion of the heat energy into the kinetic energy of motion must involve rising and descending air, but vertical movements are generally much less in evidence than horizontal ones. It is these air currents that transport water vapor from the oceans to the continents.

The extreme short-wave and long-wave parts of solar insolation do exhibit changes over periods of months and years and some are in phase with the 11-year sunspot cycle, but these do not necessarily imply changes in the solar constant. Claims for strong links between sunspot cycles, weather, and climatic conditions on earth should be considered with caution. Futhermore, regular cyclical phenomena account for only a small proportion of the total variability in climate on time scales of years and decades.

This description represents the idealized case in which all the insolation is uniformly available. The actual situation is modified by the effects of absorption and re-emission within the atmosphere, by absorption of solar radiation at the earth's surface, and by losses of radiation by reflection (Riehl, 1954; Palmen and Newton, 1969; Stringer, 1972). Over the year as a whole, latitudes polewards of about 35 degrees appear to show a net loss of energy, with the highest losses being recorded over the poles. The highest figures of average energy gain are over the north African, Arabian, and Australian deserts during their respective summers. Local or regional departures from the simple zoned pattern are caused primarily by the differential heating qualities of land, water, and vegetation cover, and by the persistence of dense cloud cover from the top of which radiation is wasted by reflection. At the local level, topography significantly affects the quantity of insolation and the duration of direct sunlight.

The amount of radiation is graded mainly by latitude and season giving an arrangement of climatic zones nearly parallel to the latitude circles and successively cooler toward the poles. The unequal receipt of heat energy geographically necessitates great lateral transfers of energy across the surface of the earth. The principal mechanism of heat exchange is the winds; at times this advected heat exceeds that coming directly from the sun, particularly in high latitudes in winter. The end products of the heat supply and its redistribution are the temperature patterns observed in the air, which in turn account for the atmospheric circulation.

In low-latitudes there are vertical circulation cells in both hemispheres, the intensity and position of which vary seasonally. Warm air near the equator rises and moves toward the poles, descends in the subtropical high pressure belts, and generates low-level flows toward the equator--the trade winds. Where the air streams from the tropical zones of the two hemispheres meet is a zone of mainly light winds, the doldrums (Fig. 2.1). Within this zone is the intertropical convergence zone which migrates seasonally north and south of the equator. Light and variable winds are also dominant in the subtropical anticyclone belt. Toward the pole sides of the subtropical anticyclones are the midlatitude westerlies, which can penetrate into the tropics during winter.

The simplicity of these zonal winds is broken by monsoon effects, by the displacement of air over mountain barriers, by land and sea breezes, and by terrain irregularities.

2.1.2 Atmospheric moisture.

The average amount of water stored in the atmosphere (about 2.5 centimeters) is only equivalent to the average rainfall over the earth as a whole during about 10 days. Because of intense influxes of moisture by air currents, rainfalls of 170 centimeters and more have been recorded in 24 hours.

High rates of evaporation and the characteristics of tropical rainfall pose particular problems. In many areas rainfall is seasonal in character, greatly limiting water availability at certain times of year. At other times, the same areas may have excessive rainfall. There is also high variability in precipitation from year to year, particularly in the semi-arid tropics, which can pose problems for human activities, partic-

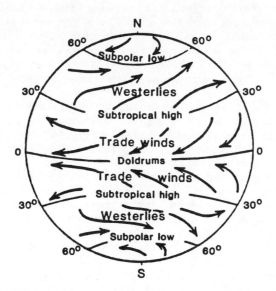

Figure 2.1 Schematic diagram of the general distribution of sea level pressure and wind zones over the world. [Source: Lamb (1972)]

ularly agriculture. It is this problem of the variability of rainfall that is important in many development projects.

In the tropics rainfall intensity tends to be high causing heavy surface runoff; much of the rainfall may be lost to agriculture. Tropical rainstorms are also often localized, creating marked differences in rainfall over short distances. Evaporation is much more constant from year to year in the tropics than is rainfall, because of the relatively small variation in key factors such as solar radiation. Annual variations in the water balance, therefore, are due more to variations in precipitation than to evaporation.

Seasonal and annual variations in rainfall and evaporation can be illustrated by water balance studies. The concept of water balance is fairly simple, but methods for estimating the water balance are quite complex. When rainfall exceeds potential evaporation, the capacity of the soil to hold moisture is reached and any further rainfall is classed as surplus. When rainfall is less than potential evaporation, soil moisture reserves are used. Once soil moisture drops below field capacity moisture deficits occur. Water balance studies are of particular importance in agriculture in determining the suitability of specific crops for particular areas and the amount of irrigation that may be required. The principal consideration is the balance between crop water requirements and available moisture.

It should be remembered that local evaporation is, in general, not the major source of local precipitation. The average residence period of a water molecule in the atmosphere is reckoned at about 10 days and during this time it can be transported thousands of miles by air currents (moisture advection). In general, the occurrence and distribution of rainfall are determined by large-scale processes and circulation patterns in the atmosphere. Whether or not it rains in Timbuktu or Delhi may depend on temperature differences between the equator and the poles, sea-surface temperatures over the oceans, and the intensity of wind circulation over continental areas. An implication of this is that the energies involved in large-scale atmospheric processes are enormous and for man to significantly modify these large-scale climatic patterns, equally large disruptions of the energy budget must be achieved.

The view of north-south movement of a rainbelt associated with a migrating low pressure trough is a great oversimplification. It is useful to start with this simple model and then modify it (Jackson, 1977a). Working from the equator toward the poles, four zones can be identified: (i) a zone with two rainfall maxima at the equator but no real dry season; (ii) another zone with two rainfall maxima but separated by a pronounced dry season; (iii) a zone with a single rain season and a single dry season; and (iv) a semiarid region with a short wet season in the cool season, under the influence of extra-tropical disturbances.

This simple pattern is modified in a number of ways--by the structure, shape, and location of certain semipermanent features such as subtropical highs and the equatorial trough; by the distribution of land and sea; and by local effects. Seasonal variations in actual distribution of rainfall worldwide (Fig. 2.2) show that average annual rainfall varies from almost zero in some deserts to over 10,000 millimeters in parts of Hawaii. At Iquique in the Chilean desert, no rain fell for one 14-year period, whereas at the other extreme 22,000 millimeters have been recorded in a single year at Cherrapunji in India.

Variability in precipitation has great significance in ecosystem stability. The precise nature of the impact depends on many factors, and it is difficult to generalize about this. Variability in semiarid areas can be devastating. In the Sahel region of Africa, a decrease in rainfall of 20 to 50 percent over a 6-year period seriously affected agriculture, livestock, inland fisheries, and entire national economies, with many social and political ramifications. The causes of such droughts, however, are not known. Whether the decrease in rainfall represents an irregular fluctuation in rainfall or is part of a long-term change in climate is not clear.

Droughts act as catalysts for meteorological and climatological research, but more research into the nature of climatic variability is needed.

For resource inventories and large-scale development projects it is important to consider the longest possible time span of climatic records available. In agriculture, knowledge of variability at shorter time spans, such as at the start and finish of the rainy season and of the probability of dry spells during the growing season, is also important. To study this rainfall reliability, analysis of daily and weekly values of precipitation, would be useful.

2.1.3 Atmospheric resources and risks.

Climate is a resource that is exploited by living organisms. Geographic variability of climate is not only a key factor explaining the

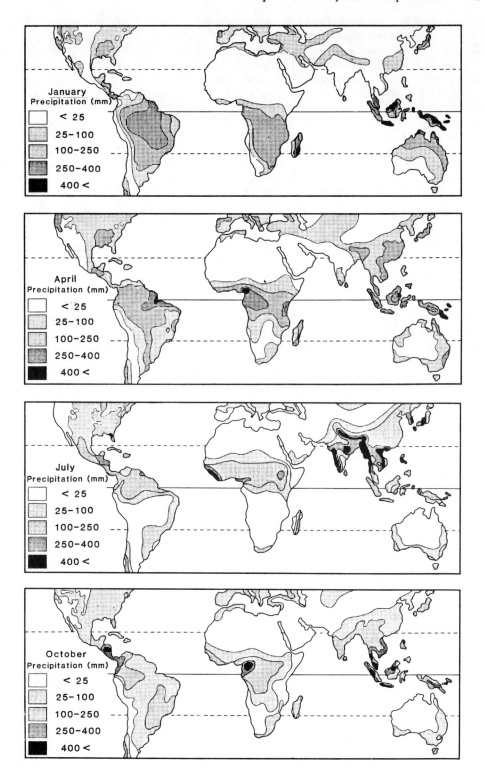

Figure 2.2 Distribution of average rainfall in January, April, July, and October. [Source: Jackson (1977)]

diversity of landscapes, but also of organisms. Weather and climatic fluctuations also pose risks or hazards, and this is particularly true in the case of extreme events such as droughts, heatwaves, frosts, or floods. It is the extreme events, irregular and infrequent, that pose a great challenge to the stability of ecosystems, and especially to humans. The natural fluctuations of climatic conditions also makes it difficult to detect with any degree of certainty the impacts of human activity on climate. In this section we will consider both the long-term average state of the atmosphere and its variations.

Although we cannot control climate, forecasting and effective use of meteorological and climatological data and information do allow us to mitigate some of the adverse impacts of extreme weather events. We can, for example, choose crops and management practices to ensure high agricultural yields under diverse conditions, design buildings for safety and energy efficiency, and manage energy and water to avoid shortages. Better use of weather and climate information should improve decision-making in many areas, with attendant social, economic, and environmental benefits (Thomas and McKay, 1978; Chandler, 1970; Pocinki, Greeley, and Slater, 1980).

The ideas summarized below are taken from a report entitled "Managing Climatic Resources and Risks" (National Research Council, 1981), and explain how to use climate data and climate information systems.

The nature of decisions for which climate information may be useful varies greatly. Decisions about activities such as crop planting, irrigation scheduling, and the short-term management of river flows are generally concerned with weather variability over time scales of days to seasons. Other decisions relate to long-term investment and design. These include the design and construction of irrigation and flood control systems and hydroelectric plants, the management of inland fisheries, and land-use planning. Here, climate over time scales of years to decades is relevant. The relevant time scale frequently is related to the magnitude of the decision: the greater the resources at stake, the longer the time span that should be considered.

There are many ways of dealing with the risks of weather. Protective measures, such as planting crop varieties less vulnerable to climatic stresses, constructing storm sewers, and restricting development on flood plains, for instance, can reduce risk. Risks may also be redistributed. Building a dam, for example, reduces the risk of frequent flooding but increases the risk of catastrophic loss should the dam fail. Managing risks associated with climate is a process that has the following components: (i) determination of risks or vulnerabilities associated with climate change or variation; (ii) assessment of sensitivity of ecosystems to climatic stress, (iii) analysis of available options and their likely outcomes, (iv) choice among the options, and (v) implementation of the option or options chosen.

In planning for the next season, the options a farmer might examine could include doing nothing, changing irrigation schedules, adopting different management practices, changing crops, or reducing the acreage planted. Individual circumstances, past experiences, and perceptions and values will influence the choice.

The context of decisions involving climate data suggest that some needs of users are particularly important. These needs, in turn, suggest some important functions that climate information systems should ideally satisfy.

1) Users need probabalistic information tailored to particular circumstances and consistent with climatological theory and practice.

2) Users often need access to a range of climatic data processed to different degrees and with different level of detail about how such data were gathered, stored, or verified.

3) Users need observations of climate parameters other than standard meteorological variables.

4) Users need climate information that can be combined easily with other data and knowledge.

5) Users need sufficient data in both space and time to ensure reliable probabalistic estimates and minimum sampling errors.

6) Users often need information on current weather and climate information.

7) Users often need special climate forecasts to permit appropriate planning and preparations.

These are the functions of an ideal climate information system against which the existing system can be judged. In the existing system, much emphasis has been placed on collecting real-time data on basic meteorological parameters such as temperature, barometric pressure, and precipitation. A by-product is a record of weather conditions spanning considerable time and space from which climatic information can be derived.

2.2 Ecosystem Properties: Geomorphology of Soils and Watersheds

The watershed is an important concept in aquatic ecosystems and must be carefully identified. Two aspects of watersheds are of interest: the first is the surface watershed and the second, the groundwater basin. In many cases these are coterminous, but in some important cases they are not (for example, in limestone regions). The surface watershed boundaries are readily established by ground topography mapping. The major watersheds and the minor watersheds of the tributary streams need to be mapped. The scale of mapping depends on the needs of the particular project or program. The topography typically needs to be surveyed in great detail, especially for most gravity irrigation and drainage projects.

In nonalluvial areas with consolidated deposits the mapping of groundwater basins is quite complex. Typically, this mapping requires extensive geophysical exploration or extrapolation from existing well data. In defining aquifers it is important to map the boundaries of water quality. In parts of Pakistan, for example, there are large pockets of saline water in the water aquifer. It is important to establish the boundaries and the areal extent of such pockets.

The size of the watershed to be considered, of course, depends on the type of development project to be undertaken. It would be difficult to imagine a case in which the whole Amazon Basin, for example, would have to be measured, analyzed, and mapped in detail. The focus will almost always be on some lower order watershed.

The gradient and the land forms, as well as the surficial geology of the basin are important considerations in defining the boundaries of the systems that require study. Steep silt-free rivers in rocky beds require significantly less data analysis than sluggish silt-laden rivers with alluvial beds. Reconnaissance by experienced geomorphologists can indicate many of the areas for which detailed environmental assessments will be required.

The soils and ground covers of the watershed are extremely important factors in understanding sediment production and transmission through a river system. Many severe problems occur after deforestation or other disturbance of shallow soils on steep slopes (Eckholm, 1976). In addition to affecting the amount of sediment runoff, the ground cover greatly influences the rate of water runoff as well. In some cases the surface runoff has been reduced to almost zero by careful forestry practices.

2.2.1 Biogeochemical cycles.

Energy flows to the earth's surface from space and returns to space when its work is done. The earth, therefore, has an open energy system but chemically it is essentially a closed system--little enters and leaves. The pathways and rates that these chemicals follow in cycling through the environment is of great concern, in particular, the cycling of the four major building blocks of living matter--carbon, oxygen, hydrogen, and nitrogen. Figure 2.3 shows the major physical transport mechanisms. Erhlich (1977) lists seven macronutrients, and thirteen micronutrients in addition to the four major nutrients. Biological components exist within these major components. And, many of the transport mechanisms are partly or totally biotic. Much of the discussion of the remainder of this section on aquatic ecosystems deals with the nutrient cycling mechanisms in different parts of the system.

Figures 2.4 and 2.5 show the biogeochemical cycles of carbon and nitrogen. Note how the hydrological cycle (Fig. 1.1 in section 1.1) intersects with the other cycles. The points at which human's intervention changes or upsets any of these cycles become points at which chains of serious environmental impact could be set in motion.

2.3 Specific Ecosystem Properties: Physical Properties

2.3.1 Climate.

We have examined what can be termed macroclimatology and large-scale meteorological processes that are relevant to understanding the

Aquatic Ecosystem Components   75

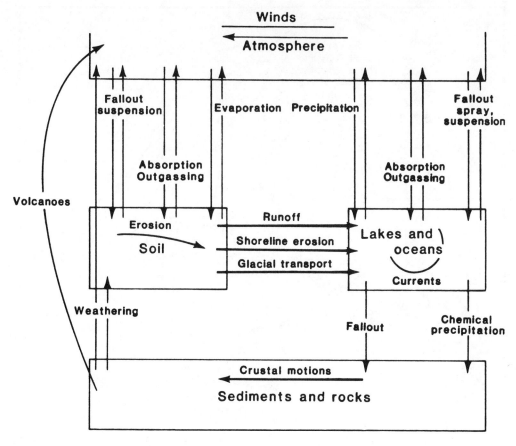

Figure 2.3 Transport mechanisms for a generalized nutrient cycle. [Ehrlich (1977)]

hydrologic cycle, the water balance, and broad geographic variations in climate (Lamb, 1972; Barry and Chorley, 1978). Now we focus on the more specific aspects of climate that a development planner will find of practical value.

There are a number of levels of concern--the country, the region, and the local jurisdiction. In conducting resource inventories, the country and region require the greatest attention. In conducting baseline studies, the focus will be on the local and sometimes the regional levels.

For national and regional studies, climatic characteristics are determined primarily by the position of the country or region with respect to the main branches of the large-scale atmospheric circulation described in section 2.1. These are modified by such factors as distance from the sea, ocean currents, altitude, and configuration of mountain ranges. Although the ocean is the major source of moisture, distance from it is not sufficient alone to account for regional variations in precipitation. Atmospheric circulation patterns and the efficiency of the precipitation mechanisms also are of great importance. Mountain ranges can act as barriers, deforming and deflecting the air flow. Of particular relevance to the hydrologic cycle and the distribution of precipitation is the fact

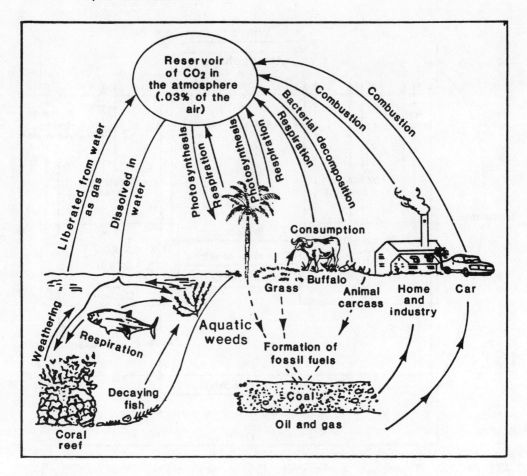

Figure 2.4 The carbon cycle. [Source: Smith (1966)]

that mountain barriers force air to rise, thus producing heavy orographic rain along the windward slopes and a dry area or rain shadow to the lee of the mountains. In the tropics, moist air seems to be concentrated in a fairly shallow layer near the surface, and maximum precipitation occurs well below the highest mountain summits.

There are two other basic types of precipitation: cyclonic (or wave) and convective. Precipitation characteristics vary according to the type of low pressure system and its stage of development, but the essential mechanism in the cyclonic type of precipitation is the ascent of air through the horizontal convergence, or coming together, of different air streams. In low latitudes such disturbances generally are embedded in the easterlies and hence move from east to west. In higher latitudes, rain-bearing disturbances generally are embedded in the westerlies and hence move from west to east. Convective cells, on the other hand, depend strongly on surface heating and convective precipitation; frequently associated with thunderstorms, they are particularly common during the warmest months. Convective rain is also associated with cyclonic disturbances when cold air undercuts warm air and forces it to rise to great heights in the atmosphere. This gives rise to large variations in amounts of rainfall over short distances.

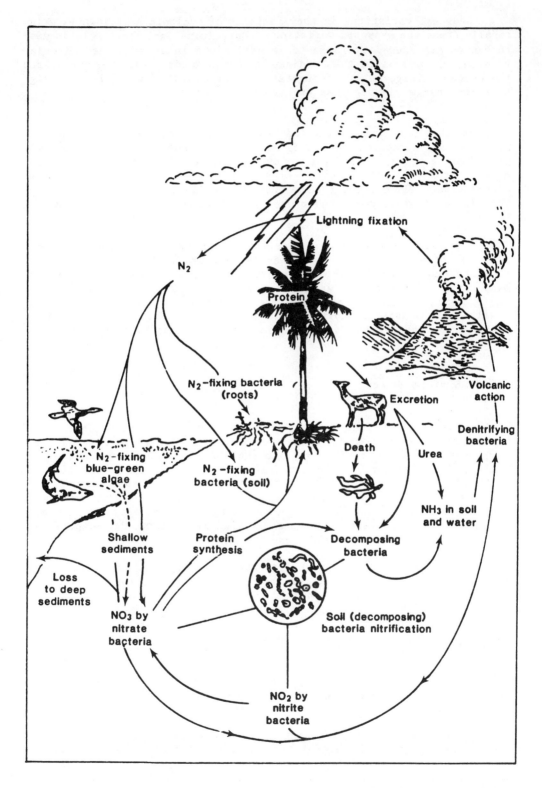

Figure 2.5 The carbon cycle in the ecosystem. [Source: Owen (1980)]

Seasonal variations in the frequency of cyclonic disturbances principally cause the seasonal variations in precipitation. The cyclonic disturbances are caused by large-scale variations in atmospheric circulation and variations in the conditions of the surface of the earth and the oceans. Large scale deforestation, for example, could lead to significant changes in regional or even global climate.

On the local scale, such factors as slope of the ground, orientation of a valley, direction of a coastline, the extent of water bodies, the nature of the surface, and proximity to urban centers will influence the climate at any particular site. On an even smaller scale, there are other factors that will determine the climate in a field, a forest, or a city. This scale of study comes under what is called microclimatology (Sutton, 1977).

The climatic elements of greatest significance in resource inventories and baseline studies on all scales are precipitation, temperature, humidity, and wind velocity. The measurement or estimation of evaporation also ranks high in priority. It is the combination of these elements and their variation over time that largely determines the amount of water available in aquatic ecosystems. Atmospheric considerations are of prime importance in the study of the structure and stability of ecosystems.

2.3.2 Surface water.

Aspects of surface water that are of concern in the management of aquatic ecosystems are the amounts of water in storage in river systems (including lakes and reservoirs) and the rates of flow at various points within river systems. These quantities vary greatly geographically and temporally. Such variations cannot be adequately described for most purposes by simple indices. Some indices, such as mean annual stream flow at a specific location, are helpful in assessing water availability, but management of the resource requires detailed records of the temporal and areal variations of stream flows and storages.

Traditionally, the variability over time of water at a fixed point in space is the primary item measured. Principal measures are long-term average precipitation, stream flow, temperature, and so on, along with average measures of seasonal variation. Thus, average quantities for the year or for all specific months are established as a reference base. Annual, monthly, or short-term quantities can then be studied in relation to the long-term averages.

Long-term average quantities are usually plotted on a map and isopleths (lines connecting points of equal value) drawn. Of particular use in water resource inventories are maps of mean annual precipitation and mean annual runoff. However, short-term variations and sequences are of critical importance in drought and flood assessment. For this reason, basic records of daily rainfall and daily stream flow at pertinent locations are essential to a water management program. Short-term variations, often hourly, are important during storms and floods. In addition, topographic mapping of drainage basins is essential in order to establish drainage boundaries, stream slopes and stream patterns.

2.3.3 Groundwater.

One part of precipitation goes to surface runoff; another percolates into the saturated portion of the aquifer and usually reappears on the surface in the form of springs and seepage, or drains directly into the ocean.

Both surface watersheds and groundwater basins perform the function of conveying water from higher elevations to lower elevations. "Depletion time" defines how long it would take for a watershed to dry out if the inputs were stopped completely. For surface water, depletion times of the order of hours to weeks are normal. If the basin includes lakes or reservoirs then the depletion time may extend from weeks to years. Depletion times for aquifers, however, are typically on the order of tens to hundreds of years.

This difference of timing leads the surface water hydrologist to study quick and often extreme events, while the groundwater hydrologist has to deal with much "slower" phenomena. This has led to the development of the two distinct specialized disciplines: groundwater hydrology (geohydrology) and surface water hydrology (hydrology).

There is also a large difference, as might be expected, in the state of the two systems. At any one time, fully 97.5 percent of the total fresh water appears as groundwater and only 1.5 percent as surface water. Of this amount, 48.7 percent is below half a mile in depth (van der Leeden, 1975). Hence, groundwater is often the most important source.

The quality of groundwater is determined both by the amount of water and the characteristics of the aquifer. In some areas of the world groundwater provides the bulk of the water supply for both municipal and agricultural supplies. In most countries groundwater plays an extremely important role in rural water supply. The relative stability of the groundwater supply, its often much higher quality in comparison with alternative surface supplies, and its location with respect to homesteads make groundwater very important.

The quantity of groundwater available has to be inferred by a variety of indirect methods based on well pumping. The quantity is difficult to estimate in the face of a variety of perched aquifers (elevated above the normal water table), or aquitards, which are often missed in hydrological investigations. Ultimately, the usable amounts of the resource are determined by the amount of annual recharge coming into the aquifer (Lloyd, 1981; and Mandell and Shiftan, 1981). This is often quite difficult to estimate given the existence of artesian aquifers and leaky influent and effluent stream beds.

The chemical parameters of quality are the important characteristics of groundwater. Groundwater tends to be more heavily mineralized than surface water, because of the leaching of salts from the ground. In many areas there are large naturally occurring salt deposits in the soils and aquifers that render the groundwater unsuitable for most, if not all, uses. In such areas there may be marked variation of water quality with depth--with the highest quality water at the top. Those sweet waters can be retrieved by judicious use of skimming wells.

2.3.4 Tides and wave action.

Along the coast, tidal rhythms determine a multitude of other physical and geological phenomena. Currents, erosion, sediment transport, and flushing of bays and harbors all are influenced by the tidal cycle. Tides determine exposure time, desiccation stress, and feeding period and affect the reproductive cycle of marine organisms. Almost every baseline study in a coastal area demands information on tidal effects. There are reliable predictive tide tables, such as those of the U.S. National Oceanographic and Atmospheric Administration (NOAA) and subsequent measured reports of actual tidal height for many areas. In underdeveloped areas of the world these data may be unavailable or available only at a few coastal sites. Since the effects of entrainment, dredging, and eutrophication often depend on water movement in and out of an area, local tidal amplitude measurements must usually be made as part of a baseline or impact study.

Wave action imparts a direct physical stress to shoreline structures and to certain natural features. Knowledge of erosion, changes in beach levels, and optimum breakwater placement depend on accurate knowledge of the position, magnitude and seasonality of wave action. Unfortunately the most intense and thus most stressful wave action is the hardest to measure. Methods range from visual estimation of wave height to sophisticated pressure or force sensors mounted on the shore. As with many structural considerations, it is often the extremes of wave action that must be planned for. Historical records of wave height during storms and attendant damage can be used to estimate these extreme conditions when such records exist.

Bottom dwelling (benthic) organisms are strongly affected both by normal wave action and by storm waves. Benthic communities can be ranked on a wave exposure scale from "protected bay" to "exposed outer coast," and the organisms themselves will be radically different across the gradient. Exposed shores, for instance, are characterized by a high splash zone and thus a broad intertidal with wide zones.

In temperate areas, winter storm waves tear up mussel beds and seaweeds providing open space for early successional species. The observed spatial pattern on such shores often includes patches, in otherwise continuous zones, that are the direct result of such wave action. The drag and shear forces produced by waves continuously select species able to withstand such stress, often at a cost of reduced size, mobility, or growth.

Currents and circulation. Knowledge of current patterns, circulation within bays, semi-enclosed bodies of water, lakes and reservoirs, and stratification of bays, estuaries, lakes, and reservoirs is necessary for the same reason as knowledge of tidal cycles. Currents along coastal shores are produced primarily by tides, although there may be some effect of offshore oceanic current at certain times of the year. Wind can also affect current strength and direction at the surface, especially in inland waters. Stratification, the development of distinct layers of water, may result in currents of different direction and magnitude at various depths.

Current tables are available for some areas (NOAA current tables), but usually not on the very local scale needed for a baseline or impact study. Current measurements must be made over a number of tidal cycles, seasons, and weather conditions.

Many coasts have permanent near shore currents that flow in the opposite direction to offshore currents. Tidal current effects are superimposed on these coastal currents.

Temperature. Water temperature has a direct influence on several aspects of both flowing and standing waters. Water temperature is the most common means of producing density layers (thermal stratification).

Temperature affects reproductive cycles of marine organisms much more that tides do. In discussing benthic community zonation and wave action, this effect is most apparent in intertidal communities. Temperature directly affects the metabolic rates of aquatic life with metabolism being accelerated to an optimum temperature above which it starts to be inhibited. An example of the range of tolerance concept is seen in Fig. 2.6. Inhibiting temperatures are rarely encountered in natural waters since the species suffering inhibition may avoid a particular habitat, but temperature may be a problem if there are new thermal inputs. Increasing temperature accelerates the rates of synthesis and of decomposition of organic matter and thus, the total biotic and nutrient cycle. It directly affects the rate of growth, development, and distribution of aquatic organisms so that some species have more generations per year in warm waters. It may also influence the time and success of spawning, and may affect incidence of disease.

Whereas the temperature-density relation apply primarily to standing waters, the metabolic aspect of temperature is significant in streams as well as lakes, reservoirs, and estuaries.

2.3.5 Salinity.

Salinity has a direct effect on both marine and fresh-water organisms. Zonation along estuaries and marshes is determined by the length of time that organisms are exposed to altered salinity and by the magnitude of the salinity differential (Fig. 2.7).

A measure of salinity is clearly a necessary part of a habitat description for estuarine biota. There are several other reasons to be concerned with salinity. It can be used to trace an effluent plume of fresh-water or low-salinity water in coastal areas. It can also be used to find boundaries between layers in stratified coastal bodies of water with fresh-water input.

Salt content also affects the quality of water used for irrigation, livestock, and human consumption. It is thus often necessary to determine groundwater salinity.

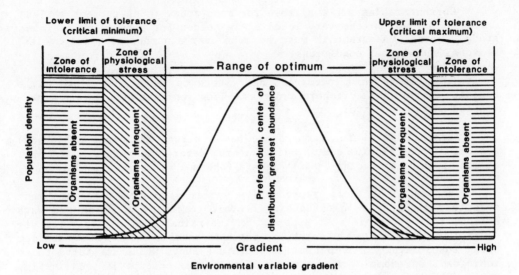

Figure 2.6 Tolerance range concept: critical minimum, maximum, and range of optimum for an environmental variable such as temperature. [Modified from Kendeigh (1974, p. 13)]

2.3.6 Density-dependent stratification.

The density of water is a function of the water temperature and of the content of dissolved solids. Water density can have a profound effect on lakes and reservoirs. Vertical stratification is the result of differences in density. Denser water forms a bottom layer, the hypolimnion. A lighter surface layer forms the epilimnion. Between these is a zone where density changes--the metalimnion (or thermocline in the older literature). The presence of this discontinuity, the metalimnion, is an important measure in lake and reservoir baseline studies.

The hypolimnion is effectively separated from the lake surface and the atmosphere. Organic materials will settle into this layer and decompose. Because of the lack of gas exchange and low light, dissolved oxygen may become depleted, drastically limiting the forms of life present. Gases such as carbon dioxide and possibly hydrogen sulfide can accumulate. In the anoxic state, further decomposition of organic matter slows, and accumulation of materials can cause the basin to fill in. The solubility of many elements is greater in these anoxic, and usually more acidic, waters, and the hypolimnion becomes a potential pool of dissolved solids.

In lakes and reservoirs, where differences in density are dependent on temperature, stratification is variable. At times of overturn, the water of the hypolimnion circulates to the surface where the gases regain their atmospheric equilibrium, and the store of dissolved solids is redistributed. Overturn occurs only when the density differences are minimal. The relation between temperature and density is nonlinear, so that a temperature differential in warm water produces a much greater density difference than an equal temperature differential in cold water (Table 2.1). Consequently, tropical lakes and reservoirs tend to stratify and destratify more readily than do waters of similar size and configuration

Aquatic Ecosystem Components   83

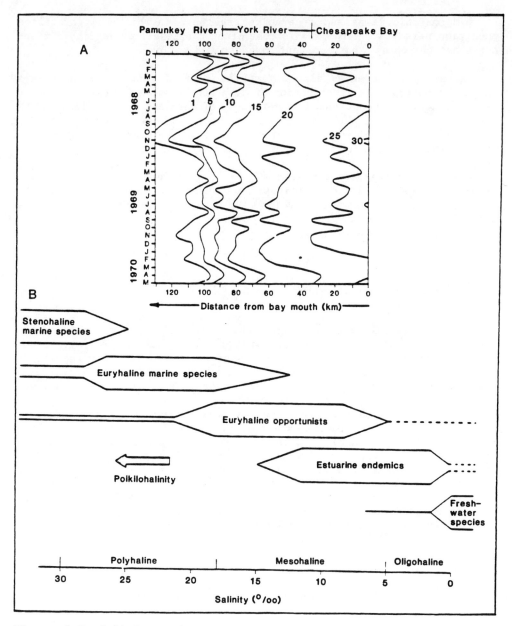

Figure 2.7 Salinity and its effects on the biota within an estuary (Chesapeake Bay).

(A) Salinity (o/oo) as a function of estuarine position and season. This is an example of how to present salinity, time, and position data in a condensed form. [Source: Boesch (1977)]

(B) Classification of species in the benthic community by tolerance and distribution. [Source: Coull (1977)]

---

in temperate regions. A small thermal gradient in warm water is much more likely to establish a significant density-dependent stratification.

Meromictic lakes and reservoirs are those in which the stratified condition persists. Meromixis is most often the result of density differences brought on by a high content of dissolved solids.

Density differences also affect the rate of settling of materials and consequently, the distribution of many life forms. For example, there is frequently a concentration of detritus, of dead and dying cells, and of organisms feeding on these materials at the first layer of density change near the upper metalimnion. Sometimes these zones of concentration are very narrow and careful sampling is needed to detect them.

Water flowing into a system may be of a different density than the receiving lake or reservoir. This produces discrete flows within the lake or reservoir, and in some reservoirs with high exchange rates, flow may be through the reservoir with little contribution to the reservoir water mass.

2.3.7 Light and transparency.

The principal importance of light to aquatic ecosystems is as the energy source to support primary production (aquatic plants). Other aspects include photoperiodic responses of some organisms for regulation of reproductive or migratory patterns. Sunlight intensity is rarely limiting to primary production in surface waters. Often less than 1 percent of sunlight impinging on the water surface is converted to photosynthetic products. Measurements of light transmission through the water column are made in order to describe the extent of the lighted or "photic" zone where photosynthesis may occur. Light is extinguished by three factors: absorption by the water itself, absorption by dissolved or suspended materials in the water, and by scattering by particles in the water. Note that there is a feedback loop with more light penetration allowing more primary production of phytoplankton which in turn both scatters and absorbs more light. Light transmission measures can be used to evaluate the productivity of the water as well as the presence of dissolved organic matter.

2.4 Specific Ecosystem Properties--Chemical Properties

2.4.1 Dissolved gases.

Dissolved oxygen enters and leaves the water through the photosynthetic or the respiratory activities (or both) of the biota and by diffusion through the surface. The net direction of diffusion is a function of the degree of saturation of the gas in the water. Solubility of oxygen in water is a function of the water temperature, with lower concentrations at warmer temperatures. Oxygen saturation at the temperatures encountered in many tropical waters (30°C to 35°C) is between 7 and 8 milligrams per liter; at near-freezing temperatures, it is approximately double that. Many species cannot tolerate exposures below 3 milligrams

Table 2.1 Changes in density in fresh water as a function of temperature. Note the great increase in stability at warm temperature. [Modified from Welch (1935)]

Temperature (C)	Density	Difference	Relative Difference*
0°	0.999868	+0.000059	7.38
1°	0.999927		
3°	0.999992	+0.000008	1.00
4°	1.000000		
4°	1.000000	−0.000008	1.00
5°	0.999992		
9°	0.999808	−0.000081	10.12
10°	0.999727		
19°	0.998432	−0.000202	25.25
20°	0.998230		
29°	0.995971	−0.000298	37.25
30°	0.995673		

*Relative to unit change from 3° to 4° or 4° to 5° C.

per liter. In tropical waters the reserve of oxygen above the critical minimum concentration is often much less than that in temperate regions. Accurate measurements of dissolved oxygen concentrations are important to any baseline study and are especially important in the tropics.

Carbon dioxide enters the water as the end product of aerobic respiration by aquatic life, by conversion of carbonates and bicarbonates dissolved in the water, or by exchange with the atmosphere. Carbon dioxide, bicarbonate ions, and carbonate ions exist in a equilibrium state so that removal of carbon dioxide, say through photosynthetic uptake, shifts the equilibrium, causing replenishment of dissolved carbon dioxide and precipitating insoluble carbonate salts. The concentration of carbon dioxide in the air over the water is usually low relative to that of the water and a net outward diffusion is expected. In well buffered waters--those with moderate or higher concentrations of bicarbonate ions--small changes in carbon dioxide concentration are difficult to detect. Thus changes in carbon dioxide content are much less sensitive indicators of metabolic activity than changes in dissolved oxygen content.

Hydrogen sulfide gas is occasionally of interest because it indicates anaerobic conditions. It is easily detected by its "rotten egg" odor. Quantitative data are usually not needed.

2.4.2 Dissolved inorganic solids.

Water is said to be the universal solvent, and water can be expected to contain most of the elements found in the earth's crust. However, of all the possible elements, a few are of special interest in baseline studies describing the chemical nature of an aquatic ecosystem: (i) those important in regulating production as basic nutrients contributing fertility to the water and (ii) those undesirable elements with potentially harmful effects on the biota. In some cases, it is important to measure concentrations of elements to compare with those of other elements because they may have synergistic effects. Also, the toxicity of some elements is a function of the degree of buffering present in the water. Thus, ratios of ions may have much interpretive value.

2.4.3 Dissolved organic matter.

Dissolved organics in water are the result of decomposition, secretions, excretions of organisms, and external introduction. Dissolved organic matter often makes up the major part (80 to 90 percent) of the total organic content of lakes and reservoirs and a somewhat less, but still significant, proportion in streams. Dissolved organic matter is a food source for microorganisms and contributes a secondary base to the food chain. In high concentration, it contributes dissolved color to the water and thus not only reduces the quantity of light penetrating to a given depth but also changes light quality by selective absorption.

Some dissolved organics present water quality problems. Tastes and odors in drinking water supplies are usually due to the presence of dissolved organics (geosmin, methylisoborneol) produced by aquatic bacteria (actinomycetes). Cyanobacteria are capable of producing toxic substances that may cause illness in livestock.

Biocides may enter the water directly through applications for control of unwanted aquatic life (weeds, insects, snails, fishes) or indirectly through application in the watershed. Several persistent (refractory) pesticides accumulate in aquatic ecosystems and undergo biological magnification through the foodchain. High concentrations may be present in species used by man. Many persistent pesticides are still widely sold for use in developing countries.

2.4.4 Inorganic particulates.

Inorganic particulates are materials held in suspension because of water turbulence, density discontinuity, or, rarely, unusual chemical conditions. Most important in baseline studies is the suspension of clays and silt. These are derived either from runoff, usually during periods of high flow, or from resuspension of shallow water sediments caused by wind-driven wave action.

These materials are important in the study of the limnology of lakes, reservoirs, streams, and estuaries. They will ultimately settle and contribute to the filling of the lake or reservoir basin. This settling may also smother the botton-dwelling life forms. This is a problem in both flowing and standing waters. Suspended materials are important in scattering sunlight, and thus reducing the depth of light penetration, and thus decreasing the productivity of the ecosystem. The feeding activities of sight-feeding fishes may also be adversely affected. In extreme cases, high inorganic turbidity may impact gill-breathers by coating the gill surfaces and reducing gas exchange.

2.4.5 Organic particulates.

Organic particulate matter consists of the plankton and bits of detritus in suspension. In the upper waters of lakes and reservoirs, its measurement provides little information because of its composite nature. In deep waters, where much of the plankton is dead, it may be used to estimate the quantity of potentially oxygen-demanding material. In streams, where most is detrital, it is used to measure the transport of potential food for bottom feeders and of the contribution of terrestrial sources to the stream or receiving lake or reservoir ecosystem.

2.4.6 pH.

The commonly used term for acidity and alkalinity of water is pH. This unit is the logarithm of the reciprocal of the hydrogen ion concentration. (The neutral point is 7.0 pH, with values less than 7.0 connoting acidity and values greater than 7.0 connoting alkalinity.) Note that a one-unit change in pH is the result of a tenfold change in the hydrogen ion concentration.

One should also remember that when processing pH data, it is inappropriate to express an average pH, because pH units are not of equal value in expressing the hydrogen ion content of the water. If a mean pH is desired, it is necessary to convert pH values to their hydrogen ion concentrations, compute the mean, and then express this mean as pH. One should exercise caution in using pH in interpreting the degree of acidity of the water. In some very acid water, such as mine wastes, the potential free acidity may be much greater than the measured pH indicates because of a reserve of undissassociated ions. Potential free acidity must be determined by titration with a strong base (a laboratory procedure).

Changes in pH are caused not only by direct additions of acid or basic substances to the water but also indirectly through changes brought about by a variety of other materials and even through the metabolic activity of the biota.

2.4.7 Conductivity.

The electrical conductivity of water is an expression of the water's capacity to conduct a current. Its value is related to the concentration of free ions in the water and to temperature. Conductivity provides a quick, convenient estimate of ionic content and thus of the fertility of the water, as well as a quick check on water suspected of having received ionic pollutants. Conductivity also gives an easy and rapid measure of salinity. A laboratory analyst may use conductivity to evaluate water before undertaking detailed mineral analyses to determine whether dilution is required.

2.5 Specific Ecosystem Properties--Biotic Properties

In baseline studies, one is often concerned with measuring the variables that define the physical environment of the biota. It is important to remember that organisms flourish within a certain range of environment conditions (Fig. 2.6). The upper and lower limits of tolerance are usually called the critical maximum and minimum, and there is also an optimum set of conditions. The tolerance range and optimum may vary with age, sex, and general condition of the organism.

2.5.1 Benthic biota.

Marine and lacustrine benthic communities are often composed of several large and relatively long-lived animals and plants. In some communities species composition and abundance change seasonally while in others, notably undisturbed coral reefs, composition changes only over long periods--decades or centuries. The benthos includes species of commercial importance, such as mussels, clams, crabs, and lobsters, as well as important prey species for fish. Often, larvae produced locally are the primary source of new recruits for benthic populations and thus local disturbance can profoundly affect the future of benthic assemblages. Other benthic populations have wide-ranging larvae and local recruitment is from a large pool of larvae derived from great distances.

Fresh water benthos includes the larvae of many insects, some of which are vectors for parasites that can infect humans. Snails, important herbivores in the natural system, also carry stages of parasites responsible for schistosomiasis and other diseases in man.

Benthic and intertidal communities can change radically on a small spatial scale (Fig. 2.8). Zonation is one of the more obvious spatial patterns, often visible at low tide on rocky shores, where plants and animals are often distinctly colored. Zonation continues beyond tidal areas and into soft-sediment communities. Zonation has distinct importance for benthic sampling regimes. Zones must be defined and individually sampled to give a complete picture of a benthic community. Intertidal and subtidal zones can change rapidly as a result of changes in wave exposure, light, or salinity and impacts from human population centers. There will be naturally occurring variability and a sampling program should take this into account.

Species chosen for detailed study are often those that are the most abundant (spatial or numerical dominants), those that are commercially important, or those known to be "indicator species" that respond to certain known perturbations. Zooplankton can be used as indicators of pollution, or its absence, as well as indicators of distinct water masses.

2.5.2 Zooplankton.

Shorelines of lakes, streams, or other bodies of water have specific, yet constantly varying, zooplankton assemblage. Quantities of zooplankton change daily as a result of tidal cycles, illumination, fresh water runoff, predation, and countless other factors. The best and most complete zooplankton sampling programs in oceanic waters have shown that zooplankton "patches" occur on the scales from a few meters to hundreds of meters. The seasonal progression of zooplankton in both fresh and salt water is well known for temperate areas; much less work has been done in tropical areas.

Because of the extreme spatial and temporal changes in zooplankton assemblages, an area's zooplankton, for all practical purposes, cannot be adequately described. Baseline or impact studies must focus on certain aspects of zooplankton assemblages designed to test specific assumptions. Zooplankton studies are usually carried out (i) to describe generally the taxa, their abundance, and their seasonal or 24-hour cycles; (ii) to estimate zooplankton as food resources as a part of fisheries studies; (iii) to determine the seasonality and abundance of larval forms of specific benthic or nektonic species in conjunction with population studies; or (iv) to determine the effects of a specific perturbation (power plant entrainment, hot water plumes from power plant discharges, changes in near-shore currents due to construction or dredging, eutrophication in lakes or estuaries, effects of specific pollutants).

For general descriptive studies, the sampling program is very complex. Samples must be taken frequently enough to measure day-night cycles, and day-to-day changes, as well as seasonality and year-to-year variability.

2.5.3 Phytoplankton.

In most lakes and reservoirs, the phytoplankton are the base of the food chain and are thus responsible for the productivity of the entire ecosystem. It is usually necessary for a baseline study to determine the kinds as well as the quantity of phytoplankton present and the rate of production. Many phytoplankton feeders are selective. In addition to their role in the food chain, some phytoplankton contribute to water quality problems, producing scums, toxins, or undesirable odors. Phytoplanktonic bluegreen algae are also important as the principal

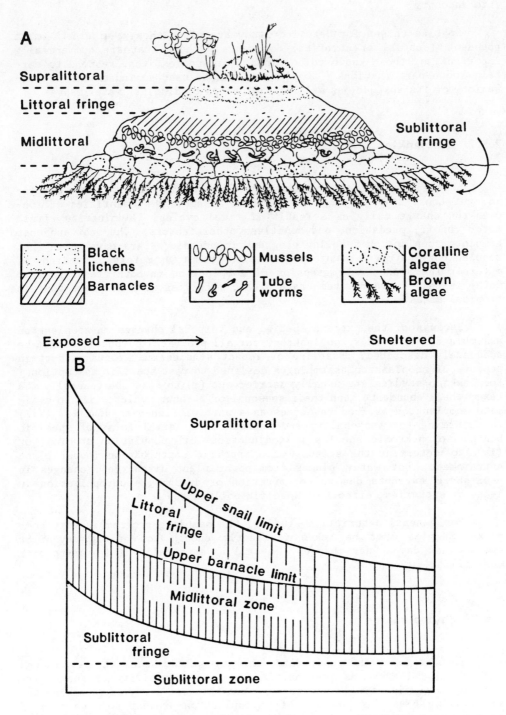

Figure 2.8 Zonation of intertidal communities. [Modified from Morton and Miller (1973)] The effects of exposure (wave energy) and intertidal elevation are illustrated. (A) Pictoral representation of zonation. Generalized elevation and exposure diagrams for a stretch of coastline from a protected bay to an exposed headline.

"fixers" of inorganic nitrogen into forms usable by other species of plant life.

Fish and fisheries. Fish are often the most prominent and best known segment of the biota of aquatic ecosystems. The species are often easily identified, and in many cases their habitat preferences are well enough known so that presence and abundance of a particular species may be used as an indicator of environmental conditions.

The abundance of fish should be recorded in as many habitats as possible, from shallow to deep waters. Where there are sources of pollution, the fish population in the affected area may be compared with that in unaffected waters.

A fish community may consist of herbivores, planktivores, insectivores, detritivores, and carnivores (predators). If the food habits are not known, the fish can often be placed in a category on the basis of structure of the mouth, gill rakers, and digestive tract.

In addition to fish, shrimp, other crustaceans, frogs, turtles, snails, and clams are often considered fishery resources and should also be included in the study of the communities.

Because fishery resources are important to the economy of most developing countries, the extent and condition of the fisheries should be recorded as part of the baseline studies. Information should be collected on the numbers and economic status of the persons involved in the industry, the equipment used, and the status of the fish populations.

2.5.4 Littoral vegetation.

Littoral vegetation is that in shallow water near the shoreline (Fig. 2.9) and consists of large plants (macrophytes) that are usually vascular species but also a few species of macroalgae.

In most lakes and reservoirs, the rooted vegetation contributes only a small and usually insignificant percentage of the total primary production even though the apparent biomass is large. However, in clear streams, ponds, and marshes, rooted vegetation is the base of the food chain. This vegetation also provides spawning and nursery grounds for many species and a substrate upon which the periphyton community develops. Some types of rooted vegetation are thought to play an important role as nutrient sources by pumping nutrients from the sediments into the water. Rooted plants help stabilize shorelines by preventing erosion and reducing turbidity.

2.5.5 Wetlands vegetation.

Areas where the soil is permanently or usually saturated with water--salt or fresh water--come under the general heading of wetlands:

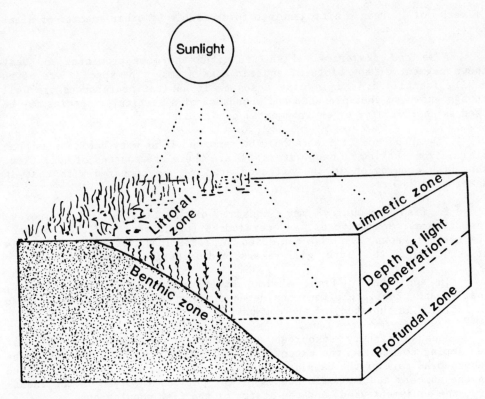

Figure 2.9 Major habitat zones of a typical lake ecosystem. [Modified from Reid and Wood (1976)]

swamps, bogs, and marshes. In some classification schemes, the intertidal zones of rocky shores or sand beaches are also considered wetlands. Seaweeds and microalgal mats are important components of saltmarsh vegetation.

Wetlands are often important habitats and are the nursery or refuge for several species. The vegetation of both fresh and salt marshes is often harvested for human use. Saltmarsh hay, reeds, and both algae and vascular plants are such used.

2.5.6 Periphyton.

The periphyton (Aufwuchs) is that community attached to the surfaces of submerged objects and is recognized as scum on rocks, pilings, plant stems, and even beer bottles. The periphyton may be considered as a special case of the benthos. It consists of microscopic species such as algae, protozoa, and bacteria; however, larger species such as snails and insect larvae are occasionally considered. Although secondary to the phytoplankton as producers in lakes and reservoirs, the phytoperiphyton are often the base of the food chain in streams. In reservoirs, inundated with timber, the periphyton community is also important.

The periphyton community is often the community most exposed to environmental variables such as pollution. Consequently, it is used to show the stress placed on the ecosystem by measuring changes in both quantity and taxonomic composition. The ratio of autotrophic organisms (primary producers) to heterotrophic organisms (consumers) is called the autotrophic index (AI), which is used to measure suspected pollution caused by organic matter such as sewage, animal wastes, and food processing wastes (Environmental Protection Agency, 1973; American Public Health Association, 1980).

2.5.7 Microbiota.

Methods used to assess microbial activity are rarely specific for individual ecosystems. Microorganisms are ubiquitous, and the same organisms are commonly found in a wide range of habitats. Unusual microorganisms or blooms of specific microbes are found in extreme aquatic environments, or under conditions of serious perturbations such as in polluted waters. Some microorganisms and microbial processes are particularly active in the environments where development projects occur. For example, high temperature in the water stimulates increased microbial activity and human pathogens play an important role in water contamination. Specific methods are needed to indicate the hazards from these pathogens, but few have been devised for specific aquatic habitats.

The structure of microbial communities can be expected to vary according to environmental conditions. However, the range of microorganisms should be similar for most of the aquatic systems we consider. Nitrogen fixation in aquatic habitats is mediated by the cyanobacteria, photosynthetic bacteria commonly found in rice paddies and in eutrophic lakes and rivers. Fixation from the atmosphere is the dominant source of nitrogen in rice paddies. It has been estimated that between 18 and 110 kilograms of nitrogen per hectare per year are fixed in rice paddies.

The process of eutrophication involves nutrient enrichment of waters, resulting in stimulation of primary productivity. Ultimately, eutrophication causes the development of anaerobic conditions in the water, deterioration of the environment, and a decrease in the diversity of the fish populations.

During development, the health hazard from microbial pathogens can be severe. The variety and density of pathogens in untreated sewage is large, and water contaminated with sewage is frequently used either for drinking or irrigation. Development projects can also cause water to become polluted with toxic chemicals--pesticides, petroleum, and industrial wastes. These chemicals can cause both severe ecological disturbances and health hazards.

2.6 Specific Ecosystem Properties--Functional Properties

2.6.1 Nutrient cycling.

The sun provides continuous energy to aquatic ecosystems and matter is constantly recycled. Matter enters the stream, lake, or reservoir from the watershed and leaves through outflow or sedimentation. The rate of matter-dependent production processes is a function of the size and rates of the flows and energy cycles operating in the system. In most aquatic ecosystems, production rates are limited by the availability of some nutrient, and not the quantity of solar energy. Typically, the element limiting primary production is either nitrogen or phosphorus. Rarely other mineral or organic materials may be limiting, but they are usually not measured first in baseline studies.

The rates of internal recycling of nutrient budgets as well as the rate of input from the watershed and output downstream or to the sediments have been described for few ecosystems. Such information is especially critical in the typically nutrient-poor tropical waters where, because of high temperatures, the rate of recycling becomes rapid and its understanding in relation to the rate of production is crucial.

2.6.2 Primary production.

Primary production is usually measured as the rate of photosynthesis and, except in unusual situations, the rate of chemosynthetic primary production is ignored. In baseline studies it is important to recognize the difference between the apparent amount of plant material present (standing crop) and the rate of primary production. For example, a large standing crop of littoral vegetation only appears to be a major component in the food chain. In reality, the greatest production is carried out by the microscopic phytoplankton.

Methods to determine primary production have been the most fully described and applied in a variety of aquatic ecosystems. They have become widely used in surveys. Primary production data may be crudely extrapolated to estimate fish and other secondary production.

2.6.3 Secondary productivity.

Often, in order to predict or assess impact, it is necessary to know which predators are most important, on which species they prey, in which zones, and so forth. Even the best sampling programs often do not address these questions. Experimental tests of predator or herbivore effects, combined with field observations or collections, are needed to determine quantitative, rather than qualitative, effects of higher trophic level species. For example, knowledge of the diet and behavior of a particular predator (keystone species) might indicate that its removal would have a greater effect on the community than its numbers alone would predict. A certain prey species may, conversely, support several specialist predators (key industry species), and its removal would disrupt the species that depend on it.

Food web diagrams illustrate the important interconnections between prey and predators. The links in a food web can provide information about the amount of energy passing along in the web. This information permits a better understanding of the relative importance of particular pathways, predators, and prey. It can also give a crude measure of the secondary productivity, the biomass or energy production at each trophic level above the primary producer level.

2.6.4 Eutrophication.

Lakes, reservoirs, and rivers occupy low places and are subject to accumulation of materials and perhaps to being filled in--a natural successional process. Eutrophication accelerates this process. Originally the term referred to enhanced production or accumulation of organic matter in the sediments because rates of decomposition were less than the rates of synthesis or input from outside the lake. However, the term has come to refer to the process of filling, including sedimentation from inorganic clays and silts.

The result of eutrophication is usually, but not always, undesirable. It may cause changes in species composition, increasing, for instance, undesirable species such as mosquitoes and midges. Floating algal mats may make boating difficult, and the drinking water may take on unpleasant tastes and odors.

Eutrophication has been recognized as a water quality problem in the United States for approximately 15 years. Several models to describe factors governing the process have been proposed (Vollenweider, 1976; Dillon and Rigler, 1974; Reckhow, 1979; Reckhow et al., 1980). A need to express quantitatively the subjective condition of eutrophication is apparent. Carlson (1977) proposed a trophic state index (TSI) which has come into wide use. However, this index has several limitations, especially in water with a high content of dissolved organic matter or suspended inorganic matter (Edmondson, 1980; Lorenzen, 1978; Megard et al., 1980).

2.6.5 Community and ecosystem descriptions.

Once the data on the composition of the community are available, certain kinds of indices or diagrams can be prepared and the information on entire communities and assemblages, including the spatial and temporal variability can be analyzed. The following approaches are commonly used: (i) graphic and tabular comparison of numbers and biomass, often including parametric or nonparametric statistical analysis, (ii) analysis of diversity, species richness, and evenness, (iii) energy flow analysis and food web construction, (iv) experimental tests of stability and resilience, and (v) multivariate statistical treatment of data to examine similarities between samples (clustering and ordination) or sources of variability (principal component analysis and multiple regression).

These analyses allow comparisons of areas, seasons, and zones, as well as descriptions of the functional or structural roles of species. It is important that the techniques applied are appropriate so that the interpretation of the data and the conclusions drawn provide the critical information.

2.6.6 Water balance.

Analysis of the water balance is a way of determining how the hydrometeorology, soil physics, and groundwater hydrology are integrated; it also provides an estimate of the total availability of water. The balance can be computed for a soil profile in a whole basin on a daily, weekly, or annual basis. The water balance method requires the computation of soil moisture, evapotranspiration, groundwater recharge, stream flow, and certain observations on the soil and vegetation.

Conceptually the water balance equates what goes in and what comes out of a hydrologic unit over a period of time. Hence, precipitation = canopy and litter interception plus actual evapotranspiration plus overland flow plus change in soil moisture plus change in groundwater storage plus groundwater runoff. But the limited availability and reliability of data often introduce large uncertainites into this simple equation.

Sampling Methods and Selection Criteria

3.1 Recommendations for Baseline Studies with Resource Inventories

In this section of the report we describe briefly baseline and resource inventory studies. We consider the impact of development projects in terms of inputs into the water, the watershed, and the atmosphere. To enable a project officer for a specific project to recommend which, if any, baselines studies should be conducted in a particular case, we have developed a system of separate recommendations for water bodies, watersheds, and air sheds (Tables 3.1, 3.2, and 3.3)

The components of aquatic ecosystems which may be selected for baseline study are used as row headings. In this matrix we have then entered one of four recommendations: A, almost always recommended; O, often recommended; S, sometimes recommended; blank, not applicable or not recommended. None of the categories implies that some one study absolutely has to be made, nor do they give any indication of the level of impact. The judgment about impact is delayed until a scientific assessment is made.

TABLE 3.1 Recommendations for baseline studies for assessing impacts on watersheds.

	Land Modification							Land Reclamation						Introduction of Chemicals			
Components of Aquatic Ecosystems	Roads	Agriculture	Settlements	Urban/industrial	Mining	Logging	Solid waste disposal	Afforestation	Agriculture	Land leveling	Water harvesting	Drainage/diking	Desalinization	Fertilizer	Pesticides	Industrial emissions	Domestic emissions
1. Physical properties																	
a. Climate																	
b. Precipitation	S	O	O	O	S	O	S	A	A	S	S	S	S	O	O	O	S
c. Surface water	O	O	O	O	O	O	O	O	A	O	A	A	S	A	A	A	O
d. Groundwater	S	S	O	O	O	S	O	S	S	O	O	O	S	A	A	O	O
e. Tides and wave action																	
f. Currents and circulation	S	S	S	S	S	S	S	S	S	S	O	S	S	S	S	S	S
g. Temperature	A	A	A	A	A	A	A	A	A	A	A	A	A	A	A	A	A
h. Salinity																	
i. Density and stratification	S	S	S	S	S	S	S	S	S	S	S	S	S	A	A	A	A
j. Light and transparency	A	A	A	A	A	A	A	A	A	A	O	A	O	A	O	O	O
2. Chemical properties																	
a. Dissolved gases	O	O	O	O	O	O	O	O	A	O	O	O	O	A	O	O	O
b. Dissolved solids																	
1) Inorganics: major nutrients	A	A	A	A	A	A	A	A	A	O	O	O	O	A	S	O	O
2) Organics	O	A	O	O	O	A	A	O	A	S	S	S	S	O	A	S	O
c. Particulates																	
1) Inorganic	A	A	A	A	A	A	A	A	A	A	O	A	O	O	O	S	S
2) Organic	S	A	S	O	S	A	O	A	A	S	S	S	S	A	S	S	O
d. pH	S	O	O	A	A	O	A	O	O	S	S	S	O	O	O	A	A
e. Conductivity	O	O	O	O	O	O	O	O	O	S	S	S	A	O	S	A	A
3. Biotic properties																	
a. Benthos	A	A	A	A	A	A	A	A	A	A	O	O	A	A	A	A	A
b. Plankton																	
1) Zooplankton	S	O	S	S	S	S	O	S	O	S	S	S	O	O	A	A	A
2) Phytoplankton	O	A	O	O	O	O	A	O	A	S	S	S	O	A	O	A	A
c. Fish and fisheries	O	O	O	O	O	O	O	O	O	O	S	S	O	O	A	A	A
d. Littoral vegetation	S	O	O	O	S	O	O	O	O	S	O	O	S	A	S	O	O
e. Wetland vegetation	S	O				S	S	O	O	A	A	A	O	A	O		
f. Periphyton	S	O	O	O	O	S	O	S	O	S	S	S	O	A	A	A	A
g. Microbiota, pathogens other	S	O	O	S	S	S	A	S	O	S	S	S	O	S	O	O	A
4. Functional properties																	
a. Nutrient cycling	O	A	O	O	O	O	A	O	A	O	S	O	O	A	S	O	A
b. Primary productivity	S	O	O	S	S	S	O	S	A	S	S	S	S	A	S	S	O
c. Secondary productivity																	
d. Eutrophication	S	A	A	A	S	O	A	A	A	O	S	O	S	A	S	S	O
e. Ecosystem indices																	
f. Water balance	S	O	O	O	S	S	S	A	A	S	O	S	S	S	S	S	S

Key: A = Almost always recommended
     O = Often recommended
     S = Sometimes recommended
     Blank = Not applicable or not recommended

TABLE 3.2 Recommendations for baseline studies for assessing impacts on water bodies.

Components of Aquatic Ecosystems	Impoundments							Withdrawal/Return Water Uses						Dredging		Introduction of Chemicals				Biotic Resource Harvest			
	Irrigation	Hydroelectric	Flood control	Water supply	Industrial	Recreational	Fisheries	Agriculture/irrigation	Livestock	Municipal/domes	Industrial	Mining	Waste disposal treatment	Estuarine	Riverine	Fertilizer	Pesticides	Industrial emissions	Domestic emissions	Fish	Shellfish	Vegetation	Aquaculture
1. Physical properties																							
a. Climate		S	S	S																			
b. Precipitation	A	A	A	S	S	S	S	A	S	S	S	S				S	S	O	S	S	S	S	S
c. Surface water	A	A	A	A	A	A	A	A	O	O	O	O	O	S	S	A	A	A	A	A	A	O	A
d. Groundwater	S	O	O	O	O	O	O	A	S	O	O	A	A	S	S	A	A	A	O	S	S	S	S
e. Tides and wave action		S	S		S	S	O							A									
f. Currents and circulation	O	O	O	O	O	O	A	S	S	S	S	S	S	A	A	A	A	O	O	O	O	O	O
g. Temperature	A	A	A	A	A	A	A	A	A	A	A	A	A	A	A	A	A	A	A	A	A	A	A
h. Salinity	A	S	S	A	O	O	A	A	A	A	A	S	S	A		S	S	S	S	A	A	S	A
i. Density and stratification																O	O	O	O	O	O	O	O
j. Light and transparency	S	S	S	S	S	S	S	A	A	O	O	O	O	A	A	A	A	O	O	O	O	O	O
2. Chemical properties																							
a. Dissolved gases	O	O	O	O	O	O	O	A	A	A	A	A	A	O	O	A	A	O	O	A	A	A	A
b. Dissolved solids																							
1) Inorganics: nutrients/others	O	O	O	O	O	O	O	A	A	O	O	O	A	A	A	A	O	O	O	A	A	O	A
2) Organics	S	S	S	S	S	S	S	O	A	O	S	S	A	O	O	A	A	O	O	O	O	O	O
c. Particulates																							
1) Inorganic	O	O	O	O	O	O	O	A	A	A	A	A	A	A	A	S	O	S	S	S	O	S	S
2) Organic	S	S	S	S	S	S	S	A	A	O	S	S	A	O	O	S	S	S	S	S	S	S	S
d. pH	O	O	O	O	O	O	O	A	A	A	A	A	A	O	O	A	O	A	A	O	O	O	O
e. Conductivity	A	A	A	A	A	A	A	A	A	A	A	A	A	O	O	A	O	A	A	A	A	O	A
3. Biotic properties																							
a. Benthos	A	A	A	A	A	A	A	A	A	A	A	A	A	A	A	A	A	A	A	A	A	O	O
b. Plankton																							
1) Zooplankton								A	A	O	O	O	O			A	A	A	A	A	S	S	O
2) Phytoplankton								A	A	A	A	A	A			A	A	A	A	A	O	O	O
c. Fish and fisheries	A	A	A	A	A	A	A	A	A	A	A	A	A	A	A	A	A	A	A	A	O	A	A
d. Littoral vegetation	O	O	O	O	O	O	O	O	O	O	O	O	O	A	A	A	O	O	O	A	O	A	O
e. Wetland vegetation	A	O	A			O	A	A	O	O	O	O	O	A	A	A	O			A	A	A	A
f. Periphyton	O	O	O	O	O	O	O	A	A	O	A	A	A	O	O	A	A	O	O	S	S	S	S
g. Microbiota, pathogens other	A	O	O	A	O	A	O	A	O	A	O	S	A	S	S	S	S	S	S	S	A	S	O
4. Functional properties																							
a. Nutrient cycling	A	A	A	A	A	A	A	A	A	O	O	O	A	A	A	S	O	O	O	A	A	A	A
b. Primary productivity	O	O	O	O	O	O	O	A	A	O	O	O	A	O	O	A	S	O	O	O	O	A	A
c. Secondary productivity	O	O	O	O	O	O	O	O	O	O	O	O	O	S	S	A	O	O	O	A	A	S	S
d. Eutrophication								A	A	O	O	O	A			A	S	S	O	O	O	O	S
e. Ecosystem indices	S	S	S	S	S	S	S	S	S	S	S	S	S	S	S	A	A	A	S	A	A		A
f. Water balance	A	A	O	O	S	S	S	A	S	S	S	S	S							S	S	O	S

TABLE 3.3 Recommendations for baseline studies for assessing impacts on airsheds.

Components of Baseline Studies	Generic & Specific Project Actions			
	Introduction of Chemicals			
	Fertilizer	Pesticides	Industrial emissions	Domestic emissions
1. Physical properties				
a. Climate			S	S
b. Precipitation	O	O	O	O
c. Surface water	A	A	O	O
d. Groundwater	O	O	O	O
e. Tides and wave action				
f. Currents and circulation	O	O	O	O
g. Temperature	A	A	A	A
h. Salinity	S	S	S	S
i. Density and stratification	S	S	S	S
j. Light and transparency	S	S	S	S
2. Chemical properties				
a. Dissolved gases	O	S	S	S
b. Dissolved solids				
1) Inorganics: major nutrients/others	A	S	A	A
2) Organics	S	A	O	O
c. Particulates				
1) Inorganic		S	O	O
2) Organic			S	S
d. pH	S		A	A
e. Conductivity	S		A	A
3. Biotic properties				
a. Benthos	S	A	O	S
b. Plankton				
1) Zooplankton		S	S	S
2) Phytoplankton	O	S	S	S
c. Fish and fisheries	S	S	S	S
d. Littoral vegetation	O	S	S	S
e. Wetland vegetation				
f. Periphyton	S	O	O	O
g. Microbiota, pathogens and other	S	O	O	A
4. Functional properties				
a. Nutrient cycling	A	S	O	O
b. Primary productivity	O	S	O	O
c. Secondary productivity				
d. Eutrophication	A	S	S	S
e. Ecosystem indices				
f. Water balance	S	S	S	S

Key: A = Almost always recommended
     O = Often recommended
     S = Sometimes recommended
     Blank = Not applicable or not recommended

II. AQUATIC ECOSYSTEMS

3.2 Methods for Physical Properties

3.2.1 Climate.

The *Guide to Climatological Practices* (World Meteorological Organization, 1960) describes most standard methods of data collection and analysis, as well as applications that are of relevant in aquatic baseline studies. Below is a list of relevant chapters in the *Guide* which we strongly recommend to our readers.

- Climatological Organization

- Climatic Elements and Their Observation
  - Definitions
  - Observations of climatic elements
  - Climatological stations
  - Instructions for observers

- Climatological Data: Collecting, Scrutiny, Storage and Supply
  - Recording procedures
  - Scrutiny of data
  - Collection of data
  - Storage and cataloging
  - Supply of climatological information

- The Use of Statistics in Climatology
  - Basic statistical theory
  - Quality control and time series
  - Correlations
  - Vector quantities

- Data Processing

- Descriptive Climatology
  - Graphical presentation of climatological data
  - Climatic maps and atlases
  - Climatological classifications
  - Descriptive texts
  - Specifications of requirements for national, subregional climatic atlases for land areas
  - Special points to be observed in the preparation of maps for different elements

- Microclimatology
  - Scope
  - Processes
  - Types of environment
  - Methods and techniques of measurement
  - Examples of microclimatological investigations
  - Recommended studies
  - Applied studies

- Publication of Climatological Data

- Application of Climatological Analysis
  - Methods of analysis
  - Agriculture
  - Engineering and industry
  - Telecommunications and energy transfer
  - Land transportation
  - Aviation

Man can interfere with the hydrologic cycle in many ways, both inadvertently and deliberately. These impacts include the direct modification of rainfall and evaporation characteristics as well as changes in factors affecting the disposition and character of surface and groundwater.

It is difficult to assess man's impact on climatic conditions in general and precipitation patterns in particular. The problems include inadequacy of data, unreliability of data, difficulties in interpreting data, and incomplete knowledge about natural conditions, processes, and variability. To determine what the effects on local and regional climatic conditions might be of constructing a large reservoir such as Lake Volta or draining extensive swamps such as the Sudd in the Sudan, one can develop a climate simulation model with data from baseline studies and then alter the model on assumptions of changed boundary conditions. This will give a theoretical estimate of the changes that can be expected, and this information will provide valuable input into project feasibility studies and environmental impact assessments.

If the project proceeds and climatic conditions are monitored, it will still be difficult, for the reasons mentioned above, to attribute changes in weather and climate to the construction project. For example, if, for 5 years, precipitation at a site 200 km from the reservoir is 20 percent higher than the average for the 30-year period before construction of the reservoir, can it be attributed to the construction of the reservoir, or is it merely part of the natural variability in climate? We do not know, but the longer our time-series for baseline studies, the greater will be our knowledge of the range of natural climatic variability and the greater will be our confidence in evaluating impacts. A long time-series of precipitation and river discharge data will add to our confidence in assessing water availability.

If a large dam is proposed and it is expected to have a life span of 50 years, it will be essential to have probability statements about the average discharge of the impounded river and the annual variations in discharge that can be expected during the life of the dam. Typically, historical records of river discharges in developing countries are short, frequently no more than 20 or 30 years, with the longest around 70 years. Such records simply are inadequate as a basis for projecting with any degree of accuracy, water availability over the next 50 years. What frequently happens is that engineers, because of the inadequacy of climatic and hydrologic data, have to be overcautious in their estimation of droughts and floods and the incorporation of this information in the design of a project can add greatly to the construction costs. The involvement of climatologists with experience and a broad perspective in reconstructing longer time-series of river discharges from, for example, precipitation records, should be considered essential in any large-scale hydrology project.

A good reference for discussions on climate variability and the design and operation of water resource systems is the paper by Schaake and Kaczmarek (1979). The World Meterological Organization has many programs concerned with improving climatological, meterological, and hydrological services in developing countries, and these should be regarded as main sources of information.

3.2.2 Surface water.

Streamflow measurements are usually made by dividing the cross-section of the stream into a number of trapezoids and measuring the flow velocity by means of a Price current meter or a pitot tube, taking a velocity measurement at 0.6 of the total depth from the surface or averaging velocity measurements at 0.2 and 0.8 of the total depth in the middle of each trapezoid. These, along with depth and with measurements of each trapezoid are used to compute the flow rate through each trapezoid. The sum of these is the total streamflow. The corresponding stage is the average of stage measurements at the start and end of the entire measurement. Details regarding streamflow measurement are contained in Chow (1974) and in World Meterological Organization (1974). The continuous stream flow record for each location is usually summarized as average flow for each day, average flow for each month and total runoff for each year.

The methods of hydrological analysis are standardized only to a minor degree, because meteorological and catchment characteristics vary greatly and are not fully understood. Judgment still plays a major role in hydrologic evaluations.

Water-supply and firm-power developments require "satisfactory" operation for a long historical period, preferably 50 years. Satisfactory can mean that substantial shortages occur occasionally for irrigation projects and rarely, if ever, for municipal, industrial, and hydroelectric projects. Simulation of project operation weekly or monthly usually supplies an adequate measure. This involves approximations of yields under each plan considered. The procedure to make approximations consists simply of applying the law of continuity of simulated operation of a proposed system. Each simulation consists of the following steps for each computation interval:

1) Computing inflow to each pertinent location in the stream system by applying ratios to recorded stream flows at gauged locations. Such ratios can be ratios of contributing drainage areas or some other factor indicating relative runoff.

2) Determining release from each reservoir required to satisfy water needs, subject to water availability.

3) Estimating evaporation from each reservoir by multiplying the average reservoir area during the computation interval by the net evaporation during that interval. Net evaporation in this sense is the difference between lake evaporation and the evapotranspiration losses that

would have taken place without the reservoir. Reservoir evaporation is best estimated from pan evaporation measurements multiplied by a coefficient, which usually averages about 0.7 and is higher in the fall and lower in the spring. Evapotranspiration losses before a project are best estimated as the difference between rainfall and runoff measured in the vicinity where no lakes exist.

4) Establishing end-of-period storages by adding inflow to start-of-period storage and subtracting outflow, net evaporation, and any other losses.

5) Establishing average flows at successive locations moving from upstream to downstream by adding reservoir releases to intermediate runoff and subtracting the difference between any diversions and return flows that occur within the intermediate area. Detailed procedures for such simulation studies are contained in the Hydrologic Engineering Center (1971-1977, vols. 1, 8, and 9). More sophisticated methods are available, such as linear or dynamic programming, but there is no evidence that results are greatly improved or study costs reduced.

Flood studies involve an index of flood severity and a frequency study of that index. Usually, the most severe value for each year is selected, values are arranged by order of magnitude and these values are plotted against recurrence interval. The recurrence interval for each observation is $(N + 1)/M$, where N is the number of years of record and M is the order number of the event, with small values of M assigned to large events. The curve drawn through the plotted points is used as the indicator of future flood frequencies. The index of severity usually refers to peak stream flow at a specified location, but other indices are used, such as runoff volumes for a specified duration, peak stage, and sediment load. A frequency function is usually computed from the mean, standard deviation, and skew coefficients of data (flows or logarithms of the flows), and functions such as the Pearson type 3 or the Fisher-Tippett extreme value function are fitted to the data using these statistics (see Chow, 1974; Hydrologic Engineering Center, 1971-1977, vol. 3).

In cases where records of stream flow are not adequate or where conditions have changed or are projected to change dramatically, flood potential can be evaluated from rainfall-runoff studies. Rainfall-runoff models range from one in which peak flow is a linear function of the average rainfall for the time of concentration, to a unit-hydrograph and loss-rate model, to the sophisticated watershed model in which all hydrologic factors are simulated. Except in simple systems, computation of flood potential from rainfall potential is highly undependable, and it is desirable to obtain stream flow data at the location of interest as early as possible.

In most water management studies, the sediment accompanying stream flows is a serious concern. In large rivers and in flat terrain, most sediment is fine and is carried in suspension (wash load), whereas in areas of high relief streams transport heavy bedloads of coarse sediment. Wash load is measured by use of a sediment sampler held at one point in a cross section (point integrating) or lowered and raised for the entire depth (depth integrating); water is collected in a sampling bottle at a rate proportional to the flow velocity. Average sediment concentrations are collected and related to flow rate over time to obtain total wash load transport. Bedload transport is computed by use of equations that relate

TABLE 3.4 Phases, problems, and techniques for groundwater supply. [Source: Mandel and Shifton (1981)]

A. Phase of exploration
Practical aims: Find groundwater in the area and supply water for specific purposes.

Problems to be solved	Techniques of investigation
1. Identification of potentially aquiferous formations.	
2. Nature and origin of porosity. Primary Fracturing Solution Weathering	Geology, geophysics (1),[a] (2), (3), (4)
3. Position of water level Elevation of natural outlet Perched horizons on impervious layers	Geohydrologic reasoning, drilling (2), (5)
4. Pumping lift. Depth of water level	Water level measurements: water level maps (7.1), (7.3); estimates of specific discharge of wells (6), (13)
5. Supply capacity of wells. Discharge-drawdown characteristics	Step drawdown tests (6.8)
6. Depth of drilling required.	Inferences from geology and water levels (5)
7. Water quality.	Chemical analyses, application of quality criteria, geochemical classification (9)

B. Phase of increasing exploitation and quantification
Practical aims: Locate more wells in area and increase groundwater supplies.

Problems to be solved	Techniques of investigation
1. Geometrical configuration of aquifer.	Geophysics, drilling, subsurface mapping (4), (5), (3)
2. Minimum distance between wells. Aquifer confined, unconfined, hydrologic constants T, S	Geohydrologic reasoning, aquifer tests (6), (11), (13)
3. Constraints on location of wells. Depth of drilling Depth of water level Poor water quality Social (land ownership, etc.)	Geologic subsurface correlations, geophysics, water-level maps, chemical methods, drilling of research holes, information on requirements and social constraints (3), (4), (5), (7), (9), (15)
4. Boundaries of groundwater systems. Geologic boundaries Hydraulic groundwater divide	Geologic maps, geologic subsurface correlations, water level maps (3), (7)
5. Mechanism of replenishment. From rain From rivers Lateral inflow of groundwater Fossil water Man-made return flow	Geohydrologic inferences, water level hydrographs and maps, observations of river flow, geochemical methods, isotope methods (7), (9), (12), (10)

TABLE 3.4 *(cont.)*

6. Mechanism of natural outflow.
    Through springs
    Seepage into rivers and lakes
    Evapotranspiration through phreatophytes, from salt marshes, and swamps
    Into oceans
	Geohydrologic inferences, water level maps, salinity maps, geochemical methods, isotope methods, air photos (11), (7), (8), (2), (9), (10), (3)

7. Estimate of average annual replenishment.	Simple water-balance methods (12)

8. Pattern of flow.
    Aquifer homogeneous isotropic
    Aquifer nonhomogeneous
    Aquifer anisotropic
    Aquifer with preferential flow channels
	Geohydrologic inferences, geologic subsurface techniques, water level maps, pumping tests, geochemical methods (2), (11), (3), (6), (7), (9)

9. Potential mechanisms of groundwater mineralization.
    Evaporites
    Seawater
    Brines of geologic origin
    Connate salts in fine-grained rocks
    Airborne salts
	Geohydrologic inferences, geochemical methods, isotope methods, drilling of research boreholes (2), (11), (9), (10), (5)

10. Model of aquifer.	Correlation and integration of all available data (3), (6), (7), (9), (8), (10), (11), (12), (14)

C. **Phase of conservation**
    Practical aims: Determine maximum sustained yield under given set of constraints.

Problems to be solved	**Techniques of investigation**

1. Compute replenishment.
    Average annual replenishment
    Annual replenishment as a function of precipitation
	Waterbalance methods (12)

2. Evaluate constraints.
    Pumping lift
    Decreased thickness of saturated section of aquifer
    Influence on spring flow and base flow of rivers
    Soil subsidence
    Economic, administrative, and social constraints
	(13)

3. Calibrate model. 4. Predict final equilibrium. Water levels for various alternatives of exploitation	Beyond the scope of this text
5. Rank constraints. 6. Determine maximum sustained yield.	(13)
7. Define observation network for monitoring and supervision.	(14)

[a] Figures in parentheses refer to the relevant sections in Mandel and Shifton (1981).

TABLE 3.5 Some recommended techniques for the analysis of major and some minor elements in groundwater. [Source: Lloyd (1981)]

Element	Analytical method	References	Analytical working range	Approximate detection limit	Minimum sample volume	Typical concentration in groundwaters	Drinking water EEC limits
Al	Complex with 8-hydroxyquinoline, extract in field into MIBK* Analysis by flame AA† (N_2O–C_2H_2)	1	5–50 μg l^{-1}	2 μg l^{-1}	500	1–60 μg l^{-1}	50 μg l^{-1} (G)
Ba	Flame AA (N_2O–C_2H_2)	—	0.1–1.0 mg l^{-1}	0.1 mg l^{-1}	2	0.01–2.0 mg l^{-1}	100 μg l^{-1} (G)
HCO_3	Potentiometric filtration with 0.01M H_2SO_4	2	100–400 mg l^{-1}	10 mg l^{-1}	25	50–100 mg l^{-1}	—
Br	Neutron activation analysis	3	0.5–100 mg l^{-1}	0.1 mg l^{-1}	1	50–500 μg l^{-1}	—
Ca	Flame AA (Air–C_2H_2)	—	5–40 mg l^{-1}	<1 mg l^{-1}	2	5–500 mg l^{-1}	100 mg l^{-1} (G)
Cl	$AgNO_3$ titration using Ag/AgS ion selective electrode as monitor	—	5–800 mg l^{-1}	1 mg l^{-1}	10	10–200 mg l^{-1}	25 mg l^{-1} (G)
Cu	Flameless AA—direct	4	1–50 μg l^{-1}	1 μg l^{-1}	1	1–100 μg l^{-1}	3 mg l^{-1} (M)
F	Ion selective electrode	—	0.1–10 mg l^{-1}	0.05 mg l^{-1}	10	0.1–5 mg l^{-1}	1.5 mg l^{-1} (M)
Fe (total)	Flameless AA	5	2–100 μg l^{-1}	3 μg l^{-1}		10–20,000 μg l^{-1}	50 μg l^{-1} (G)
Fe^{2+}	Spectrophotometric with 2,2'-dipyridyl	6	0.08–1.0 mg l^{-1}	0.01 mg l^{-1}	25	0.01–20 mg l^{-1}	50 μg l^{-1} (G)
Li	Flame AA (Air–C_2H_2)	7	20–200 μg l^{-1}	5 μg l^{-1}	2	2–60 μg l^{-1}	—
Mg	Flame AA (Air–C_2H_2)	—	0.5–6.0 mg l^{-1}	0.05 mg l^{-1}	2	1–200 mg l^{-1}	30 mg l^{-1} (G)
Mn	Flameless AA—direct	8	1–50 μg l^{-1}	0.5 μg l^{-1}		1–50 μg l^{-1}	20 μg l^{-1} (G)
Ni	Flameless AA—direct	4	10–100 μg l^{-1}	5 μg l^{-1}		1–50 μg l^{-1}	50 μg l^{-1} (G)
NO_3–N	Spectrophotometric (UV)	9	0.1–1.0 mg l^{-1}	0.01 mg l^{-1}	1	1–100 mg l^{-1}	5.6 mg l^{-1} (G)
HPO_4	Spectrophotometric—molybdenum blue	10	0.06–3.0 mg l^{-1}	8 μg l^{-1}	40	0.01–0.5 mg l^{-1}	0.35 mg l^{-1} (G)
K	Flame AA (Air–C_2H_2)	7	0.5–10 mg l^{-1}	0.1 mg l^{-1}	2	0.2–30 mg l^{-1}	10 mg l^{-1} (G)
SiO_2	Spectrophotometric—molybdenum blue	11	0.5–4.0 mg l^{-1}	0.05 mg l^{-1}	25	1–10 mg l^{-1}	—
Na	Flame AA (Air–C_2H_2)	7	0.25–400 mg l^{-1}	0.1 mg l^{-1}	2	5–1,000 mg l^{-1}	20 mg l^{-1} (G)
Sr	Flame AA (Air–C_2H_2)	12	0.05–1 mg l^{-1}	0.02 mg l^{-1}	2	0.05–10 mg l^{-1}	—
SO_4	Titration with $Ba(ClO_4)_2$ using thorin as indicator	13	10–100 mg l^{-1}	10 mg l^{-1}	70	10–500 mg l^{-1}	25 mg l^{-1} (G)

*MIBK—methyl isobutyl ketone   †AA—atomic adsorption

[References: 1, Barnes (1975); 2, Barnes (1964); 3, Cawse and Piersen (1972); 4, Boyle and Edmond (1977); 5, Segar and Cantillo (1975); 6, Rainwater and Thatcher (1960); 7, Ure and Mitchell (1975); 8, McArthur (1977); 9, Miles and Espejo (1977); 10, Murphy and Riley (1962); 11, Fanning and Pilson (1973); 12, David (1975); 13, Fritz and Yamamura (1955)]

flow velocities to the size distribution of the load after samples are taken from the stream bed at many points in a grid system.

Total transport is the sum of the suspended and bed loads transported. As stream channel characteristics and flow change from location to location and from time to time, sediment transport capacity also changes. Changes occur particularly at lakes and reservoirs, where velocities are greatly reduced, bed load drops, and much of the suspended load settles out. Detailed procedures for sediment measurement and computation are given in Chow (1974).

3.2.3 Groundwater.

Table 3.4 (Mandel and Shiftan, 1981) outlines the various phases and problems encountered in developing groundwater as well as investigative techniques that can be applied. There are many excellent texts available on these techniques in addition to Mandel and Shiftan (1981); for example, Chow (1974), Todd (1980), Freeze and Cherry (1979), and Lloyd (1981). Methods for chemical analyses of groundwater are outlined in Table 3.5 (Lloyd, 1981).

It is unfortunate, but there has been little evaluation of how applicable these techniques are in developing countries. Some general observations may be made, however, about manpower training, equipment, and the time frame for analysis.

1) Training. Geohydrological work requires more skill and education than that required for surface hydrology. In most developing countries few civil engineers have good training in groundwater hydrology; at least in part because there is little role for professionals in this area. Foreign engineers will have to be consulted initially until training programs are instituted that emphasize groundwater equally with surface water.

2) Equipment. The equipment required for these technical studies does not have to be elaborate. Small mobile drilling rigs are of great use.

3) Time frame. Frequently data on groundwater is meager. Gathering new data is expensive and time consuming, especially if wells have to be drilled. Even after an area is sampled adequately once, water level fluctuations during several years should be measured monthly or seasonally to estimate aquifer yield with any confidence.

3.2.4 Tide measurements, waves, and estimation of intertidal elevation.

Tidal fluctuations are affected by local geography. The topography of bays, estuaries, straits, and channels can amplify, dampen, or delay tides. There are predicted tide levels available for many coasts of the world prepared from data on lunar cycles, local conditions, and past tidal measurements, but up to 65 separate factors govern local tidal cycles (Pond and Pickard, 1978). Actual tides at a research site may be quite different from published information for a nearby location. If tides and currents are important factors in a particular study, they must be measured on site.

Tidal categories. Tides may be classified as: (i) diurnal, a single low and high tide per day (not common); (ii) semidiurnal (two low and two high tides per day) and of equal amplitude; or (iii) semidiurnal and of unequal amplitude. One type of tide is usually characteristic of a given large stretch of coastline. Both diurnal and semidiurnal equal tides have a daily high water (HW) and a low water (LW) while semidiurnal unequal tides have a higher high water (HHW) and a lower high water (LHW) as well as a lower low water (LLW) and a higher low water (HLW). Superimposed on the daily cycle is a lunar monthly cycle from extreme (spring) tides to those of least amplitude (neap tides).

The tide datum. At any point along a coast, the tide must be measured with reference to a presumed fixed point so that tides on different parts of the coast can be compared. This point is referred to as the "datum." For most of North America the datum is the mean low tide or the mean lower low tide mark, the mean of all daily lower low tides, or low tides, for a year. Other coasts use mean monthly lowest low tide, or the most extreme low tide ever observed. Each study area must have its own calculated datum. If there is to be a long-term record of tidal levels, a mean low or lower low water can be determined. If tidal heights are to be taken only sporadically, the lower low tide of a few days can be compared to the predicted level for that day. This gives an average estimated tidal height for any fixed point when taken repeatedly and averaged. (Pugh, 1978; Pond and Pickard, 1978; Pickard, 1979).

Sampling, analysis, and selection. The most common measuring and recording techniques for tidal height are the following three. (i) For sighting and manual recording, a permanent marker is used as a reference point with either a fixed graded pole or one that can be set up for a measuring interval. Tide height relative to the fixed structure is noted at predetermined intervals. Once a series of low tides has been recorded, a zero-tide datum can be determined (see below). (ii) Floating tide gauges are usually composed of a recording device and a float surrounded by a wave-damping well. The float moves up and down with the tide but is not affected by waves or wind chop because of the well. The recording device can be electronic or use pulleys and lines from the float to a pen and chart (Pickard, 1979). (ii) Pressure gauges are located on the bottom and sense the height of the column of water above. Readings will be affected by large waves and, in fact, such a gauge can act as a wave gauge as well as a tide gauge. The pressure sensor is connected either by pipe or cable to a recording device on shore. In situ recorders are also available (Pugh, 1978; Pickard, 1979).

Determining intertidal elevation. To describe intertidal biological zonation and the physical profile, accurate estimations of intertidal elevation must be made. Two simple methods can be used.

1) In the pole-and-transit method, a fixed point above the highest tide level is chosen as a reference (an identifiable rock or a concrete or metal marker). The surveyor's transit (telescope with leveling device) is

positioned over this point and the elevation of the eyepiece is measured. A second person with a wide pole marked with 1-, 10-, and 100-centimeter bars places the pole on each point to be measured. The reading is taken through the transit and is the vertical distance of the measured point below the transit telescope center. If the pole is too far away to be read easily through the transit, the person with the pole can move a large marker up and down until the transit user signals that the marker is level with the transit. The reading is then taken by the person with the pole.

2) The pole and horizon method is similar except that the transit is replaced by an observer's eye above a similar high reference point. Sighting along the top of a meter stick above the reference point is convenient. The point at which the horizon (preferably over water on a clear day) intersects the pole is the difference in elevation between the eye and the measured point. The person with the pole moves a marker (finger or card) until signaled to stop by the observer. Alternatively, the observer can use binoculars and take the reading himself. This method is at least as accurate as the transit method when the horizon is visible. It can be used anywhere and does not require expensive instruments.

Wave characteristics. Important descriptive characteristics of waves include the amplitude (height from trough to crest), period (time between successive crests), compass direction, and variation in both amplitude and period (Bascomb, 1964). Five methods that can be used to make measurements under various conditions are the following.

1) Pressure gauge and transducer--a permanent record of pressure from a bottom-mounted transducer gives the best long-term record of wave amplitude and period (Pugh, 1978; Pickard, 1979). From such data, mean amplitude and period can be calculated as functions of tide, season, and wind velocity.

2) Electronic wave gauges--electronic methods use pairs of bare wires to complete a circuit, a bare resistance wire that is covered more or less as waves rise and fall, or an insulated wire hung vertically to determine the capacitance between the wire and the changes in sea water with emersion during wave passage (Pond and Pickard, 1978).

3) Direct sighting--wave amplitude can be estimated from direct observation of a series of waves passing a rock, pier, or jetty of known dimension. A painted scale located at the point of observation will help. The observer usually records the mean of the largest waves passing the point in a given time (for example, for the largest third of the waves this is termed the significant wave height). The period of the waves can be estimated by timing a succession of waves passing the reference point.

4) Photographic wave monitoring--a lateral view of waves passing a structure of known dimension can be photographed to record the position of a series of wave troughs and crests, or stereoscopic photographs of the sea surface from above (aerial photography) with a ship, pier, or piece of coastline for scale can be taken. Height of a series of waves is determined from the lateral difference between peaks in the two photographs. Wave direction is easily identified from aerial photographs.

5) Aerial remote sensing--airplanes flying at relatively constant altitude can use a narrow radar beam to measure wave heights over long

transects of sea surface and record them on chart paper or digitally on magnetic tape (Pond and Pickard, 1978). This method is more useful in open ocean oceanographic studies than along shorelines.

3.2.5 Current measurement.

Surface currents along many coasts have been measured as a function of tidal period and amplitude. Such data are available for general use in coastal navigation (e.g. NOAA current tables). These measurements are rarely taken on a fine enough scale for use in studies of a particular part of a bay, estuary, or open coast. In many cases, currents are locally dependent on weather (wind, rainfall) especially in or near rivers. In addition, most surveys have measured surface currents only and such data cannot be used to accurately calculate volume transport or flushing rates for a given body of water. Because of turbulence, mixing, and tidal effects, accurate and detailed description of current in all locations and at all times are extremely difficult to obtain. Although there are now excellent instruments available for current measurement and long-term automatic recording, they are very expensive and still sample at only one point each. (See LaBarbera and Vogel, 1976; Maragos, 1978b; Pond and Pickard, 1979; and NOAA Current Tables.)

Sampling and selection criteria. Current measuring techniques are divided into "Lagrangian" (movement path of a water mass over a distance) or "Eulerian" (speed and direction at a fixed point) methods. Several of both types of methods, their limitations, cost, and relative ease of use are discussed.

1) Neutrally buoyant objects and fluids (Lagrangian). Any neutrally buoyant matter should move with the same direction and speed as particles of a moving fluid. Fluorescein dye patches are often used to trace water movement. On the surface of a body of water, a dye patch moves and spreads, and the center can be used to estimate mean velocity, speed and direction. In faster moving water, photographs or direct observations of a dye "front" give the approximate mainstream velocity. This technique also helps define a boundary layer above a surface, to estimate its depth, and to detect turbulent eddies. Other fluids such as pollutant plumes, fresh-water plumes, water with suspended material, and water of a different temperature can be used to estimate velocity and direction. For example, hot water plumes from industry and power plants show up quite well in aerial infrared photographs. Velocity of a given patch of fluid can be estimated from successive photographs a few seconds or minutes apart.

Finally, neutrally buoyant objects (such as plastic strips or water filled balls) can be used. In situ underwater photography with timed exposures, or direct observation from a dock or anchored boat can also be used. The advantages of such techniques are low cost and simplicity. The disadvantages are that they must be employed in situ, do not offer much potential for remote or automated monitoring, and consume relatively large amounts of time per flow measurement.

2) Drift bottles and drogues (Lagrangian). Drift bottles are glass or plastic vessels partly filled with water (or weighted), containing an

addressed stamped postcard for return to an investigator. They are designed to drift for long periods with the current until they wash up on a beach. Although they were of some use in early oceanographic studies, there are much better methods now available and their continued use is not advised where improved methods can be employed. Drift drogues consist of a lightly weighted biplanar drogue one to tens of meters below the surface and a cable or rope to a small float at the surface. The float has either a bright colored rod, flag, or light attached and is tracked from shore, boat, or airplane. Drogues are cheap to make, easy to deploy, and easy to track by triangulation or navigational methods. A recent improvement is the addition of lightweight animal-tracking radios which transmit at a narrow frequency (150-151 hertz) for up to a year or more. The drogues are tracked from two points on shore and the distance and direction determined (daily or more frequently) by triangulation. Drogues have been used to map eddy currents behind the island of Hawaii, and their position was mapped up to 60 miles from the receiving station (Lobel, 1981). This technique has a relatively modest cost ($100 to $200 per radio, $20 to $30 if not assembled, $800 to $1200 per receiver) and provides useful data on surface or near-surface water mass movements. Some of the available radios transmit pulses whose frequency changes with temperature so that this information can also be obtained remotely.

Satellite-tracked drift buoys have been used in large-scale oceanographic studies. Their expense and logistical complexities make them prohibitive for small-scale coastal studies, however.

Neutrally buoyant floats (Swallow floats) can be constructed so that their density is equal to that at any desired depth. Sonic pulses transmitted from the float are monitored from a ship which is taking precise longitude and latitude readings concurrently (Pickard, 1979). These are also expensive and difficult to use in any except large-scale studies on the open ocean.

3) Drag angular deflection (Eulerian). A weighted biplanar surface hung by a fine wire or monofilament line (less than 1 to 20 meters below the surface) from a fixed support above the water (Chesapeake Bay Institute Drag) will deflect the line by an angle related to the water velocity. This is an inexpensive and simple technique for application in relatively shallow water where a firmly fixed platform (pier, wharf, or pole) is available. It must be recorded manually although resistance wire or a variable resistor (rheostat) might be used above water at the point where the line is attached for remote sensing. Long-term recording versions have used a movie camera in a housing to record the deflection of a float and a compass direction on film. These units are moored to the bottom and the entire housing floats upright.

4) Rotor and propeller current meters (Eulerian). Most early current meters consisted of a multiblade propeller (Eckman current meter) responding linearly in the range of 2 to 25 centimeters per second (Pickard, 1979). The Eckman meter also measures current direction with weights that are dropped into eight slots in a circular array. Propeller meters can record flow directly on counting wheels or remotely by means of a magnet in the propellor and a reed switch in the mount. A cable connects the propeller unit to the shore, ship, or buoy. For an inexpensive remote-sensing meter, the propeller meter with magnetic pick-up can be

connected to a small VHF radio on a buoy, which can transmit data for up to a year on one set of batteries.

A Savonius rotor is a two-tiered rotor which responds equally to flows from various direction and of various speeds (2 centimeters per second up to several meters per second). This type of rotor has been combined with electronic recording (enclosed chart recorder or digital recording on tape), allowing deployment for months although data collection occurs at fairly wide intervals. Fouling on the rotor and entanglement with seaweeds are problems with all rotor flow meters. Weekly examination and cleaning are sometimes necessary.

5) Electromagnetic current meters (Eulerian). Several types of meters work on the principle of moving a conductor (sea water) past electrodes to generate a current of several millivolts. One version requires electrodes to be positioned along the flow axis tens of meters apart (Pickard, 1979) and is useful primarily for unidirectional nonturbulent flow. Another uses a coil of wire placed parallel to the flow. A third type has a spherical sensor only a few centimeters in diameter. This can be used with a cable and a recording device on shore or on a ship. Another version uses digital or chart recording equipment within a housing. The electromagnetic meters have the advantage of operating without moving parts and therefore not affected by fouling. They are, however, the most expensive units now available ($5,000 to $11,000).

6) Thermistor flow meters (Eulerian). Thermistor flow meters measure the electrical current needed to keep a small thermistor bead at a constant temperature. The faster water moves past the thermistor, the more current is used, (LaBarbara and Vogel, 1976). The advantages of this meter are that low velocities can be measured accurately and flow can be measured within millimeters of a surface. Since ambient temperature must be known for calculation, long-term remote deployment is not possible now.

Current velocities are often determined in order to calculate the total amount of water flowing through a given area per unit time (volume transport). This measure depends on (i) the velocity profile away from the substrate, (ii) the bottom profile and nature of the boundary layer and, (iii) stratification and semi-independent movement of water layers. The calculations are easiest to make over a smooth substrate at velocities that produce a thin boundary layer (Maragos, 1978b; Pond and Pickard, 1979).

3.2.6 Temperature.

Subsurface water temperatures are usually determined by a thermistor thermometer with a weighted probe and cable of sufficient length to reach the bottom. This thermometer may be a single instrument or a component of submersible multivariable analyzers. Measurements are usually recorded at 1-meter intervals and more frequently if large termperature differences are encountered. The instruments are rugged and reliable, and their accuracy is adequate when properly calibrated. Weighted minimum-maximum thermometers provide an inexpensive alternative for in situ measurements. However, their sensitivity will not be adequate in many studies, especially in warmer waters. It is also possible to use a water sampler and collect samples from discrete depths and determine the temperature of the samples. This method is less preferred because the temperature may change during sampling and the number of discrete depths sampled is usually too few because of the labor required.

Temperature variations of less than 1 centigrade degree of surface waters may be detected from the infrared scanner data obtained from satellites. This is a valuable tool in assessing patterns of thermal discharges, currents and upwellings, and distributional patterns of inflowing waters in lakes and reservoirs. (Lind, 1979).

Sampling and analysis. The principal measurements are accuracy and completeness, and these may vary by season and water type. In the tropics with usually high temperatures, a small change in temperature has a great effect on water density and thus presence or absence of stratification (see section 2.3.6). In such situations, changes of as little as 0.1 centigrade degree may be needed, whereas in temperate waters, reading to 0.5 centigrade degree is adequate. Also in the tropics, the small temperature but great density differences produce fine scale thermal structure; thus measurements may be required at much less than 1-meter intervals. The sampling design should also allow for extensive measurements through the year to document such events as the onset and breakup of stratification. Surface temperatures and temperatures of shallow waters and of slow flowing streams may vary greatly in the daily cycle. Samples of both the predawn minimum and the midafternoon maximum are needed.

Temperature data are often presented as depth-temperature graphs (Fig. 3.1). This type of graph is used extensively in reporting aquatic data because of the ease in conceptualizing patterns within the lake, reservoir, or sea.

Selection criteria. Temperature measurements are central to almost every limnological study. They should be the first measurement made because much of the subsequent sampling program depends on these data. Skills required are minimal. Thermister thermometers are among the least expensive and are often the most reliable of electronic field instruments.

3.2.7 Salinity.

The salinity of oceanic and estuarian water reflects the abundance of several different salts (whose ratios remain constant) and the degree of mixing of fresh water and runoff with saltier water (Fig. 2.7). In lakes, several salts are derived from the surrounding soil and rocks or are brought in by surface or below surface inflows. Salinity is usually expressed as parts per thousand by weight of all solid material excluding carbonate, bromine, iodine, and organic matter. It averages 35 grams per kilogram in open ocean water. The ratios of the various elements that comprise the salts are not nearly as constant in salt lakes, and often individual ions must be determined by chemical methods or elements by atomic absorption spectroscopy. The following methods are commonly used for salinity determinations.

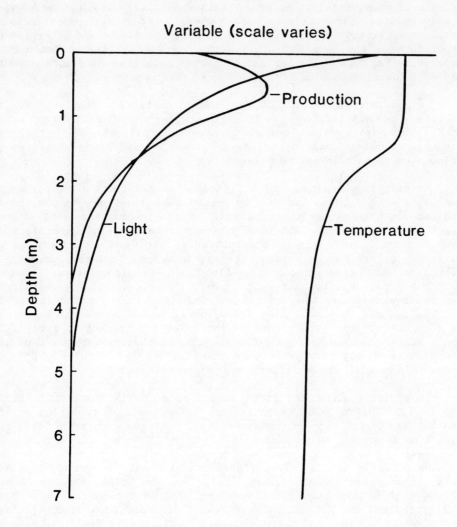

Figure 3.1 Temperature, light, and primary production rate plotted with depth.

1) Determination of chlorinity by silver nitrate titration and subsequent calculation of salinity (Knudsen method). Salinity is calculated as: s=1.80655 chlorinity (parts per thousand) with an accuracy of ±0.02 parts per thousand in routine measurement (Pickard, 1979). Strickland and Parsons (1974) note that titrations must be made at constant temperature.

2) Salinity by electrical conductivity. Electrical salinometers measure electrical conductivity of the water and compensate for changes in conductivity with temperature. This method has an accuracy of 0.003 parts per thousand salinity (Pickard, 1979). Salinity values corresponding to conductivities at known temperatures are given in the International Oceanographic Tables (1966) and must be used when a compensating salinometer is not available (Strickland and Parsons, 1974). This method is approximately ten times as rapid as the titration technique (see section 3.3.1).

3) Conductivity, temperature, depth instruments. Salinometers that measure conductivity, temperature, and depth in situ are desirable in the open ocean where small salinity and temperature differences must be determined for a wide range of depths. Data is transmitted by cable or recorded in the instrument. Although these instruments may not be feasible for small studies, they can greatly increase the quantity of data collected when they are used.

3.2.8 Density and stratification.

Density is a function of temperature, depth, and the amount of dissolved material in the water (salts, organics, and so forth). At equilibrium, stratified water masses will go from low density at the surface to high density at the bottom. If this is not the case, equilibrium has not been reached and water of different density will tend to move vertically. Density in the open ocean ranges from 1021.00 kilogram per cubic meter at the surface to 1070.00 at a depth of 10,000 meters (Pickard, 1979).

Distinct differences in temperature, salinity, or dissolved gases and chemicals indicate stratification. In lakes, temperature is a reliable indicator of various strata. In coastal areas, temperature and salinity are useful indicators where there is fresh-water discharge nearby.

Sampling and analysis. For sea water brought to the surface, in situ density can be calculated from known conductivity, depth, and in situ temperature. Volumetric determination of water density (weight of accurately measured volume) is also possible with analytical balances in the laboratory. This technique is often used at coastal stations or for fresh water.

Although density in fresh waters may be measured by a simple specific gravity hydrometer, it is most commonly inferred from indirect measurements of water temperature, dissolved solids content, or both.

3.2.9 Light and transparency.

Light measurements usually are made to determine the penetration of light into the water, and thus, the transparency. Light transmission is measured with a submarine photometer, which operates on the same principle as a common photographic light meter. In addition to the photocell, other detectors are available such as the quantum cell, which measures light flux on an area basis. Transparency can be characterized by relating light intensities at given depths to that of the surface. A regression of light extinction on depth describes a line, the slope of which is the extinction coefficient. This value is often used to compare the transparency of different waters. Transparency, as visibility, may be determined by the use of a simple Secchi disk. This is a 25-centimeter black and white disk that is lowered into the water, and the depth at which it disappears from view is recorded. Although apparently crude, it produces remarkably good results. Use of the Secchi disk and the submarine photometer are described by Lind (1979, pp. 23 and 28) and Vollenweider (1969, pp. 158 and 171).

Sampling and analysis. Obtaining a representative sample is the primary concern. Much spatial variation in transparency is expected. Lakes and especially reservoirs often have much greater concentrations of suspended solids near points of inflow and near shorelines. Dissolved organic color is also usually greater near inflows and near extensive beds of vegetation. In some tropical waters high concentrations of dissolved color are encountered as warm, slow moving, waters leach soluble organics from the abundant vegetation, producing the so-called "black water" that is found in tributaries of the Amazon. Much seasonal variation is also expected in light penetration because of factors such as seasonal plankton blooms, rainfall runoff events, stratification, seasonal wind patterns, and the angle of incidence of sunlight. However, little difference in light penetration measurements is found if sampling occurs between 0900 and 1500 hours.

Selection criteria. Studies of light transmission should be part of all lake and reservoir baseline studies and of studies of larger rivers or of those with obviously high color or suspended matter. Skill levels required are minimal. Photometers are expensive, and although made for field use, require some special care, such as protection of the photocell from extreme heat. The simplicity and reliability of properly taken Secchi disk data requires that it be a part of all studies.

3.3 Methods for Chemical Properties

3.3.1 Dissolved gases.

We will look at the two major dissolved gases--oxygen and carbon dioxide.

1) *Dissolved oxygen.* Two methods of measuring dissolved oxygen are acceptable: the modified Winkler chemical method and the polarographic

method, which uses a Clark-type electrode either in a bottle or directly in the stream, lake, or reservoir. Advantages of the Winkler method are its proven reliability and relative simplicity. Its disadvantages are the relatively high level of skill needed to prepare and use the reagents, the instability of chemical reagents (primarily the iodide reagent and the thiosulfate titrant if it is used), and the fact that discrete samples are used which may reduce the number of depths and sites sampled. The samples must be transported to a laboratory for titration, and should not receive prolonged exposure to direct sunlight. In a few waters, the presence of unusual quantities of highly oxidizing or reducing substances may produce interferences resulting in either over- or underestimation of the true concentration.

The polarographic oxygen meter method has the advantage that no water samples are taken or transferred. The oxygen content is measured in situ with an electrode on a weighted cable. The meter is relatively free from interference although the membrane on the electrode may require occasional replacement. Relatively minimal skill is required to calibrate the meter and operate it in the field and only moderate skills are required for membrane replacement and original calibration in the laboratory. Because of its simplicity, it encourages the taking of adequate numbers of measurements. The disadvantage is primarily in original cost. Most instruments are relatively rugged and can withstand normal field use, but as with most electronic instruments, failure will probably be more frequent in the humid climates.

The Winkler method is described by the American Public Health Association (1980, p. 390); the Environmental Protection Agency (1979a, p. 360.2); Lind (1979, p. 72); Mackereth, Heron, and Talling (1978, p. 24); and Strickland and Parsons (1974, p. 21). The polarographic method is described by the American Public Health Association (1980, p. 395) and the Environmental Protection Agency (1979a, p. 360.1).

Sampling and analysis. Sampling by the Winkler method requires special care to avoid exchange of the gas with the atmosphere. Turbulence during collection or transfer will cause losses. Oxygen samples must always be taken with a sampler such as the Kemmerer, Van Dorn, or Niskin water bottles, never with submerging bottles or dipping and pouring. Because the concentration of dissolved oxygen in the water is often a direct function of metabolic activity, as well as physical factors, a sampling program must include sampling during the daily periods of maximum community oxygen production and accumulation (late afternoon) and maximum community oxygen consumption (predawn). These deil cycles are often large in the tropics. Sampling must also take into account seasonal variations, which result from temperature changes and changes in the rates of metabolism of the biota. In order to detect conditions possibly stressful to aquatic life, sampling must occur at that time when all forces coincide to produce the minimum oxygen concentration encountered by the organisms. Large vertical gradients in oxygen concentration can be expected in most lakes and reservoirs during stratification, and samples must be taken from the metalimnion and hypolimnion as well as the epilimnion.

Selection criteria. A detailed knowledge of the content and temporal and spatial variation in oxygen levels is required of most baseline studies. Only in moderately to rapidly flowing streams where turbulence will maintain the oxygen content in equilibrium with the atmosphere and where no known sources of organic pollution occur, can this variable be omitted from baseline studies. Skill levels required are low to moderate depending on the technique selected. Costs for the Winkler method are relatively low, and oxygen meters, although moderately expensive, often are combined with a temperature thermistor and occasionally other sensors.

2) *Carbon dioxide.* Free carbon dioxide may be determined by titration. However, because of the danger of loss to the atmosphere during handling and titration, this method is rarely used. A preferred method is to use the known equilibrium of carbon dioxide with other forms of inorganic carbon in relation to the pH and temperature of the sample as a basis for calculating the concentration of the free gas present (American Public Health Association, 1980, p. 268; Lind, 1979, p. 62; Strickland and Parsons, 1974, p. 33).

Sampling and analysis. As with sampling dissolved oxygen, precautions must be taken to avoid loss during collection and handling of the sample. In some tropical areas with well-leached soils, the carbonate and bicarbonate ion content is low; the buffering capacity permits metabolic activity, adding or removing carbon dioxide, to make significant changes in concentration. In such regions, daily cycles, which may be large in the tropics, often show maximum carbon dioxide content before dawn and minimum in late afternoon. Because much of the carbon dioxide content is derived from the decomposition of organic matter, it usually increases with depth and may reach high concentrations in the hypolimnion.

Selection criteria. Determination of free carbon dioxide may not be needed in alkaline waters such as those found in regions of limestone substrate. In such, most inorganic carbon will be in the form of bicarbonates and carbonates. However, because the measurement of pH, temperature, and alkalinity (bicarbonate plus carbonate) is usually done in a baseline study, all the necessary information is available for calculating free carbon dioxide. In more acidic waters such as the organic rich waters of humid tropical areas, or in hypolimnetic waters of stratified lakes, the free carbon dioxide content and its variation should be calculated. Skills required for calculation are moderate--the ability to determine pH and perform simple titrations. Costs are small because only a pH meter, simple glassware, and inexpensive reagents are needed.

3.3.2 Dissolved inorganic solids.

Although analyses differ, sampling procedures for most dissolved elements are similar. Samples must (i) represent the water mass in time and space, (ii) must be collected and stored without contamination, and (iii) must be stored or preserved to prevent changes in form of the element and to prevent loss by adsorption. Sampling in lakes and reservoirs must be conducted in all seasons and at all depths. In rivers and in riverine portions of lakes and reservoirs, sampling must be on an event (rainfall-runoff) basis and may require sampling at frequent (hourly) intervals during the rise and fall of the hydrograph. In temperate regions especially, a large percentage of the annual contribution of dissolved solids enters during rainfall-runoff periods. In the humid tropics, the leaching of the watershed tends to reduce the magnitude of contribution by single events. Annual contribution or "loading" of some element is not only a function of its concentration in the water but also of the volume of water received. Consequently, an adequate sampling program requires hydrographic data on volumes received.

Because many elements are normally present in small concentrations (measurements in parts per billion), it is important to avoid contamination from the sampler, disturbed substrates, airborne dusts, and the sample bottles.

Sample storage and preservation is critical (Table 3.6). If the samples are not to be analyzed within a prescribed time, as may occur in remote areas, it may be necessary to filter to avoid release from or adsorption to particles. Because of different preservation requirements for different elements, a baseline survey will require several determinations from the sample, each being preserved differently.

Skill requirements for analyses of dissolved materials vary from moderate to high (Environmental Protection Agency, 1979b, chapter 9). Costs are greatly variable and depend on the element, its concentration, and any preliminary treatment required. All laboratories must be in compliance with standards such as those prescribed by the Environmental Protection Agency (1979b, chapter 6).

Depending upon specific potential or realized water quality problems, analysis for several different elements may be required. Only those needed to characterize an aquatic ecosystem through baseline studies are discussed below.

1) *Phosphorus*. Phosphorus analyses must differentiate between total phosphorus and inorganic (ortho) phosphate phosphorus (sometimes called soluble reactive phosphorus). Routine analysis is by colorimetric measurement of an antimony-phospho-molybdate complex. Total phosphorus is determined after digestion with persulfate and heat converts other forms to the soluble reactive form. Concentration of inorganic phosphorus in many natural waters is sufficiently low so that determination by simple instruments, such as the Spectronic 20, is not possible (American Public Health Association, 1980, p. 409; Environmental Protection Agency, 1979a, p. 365.2; Lind, 1979, p. 77); Mackereth, Heron, and Talling, 1978, p. 83; and Strickland and Parsons, 1974, p. 49).

Sampling and analysis. There should be precautions to avoid contamination, especially from detergents in sample bottles, in the laboratory glassware, and even airborne traces in the laboratory. No phosphate-bearing materials should be used in the laboratory. Also special precautions in storage (24 hours preferred) and preservation (on ice for inorganic; acidified for total) are required to prevent conversion of forms or biotic or adsorptive uptake of the inorganic form.

Selection criteria. Phosphorus is essential to the productivity of any aquatic ecosystem, and worldwide, is the element that most often limits the production of living matter. Complete phosphorus budget information is required in all baseline studies to adequately describe the production potentials of the ecosystem. Skill levels required are moderate to high, especially where trace quantities are to be measured and possibly extraction-concentration procedures are required. If an equipped chemical laboratory is available, costs for reagents and filters are not

TABLE 3.6 Recommendation for sampling and preservation of samples. [Source: Modified from the Environmental Protection Agency (1979a, p. xv)]

Measurement	Preservative	Holding Time
Hardness	Cool, 4°C or HNO_3 to pH 2	6 Mos.
pH	Determine on site	6 Hrs.
Particulates		
Inorganic	Cool, 4°C	7 Days
Organic	Cool, 4°C	7 Days
Temperature	Determine on site	No Holding
Turbidity	Cool, 4°C	7 Days
Metals	Filter on site then HNO_3 to pH 2	6 Mos.
Nitrogen		
Ammonia	Cool, 4°C H_2SO_4 to pH 2	24 Hrs.
Kjeldahl, Total	Cool, 4°C H_2SO_4 to pH 2	24 Hrs.
Nitrate plus Nitrite	Cool, 4°C H_2SO_4 to pH 2	24 Hrs.
Dissolved Oxygen		
Probe	Determine on site	No Holding
Winkler	Fix on site	4–8 Hrs.
Phosphorus		
Orthophosphate	Filter on site Cool, 4°C	24 Hrs.
Total	Cool, 4°C or H_2SO_4 to pH 2	24 Hrs.
Sulfate	Cool, 4°C	7 Days
Organic Carbon	Cool, 4°C H_2SO_4 of HCl to pH 2	24 Hrs.
Pesticides	Cool, 4°C	7 Days

great. This analysis is slightly more labor intensive than some in maintaining the cleanliness required.

2) *Nitrogen.* Nitrogen is the second most frequently limiting element to biological production in fresh waters. And, in some tropical regions, it is the primary limiting element. The nitrogen cycle in fresh waters may be complex with episodes of high nitrogen fixation by bluegreen algae, inflows of fixed nitrogen from the watershed, and bacterial denitrification. Thus, it is important that all forms be determined in order to provide a total nitrogen balance for the ecosystem. Laboratory analyses by the Kjeldahl method should be for organic nitrogen, ammonia, nitrate, and occasionally, nitrite nitrogen. Kjeldahl procedures involve a digestion step to convert the organic nitrogen to ammonia. Ammonia, either free in the water or derived from Kjeldahl digestion may be determined either by a potentiometric method with an ammonia-sensitive electrode, or colorimetrically by the phenate method. The latter is necessary for low concentrations. Direct methods of determining nitrate nitrogen have low sensitivity; thus for most water, nitrate is reduced to nitrite by the use of cadmium metal, and nitrite determined by diazotization. If diazotization is used on the cadmium-reduced sample, the result is both nitrate and nitrite nitrogen. Nitrite alone is calculated by the difference between the nitrite concentration in the reduced sample and that of a non-reduced sample.

Kjeldahl procedures are described by the American Public Health Association (1980, p. 386), the Environmental Protection Agency (1979a, p. 351.4), and Strickland and Parsons (1974, p. 143). The phenate method for ammonia nitrogen is described by the American Public Health Association (1980, p. 360), Lind (1979, p. 84) and Strickland and Parsons (1974, p. 87). Ammonia nitrogen by the selective ion electrode method is described by the American Public Health Association (1980, p. 362), the Environmental Protection Agency (1979a, p. 350.3), and in literature by the electrode manufacturer. The cadmium reduction procedure for nitrite nitrogen and the diazotization procedure for nitrite nitrogen are described by the American Public Health Association (1980, pp. 370 and 380), Environmental Protection Agency (1979a, pp. 353.3 and 354.1) and Strickland and Parsons (1974, p. 71).

3) *Alkalinity.* Analysis of total alkalinity, or its components, is relative simple. A titration with strong acid to a specified pH is used. The pH endpoints may be determined with a pH meter or color indicator. (American Public Health Association, 1980; Environmental Protection Agency, 1979a; Lind, 1979).

Sampling and analysis. Alkalinity of waters is a function of the bicarbonate, carbonate, and rarely the hydroxyl concentration. A shift in pH, such as that brought on by high rates of photosynthetic activity, may cause carbonates to precipitate. Sampling programs should include vertical distribution to document any increase with depth. This precipitation will also follow a daily pattern which will require notation of time of sampling. High periods of rainfall will usually dilute the alkalinity of waters; thus seasonal patterns and runoff events should be considered.

Selection criteria. In baseline studies, alkalinity is used in general characterization of the waters and their buffering capacity. Sufficient sampling to reveal general patterns is required. Skill levels for the titration involved are low to moderate; the reagents used are relatively stable; costs are low.

4) *Sulfate.* Sulfate is common in natural waters and is usually the anion of secondary importance to the carbonate system. Sulfate concentrations are determined either by a turbidimetric or gravimetric method. In either case, sulfate is converted to a barium sulfate suspension which is filtered and weighed or mixed and turbidity measured by nephelometer or colorimeter. The gravimetric method is described by the American Public Health Association (1980, p. 438) and the Environmental Protection Agency (1979a, p. 375.3). The turbidimetric method, which may use a prepared reagent such as SulfaVer (Hach Chemical) is described by the Environmental Protection Agency (1979a, p. 375.4) and Lind (1979, p. 91).

Sampling and analysis. Under anaerobic conditions, sulfate undergoes reduction to sulfur and hydrogen sulfide. Thus a decrease may occur in deep waters or in waters with heavy organic pollution loads. In baseline studies, sulfate, like alkalinity, is used to characterize the waters; thus an annual survey with sampling at approximately monthly intervals at representative depths is adequate. Skill levels required are low to moderate for the gravimetric method and low for the turbidimetric method. If the laboratory has a drying oven, balance, and colorimeter or nephelometer, costs are low.

5) *Calcium and magnesium.* Water with high concentrations of calcium and magnesium, and rarely other polyvalent ions, is referred to as hard water. Analysis of calcium and magnesium, either individually or in combination, is by complexometric titration. The endpoints of these titrations are determined by color indicators. (American Public Health Association, 1980, pp. 185 and 213; Lind, 1979, pp. 69 and 71).

Sampling and analysis. Since in most waters the calcium and magnesium are associated with the bicarbonate, carbonate, and sulfate ions, they covary with them, and sampling considerations for the all ions apply. Skill levels required for the preparation of the reagents, which may be purchased ready-to-use, and for the titration are low. Only simple laboratory titration glassware is required and costs are low.

6) *Iron.* Iron is not only important as a constituent of plant and animal life, it may also play a role in the availability of other elements of interest, such as phosphorus. In natural waters iron may exist in both the oxidized ferric and reduced ferrous states. Analysis is usually of total iron by atomic absorption spectroscopy, although a colorimetric phenanthroline indicator is available. Because of its simplicity, the atomic absorption method by direct aspiration for moderate to high concentrations or by graphite furnace for low concentrations is preferred when the instrumentation is available. The atomic absorption method is described by the Environmental Protection Agency (1979a, p. 236.1); the phenanthroline method is described by the American Public Health Association (1980, p. 201) and by Strickland and Parsons (1974, p. 107).

Sampling and analysis. Waters with high concentrations of iron are often recognized by an orange ferric iron precipitate near the shoreline. Occasionally orange colonies of iron oxidizing bacteria are visible. Because of its insolubility in the ferric state, little iron is usually in well-oxygenated waters such as streams and the epilimnion of lakes and reservoirs. High concentrations may occur in the hypolimnion because of precipitation and solution under low oxygen conditions. Samples must be preserved by acidification to prevent precipitation or adsorption during storage. Ferric iron often forms organic complexes and may be partially removed by filtration; thus unfiltered samples should be analyzed. The skill level to use atomic absorption is moderate and for the phenanthroline method, moderate to high. If an atomic absorption spectrometer is available, costs by this method are low. Costs for the phenanthroline method are low to moderate.

7) *Other metals.* Most of the other metals are not routinely analyzed in baseline studies unless pollution is suspected or water with low concentrations of a metal are of interest. Samples for other metals are preserved by acidification, and analysis is by atomic absorption spectroscopy, by direct aspiration, or for greater sensitivity, by graphite furnace. Properly preserved samples for metals analyses by atomic absorption require relatively small volumes and may be held for several weeks, allowing transport to a laboratory with the necessary instrumentation. Atomic absorption methods are described by the American Public Health Association (1980, p. 147) and Environmental Protecton Agency (1979a, p. 200.0).

3.3.3 Dissolved organics.

Total dissolved organic matter is measured from filtered water samples by gravimetric methods in which an evaporated water sample is weighed before and after ignition in a furnace to drive off all organics. Or, it is determined as organic carbon in which the carbon dioxide liberated from the complete digestion of organic matter is determined with an infrared analyzer. The latter procedure is preferred.

Pesticides may be measured in the water, the sediments, and in the tissues of the aquatic organisms. Because they concentrate in the sediments and in tissues, analysis of these will often detect pesticides where concentrations in the water will be below detection limits or where pesticides have been, but are no longer present.

The gravimetric procedure is described by the American Public Health Association (1980, pp. 94 and 95) and Lind (1979, p. 98). The infrared analyzer method is described by the American Public Health Association (1980, p. 471), the Environmental Protection Agency (1979a, p. 415.1), and Wetzel and Likens (1979, p. 129). Methods for detecting the common insecticides and herbicides are described by the American Public Health Associaton (1980, p. 493 and supplement p. S51).

Sampling and analysis. Standard sampling techniques for representative collections in time and space are followed. Samples should be filtered immediately to prevent conversion from particulate organics. Samples must be analyzed soon or preserved to stop bacterial degradation of organics. Skill levels required are low for the gravimetric method and moderate for the infrared method. Costs for the gravimetric method are low, and moderate (for ampoules and gases) for the infrared method. Skill levels and costs for the extraction and analysis of pesticides are high. A laboratory with a gas chromatograph having a variety of detectors is required.

3.3.4 Inorganic particulates.

Sources of inorganic particulates are nonvolatile filterable residues, primarily clays and silts. The method is gravimetric determination of materials retained on a glass-fiber filter after ignition in a furnace (American Public Health Association, 1980, pp. 92 and 95; Environmental Protection Agency, 1979a, pp. 160.2 and 160.4); Lind, 1979, pp. 95 and 98).

Sampling and analysis. The content of suspended inorganics varies in time and space. High concentrations occur near where rivers flow into lakes and reservoirs during high water periods, near the shoreline during wave action, and in rivers after rain. Because one of the principal concerns with this material is restriction of light penetration, measurements of turbidity and water transparency are often used when muddy water is present, even though lack of transparency results not only from clays and silts but suspended organics as well. Thus inorganic suspended solids are usually only determined in baseline studies where there is particular concern about watershed erosion and transport to the lake or reservoir.

Skills required are low. The requirement for a muffle furnace to operate at 55°C may restrict the method to better-equipped laboratories. Costs of analyses are low.

3.3.5 Organic particulates.

This is the same as suspended volatile solids and is determined gravimetrically by weight loss of a sample on a glass-fiber filter after ignition in a muffle furnace usually at 55°C. Alternately, digestion and infrared analysis may be used in conjunction with dissolved organic matter determination (see section 3.3.3). Organic particulates consist of plankton and detrital particles, except the inorganic parts of these, such as diatom tests.

Sampling and analysis. Quantities of detrital organic material will be greater in the riverine portion of lakes and reservoirs and near shoreline with vegetation. Because of their nearly neutral bouyancy, some small organic particles may float at the surface and others may sink until an increase in water density is reached. This often produces a thin layer at the upper limit of the metalimnion. Sampling designs must take this heterogeniety into account. Sample preservation is important in preventing loss in the breakdown of particles. A problem that may arise in analysis is the possible breakdown of carbonaceous materials at high temperatures and their loss which would be interpreted as organic matter loss. This is a special problem in highly alkaline waters where precipitation of carbonates is probable. It is minimized by using the lowest possible furnace temperature. Skills required are low. The requirement for a muffle furnace may prohibit the procedure for some laboratories. Costs are low.

3.3.6 pH.

The pH is determined either with potentiometers or with indicators that take a unique color for a specific pH value. The meter, preferred for almost all studies, is calibrated before each use with at least one buffer solution of known pH (American Public Health Association, 1980, p. 402, Environmental Protection Agency, 1979a, p. 150.1; Lind, 1979, p. 55).

Sampling and analysis. Measurements of pH are only as reliable as the buffers used for calibration. Buffers deteriorate, but buffer concentrates that are diluted with distilled water before use are more stable and are preferred in warm climates and remote areas. Because of this deterioration problem, and to check on the linearity slope of the electrode, two buffers of different pH should be used. Buffers should always be discarded after use. Special care must also be given to the maintenance of the electrode. Coatings of residues from dirty water, especially organics and plugged fiber junctions will cause sluggish response and drift. An electrode may be damaged if the instrument is turned on when the electrode is not immersed.

Because changes in water pH are often due to the metabolic activity of aquatic life, the sampling design should be similar to that required for dissolved oxygen or carbon dioxide. pH determinations should be made on site if possible and if not, immediately upon return to the laboratory. Samples must not be preserved, but may be held on ice for several hours. They should be allowed to return to nearly buffer temperature before measurements are made. Skills required are low. Costs are low except for the initial cost of instrumentation. Some multiple analyzers are now available that incorporate electrodes for pH along with those for oxygen, temperature, and conductivity.

3.3.7 Conductivity.

Conductivity is measured with potentiometers that have a dip-probe with two electrodes. The conductance of an electric current between these is read on a meter. The electrical conductance is proportional to the content of ionic materials in the water. As conductivity is temperature dependent, water temperature must be measured to permit correction to a standard temperature, usually 25°C (American Public Health Association, 1980, p. 70; Environmental Protection Agency, 1979a, p. 120.1; Lind, 1979, p. 93).

Sampling and analysis. Specific conductance is a good survey tool for baseline studies. It is very inexpensive, and large numbers of samples may be processed in a short time. One may use it to detect general patterns in the vertical and horizontal distribution of total ions. This information can be used as a guide to areas needing more detailed sampling. The electrode calibration should be checked periodically because through use, coatings are lost and residues may build up, changing the "cell constant." Skills required are low. Cost is very low after initial investment.

3.4 Methods for Biotic Properties

3.4.1 Benthic community analysis.

Benthic organisms occur on surfaces (epibenthos or epifauna) or within sediments (infauna). The type of substrate determines what sampling techniques will be used. Major substrate categories include solid rock, coral and coral rock, tree roots, boulder fields, cobbles, gravel, sand, mud, clay, and various mixtures. A tropical shore site might include, within a few hundred meters, a coral reef, sand channels, a muddy bay, sea-grass meadows, shell gravel, a rocky intertidal zone, and more, each requiring a different sampling program. On sand beaches, for instance, zonation from high to low intertidal zones is complicated by vertical zonation within the sediment.

Sampling of benthic communities should cover the various zones during all the seasons and from year to year as well as provide as complete a taxonomic breakdown as possible. General community descriptions are often inadequate in baseline studies and impact assessments, because few taxonomists are available who can identify certain groups, such as polychaete worms, amphipods, and other small crustacea, insect larvae, sponges, hydroids, bryozoans, seaweeds, or microalgae. Often they are sent much more material than they can possibly identify. The caveat to persons planning baseline studies is to weed out common species and to send only a few, well-sorted specimens at a time to experts who have agreed, in advance, to examine material. The need for expert identification in many taxa cannot be overemphasized, and there are not enough qualified people to go around. Finally, it is often better to leave a species unidentified than to misidentify it. Training and supervision of community personnel in identification can be very important (Whyte, 1977).

Benthic sampling of streams (lotic waters) presents difficulties because of the continuous flow. However, downstream nets can be used to catch organisms scraped off of surfaces or dug out of sediments upstream. The Surber stream sampler is described in Lind (1979).

Temporal sampling, recruitment and succession. Natural catastrophes, seasonal perturbations, or massive pollution can disturb soft substrates, opening them up for colonizing species and a new succession of species (Woodin, 1974 and 1978; Whitlatch, 1977; Simon and Dauer, 1977; Grassle and Grassle, 1977; Sanders et al., 1980). Such changes can be monitored with transect or quadrat sampling. Because there may be some sampling overlap with areas that have been dug up for analysis, the sampling area should be large enough to make this unlikely.

Experimental defaunation by sieving and baking can be used to create macrofauna-free sediment boxes that can be set out at intervals to see which species have larvae in the plankton capable of settlement. Larger experiments can be used to examine colonization and succession.

General methods Methods devised to quantify organisms depend first on the type of substrate and then on the orientation (epifaunal or infaunal). The outline below provides a summary appropriate to each.

I. Hard substrates. Organisms are either encrusting or erect. Erect forms can produce a second level (canopy) above the substrate often obscuring those below. Furthermore, mussels, oysters, and seaweeds often act as secondary substrate for other assemblages of both plants and animals.

    A. Destructive sampling (analysis of numbers and biomass).

        1. Scraped samples from intertidal quadrats (Dayton, 1971).

        2. Scraped samples from subtidal rock collected with airlift suction devices (Harris, 1976).

    B. Non-destructive sampling (analysis of numbers or percent cover).

        1. Direct counts of organisms along transects (Porter, 1972a and b, Menge, 1976).

2. Direct counts of organisms in randomly placed quadrats (Dayton, 1971 and 1975; Connell, 1966a and b).

3. Photographs of organisms in permanent quadrats (Connell, 1972).

4. Photographs of quadrats taken at random or at even intervals along transects (Sebens, in press).

C. Determinations of percentage cover.

1. Planimeter or digitizer--perimeters of organisms and colonies on photographs, or tracings on plastic, are outlined using a planimeter or a digitizer connected to a computer; either method gives a good estimate of space cover for each entity measured (Buss, 1980).

2. Dot patterns--the percentage of cover of a given species on a surface will be directly related to the fraction of random dots on that species or its photographed image. This is also true for a grid of evenly placed dots, although even spacing could place some organisms off register and thus cause them to be underestimated. Randomly generated dot patterns (from random number table or computer programs) are thus prefered.

Such dot patterns, often a dark or light circle with a clear center, are drawn on clear plastic sheets, flexible for irregular surfaces, which is placed over a surface in the field (usually intertidal); the identity of organisms under each dot is determined. Because 100 dots per quadrat give standard deviations of 5 to 10 percent (Connell, 1970; Dayton, 1971), 100 or more dots per quadrat should be used (Fig. 3.2). Species found in the quadrat but not in contact with any dot are recorded as trace species (Menge, 1976; Lubchenco and Menge, 1978).

When it is not possible to lay out such quadrats in the field, photographs of marked permanent quadrats or random quadrats along a transect can be projected onto paper with similar random dot patterns (Connell, 1970). A random pattern should be used only once on a quadrat. Note that unless all percentage data fall between 30 and 70 percent, an arcsine transform of percent should be calculated (Sokal and Rohlf, 1969) before statistical treatment. Standard deviation or 95 percent confidence intervals for the mean can be calculated on such transformed data, and reported after being transformed back (Lubchenco and Menge, 1978).

II. Soft substrates. Organisms are either epibenthic (and methods for hard substrates can be used) or infaunal and often not visible from the surface or in photographs. Grabs, dredges, and suction devices are used to sample such assemblages. There are really no non-disruptive methods although Peterson and Andre (1979) found that they could dig up bivalves, measure them, and rebury them successfully.

A. Dredging. Epibenthic dredges sample only the first few centimeters of sediments and the organisms in or on that layer (Sanders et al., 1965). Other benthic dredges bite deeper into the sediments and sample organisms that burrow deeper. Since some bivalves and shrimp can burrow more than a meter into sediments, dredging almost never provides an equally quantitative sample of all species. The real area and depth sampled are never known for certain.

B. Grab samples. Grabs, of which there are numerous designs, dig into the sediment by their weight when dropped or when pushed by a diver. It is usually obvious when a grab dropped from a ship has functioned properly and will contain a full measure of sediment. Diver operated grabs are most certain to penetrate to the desired depth. Volume and depth of sample are known more accurately for grabs than for dredges (Sanders, 1958 and 1960). Grabs are the most common methods used for sampling lake benthos (Lind, 1979).

C. Box cores. Intertidally, metal square or circular frames are driven into the sediment, allowing deeper penetration than the methods mentioned above. Sediment is dug out of the core and put through a set of graded mesh sieves. Subtidally, airlift suction hoses are used instead of shovels to move sediment out of the core and through a mesh bag in the water or a strainer on a boat (Thomassin, 1978; Harris, 1976).

D. Coarse sediments. Shell beds, coarse gravel, or cobbles are not adequately sampled by most of the methods mentioned above. A combination of box core or quadrat, digging or airlift, and sieving can separate the mobile fauna. The surface of the shells or rock must also be examined for attached organisms.

E. Sample preservation and sorting. Once samples have been collected, animals must be fixed, stained, identified, counted, measured, and often weighed. They can be fixed initially in 7 to 10 percent formalin in sea water for a few days, stored in 70 percent alcohol, and stained. Note that weights of materials preserved in alcohol may be low because of lipid dissolution.

F. Sediment analysis. Descriptions of soft-substrate communities rely on measurements of sediment grain size, water content, and nutrients (protein, nitrogen, or carbon). Sediment particle size determines nomenclature (clay, silt, sand, gravel, and so on) (Fig. 3.2) (Eltringham, 1971; Sanders, 1956, 1958, and 1960; Fenchel, 1969; Stoddart, 1978c). In many cases sediment type is a result of the activity of the deposit feeders (Rhoads and Young, 1970; Levinton, 1977; Rhoads et al., 1977). Water content or porosity is a measure of free space within sediments, and is sometimes related to the presence of types of meiofauna (microscopic infauna between grains) (Jansson, 1967; Morgan, 1970). Depth of the anoxic (black) layer characterizes some sediment with respect to the distribution of both macro-organisms and bacteria (Fenchel, 1969; Fenchel and Riedl, 1970).

Organic carbon (Riley, 1970; Newell, 1970), total nitrogen (Newell, 1970) and protein content (Meadows and Campbell, 1972; Watling et al., 1979) are all related to the biomass of bacteria, surface microalgae, and detritus in sediment and may reflect their suitability for deposit feeders. The relation is not direct and depends to a great extent on turnover rates of both carbon and nitrogen (Levinton, 1979). Direct counts of bacteria by epiflourescence microscopy and analysis of chlorophyll content are probably better estimators of food availability, but experiments with density manipulation must be carried out to determine actual rates of energy flux (Tenore, 1977).

Whole-community metabolism (rate of oxygen consumption per unit of substrate area) has been used as an index of biological activity (Pamatmat, 1968, 1975, and 1977; Pamatmat and Fenton, 1968; Smith, 1973 and 1974; Smith et al., 1973), but is a composite of many causes and thus difficult to interpret or use as an indicator of community condition.

Benthic sampling-transect and quadrat methods. The theory behind transect sampling is that species abundances are proportional to their occurrence along a randomly placed line through a two-dimensional system. A series of transects can also be laid out in a grid pattern and across various zones. Changes in species abundances along transects show variations in zones and transitions between zones. Transects are either sampled at even intervals or at random points (Strong, 1966; Porter, 1972a and 1972b; Loya, 1978; Quinn and Gallucci, 1980; Poole, 1974). As a general guide, a system of equally spaced transects perpendicular to shore as shown in Fig. 3.3 which was prepared by the panel and sampled regularly is appropriate for a description of zonation and some types of spatial heterogeneity. Species at the sampling points (below the point or in a quadrat at that point) or species below each link of a chain transect are counted.

Abundance of species or types of "cover" within an area may be described at a finer scale by having evenly spaced transects laid parallel to shore within zones. Samples should then be taken at all points or at randomly chosen points along the line to avoid problems associated with species that have regular spacing (Fig. 3.3). If zonation is not evident, there is no reason why transects laid perpendicular to shore could not also be sampled at random.

Quadrat sampling is another way of sampling along a transect, gathering information on abundance, biomass, or percentage of cover at each point. Random dot quadrats for epibenthos have already been described. Grid quadrats are sometimes used, but the point method is preferred and is no more difficult. Quadrats can be done with box cores with a substrate sample taken at each selected point on a transect. Some investigators have used a "blind toss" of a quadrat within a prescribed area but this is not random sampling and should be avoided. It is easy to make a transect, generate lists of random numbers, and set out the transect several times at regular intervals across the study site (stratified random sampling). Even better is a grid of transects at right angles. With this arrangement, the Cartesian coordinates can be chosen at random and quadrats placed completely at random within the site (Fig. 3.3-I).

If permanent quadrats are constructed, their positions should be chosen at random. Variation between areas may be high even on a fine scale and in such cases, permanent quadrats ensure that the same

130   II. AQUATIC ECOSYSTEMS

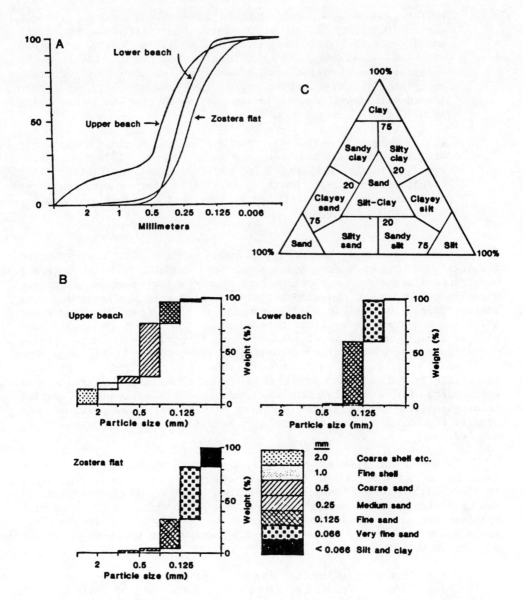

Figure 3.2 Soft-sediment habitats by size distribution. (A) Cumulative composition (percent) of sediment by particle size for three habitats. [From Morton and Miller (1973)] (B) Size-frequency distribution of particles for the three habitats in (A). [From Morton and Miller (1973)] (C) Classifications of sediments as mixtures of sand, clay, and silt.

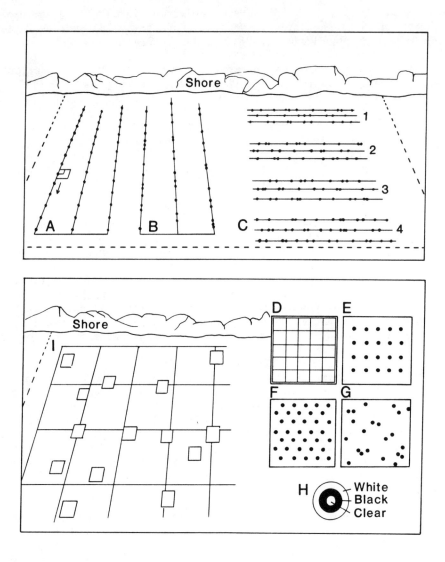

Figure 3.3 Examples of transect and quadrat sampling layouts.

microhabitats are monitored repeatedly (Connell, 1970, Sousa, 1979a and 1979b, Polderman, 1980).

Special problems of coral reef surveys. Coral reef benthos present several problems, especially the variety of microhabitats for plants and animals. Simple chain transects have been used to survey coral species diversity and depth zonation (Porter, 1972a and b) and more sophisticated methods have been developed that are based on those used in plant ecology (Loya, 1978; Scheer, 1978). Permanent quadrats can be established and used to follow changes in the percentage of cover and growth or damage to individual colonies over several years (Connell, 1973). General surveys and baseline studies have been carried out on several reefs in the Caribbean (Birkeland et al., 1976a), the Pacific Coast of Central American (Glynn, 1976) and the South Pacific (Amesbury et al., 1975a, 1975b, and

1976; Tsuda et al., 1975; Smith and Henderson, 1978). These surveys use such techniques as photographic mapping (Rutzler, 1978) and descriptions of reef physiography (Pichon, 1978; Stoddart, 1978a and 1978b).

Pollution from sewage outflow, oil spills, and sedimentation can severely damage reefs and corals (Loya, 1975; Birkeland et al., 1976a and b). Most information has come from studies of existing perturbations.

Grigg and Maragos (1974) examined reef development on old submarine lava flows and found that 150 years or more was needed for a reef to return to a normal coral diversity and more than 80 years to reach normal percentage of coral cover. Damage to reefs may last a century or more, whereas in temperate intertidal areas, damage may be removed in a few decades (Sanders et al., 1980).

Fouling organisms. Encrusting benthic invertebrates and algae often settle on and "foul" pilings, cement, boats, pipes, and other artificial structures in the water. These organisms are usually a normal part of the local hard-substrate benthos. Often, they are the shorter-lived seasonal component when the structures in question are periodically cleaned or replaced. On softer surfaces, such as wood, plastic, and rubber, several fouling species may bore into the material and cause extensive damage. Human activities that change the temperature, salinity, or flushing rates of coastal areas can have a large impact on the community of fouling organisms.

No single substrate is sufficient as an experimental settling panel to test for the presence of larvae and the growth rate of the fouling community. In some areas it is possible to use fragments of the natural substrate (Osman, 1977 and 1978, Sousa, 1979a and b; Sutherland, 1980), but it is often simpler to use artificial panels (plastic, slate, asbestos board, wood) since the flat surface makes analysis easier. The test substrate used should depend on the effect being tested--the effect on the natural local encrusting community on organisms that may potentially foul wooden, concrete, or other structures.

General discussion of methods for estimating abundances and percentage of cover can be found in Dayton (1971), Porter (1972a and b), Menge (1976), Done (1977), Loya (1978), Lubchenco and Menge (1978); Watling et al. (1979), John et al. (1980), Buss (1980), Jones et al. (1980), George (1980), and Russell (1980). Coral reef methods are discussed by Porter (1972a and b), Connell (1973), Grigg and Maragos (1974), Loya (1978), Glynn (1976), and Stoddard (1978a and b). Soft-substrate biotic community methods are given by Sanders (1956, 1958, and 1960), Sanders et al. (1980), Fenchel (1969), Woodin (1978), Levinton (1977 and 1979), and Whitlatch (1977). On the overall design of benthic monitoring studies, see Odum et al. (1974), Lewis (1976, 1978a, 1978b, and 1980), Erickson (1979), Cairns et al. (1979), Clokie and Boney (1980), Hiscock and Mitchell (1980), Jones et al. (1980), Knight and Mitchell (1980). Finally for discussions of methods for monitoring the effects of pollution on benthic organisms, see North (1974) Wood and Johannes (1975), Birkeland et al. (1976a), and Pearson and Rosenberg (1978).

Population structure, reproductive cycles, and growth of benthic organisms (including coral). Population characteristics can be determined using transect data or quadrat surveys. Histograms can be used to illustrate size-frequency distributions and changes in the distributions with time. In some species distinct year classes are evident from such histograms, and the magnitude and size of recruitment events can be studied. Population size and density estimates can also be made from transect and quadrat results except for mobile benthic organisms (fish and crustaceans), where visual counts, or mark and recapture techniques may be appropriate. In population studies, densities, sizes, and distribution among size classes are often compared (Patil et al., 1971a and 1971b; Scheer, 1978; Chapman and Gallucci, 1979; Patil and Rosenzweig, 1979).

Reproductive cycles are quantified by gonad indices (Giese and Pearse, 1974) from collected animals. Measurements of gonad weight and volume (Seed, 1969) or cross-sectional area (Sebens, 1981) are taken for organisms with amorphous and numerous gonads. Some animals can be brought alive to the laboratory and release of larvae or gametes observed. Certain crustaceans and fishes can be observed in the field and egg masses on the body (crabs and lobsters) or brooded on the substrate (some fish) can be counted and measured.

Growth is a sensitive measure of environmental quality but depends on crowding, food supply, and physiological state. Some species inhabiting the best areas may get so crowded that individual growth and reproductive success are very low. To complicate matters, growth in many species is dependent on habitat. In such species, comparisons across habitats for one size are inadequate. Fitting growth increments to growth models (Jones, 1976; Kaufman, 1981; Sebens, 1982a) or statistical comparison at all size classes (Sebens, 1982b) are needed. Marked individuals are usually used in such studies.

Coral growth, primarily calcium carbonate deposition, presents another problem since moving the colony is usually detrimental. Weights of colonies in sea water (bouyant weight) (Jokiel, Maragos and Franzisket, 1978), or estimates of volume in situ (Maragos, 1978) can be made over time. Coral also lays down alizarin red dye in bands when incubated with the dye in bags in the field. Growth after that banding (or a series of bands) can be measured at the end of a study by killing and slicing the coral (Lamberts, 1978).

The record of daily, seasonal, and annual growth in coral is visible in density bands from X-rays of cut sections (Buddemeier, 1978; Hudson, 1977). This technique has great potential in comparisons of sites, habitats, and depths for control and impacted areas in the tropics. Even areas without well-established reefs usually have a few coral species to provide growth data for many years before a given project is initiated (Shinn, 1966; Weber et al., 1975; Glynn, 1976).

3.4.2 Zooplankton sampling.

Zooplankton includes: resident species, which are always planktonic, (holoplankton), the larvae of benthic invertebrates (meroplankton); and the larvae of fish (ichthyoplankton). They are categorized by size into megaplankton (>2 mm), macroplankton or mesozooplankton (0.2 to 2 mm), microplankton (20 to 200 µm), and nanoplankton (2 to 20 µm). Ultrananoplankton (<2 µm) are primarily bacteria. Almost all zooplankton are >2 µm (net plankton); however, early plankton studies and those associated with recent environmental surveys have used nets with 200 μm mesh or larger, thus omitting the largest biomass fractions and most benthic larvae. For each size fraction there are specific methods of sampling, preservation, and analysis. On general zooplankton sampling, see Tranter and Fraser (1968), Steedman (1976), Jacobs and Grant (1978), Lind (1979), North (1974) (entrainment). On coral reef plankton, see Emery (1968), Hobson and Chess (1976 and 1978) Porter, Porter, and Olhorst (1978), Porter et al. (1977), Johannes (1978), Rutzler et al. (1980).

Sampling and analysis. The outline below is condensed from the 1968 UNESCO study of zooplankton sampling methods and the 1976 UNESCO publication on methods of zooplankton preservation (Tranter and Fraser, 1968; Steedman, 1976).

Megaplankton contains fast swimming, sparsely distributed forms. Problems of net avoidance and sampling scale are severe (see above references).

Large mesozooplankton (>1 mm) includes larger crustaceans, eggs, ichthyoplankton, medusae - includes some megaplankton).

1. The plankton net must be towed at fairly high speeds (2 to 3 knots for a simple net or 6 knots for an encased sampler). The suggested net (1 mm mesh nylon) has a cylindrical front section and conical rear section.

2. Preservation, counting, and biomass determination are carried out as for the small mesozooplankton (see below).

3. Special consideration must be given to two known biases. The first is that larval fish can avoid the nets as can certain larger crustaceans. The second is that gelatinous zooplankters are often broken up during sampling. These can be adequately sampled only by in situ counting or by photography using scuba equipment.

Small mesozooplankton (200 to 1000 μm includes larvae, eggs, and most crustaceans.

1. Plankton nets (200 μm aperture nylon) with a cylindrical fore-section and a conical body [WP-2 net, (Tranter and Fraser, 1968); Wisconsin plankton net (Lind, 1979)] are suggested for these zooplankton. The cylindrical section sheds captured material and avoids clogging for longer periods. Vertical tows for only a few minutes each are also necessary to prevent clogging (for example, 4.5 minutes at 45 meters per minute hauling speed). In shallow water or where a specific depth is to be sampled, horizontal towing may be used.

2. Flow meters should be used with the net to determine the water volume sampled. The flow meter (propeller, nonreversing) is placed midway between the rim and center of the net mouth. A second meter, outside the rim in mainstream flow, can be added to calculate filtration efficiency--less than 85 percent efficiency indicates clogging.

3. Automatic closing methods can be used to stop plankton collection at a known depth (for example, lead messenger weights).

4. Marine plankton should be preserved in nine parts sea water with one part buffered formalin solution (concentrated 38 to 40 percent formaldehyde in solution). The concentrated formalin is buffered with borax, or other suitable buffer, to an approximate pH of 8 for sea water. Fresh water zooplankton are preserved in 4 to 5 percent buffered formalin in fresh water (Lind, 1979).

5. Biomass can be determined on a split fraction of the sample by weighing the plankton sample which has been deep frozen and dried at 60°C; it is weighed again after being heated at 500°C to give both dry weight and ash-free dry weight. Settled volume is also used as a rough measure of plankton abundance.

6. Counting and identification are carried out in a Sedgwick-Rafter cell.

Microzooplankton (<200 µm) includes protozoa, eggs, larvae, and smaller copepods.

1. Water samples are collected in 10 liter bottles--for zooplankton <150 µm), with membrane filters; for larger zooplankton (75-200 µm) with fine nets (40 µm aperture). Nets of such a fine mesh clog easily.

2. Preservation is usually with buffered formalin (4 to 5 percent in sea water) although direct examination of living material is preferred. Rhode's iodine fixative is useful for the naked protozoan plankters.

3. Sedimentation is the preferred method for concentrating microzooplankton.

4. Counting is done on a Sedgwick-Rafter cell for the larger animals and on a hemocytometer for the smaller ones. A square counting cell the size and thickness of a cover slip and retaining 0.5 ml is useful for most sizes of microzooplankton.

Sampling designs and data analysis for zooplankton. Most sampling designs for zooplankton attempt to include daily cycles, depth distribution, seasonal changes, and spatial distribution over the study area. In addition, areas with problems of water entrainment (power plants or locks), may require that plankton samples be taken before and after entrainment. It is usually necessary also to determine living and dead zooplankton. For most applications, both holoplankton and meroplankton must be quantified and thus sampling should include microzooplankton (many benthic larvae are < 200 µm), macroplankton, and megaplankton. Plankton are known to change drastically from day to day in coastal systems, and certainly from week to week. If at one station, three samples are taken twice a day and biweekly throughout a year, there are 138 samples to analyze. In order to sample even 10 stations, that number rises to 1380 samples in a year. For practical reasons, areas for sampling programs should be identified carefully. If it is necessary to examine plankton only upstream and downstream of an entrainment project (North, 1974), then the sampling program can be limited. In coastal systems where information is desired on larval abundance at different times of year, a single station may suffice (Jacobs and Grant, 1978).

Coral reef zooplankton reside on the substrate during the day, move upward at dusk, and settle again at dawn. Different substrate types support different zooplankton assemblages. While water near a stretch of ocean sand beach or rocky shore may have a fairly homogeneous zooplankton distribution, an estuary, inlet, or deep channel will produce a zonal or patchy structure to the zooplankton. Each habitat must be sampled as well as the abundance and distribution of specific larvae. If it is known when larvae are present, the study can concentrate on those periods.

Sampling for the effects of a specific perturbation can sometimes be done in the field but often requires laboratory studies or studies within field enclosures. Since the conditions under which zooplankton function, feed, and grow are often very narrow, studies must be carefully controlled. In sufficiently large field enclosures, zooplankton responses to light, temperature, eutrophication, density, and other variables can be investigated experimentally and compared to a control. This technique has been used successfully in temperate lakes (Kerfoot, 1980) but is less used in the ocean because of large circulation patterns, long vertical migration, and other factors (Grice and Reeve, 1981).

Problems associated with taxonomic identification are legion. Often studies must be limited to a few large, well-known species.

Zooplankton mortality during entrainment. Zooplankton that are trapped within a body of water with high temperature or salinity (near a power plant outflow, for instance) often suffer heavy mortality (North, 1974). This is more likely to be a problem for the larvae of inshore fish and invertebrates than for the wide-ranging open water zooplankton. The most important consideration is that effects of entrainment or effluent are severe in estuarine and shallow littoral areas. Studies of entrainment effects use the techniques decribed for zooplankton and require the calculation of volume transport.

Demersal plankton. Coral reef plankton and that in other coastal locations, with a component that migrates to the surface at night, can be sampled with standard net tows from boats or divers near the bottom and near the surface (Emery, 1968; Hobson and Chess, 1976 and 1978; Johannes, 1978; Porter, Porter, and Ohlhorst, 1978). Plankton traps (Porter et al., 1977) or suction devices (Rutzler et al., 1980) on the bottom can also be used to find out which zooplankton are truly demersal. Mesh sizes for nets will vary for the desired size category to be sampled.

Plankton traps consist of inverted nets or plastic cones leading into a floating bottle filled with anaesthetic in sea water. Zooplankton migrate from the substrate up the funnel, and into the trap. Suction devices include pumps to the surface (Porter et al., 1977) and submersible pumps that pull water through a net (Rutzler et al., 1980). The latter has the advantage of not passing plankton through a propeller and of sampling plankton near the bottom that do not migrate vertically. Measurement of plankton upstream and downstream across a reef can be used to show depletion of plankton (by coral and other predators) in natural or perturbed systems (Glynn, 1973).

3.4.3 Phytoplankton.

A baseline study of a lake or reservoir should include data on the kinds of phytoplankton present, their distribution in time and space, and their rate of production. Also because of their small size, most will pass through the mesh of commonly used plankton nets, and other techniques must be used for collection. Because of their small size and high concentrations, small volumes are usually required. Samples from discrete depths may be taken by water sampler or by pump, preserved, and taken to a laboratory for analysis, where they are concentrated by centrifugation, filtration, or settling.

Identification of species usually requires high magnification and high resolution microscopy (magnification, 1000). Counting of the concentrate is often done at somewhat lower magnifications in special chambers (Palmer cells or hemocytometers), on gridded, cleared membrane filters, or by inverted microscopy.

Phytoplankton biomass is often estimated indirectly by spectrophotometrically or fluorometrically determining the quantity of chlorophyll present. Ratios of the different chlorophylls give an indication of major divisions of algae present. Ratios of chlorophylls to their degradation products (phaeophytins) give an indication of the "health" of the phytoplankton community (American Public Health Association, 1980, p. 931; Environmental Protection Agency, 1987; Lind, 1979, p. 115, 129; Strickland and Parsons, 1974, p. 185).

Sampling and analysis. Terminology may cause confusion. The phytoplankton are the plant members of the larger plankton community of lakes and reservoirs. Usually, but not always, they are microscopic. The term nanoplankton refers only to the size of the organism (small enough to pass through a plankton net with a mesh opening of 64 µm). Most phytoplankton are nanoplankton, but some colonial species are larger. Also, some animals such as protozoa and rotifers are included in the nanoplankton. Because phytoplankton populations may rise or crash in a few days, frequent sampling is important. Population growth changes may be very localized, and populations may be hyperconcentrated or hyperdispersed by currents and Langmuir circulations. Because of light dependency, phytoplankton are found in the upper illuminated layer of the water, the photic zone. Here they often stratify on a fine scale, apparently in response to the light gradient, to nutrient availability, and to grazing pressure. Consequently the sampling design should permit the collection of many samples in both horizontal and vertical directions. When many discrete samples may not be taken, a compromise suitable for baseline surveys is to take an integral sample from throughout the photic depth with a pump (easily made from a garden hose, inexpensive bilge pump, and a 12 volt battery).

Among the phytoplankton are delicate species with flagella and soft gelatinous sheaths. Many sampling, preservation, and concentration techniques may damage such cells beyond recognition. Gentle preservation is required (for example, Lugols iodine and concentration by sedimentation). For routine baseline surveys, the membrane filter method of concentration is often appropriate. In tropical waters, because of the low buoyancy of the warm waters, the average organism size may be somewhat smaller than that in comparable temperate waters. To improve buoyancy, some organisms secrete gelatinous sheaths. Thus, in tropical waters a greater proportion of the community may be expected to have these delicate structures. Because of their small size, decomposition of phytoplankton cells is especially rapid in warm waters. Thus, special care should be given to prompt preservation.

Selection criteria. Because identification of most phytoplankton requires a high degree of taxonomic skill and high quality microscopes, samples from many developing countries will require shipment to outside authorities. Identification to taxonomic levels higher than species provides little usable data and is often misleading. Preserved samples may be shipped in polyethylene bottles that are completely full (no bubbles). Costs of adequate sampling will vary widely. Laboratory costs are low, but the costs for identification and enumeration by specialists are high.

3.4.4 Fish and fisheries.

The fish fauna need to be sampled in most baseline studies. A species list by habitats with information on whether a species is rare, uncommon, common, or abundant may be adequate. Reductions in numbers of species or major shifts in species composition may indicate the effects of pollution, construction, or other environmental stresses. Low species diversity indices can indicate stressed conditions.

Sampling can be difficult because the equipment used to catch fish is somewhat selective of the species and sizes and certain species may be in habitats that can not be efficiently sampled by available equipment. (Bagenal, 1978, chapters 2 and 3; Brandt, 1972; Hocutt and Stauffer, 1980; Nedelec, 1975; Saila and Roedel, 1979).

Sampling and analysis. Habitats should be identified before sampling starts and will to a considerable extent indicate the types of sampling needed. Many habitats can be recognized by a quick field survey. Maps with soundings or depth contours indicated can help in locating sampling sites. The habitats in a lake or reservoir will include littoral (or shallow-water) areas, which are classified by bottom types or the presence of vegetation, bays, mouths of tributary streams, and open water; or limnetic areas where sampling may be done at several depths. In lakes with oxygen deficient hypolimnions, sampling is usually not productive below the thermocline. In streams, riffle and pool areas should be sampled at several points from headwaters to the mouth. Larger rivers and extensive flooded areas or marshes are often difficult to sample.

The methods for capturing fish will have to take into account water depth, bottom type, presence of vegetation or other obstructions, and current. It should be recognized that any method short of total kill (for example, rotenone poisoning) is selective and may bias a sample. For most purposes, gear similar to that used by local fishermen may be most suitable. Often data may be collected from the fishermen's catch, and fishermen may provide useful information (Acheson et al., 1980).

In order to take species and sizes not usually sought by fishermen and to quantify the abundance estimates more precisely, special techniques may be used--standardized trap nets, trawls, seines, or gill nets.

Experimental gill nets have sections with graded sizes of mesh to catch fish of all sizes. These nets may be set for specified periods or the trawls or seines dragged to cover specified areas so that the catch-per-effort can be used as a measure of abundance. Because fish activity and location differ with time of day or season, sampling should represent all time periods, or at least comparable time periods so that comparisons can be made.

Electric shocking and poisons (usually rotenone) are two methods for collecting fish not generally available to the local fishermen but of possible use for baseline studies.

Species readily identified in the field may be recorded without removing the fish, but a few specimens of each species should be preserved for record purposes and for verification of identifications. Several specimens of species not positively identifed should be preserved and labeled. Formalin (10 percent) or formaldehyde (4 percent) neutralized with borax is satisfactory as a preservative. Most specimens can be identified readily by available taxonomic keys and then may be checked by experts.

Field records should include full information on the equipment used, length of time it was set or area covered, description of habitat and depth where the catch was made, location, date, and numbers of fish of each species caught. Lengths of each fish should be recorded unless several hundred are caught and then a representative sample should be measured. Lengths may be standard lengths (to end of vertebral column), fork lengths (to end of central portion of caudal fin), or total lengths (to end of caudal fin with the two lobes compressed to give maximum length), but the measurement used should be recorded and should be consistent through the study.

Large numbers of fish may be measured conveniently in the field with special measuring boards with graph or evenly lined paper under a thin sheet of plastic. With the snout of the fish against an endboard, the position of the end of the caudal fin can be recorded with the plastic and graph paper. The numbers of holes in each length interval can be recorded in the laboratory. This method eliminates much of the writing in the field. Separate sheets are used for each species and properly labeled.

Lengths and weights should be measured and recorded from a reasonable number of fish, perhaps 24 of each species from each catch. In many cases, scales, otoliths (calcareous ear deposits), fin rays, or some bony structure should also be collected from these fish to determine age and growth rates. These structures can usually be conveniently stored in individual envelopes on which measurements are recorded, and if possible the sex, sexual maturity and stomach contents.

One measure of the relative health of a fish is the condition factor: $K = W \times 10^2 / L \times 10^3$, where W is weight in kilograms and L is length in millimeters. Condition factors should be analyzed by size classes to see whether the relative weight increases or decreases with length of the fish. The numerical value of K means little except in comparison with K values from other populations of the same species (in the same geographical locations).

Condition can also be described by the weight-length regression: $\log W = a + b \log L$. These condition factors can be compared by analysis of variance of the regressions from different populations.

Age and growth of fish are often used as indicators of the condition of the population (Bagenal, 1978). Fish grow when conditions are favorable and can maintain themselves for relatively long periods without growing. When regular seasonal changes are pronounced, as in temperate and arctic climates, annual marks on scales and bony structures can often be interpreted to determine the age of fish. Past growth can also be estimated on the assumption that growth of the scale or bone is proportional to growth in length of the fish. Then, $L_i = S_i \times L_c/S_c$ where
L_i = length of fish at previous year i,
L_c = length of fish at capture,
S_i = scale measurement to annulus i, and
S_c = scale measurement at time of capture.

Since the growth may not be directly proportional, corrections can be made through study of the length-scale relation, but direct proportion is often accurate enough.

Otoliths, spines, or bones are frequently more reliable indicators of age than scales for fish in the tropics, but even these structures may not show annual rings where seasons are poorly defined. Otoliths are collected from the inner ear near the base of the skull, then split or sectioned, and examined with a microscope (magnification of 400x may be needed).

Age may be estimated by a plot of the number of fish of each length measurement. Fish of similar age or similar length and their frequency appears as a peak or mode on the length frequency graph. The modes may indicate the average size of several age groups. Because spawning may occur during most of the year in the tropics, this methods will not give satisfactory estimates of age.

The baseline report of species distribution and abundance should be compared with earlier data on the same waters if possible and with regional lists or other comparable data. Species which are no longer present or which would be expected to be present on the basis of the available habitats may provide clues about stresses on the population. For example, the dominance of species capable of living in waters with low oxygen such as the anabantids would suggest oxygen stresses.

Species diversity indices (similar to those used with other organisms) are convenient measures for making comparisons but are not always readily interpreted. Sale (1977) proposed that in reef fish communities, community structure may be unstable--that is, species composition at a given site may not recover after a disturbance but a new species composition may develop as the result of chance recruitment--and that the diversity may be directly related to the rates of small-scale unpredictable disturbances. The situation in tropical inland waters may be similar to that on reefs. Inland waters may be subject to frequent small-scale disturbances and the variety of species replacing others may be fairly high in the tropics. Species diversity may decrease if the number of predators are reduced since predators keep more of the environment available for invasion.

Fishes in various habitats should be identified as herbivores, planktivores, and so on; many species may be in different roles at various stages in their growth.

A convenient ratio to represent a population is the weight of the forage fish (F) to the weight of the carnivores (C). Values of 3 to 6 in the ratio of F to C indicate balanced populations in U.S. bass-bluegill ponds, and such values apply to other situations since predators need about 3 to 6 grams of food for each gram of growth. One complication is that a significant part of the population may be composed of neither carnivores nor fish of a size that can be eaten by the average carnivore.

Selection criteria. The cooperation of fishery office personnel is needed first, and they can often provide valuable information on the water and its fishery. If rotenone or electric shock methods are to be used for collecting, the attitude of the fishery officer and of local fishermen should be considered because these methods are usually banned, but exceptions may be made for scientific collecting. In general, use of rotenone or electric shock in limited areas have no significant effect on fish populations or future fishing. In very soft waters, electric shock is not an effective method of sampling because of the poor conductivity of the water.

For sampling in shallow water, seines, cast nets, electric shockers, and trap nets are usually used, but abundant vegetation may rule out the use of seines and cast nets. In open waters, gill nets and trawls are most commonly used. Local modifications of the gear may be more efficient. For comparative catch-per-effort data, however, standard gear should be used.

Field sampling is labor intensive and can often be handled by using local personnel (fishery staff, fishermen, and biology studies) under supervision of a trained scientist. The trained scientist is needed for planning, directing the field work, analyzing data, aging of fish, and writing up findings. Expert identification of some fish may be necessary but the amount of this is usually minimal.

Fisheries. Most governments collect statistics on their fisheries and these may be of value in baseline studies. However, statistics are meager or lacking on the small fisheries found on many inland waters. While these fisheries may be smaller than the marine and coastal fisheries usually studied, they are important sources of food and employment to the local people. A fishery survey should include study of (i) the fishery resource, (ii) capture and harvest, (iii) processing, distributing, and marketing, and (iv) consumption (Bazigos, 1974; Gulland, 1969; Saila and Roedel, 1979).

Sampling and analysis. A major problem in assessing inland fisheries is that they are usually composed of many species of fish, harvested by a large number of people with a variety of gear, and are locally consumed or marketed at several small markets. Securing an adequate baseline for a fishery thus requires a well-designed sampling scheme and a great deal of sampling.

Methods for the survey of the fishery resource have been discussed above but additional information, particularly length-frequency distributions, weight-length ratios, age, and growth can be secured from the fish caught by the fishermen and sold in the markets. These data are needed in assessing the effects of fishing on the resource. When the catch is composed of small fish in comparison with the same species in natural populations it usually means that the fishing pressure has been great, possibly excessive.

After a preliminary survey to identify the numbers and types of fishing practices used in various areas (Bazigos, 1974), a sampling schedule is established to record the catches by contacting the fisherman at the time particular gear is lifted or when the catch is brought to shore. The sampling should record amount of gear, time fished, and species, numbers, and sizes of the fish. The size and species collected by each type of gear are particularly important and sampling should be designed to evaluate the catch from each type of gear at all periods throughout the annual fishing cycle. From the estimates of the number of gear units and of the average catch-per-unit, total harvest may be calculated.

Once the location and number of markets is known, they should be sampled on a regular schedule. The amounts of fish of each species brought to the market each day should be recorded as well as whether all fish are sold and prices. The proportion of fish which are locally consumed before reaching the market, sent to other markets, or processed by salting or drying should be estimated. Samples of the fish in the market should be measured. The marketed fish may differ in species composition and size frequency from those caught. Small fish may be eaten by fishermen and local people and only large fish selected for shipping. Data on large fish are particularly important since these are most often affected by overfishing or other disruptions. The percentage of fish in the diet of residents should be estimated. The social and economic condition of the fishermen, marketers, and consumers may also be of interest. Local personnel will usually be needed to collect these data.

The morphoedaphic index, which is determined by dividing the total dissolved solids by the mean depth of the lake or reservoir, has been used in estimating potential yield (Henderson et al., 1973; Ryder et al., 1974). This index is not useful in large shallow bodies of water, however, since mean depth is not a significant factor.

Selection criteria. Contact and cooperation with the fishery agency and local fishery officers is essential in this phase of the study. Data collection is again labor intensive and local labor specially trained and supervised by an expert is most efficient. A baseline study on fisheries should include at least a full annual cycle of data collection. As much data as feasible should be recorded with precise description of the methods used in collection because such spot data may be valuable in interpreting change in a fishery when more complete data are lacking.

3.4.5 Littoral vegetation.

Analysis will require collection, washing, identification, and if quantitative data are desired, drying and weighing of samples. Dried and pressed herbarium reference specimens may be retained. Sample collection for qualitative evaluation is by plant grapple for submerged species and simply by hand for emergent and floating species. Quantitative sampling requires use of a metal grid (1 square meter) placed over the plants and all above ground parts are harvested. For submerged species, scuba equipment may be needed. A sonar depth-finder may be used to locate submerged beds (American Public Heath Association, 1980, p. 983; Lind, 1979, p. 176; Westlake, 1969, p. 25). Aerial color photography, from small aircraft and hand-held camera, is a valuable tool in mapping the extent of emergent and floating plant beds. Use of filters to provide a "false color" image permits detection of some submerged beds (Benton and Newman, 1976).

Sampling and analysis. A baseline study should include, at minimum, a survey and classification of the aquatic vascular plants and macroalgae. In temperate regions there is significant variation both in varieties and quantities present at different seasons, and samples should be taken at least monthly during the growing seasons. In the tropics, populations are more constant and quarterly sampling may be adequate. There is spatial variation with major forms growing in concentric rings around lake and reservoir shorelines. Emergent species near the shore and floating species may be obvious, but the innermost ring of submerged species, which may make up the largest portion of the macrophyte community, must be sampled adequately. Aquatic macrophytes, especially submerged species, have a high percentage of water in the tissues. This encourages rapid decomposition, especially in warm climates, unless prompt attention is given to preservation. An effort should be made to collect plants with flowering parts for ease of identification. Most collections do not include roots and tubers, the extent of which vary from species to species; reports must always note this.

Selection criteria. Macrophyte surveys should be made in all waters with special attention paid to places where these plants are abundant, such as in shallow lakes and reservoirs, clear rivers, and the tropics. Skills required for collection and preservation are low and those for identification are moderate; many species can be identified by aquatic biologists with moderate taxonomic experience. Some species will require confirmation by authorities. Costs of collection and preservation are low. Many authorities will confirm occasional specimens free in exchange for herbarium-quality materials, especially from less studied regions.

3.4.6 Wetlands vegetation.

Wetlands include areas permanently, seasonally, or occasionally covered by water or underlain with water-saturated soil. Vegetation in such areas along sea coasts includes marsh grasses, which are able to withstand high salt concentrations and long immersion, and other salt-tolerant vascular plants. Dune and beach areas have another salt and spray-tolerant flora above the tide level. Fresh-water lakes, ponds, reservoirs, and rivers are often bordered by wetlands, swamps, and periodically flooded lowlands (Odum et al., 1974; Cowardin et al., 1979). Methods for describing wetlands vegetation will be nearly identical to those used in any terrestrial plant study (see part IV).

3.4.7 Periphyton.

Samples of this community are scraped from the substrate and preserved before identification. Quantitative samples are difficult to take from natural substrates. Consequently, artificial substrates are usually used. The most common is the microscope slide which is submerged for several weeks to allow colonization on it. The community is then examined directly on the slide (American Public Health Association, 1980, pp. 964-965; Environmental Protection Agency, 1973; Lind, 1979, p. 168).

Sampling and analysis. The type and abundance of periphyton is a function of the of substrate. For a baseline study, substrates must be identified. The periphyton community, as a whole, grows more slowly than phytoplankton. Samples adequate to describe seasonal variation may be taken monthly. Although artificial substrates are of great value, colonization on these will differ from comparable natural substrates. Thus these data are of value only for comparisons of ratios of species or functional groups of species on a substrate. Because many members of this community are soft-bodied and may have gelatinous sheaths, prompt but gentle preservation is essential (4 percent, neutral pH, and osmotically balanced, formalin works well).

Skills for sampling and preservation are low, but those for identifying members of this diverse community are high. An aquatic biologist will probably not be able to identify most organisms to species level, and specimens will need to be sent to authorities. Skills required for non-taxonomic studies of this community such as biomass determination, and pigment quantities and ratios are moderate. Costs of sampling are low to moderate and depend on the use of and type of artificial substrates. Laboratory costs are low to moderate except for pigment and biomass studies. The use of authorities for taxonomic work may be expensive.

3.4.8 Microbiota.

Bacteria in aquatic habitats are usually identified by a set of physiological tests. Specific tests are used to determine whether unusual bacteria are developing in aquatic habitats near development projects. This identification process is less complex for algae, fungi, and protozoa, since identification of these organisms is based on morphology and life cycles. Quantitative assessment of bacteria is normally based on dilution methods and colony counts on agar (Stanier et al., 1976). Protozoa, algae, and fungi are counted under a microscopic. Microbial activity in aquatic habitats is normally assessed by a variety of biochemical techniques. These include radioisotope assays, measurement of adenosine triphosphate, and other methods to indicate specific metabolic activity (Roswall, 1972).

Community structure. Methods for studying microbial communities, and in particular bacterial communities, by biochemical techniques have recently been described by Bobbie and White (1980). More classical approaches to the understanding of community structure of fresh-water microbial communities have been described. Methods for the determining algal community structure, have been well described. Algae are frequently used to indicate water quality. A shift in algal species dominance and a decrease in diversity in the number of species frequently occurs in aquatic habitats that are under stress. The use of algal community structure to determine water quality in development projects would be useful. Methods available for this purpose have been discussed by Patrick (1973).

Rapid in-stream monitoring of aquatic quality has been achieved with a sequential comparison index in which stream quality is expressed in terms of community structure and particularly the changes in diversity of diatoms. A typical slide community from a healthy stream would contain about 25,000 diatoms representing about 65 species. Laser holography is used to determine the effect of stress on these communities (Cairns et al., 1973). This in-stream monitoring system, which uses species diversity of algae, would be particularly useful in assessing changes in community structure in aquatic ecosystems near development projects.

Estimates of biomass. For estimating microbial biomass, assays of cellular constituents such as adenosine triphosphate, protein, or nucleic acids are used (Ellwood et al., 1981). Frequently it is important to determine microbial activity rather than biomass. Typically, this information is obtained by analysis of uptake of radioactive organic substrates (Bott, 1973). More accurate estimates made with maramic acid have also been developed (Moriarty, 1978). Because metabolic activities can be expected to be significantly affected in rivers and lakes perturbed by pollutants produced during development, it is important to use the most modern accurate biochemical methods to assess these effects on the microflora.

Nitrogen fixing microorganisms. Nitrogen fixation (Horne, 1978) plays a major role in the maintainance of algal productivity, and therefore of fish productivity, in tropical lakes. It is important to monitor the effects of development on this fixation process continuously. The use of the stable isotope nitrogen-15 and of the reduction of acetylene to ethylene gas by the enzyme nitrogenase permits accurate analysis of nitrogen fixation in natural waters. Populations of cyanobacteria also need to be monitored. Nitrogen-fixing cyanobacteria include *Anabaena* and *Nostoc*; taxonomy is normally determined morphologically. In Lake George, Uganda, nitrogen fixation by these bacteria supplies 30 percent of the daily nitrogen consumption, and more than 50,000 *Tilapia* per day are harvested. Perturbations caused by development, particularly toxic chemicals, can be expected to inhibit nitrogen fixation.

Determination of the nitrogen balance is affected by denitrification and by grazing. Methods for assessment of nitrogen balances in traditional Asian low input rice cropping systems have by discussed by Wetsclaar (1981).

Treatment of waste water. Development can be expected to be accompanied by increases in human population density and therefore in the number of sewage treatment plants. In developing countries, biological treatment of domestic waste is typically achieved either with trickling filters or oxidation ponds. In most cities and rural areas there is no organized waste treatment. An excellent set of 12 volumes dealing with appropriate technology water supply and sanitation has been prepared by the World Bank (1978-1982, 1982a and b). Conventional methods for assessing effectiveness of treatment before release of the treated water to rivers and lakes can be applied. These methods include the biochemical oxygen demand (BOD) test and coliform bacteria counts as well as assessment of chlorine demand (American Public Health Association, 1975).

Pathogenic microorganisms. Coliform bacteria are used as indicators of the health hazard from pathogenic microorganisms in water (Cabelli, 1979). In developed countries, the coliform test provides effective protection against water-borne disease. Viruses may be present in natural waters in which coliform bacteria are absent, however, and may be capable of causing human disease. In recent years specific tests for viruses in water have been developed (Mandell et al., 1979).

Development projects may help create severe the human health hazards. The pool of pathogens in sewage under these conditions is large, both in variety and density. Water contaminated with sewage is frequently used either for drinking or irrigation. The standard coliform count used in developed nations to indicate the presence of human sewage is rarely sufficient for the analysis of water in developing nations. Additional tests are required for pathogens that are protozoan, viral, or helminthic in origin. Methods for analysis of these microorganisms have been described by Mandell et al. (1979).

Self-purification. Elimination of organic matter carried in sewage and other waste is conventionally measured by the BOD test. A low BOD indicates the absence of a high concentration of organic matter available as a substrate for microbial growth. Rivers and streams have the capacity for self-purification--that is microbiological elimination of organic matter. However, frequently the BOD test does not provide an adequate measure of this capacity. Wuhrmann (1974) has developed methods to measure more directly the response of the microbial community to loading with organic matter and the capacity of the microflora to purify the water, thus providing a more accurate measure of self-purification.

Wuhrmann's methods are particularly appropriate for development projects. The standard BOD test is carried out at 25°C and is a measure of oxygen demand in temperate waters. The BOD test does not provide an assay for the more rapid metabolism that occurs at the higher temperatures found in countries where most development projects have their impact. In addition the specific substrates in these environments may degrade at rates that are not assessed by the BOD test.

Toxic chemicals. Detoxification of toxic chemicals in the water will occur by microbial action, which is temperature dependent. Methods are available to determine the rates of microbial biodegradation of toxic chemicals in water (Giger and Roberts, 1978). The traditional and simplest method to assess microbial degradation of toxic chemicals is to determine the increase in bacteria, but results are inaccurate because they do not necessarily reflect metabolic activity. Radioisotopic assays or oxygen utilization measurements more accurately measure biodegradation of toxic chemicals. However, chemical assays with chromatograpy and spectroscopy provide the most commonly used methods (Giger and Roberts, 1978).

It is not likely that such sophisticated methods will be available in the regions where development projects are being undertaken. The hazard of ecological damage by toxic chemicals may be determined less accurately from bioassays. The response of specific protozoan and algal communities to stress by toxic chemicals has been used to indicate the danger of ecological disturbance by toxic chemicals (Cairns and Dickson, 1971). Blockage of microbial chemoreceptors has also been used to indicate behavioral aberrations caused by toxic chemicals in aquatic environments (Chet and Mitchell, 1976).

3.5 Methods for Functional Properties

3.5.1 Nutrient cycling.

For studies of nutrient cycling, measurements of inputs and outputs of the aquatic ecosystem must be made as well as rates of recycling of nutrients within the ecosystem (see Fig. 2.3). This is a complex study where no single technique applies. For example, a nitrogen budget of a reservoir requires frequent monitoring of nitrogen entering in water flowing in from stream or from surface runoff and that leaving through outflow. Measurements are required of nitrogen entering from rainfall and from nitrogen fixation by organisms, as they are from loss to bacterial denitrification and to the sediments. Changes in the relative concentrations of nitrogen as ammonia, nitrate, and nitrite within all parts of the ecosystem should also be measured.

Sampling and analysis. Possibly no other baseline studies are as crucial to understanding the functioning of aquatic ecosystems as the nutrient budgets. Nitrogen and phosphorus budgets, and if possible, complete descriptions of all aspects of nutrient cycles, should be strongly considered in planning a baseline study. Nutrients should be sampled both on a regular time schedule and in conjunction with rainfall-runoff events. It is necessary to have accurate measurements of volume discharge into and from the ecosystem so that measured concentrations may be extended to total loads or losses. This may require installation of a hydrograph on the principal streams, or if labor is available, frequent flow-meter measurements at a stream segment of known cross-sectional area. The measurement of biological activities (nitrogen fixation and denitrification) is often seasonal and temperature-, light-, oxygen-, and nutrient-dependent. Careful monitoring for conditions favoring these processes must be made and measurements of these processes carried out intensively during periods of activity.

If nutrient budgets and nutrient cycles are to be measured accurately, highly skilled personnel must design the study. The sampling program is labor intensive but skill requirements are low. Some of the analytical procedures require only moderate skill in analytical chemistry, but other aspects, such as nitrogen fixation or denitrification studies, require considerable expertise. Costs for much of the routine chemical analysis (see methods for specific elements) are low, but procedures for nitrogen fixation require a gas chromatograph and bottled gases.

3.5.2 Primary production.

To observe primary production, measurements should be made of phytoplankton production in all lakes and reservoirs, macrophyte production in streams, lakes, and reservoirs when these plants are present in obvious quantities, and periphyton production in streams. Primary production is measured by one or more of several methods: uptake of radioactive inorganic carbon, evolution of oxygen into the water, removal of carbon dioxide from the water (by change in pH), and change in the standing biomass from the annual minimum to the annual maximum. In some cases, changes in the extent of the macrophyte beds may be estimated from aerial photography or satellite imagery. Radioactive carbon (carbon-14), biomass, and oxygen methods are described by the American Public Health Association (1980, pp. 957, 969, and 992), Lind (1979, p. 134), Vollenweider (1969, pp. 62 and 100), and Strickland and Parsons (1974, p. 261).

Sampling and analysis. The radioactive carbon method is the most sensitive and should be used where low rates of production are expected. Originally developed for phytoplankton production, it is easily adapted to periphyton production. Results are usually considered as "net" primary production. The oxygen method is suitable for more productive waters with relatively high rates of primary production. It is suitable for measuring the production of phytoplankton, periphyton, and macrophyte (with certain limitations). Data from the oxygen method can be used to calculate either net or gross primary production. The sensitivity of the carbon dioxide method varies because of differing buffering capacities of water. In well-buffered alkaline waters, high rates of primary production are needed to produce a measurable change in pH. Consequently, this method is rarely employed. The change in biomass method is the most common method used for macrophyte production and gives net production but does not account for tissue loss between measurements. One must also specify if below-ground parts are included.

The sampling program must be designed to take into account the light and nutrient dependency of primary production. Samples should be taken throughout the photic zone and all year in tropical and subtropical regions, and during ice-free times in temperate regions (subice production is usually inconsequential to the annual total). Macrophyte monitoring by photography or imagery is hindered by problems of dust in arid regions and of water vapor and cloud cover in more humid areas.

Because of the expense of radioisotopes and instrumentation, as well as possible legal problems in their use and disposal, the oxygen method may be more appropriate for use in developing countries, once it is shown to have adequate sensitivity. Skills required for use of the radiocarbon method are moderate if the isotope solutions purchased are calibrated and ready to use. Skills for the dissolved oxygen method and biomass change method are low to moderate, although scuba diving may be required to measure macrophyte production. High skills are required in designing the program and in interpretating the results.

3.5.3 Secondary productivity.

Functional roles, food webs, and predator-prey interations are important considerations, but most baseline studies and resource inventories omit analysis of functional roles in the communities under study. Sometimes there is published information on trophic position, prey choice, and energetics of a few species. It is usually not possible to obtain such information for all species, but it is possible, although not easy, to identify species whose effects on the community are more significant than might be predicted by their abundance alone. Such species have been called "keystone predators" (Paine, 1966) if they affect the relative abundances of important prey species or "key industry species" (Birkeland, 1974) if they are common prey species supporting numerous predators.

Dayton (1971 and 1975) proposed the following protocol for community level studies of marine benthos: (i) identify the species that are most abundant either numerically, by biomass, or by percentage of cover; (ii) determine the competitive interactions among these species, including life history information such as recruitment rate, mortality, and growth rate; (iii) examine competition directly by experimental removal or addition of each species in contact with others; (iv) determine the important herbivores, predators, and higher trophic level species by observation and by experiment.

Caging, removal, and density manipulation of potentially important animals can be used to verify their impact (see, for example, Connell, 1966a and b, 1970; Paine, 1969, 1974, 1977; Woodin, 1974; Menge, 1976; Virnstein, 1977). This technique concentrates the investigation of functional roles on community dominants and important consumers. Potential impact can then be examined by its effect on these species and the system respond to change in them. Levins (1976) proposed a "loop analysis," which can be either a qualitative or quantitative diagram of interactions between species.

Classically, such information has been displayed in a "food web." Cohen (1978) breaks this into a "source web," a food web beginning with particular prey species and progressing to higher trophic levels, and a "sink web" which begins with the dominant consumers and identifies as many of their prey as possible. A "community web" includes all significant species or groups and identifies links between them (Fig. 3.4). The links in a food web can be quantified with either numerical, biomass, or energetic measures of consumption or "energy flow" (Paine, 1971 and 1980). However, the simplest and most feasible approach is that of Dayton (1975).

Limitations of time, funds, and manpower will likely restrict the study to information on the important lower trophic level of species and their interactions followed by a determination of which species have the greatest controlling effects at higher levels. This should always include the existing human impact.

150   II. AQUATIC ECOSYSTEMS

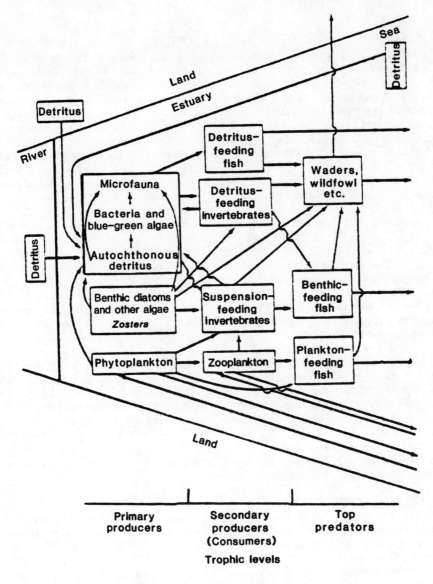

Figure 3.4 A food web for an estuarine community, an example of a 'community' food web, identifying major tropic groups, rather than individual species. [Source: Barnes (1974, p. 37)]

3.5.4 Eutrophication.

In lakes and rivers, phosphorus is the limiting nutrient that controls the eutrophication process (Stumm and Stumm-Zolinger, 1972). A model has been developed for relating additions of phosphorus to changes in algal abundance and transparency (Lorenzen, 1978). This model would be particularly useful for development projects where there is a serious danger of nutrient enrichment of aquatic habitats because the methods permit assessment of the critical factors controlling deterioration of aquatic habitats. These models should be interpreted with caution because of the quality of data used in the modeling process.

There is no discrete method to measure the rate of eutrophication in waters. Several methods may be used to measure changes symptomatic of the aging process. Obvious symptoms such as floating mats of vegetation or scums of algae permit measurement by collection and weighing of the material, or by changes in the average chlorophyll content of the water over several years. Shifts in ratios of fish (or other organisms), determined from the historical record, may be used. Data should be gathered from lakes and reservoirs in less advanced stages of eutrophy on the nitrogen and phosphorus budgets, on the average annual and average summer chlorophyll content, on reductions in transparency (by Secchi disk), and on rates of oxygen depletion in the deep waters. The historical record should be studied from sediment cores for changes in species composition through direct examination of undecomposed hard parts, pigment degradation products, and changes in the deposition of phosphorus. Wetzel and Likens (1979, pp. 306 and 317) describe laboratory methods for sediment analysis and for depletion of hypolimnetic oxygen (p. 317). But see also Lind (1979, p. 149). Otherwise, refer to specific methods for nutrients, biomass, productivity, and so on.

Sampling and analysis. Eutrophication studies are long-term projects requiring many skills. A careful search of the research literature, reports, fishery projects, and water supply records will be helpful. Measurement of active processes such as nutrient budgets and production must be repeated for several years because of great year to year variation. Sedimentation studies avoid the possible bias by sampling only regions of a basin that serve as sinks with great accumulations or those that are scoured by currents. High skills are required for proper study design. Many of the analytical procedures require only moderate skills. Some taxonomic aspects will require skills of a taxonomic authority. The duration of a eutrophication study will make it relatively expensive although much of the discrete data can be gathered inexpensively.

3.5.5 Ecosystem indices.

Continued high diversity is often assumed to be a desired characteristic of natural communities, since it means that there is a maintenance of rare species, for their own sake as well as for possible sources of genetic material in the future. Detrimental impact can result in decreased diversity as well as decreased abundance of organisms. It is often useful to measure diversity of communities or to monitor changes in diversity after some impact.

The term diversity is used to describe the observations that (i) some assemblages are composed of small sets of species and (ii) some are dominated by one or a few species, even though there may be many rare ones. Low diversity occurs when either of the above conditions are true, although the two are very different. Samples could be compared strictly by the number of species (species richness), by the way in which individuals are distributed among species (evenness) (Fager, 1972; Peet, 1974; Pielou, 1977) or by species equitability (Lloyd and Ghelardi, 1964).

Diversity indices attempt to combine concepts of species richness and evenness (Fager, 1972; Peet, 1974; Pielou, 1977; Grassle et al., 1979). The simplest is Simpson's index which measures the probability that two individuals drawn at random from a population are of the same species. Most other measures of diversity quantify information on a system or sample. The common ones are as follows: (i) Brillouin's index used for an entire population of individuals, not a random sample from a larger population; and (ii) Shannon-Weaver diversity index, used for a random sample of an effectively infinite population; it gives more weight to the presence or absence and abundance of rare species than does Simpson's index. In calculating the Shannon-Weaver index, it is assumed that all species in the population also occur in the sample. This is not usually true and a correction must be made (Peet, 1974).

There are two considerable problems with diversity measures: scale and significance. As sample size and number of samples increase the estimate of true diversity will reach a plateau. It is useful to find out, for each assemblage or community, at what values this occurs (Loya, 1978; Weinberg, 1978). By so doing the investigator can validate his sampling procedures.

Statistical comparison of diversity between areas or samples is difficult. One technique (Patil and Taillie, 1979) is to calculate a series of related indices of diversity to determine whether they all show differences in the same direction. Species richness (number), relative abundances of particular species, and individual species population densities are important components of diversity and can be compared by standard statistical techniques (analysis of variance or nonparametric methods) (Sokal and Rohlf, 1969).

Multivariate comparisons between samples. Clustering techniques allow comparison of similarities in samples taken at different times and locations. For example, if a sampling program for describing benthic faunal assemblages in a specific depth range and geographical area includes monthly or more frequent sampling, it is often of interest to know whether or not there are distinct habitat patches within the area and whether or not there are discrete temporal assemblages such as a characteristic summer or winter fauna or a wet and dry season fauna. It might also be important to determine whether or not samples from a perturbed area show clustering effects or natural variability, as shown by samples from outside the affected region. It must be noted that large inherent variations in faunal density and species composition will make the detection of differences due to some specific impact difficult to determine. The number of replicate samples needed to find such differences can be prohibitively large (Kaufman et al., 1980).

The applications of the above types of multivariate statistical methods are described by Gauch and Whittaker (1972), Sneath and Sokal (1973), Rohlf (1972), and Poole (1974). A brief description of some statistical tests and techniques, suggested applications, and references in which the techniques was used follows.

1. *Multiway analysis of variance (ANOVA).* This technique can be used in conjunction with various clustering methods to test for statistically significant differences between groups of samples differing in a number of characteristics (two-way ANOVA) or between experimental treatments and controls (three-way ANOVA) as well as for significant interactive

(synergistic) effects between variables affecting sample parameters (Sokal and Rholf, 1969; Zar, 1974; Kaufman et al., 1980). Three-way ANOVA was used to study effects of different petroleum products on clam growth at several salinities and temperatures. Two-way was used to study densitites of clam species in samples taken at several depths and from several sediment types.

2. *Multiple regression analysis.* This is a type analysis of variance in which variables are continuous or almost continuous and is used to test the effect of independent variables on a dependent variable. Regression analysis is used to study the effect of each independent variable on total variance and additivity of effects is assumed. For example, the effects of depth and salinity on population density or mean size of samples of a clam population collected along a depth gradient where salinity varies both within and across depths might be tested (Sokal and Rohlf, 1969; Zar, 1974; Kaufman et al., 1980).

3. *Analysis of covariance.* This analysis is similar to the multiple regression except that no assumptions are made about dependence or independence of variables. Tests of correlation are used to sort out the relation of each variable to another, with the variance due to each factored out. In many cases, use of this technique can minimize sample sizes necessary to detect small significant differences due to any variable. This technique could be used in studies of biomass of infaunal organisms collected at various depths at which sediment grain size varies. The relation between biomass and depth, biomass and grain size, grain size and depth, can all be analyzed. The example given in multiple regression analysis above could also be analyzed by this method (Sokal and Rohlf, 1969; Zar, 1974; Kaufman et al., 1980).

4. *Similarity indices and dendrograms.* Similarity indices can be used to compare relative or absolute abundance (number, biomass, or physical parameters) in samples and indicate groupings or clusters of samples that are similar in biotic or abiotic characteristics. The basic technique is (i) calculate similarity between pairs of samples by one of a variety of indices, (ii) construct a matrix of similarity indices, (iii) combine groups that are most similar and generate a new matrix for combined groups, and (iv) construct a dendrogram from the values obtained in (ii) and (iii); vertical distance on the chart is a measure of dissimilarity (more similar samples branch lower on the diagram) (Fig. 3.5). In this figure, those sites or groups of sites which branch closest to the bottom of the diagram are the most similar in species composition. Similarity was calculated using Sorenson's index (Jokiel and Marigos, 1978, p. 81).

Although this technique produces an easily interpreted visual presentation of sampling results, statistically significant differences between samples must be analyzed by appropriate parametric or nonparametric tests.

Clustering methods have been reviewed by Sneath and Sokal (1973); Clifford and Stephenson (1975); Neff and Marcus (1980). Examples of their use in ecological studies can be found in Stephenson and Williams (1971), Boesch (1973 and 1977a); Whitlatch (1977), Jokiel and Maragos (1978), Hartzband and Boesch (1979), and Sutherland (1980).

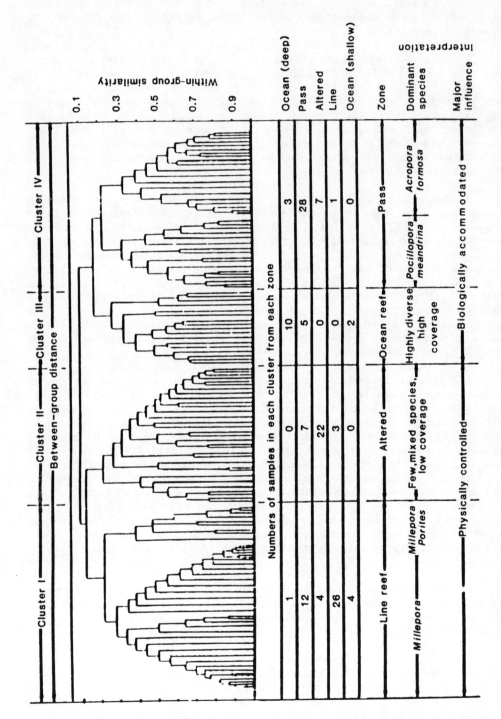

Figure 3.5 Dendrogram (cluster diagram) for comparing similarities in biotic and abiotic sample groupings derived from sites along a coral reef transect. [Source: Jokiel and Maragos (1978)]

5. *Ordination techniques - continuous multivariate methods.* Noy-Meier and Whittaker (1977) point out two main reasons for carrying out multivariate analyses on community samples. The first is to identify groups of related sites (or species) or typical groups without discrete boundaries between them. The second is to identify and define sequences of community composition (coenoclines) that vary along identified environmental gradients. Ordination includes any method of arranging sites (samples, quadrats), or species, by similarities and in relation to one or more environmental (or succession) gradients. In the broad sense, arrangement of such samples along any set of defined axes can be an ordination. The steps are to (i) set up a community composition matrix for species scores (presence, abundance, cover) by sites (samples); (ii) assign sites and species to one of several continuous variables (axes, components, dimensions, gradients); and (iii) present compositional similarities in condensed form. Any technique must take a large number of correlated variables and reduce them to a small number of significant but uncorrelated variables (Poole, 1974).

Two types of sample comparisons are recognized. In the first (R type), a matrix of similarities between species is defined, such as species characters in numerical taxonomy, species occurrences, or ecological characteristic. The matrix can then be used in comparisons of sites or samples. In the other method (Q type) a matrix of similarities between sites or samples is based on abundances (for example, presence and absence or percentage cover) or species in each of the sites or examples. For example, quadrats in a salt marsh may be quantified by percentage of cover of each plant species. Similarity between samples is calculated and plotted on axes that represent sample extremes along intertidal elevation and salinity gradients. For comparisons between methods, see Whittaker and Gauch (1973); Kessell and Whittaker (1976); Noy-Meier and Whittaker (1977). Computer programs are available for ordination (Green, 1980). The following methods have been used in community ordination.

Polar ordination (Wisconsin polar or Bray-Curtis polar). Polar ordination is used in comparing sites or samples taken along a continuum of one or more environmental gradients. In the unmodified analysis, the most dissimilar samples are taken as the end points of the first ordination axis. Similarities are calculated by methods such as a coefficient of dissimilarity, a percent dissimilarity, and the Euclidian distance. Kessell and Whittaker (1976) used all these techniques on a computer-generated community along a gradient with normally distributed species and found that the coefficient of dissimilarity gave the most consistent match between calculated ordination and the real position along the coenocline.

One major limitation of the technique is dependence on "outliers" or on sampling accidents to set the axes. Whittaker and Gauch (1973) suggest that end points of samples from known environmental gradients be used to set the axes rather than the most different samples. This may introduce some error but will eliminate the arbitrary nature of the standard axes. For this reason, the use of reciprocal averaging in stream surveys has been suggested. Kessell and Whittaker (1976) found polar ordination to be the most useful ordination technique for samples from a coenocline; they also report it to be the least subject to distortion from high between-sample diversity and high variability between repeated sampling or trials (noise) in comparison with principal components analysis and discriminant function analysis. Noy-Meier and Whittaker (1977) note that there are several recent modifications of the methodology for assigning axes in polar ordination that make the technique less subject to outlier samples and that might improve its general usefulness. (Poole, 1974; Kessell and Whittaker, 1976; Noy-Meier and Whittaker, 1977; Whittaker, 1978; Green, 1980.

Principal components analysis. Principal components analysis is a technique to find variables (principal components) that can account for most of the variance in a set of samples. These principal components will probably not correspond to real environmental gradients but to transforms of sets of species abundance data for community samples. The species abundance composition of each sample is used as a point in a space where each species is a dimension. Obviously, this is too difficult to either work with directly or to display in any meaningful way. The principal components are thus defined such that the first one is as variable as possible, the second one is "orthogonal" to the first, and the second is a variable as possible given the orthogonal constraint. Other principal components are defined in a similar way. Usually, the first two or three principal components will account for 99 percent of the sample variance and thus define the important relations between samples. A two-dimensional plot of sample positions along axes corresponding to the first two principal components allows comparison of groups and relationships in a space that accounts for a large proportion of the variance.

Kessell and Whittaker (1976) suggest that this technique does not work well for community sample ordination with any level of between-sample diversity or variability between replicates and that it is most appropriate for R type (species-based) comparisons. They note that it has been used successfully to break down sets of samples into recognizable "community types" and that it is useful for such "clustering" applications. (Poole, 1974; Kessell and Whittaker, 1976; Noy-Meier and Whittaker, 1977; Whittaker, 1978).

Canonical correlation analysis. This technique can be used to make clusters by indicating similarily between samples along axes of "canonical variables" (calculated to describe contribution to variance between samples). Variables are identified in the analysis that contribute most to between-station differences relative to within-station differences when the sampling design is nested in that fashion. The technique is probably not appropriate to Q-type community sample ordination according to Noy-Meier and Whittaker (1977) although it can be used to determine clusters of similar samples (Noy-Meier and Whittaker, 1977; Whittaker, 1978; Jokiel and Maragos, 1978).

Discriminant function analysis. In this analysis individuals (representing categories) are compared; attributes of individuals in each category are assumed to be normally distributed. It provides a set of between-sample distances that can be used or ordination, although it is better suited to species comparison (R-type) than to the ordination of community samples, according to Kessell and Whittaker (1976). They note that the technique produces a poor fit between the ordinated position of samples and their real position along a coenocline generated in a computer simulation study. Although the method is mathematically sophisticated in comparison with polar ordination or principal components analysis, its usefulness in ecological studies has not been shown (Sneath and Sokal, 1973; Kessell and Whittaker, 1976; Noy-Meier and Whittaker, 1977; Whittaker 1978; Green, 1980).

Reciprocal averaging. This relatively new technique allows sorting of samples according to species composition by repeated application of a ranking and averaging technique. It uses neither gradients nor axes and is thus easy to use on even small computers. It has gained acceptance and increased application over the past few years and may replace polar ordination (Whittaker, 1978).

3.5.6 Water balance.

All of the many methods for measuring water balance have limitations of which one of the most serious is the availability and reliability of data. Direct measurements of water loss are difficult because it is often not possible to replicate real world conditions, and indirectly, measurements are often incomplete or inacurate. In the tropics, sparse meteorological networks, manpower, and financial limitations add to the problems.

Methods of assessment can be classified as follows: (i) direct measurements, which use evaporation pans, atnometers, and lysimeters; (ii) meteorological formulas, which use aerodynamic methods, energy budget methods, a combination of these two, or empirical formulas; and (iii) moisture-water budget methods.

Jackson (1977a) reviews the general principles and limitations of the most commonly used methods. He notes that the degree of simplification of water balance studies is related to the purpose of the study, the available data, and the facilities available to conduct analyses. He also presents some of the major applications of water balance studies, which can be summarized as follows:

1) To provide a general overview of the water conditions in an area.

2) To form part of a model for investigating rainfall-runoff interactions and stream flow predictions from climatic data.

3) To assess the suitability of an area for a particular crop or the suitability of a crop to an area.

4) To assess irrigation requirements, both quantity and interval.

5) To examine water-yield relationships.

6) To assess water use by a particular vegetation or crop type.

7) To assess man's impact on an ecosystem.

3.6 Methods for Water Resource Demands

The principal measures of human water use are water withdrawals (the water taken from a surface or groundwater source for off stream use) and consumption (the water withdrawn which is not returned to a surface or groundwater source). These two measures give very different indications of the demands a given use places on water supplies. In the United States, for example, less than one-third of total fresh-water withdrawals are consumed, but the range varies from less than 2 percent for stream electric to 100 percent for livestock use. Although both measures provide useful information about human water use, they have limited value for determining the pressures human activities place on water resources. Both measures ignore water uses such as navigation, preservation of riparian plants and animals, hydroelectric power generation, and waste disposal, all of which do not involve water withdrawals or off stream consumption but may be competitive with other uses. Moreover, neither measure takes into account the qualitative impacts of human uses. In particular, if the wastes delivered to a water body exceed the assimilative capacity of the system, the value of the water for other purposes will be degraded.

Estimating water use for project planning traditionally has been a relatively simple process which ignores the complexities of human impacts on water supplies and the opportunities for influencing use through alternative pricing structures or management institutions. The usual approach is to assume a given quantity of water is required for a particular purpose such as irrigation or domestic supplies. Total projected water use simply multiplies the per unit (for example, per acre or per person) requirements by the appropriate number of units to be serviced by the project. This approach implicitly assumes the ratio of water to output is insensitive to changes in the prices of water, other inputs, or the final product. Furthermore, the required quantity of water is commonly determined as the amount that would be demanded and used efficiently only if it were a free commodity. Indeed, when water is not metered (which is frequently the case), it is viewed and treated as a free good by the user. With irrigation projects, for example, water requirements commonly are estimated as the quantity of water needed to maximize the yield of a particular crop (on the basis of local growing conditions and existing water distribution methods). Since farmers often are not charged by the quantity of water received for irrigation and may not even have any control over that quantity, they have no incentive to apply less than the yield-maximizing quantities. Yet, if water is scarce, an efficient allocation and management of water supplies would likely call for less water use per acre (enabling limited supplies to be spread over more acres or diverted to other uses) and for some substitution of labor and capital resources for water resources to improve distribution efficiency.

The concept of a demand for water that varies with price is rarely introduced into water planning. This planning deficiency stems in part from some of the features of water that complicate the tasks of using it efficiently and of estimating the demand for it. Complicating features include: some water uses can not be priced and, therefore, do not show up in usual demand estimates; most uses are either not consumptive or only partially consumptive of the water used, but they may alter the value of the return flows for other potential users; water is often a fugitive resource flowing from one property to another making it difficult to establish clear property rights to such resources; and transfer of water among users tend to have important impacts on those who are not direct parties to the transaction. All these features limit the usefulness of an unfettered market system for allocating scarce water resources and make it difficult to determine the impacts of alternative charges on water use.

Within developing areas, efforts to estimate the sensitivity of water demand to price have been notably unsuccessful because of the combination of the complicating features mentioned above and the inadequacies of the data on water use. The use of questionnaires to estimate the willingness to pay for water also has been useless (Sanders and Warford, 1976).

Despite the apparent futility of obtaining reliable estimates of the responsiveness of water demand to price, it is important to recognize that water use does depend on price or the incentives users have to conserve, and the efficient use of the resource depends on the existence of appropriate incentives. Where water is viewed as a scarce resource, some effort should be made to determine the response of the principal water users to alternative water charges and the institutional arrangements required to provide the incentives and opportunities for efficient use. Even if the project does not include water management institutions, which permit users to control their water withdrawals and costs, demand analysis might indicate the benefits of delivering less than the yield-maximizing quantities of water to large consumers.

There are established techniques for analyzing the demand for offstream consumptive uses of water for domestic, industrial, mining, and agricultural purposes. The bibliography lists sources that describe and evaluate alternative demand estimation techniques. Two references (Wollman and Bonem, 1979; U.S. Water Resources Council, 1978) project national and regional water use for the United States by estimating requirements. Others describe the complexity of and the data requirements and methodologies for relating water use to water costs or prices. Bohi (1971, chapters 1 and 2) provides a good general discussion of the issues and methodologies associated with estimating resource demand, and several of the other references describe studies relating specific water uses to price (see Young and Grey, 1972; Ruttan, 1965; and Howe and Linnaweaver, 1967).

Several investigators have examined the opportunities for and costs of altering the water use associated with a given industry. Cootner and Lof (1965) identify the technological opportunities for and costs of water recirculation and water quality adjustment in thermal plants; Lof and Kneese (1968) examine the economics of water use in the beet sugar industry; and Russell and Vaughan (1976) examine water use in steel production. These studies indicate the importance of engineering and institutional factors in determining the pressures placed on aquatic ecosystems.

Special Methods: Current Research and Expected Future Methods

4.1 Introduction

It has been emphasized throughout this discussion that the hydrological cycle provides a framework for analyzing aquatic ecosystems. It has also been emphasized that analysis at varying scales is required and that new methods, some continental in scope and others at microscopic levels, will offer important new data sources. Of equal importance is work now underway on modeling of aquatic ecosystems and the complex relationships involved. For example, in atmospheric science, advances are expected to be made in such subjects as mesoscale meteorological modeling, climate modeling, air chemistry, and weather and climate forcasting. Only some of the new methods and data sources are discussed briefly in this section, including weather modification, use of remote sensing, modeling of the relationships between climatic factors and economic variables, and aquatic microbiology. Our knowledge of physical and biotic components of ecosystems should continue to improve and eventually expand on our knowledge of the sensitivity of ecosystems to human involvement.

Weather and climate forecasting will advance slowly, but even now existing data and information are underutilized. Closer communication between experts in different disciplines could result in significant improvements in resource management and development planning. In many cases, the constraints are human, institutional, and political, and we do not have to wait for further technological advances to do a better job.

4.2 Remote Sensing

Remote sensing deals with more than data just from satellites. In many cases low altitude aerial obliques are a cost effective approach.

For physical mapping of watersheds the aerial and satellite methods are already cost effective in many areas. For resource assessment, phenomena such as floods and snow cover are also conveniently mapped in aerial extent and current research is underway to enable quantitative evaluation of depths and water yields. Excellent summaries of recent work in satellite hydrology are given in Deutsch (1981) and specifically for rainfall monitoring, by Barrett and Martin (1981). Table 4.1, drawn from Barrett and Martin (1981), gives the current and the future expected status of climate parameter monitoring according to NASA. Eight case studies in marine environmental management are given by Behie and Cornillon (1981).

4.3 Weather Modification

One area in which environmental project officers may find themselves involved is weather modification, particularly attempts at the artificial stimulation of rainfall. It is not intended here to present a review of the various methods that have been and currently are used to modify weather, and indeed climate; these are covered in Hess (1974) and Wegman and DePriest (1980). The point of emphasis is that such attempts, usually extremely costly, should be regarded as experimental. Particularly in developing countries, results of experiments are extremely variable and interpretation is difficult. In some areas increases in rainfall of 10 to 20 percent at an acceptable level of statistical significance are recorded, but in others there has been no significant increase, and decreases have sometimes been found. It is recommended, therefore, that project officers exercise caution both in advocating weather modification

TABLE 4.1 Status of climate parameter monitoring, according to NASA. [Source: Barrett and Martin (1981)]

		Climate type			Can requirement be met by...		
No.	Parameter	A	B	C&X	Current systems?	Approved future systems?	Climate observing system (1980's)
Weather Variables							
1.	Temperature Profile	↑	↑	↑	[Standard Weather station obsns.]	Yes	Yes
2.	Surface Pressure		Basic			(No remote technique)	
3.	Wind Velocity		FGGE Measurement			(from cloud motions)	
4.	Sea Surface Temperature					Yes	Yes
5.	Humidity	↓	↓	↓		Yes	Yes
6.	Precipitation	✓	✓	✓		No	No
7.	Cloud Cover	✓	✓	✓		Yes	Yes
8.	Boundary Layer Stability	✓					
Ocean Parameters							
4a	Sea Surface Temperature	✓	✓	✓	No	Yes	Yes
9.	Evaporation		✓	✓	No	No	No
10.	Surface Sensible Heat Flux		✓	✓	No	No	No
11.	Wind Stress		✓	✓	No	No	No
12.	Sea Surface Elevation			✓	No	Maybe	Maybe
13.	Upper Ocean Heat Storage			✓	No	No	No
14.	Temperature Profile			✓	Partially	Partially	Partially
15.	Velocity Profile			✓	Partially	Partially	Partially
Radiation Budget							
7a	Clouds (Effect on Radiation)		✓	✓	Almost	Yes	Yes
16.	Regional Net Radiation Components		✓	✓	No	No	Yes
17.	Equator-Pole Gradient		✓	✓	No	No	Yes
18.	Surface Albedo		✓	✓	Yes	Yes	Yes
19.	Surface Radiation Budget	✓	✓	✓	Partially	Partially	Partially
20.	Solar Constant		✓	✓	No	Yes	Yes
21.	Solar Ultraviolet Flux		✓	✓	No	Yes	Yes
Land, Hydrology, and Vegetation							
6a	Precipitation	✓	✓	✓	No	No	No
18a	Surface Albedo		✓	✓	Yes	Yes	Yes
22.	Surface Soil Moisture	✓	✓	✓	No	Partially	Yes
23.	Soil Moisture (Root Zone)	✓	✓	✓	No	No	No
24.	Vegetation Cover (Non-Forest)	✓	✓	✓	Almost	Yes	Yes
25.	Evapotranspiration	✓	✓		No	No	No
26.	Plant Water Stress	✓	✓		No	No	No
Cryosphere Parameters							
27.	Sea Ice (% Open Water)	✓	✓	✓	Yes	Yes	Yes
28.	Snow (% Coverage)	✓	✓	✓	Yes	Yes	Yes
29.	Snow (Water Content)	✓	✓	✓	No	Yes	Yes
30.	Ice Sheet Surface Elevations			✓	No	Maybe	Yes
31.	Ice Sheet Horizontal Velocity			✓	Yes	Yes	Yes
32.	Ice Sheet Boundary		✓	✓	Yes	Yes	Yes
Atmospheric Composition							
21a	Solar Ultraviolet Flux			✓	No	Yes	Yes
33.	Stratospheric Aerosol Optical Depth	✓			No	Almost	Almost
34.	Tropospheric Aerosol Optical Depth	✓		✓	No	No	No
35.	Ozone	✓		✓	No	Yes	Yes
36.	Stratospheric H₂O			✓	No	Yes	Yes
37.	H$_2$O, NO$_x$	✓		✓	Yes	Yes	Yes
38.	CO$_2$			✓	Yes	Yes	Yes
39.	CFM's			✓	Yes	Yes	Yes
40.	CH$_4$			✓	Yes	Yes	Yes

projects and in evaluating the claims of weather modification consultants. Besides doubts as to suitable technology, there are also many uncertainties as regards the economics of weather modification and major legal issues remain unresolved.

4.4 Climate-Economic Modeling

An emerging practice is the development of models to project the effects of climate fluctuations on national economies. Water resources are important functions of climate and play an important role in these econometric and input-output models. The models do make it possible to explore relationships between environmental and economic variables. Fuller understanding of these relationships could assist in the design of programs to redress, for example, the economic hardships associated with droughts and floods (Cooter, 1980).

4.5 Aquatic Microbiology

New methods in aquatic microbiology are developing rapidly. Sophisticated biochemical techniques are being used to determine metabolic processes mediated by microorganisms. Radioisotopic techniques, and particularly use of tritium-labeling, are increasingly being used. High pressure liquid chromatography permits analysis of very low concentrations of organic toxicants. This method is also beginning to be applied in assessing quantitatively specific microbial populations both in sediments and in the water column. New methods are also being developed for the analysis of pathogens. Rapid identification kits, with prepackaged test chemicals, are beginning to appear on the market and should facilitate identification of pathogens.

Analysis of microbial processes is dependent on detection of very small quantities of chemical products. The development of microprocessors and other computer-assisted techniques has sharply increased the accuracy and range of these methods. It is probable that new methods that use advanced physical techniques, such as laser spectroscopy, will permit much better analysis of microbial processes in aquatic ecosystems.

References Cited

Acheson, J.M. et al. 1980. *The fishing ports of Maine and New Hampshire: 1978.* Orono, Maine: Univ. of Maine Sea Grant Publications.

American Public Health Association. 1975. *Standard methods for the examination of water and wastewater.* 14th ed. Washington, D.C.: American Public Health Association.

American Public Health Association. 1980. *Standard methods for the examination of water and wastewater.* 15th ed., with supplement. Washington, D.C.: American Public Health Association.

Amesbury, S.S., R.T. Tsuda, W.J. Zolan, and T.L. Tansy. 1975a. *Limited current and underwater biological surveys of proposed sewer outfall sites in the Marshall Island District.* Ebeye, Kwajalein Atoll. University of Guam Marine Laboratory Tech. Report No. 22. 30 pp.

Amesbury, S.S., R.T. Tsuda, W.J. Zolan, and T.L. Tansy. 1975b. *Limited current and underwater biological surveys of proposed sewer outfall sites in the Marshall Island District.* Darrit-Uliga-Dalap area, Majuro Atoll. Technical Report No. 23. Agana, Guam: University of Guam Marine Laboratory.

Amesbury, S.S., R.T. Tsuda, R.H. Randall, C. Birkeland, and F. Cushing. 1976. *Limited current and underwater biological surveys of the Donitsch Island sewer outfall site, Yap Western Caroline Islands.* Tech. Report No. 24. Agana, Gaum: University of Guam Marine Laboratory.

Babcock, R.H. and J.J. Matera. 1973. *The two books on the water supply of the city of Rome of Sextus Julius Frontinus.* Boston: New England Water Works Association.

Bagenal, T., ed. 1978. *Methods for assessment of fish production in fresh waters.* 3rd ed. I.B.P. Handbook No. 3. Oxford: Blackwell Scientific Publ.

Barnes, I. 1964. Field measurement of alkalinity and pH. *Wat.- Supply Irrig. Pap. Wash.* 1535-H.

Barnes, R.B. 1975. The determination of specific forms of aluminium in natural waters. *Chem. Geol.* 15:177-191.

Barnes, R.S.K. 1974. *Estuarine biology.* London: Edward Arnold.

Barrett, E. and D. Martin. 1981. *The use of satellite data in rainfall monitoring.* London: Academic Press.

Barry, R.G. and R.J. Chorley. 1978. *Atmosphere, weather and climate.* 3rd ed. London: Methuen.

Bascomb, W.J. 1964. *Waves and beaches.* Garden City, N.Y.: Doubleday.

Bazigos, G.P. 1974. The design of fisheries statistical surveys--inland waters. *FAO Fish. Tech. Paper 133.* Rome: FAO.

Behie, G. and P. Cornillon. 1981. *Remote Sensing, a tool for managing the marine environment: Eight case studies.* Ocean Engineering, NOAA/SEA Coast, University of Rhode Island Marine Technical Request 77.

Benton, A.R., Jr. and R.M. Newman. 1976. Color aerial photography for aquatic plant monitoring. *J. Aquatic Plant Management* 14:14-16.

Berg, G., ed. 1978. *Indicators of viruses in water and food.* Ann Arbor, Mich.: Ann Arbor Science.

Birkeland, C. 1974. Interactions between a sea pen and seven of its predators. *Ecol. Monogr.* 44:211-232.

Birkeland, C., A.A. Reimer and J.R. Young. 1976a. *Survey of marine communities in Panama and experiments with oil.* PB 253 409/7GA. EPA Report 600/3-76/028. Springfield, Va.: National Technical Service.

Birkeland, C., R.T. Tsuda, R.H. Randall, S. S. Amesbury, and F. Cushing. 1976b. *Limited current and underwater biological surveys of a proposed sewer outfall site on Malakal Island, Palau.* Tech. Report No. 25. Agana, Guam: University of Guam Marine Laboratory.

Bobbie, R.J. and D.C. White. 1980. Characterization of benthic microbial community structure by high resolution gas chromatography of faty acid methyl esters. *Appl. and Environ. Microbiol.* 39:1212

Boesch, D.F. 1973. Classification and community structure of macrobenthos in the Hampton Roads area, Virginia. *Mar. Biol.* 21:226-244.

Boesch, D.F. 1977a. Application of numerical classification in ecological investigations of water pollution. *Special Scientific Report* No. 77, Virginia Institute of Marine Science, Grant No. R803599-01-1. Corvallis, Ore.: U.S. Environmental Protection Agency.

Boesch, D.F. 1977b. A new look at the zonation of benthos along an estuarine gradient. In *Ecology of marine benthos*, ed. B. C. Coull, pp. 245-266. Columbia, S.C.: Univ. of S. Carolina Press.

Bohi, D.R. 1971. *Analyzing demand behavior: A study of energy elasticties.* Baltimore, Md.: The Johns Hopkins Univ. Press for Resources for the Future.

Bott, T.L. 1973. Bacteria and the assessment of water quality. In *Biological methods for the assessment of water quality*, eds. J. Cairns, Jr. and K. L. Dickson, pp. 61-75. Proceedings of a symposium held 26-29 June 1972 in Los Angeles, Calif. ASTM/STP 528. Philadelphia: American Society for Testing and Materials.

Bourne, P. 1982. Clobal eradication of guinea work, editorial. *J. Royal Society Medicine* 75:1-3.

Boyle, E.A. and J.M. Edmond. 1977. Determination of copper, nickel and cadmium in sea water by APDC chelate precipitation and flameless atomic absorption spectrometry. *Analyt. Chim. Acta* 91:189-97.

Bradley, D.J. 1977. Health aspects of watersupplies in tropical countries. In *Water, wastes and health in hot climates*, eds. R. Feachem, M. McGarry and D. Mara, pp. 3-17. New York: Wiley-Interscience.

Brandt, A. von. 1972. *Revised and enlarged fish catching methods of the world.* London: Fishing News (Books).

Briscoe, J. 1978. The role of water supply in improving health in poor countires (with special reference to Bangladesh). *Amer. J. Clinical Nutrition* 31:2100-2113.

Buddemeier, R.W. 1978. Coral growth: retrospective analysis. In *Coral reefs: Research methods*, eds. D. R. Stoddart and R.E. Johannes, pp. 551-572. Paris: UNESCO.

Buss, L.W. 1980. Competitive intransitivity and the size-frequency distributions of interacting populations. *Proc. Natl. Acad. Sci.* 77:5355-5359.

Cabelli, V.J. 1979. What do water quality indicators indicate? In *Aquatic microbial ecology*, eds. R.R. Colwell and J. Foster, pp. 305-336. Proceeding of American Society of Microbiology Conference. UM-TS-80-03. College Park, Md.: University of Maryland Seagrant College Program.

Cairns, J., Jr. and K.L. Dickson. 1971. A simple method for the biological assessment of the effects of waste discharges on aquatic bottom-dwelling organisms. *J. Water Pollution Control Federation* 43:755-772.

Cairns, J., Jr., K.L. Dickson and G. Lanza. 1973. Rapid biological monitoring systems for determining aquatic community structure in receiving systems. In *Biological methods for the assessment of water quality*, eds. J. Cairns, Jr. and K.L. Dickson, pp. 148-163. ASTM/STP 528. Philadelphia: American Society for Testing and Materials.

Cairns, J., Jr., G.P. Patil, and W.E. Waters. 1979. *Environmental biomonitoring, assessment, prediction, and management.* Certain case studies and related quantitative issues. Statistical Ecology Series. Vol. 11. Burtonsville, Md.: International Co-operative Publishing House.

Carlson, R.E. 1977. A trophic state index for lakes. *Limnol. and Oceanogr.* 22:361-369.

Cawse, P.A. and D.H. Pierson. 1972. An analytical study of trace elements in the atmospheric environment. *Report R.7134*. Harwell, England: Atomic Energy Research Establishment.

Chandler, T.J. 1970. *The management of climatic resources.* London: Lewis and Co. for the Univ. of London.

Chapman, D.G. and V. Gallucci. 1979. *Quantitative population dynamics.* Statistical Ecology Series. Vol. 13. Burtonsville, Md.: International Co-operative Publishing House.

Chet, I. and R. Mitchell. 1976. Petroleum hydrocarbons inhibit decomposition of organic matter in seawater. *Nature* 261:308.

Chow, V.T. 1974. *Handbook of applied hydrology.* New York: McGraw Hill.

Clifford, H.T. and W. Stephenson. 1975. *An introduction to numerical classification.* New York: Academic Press.

Clokie, J.J.P. and A.D. Boney. 1980. The assessment of changes in intertidal ecosystems following major reclamation work: framework for interpretation of algal-dominated biota and the use and misuse of data. In *The Shore Environment*, Vol. 2, eds. J.H. Price, D.E.G. Irvine, and W.F. Farnham, pp. 609-676. London: Academic Press.

Cohen, J.E. 1978. *Food webs and niche space.* Monographs in Population Biology No. 11. Princeton, N.J.: Princeton Univ. Press.

Connell, J.H. 1966a. Effects of competition, predation by *Thais lapillus* and other factors on natural population of the barnacle *Balanus balanoides*. *Ecol. Monogr.* J1:61-104.

Connell, J.H. 1966b. The influence of interspecific competition and other factors on the distribution of the barnacle *Chthamalus stellatus*. *Ecology* 42:710-723.

Connell, J.H. 1970. A predatory-prey system in the marine intertidal region. I. *Balanus glandula* and several predatory species of *Thais*. *Ecol. Monogr.* 40:49-78.

Connell, J.H. 1972. Community interactions on marine rocky intertidal shores. *Ann. Rev. Ecol. and Sys.* 3:169-192.

Connell, J.H. 1973. Population ecology of reef-building corals. In *Biology and geology of coral reefs*, Vol 2, eds. O.A. Jones and R. Endean. New York: Academic Press.

Cootner, P.H., and G.O.G. Lof. 1965. *Water demand for steam electric generation: A economic projection model.* Baltimore, Md.: The Johns Hopkins Univ. Press for Resources for the Future.

Cooter, W. 1980. Possible applications of input-output models in climatic impact assessments. In *The economic impact of climate*, Vol. I, pp. 167-191. Proceedings of two workshops on the structure of economic models, held in May 1980, Oklahoma.

Coull, B.C., ed. 1977. *Ecology of marine benthos.* Columbia, S.C.: Univ. of S. Carolina Press.

Cowardin, L.M., V. Carter, F.C. Golet and E.T. LaRoe. 1979. *Classification of wetlands and deepwater habitats of the United States.* FWS/OBS-79/31. Washington, D.C.: U.S. Dept. of the Interior, Fish and Wildlife Service.

Critchfield, H.J. 1974. *General Climatology.* Englewood Cliffs, New Jersey: Prentice Hall.

David, D.J. 1975. Magnesium, calcium, strontium and barium. In *Flame emission and atomic spectroscopy*, Vol. 3, *Elements and matrices*, eds. J.A. Dean and J.C. Rains, pp. 33-64. New York: Marcel Dekker.

Dayton, P.K. 1971. Competition, disturbance, and community organization: the provision and subsequent utilization of space in a rocky intertidal community. *Ecol. Monogr.* 41:351-389.

Dayton, P.K. 1975. Experimental evaluation of ecological dominance in a rocky intertidal algal community. *Ecol. Monogr.* 45:135-159.

Deutsch, M. 1981. *Satellite hydrology.* Minneapolis, Minn.: American Water Resources Assoc.

Dillon, P.J. and R.H. Rigler. 1974. A test of a simple nutrient model predicting the phosphorus concentration in lake water. *J. Fisheries Res. Board, Canada* 31:1771-1778.

Done, T.J. 1977. A comparison of units of cover in ecological classifications of coral communities. In *Proceedings: Third Int. Coral Reef Symp.* Vol. 1, ed. D. L. Taylor, pp. 9-14. Miami, Fla.: University of Miami Rosensteil School of Marine and Atmos. Science.

Dunne, T. and L.B. Leopold. 1978. *Water in environmental planning.* San Francisco: W.H. Freeman

Eckholm, E. 1976. *Losing ground.* New York: Norton.

Edmondson, W.T. 1980. Secchi disk and chlorophyll. *Limnol. and Oceanogr.* 25:378-379.

El-Ashry, M.T. 1980. Groundwater salinity problems related to irrigation in the Colorado River basin. *Groundwater* 18:37-45.

Ellwood, D.C., J.N. Hedger, M.J. Latham, J.M. Lynch and J.H. Slater, eds. 1981. *Contemporary microbial ecology.* London: Academic Press.

Eltringham, S.K. 1971. *Life in mud and sand.* London: The English Univ. Press.

Emery, A.R. 1968. Preliminary observations on coral reef plankton. *Limnol. and Oceanogr.* 13:293-303.

Erickson, P.S. 1979. *Environmental impact assessment. Principles and applications.* New York: Academic Press.

Erhlich, P.R. 1977. *Ecoscience: Population resources in environment.* San Francisco: W. H. Freeman.

Environmental Protection Agency. 1973. *Biological field and laboratory methods for measuring the quality of surface waters and effluents.* EPA-670/4-73-001. Environmental Protection Agency. Springfield, Va.: National Technical Information Service.

Environmental Protection Agency. 1979a. *Methods for chemical analysis of water and wastes, 1978.* EPA-600/4-79-020. Environmental Protection Agency, Cincinnati, Ohio. Springfield, Va.: National Technical Information Service.

Environmental Protection Agency. 1979b. *Handbook for quality control in water and wastewater laboratories.* EPA-600/4-79-019. Environmental Protection Agency, Cincinnati, Ohio. Springfield, Va.: National Technical Information Service.

Fager, E.W. 1972. Diversity: a sampling study. *Amer. Nat.* 106:293-310.

Fanning, K.A. and M.E.Q. Pilson. 1973. On the spectrophotometric determination of dissolved silica in natural waters. *Analyt. Chem.* 45:136-140.

Feacham, R., M. McGarry and D. Mara, eds. 1977. *Water wastes and health in hot climates.* New York: Wiley-Interscience.

Fenchel, T. 1969. The ecology of marine microbenthos. IV. Structure and function of the benthic ecosystem, its chemical and physical factors and the microfauna communities with reference to the ciliated protozoa. *Ophelia* 6:1-182.

Fenchel, T. and R.H. Riedle. 1970. The sulphide system: a new biotic community underneath the oxidized layer of marine sand bottoms. *Mar. Biol.* 7:255-268.

Freeze, R.A. and J.A. Cherry. 1979. *Groundwater.* Englewood Cliffs, New Jersey: Prentice-Hall.

Fritz, J.S. and S.S. Yamamura. 1955. Rapid microtitration of sulphate. *Analyt. Chem.* 27:1461-1464.

Gauch, H.G., Jr. and R.H. Whittaker. 1972. Comparison of ordination techniques. *Ecology* 53:868-875.

George, J.D. 1980. Photography as a marine biological research tool. In *The shore environment*, Vol. 1. *Methods*, eds. J. H. Price, D. E. G. Irvine and W. F. Farnham, pp. 45-116. London: Academic Press.

Giese, A. and J.S. Pearse, eds. 1974. *Reproduction of marine invertebrates.* Vol. 1. *Methods.* London: Academic Press.

Giger, W. and P.J. Roberts. 1978. Characterization of persistant organic carbon. In *Water Pollution Microbiology*, Vol. II, ed. R. Mitchell. New York: Wiley.

Glynn, P. 1973. Ecology of Caribbean coral reef. The *Porites* reef-flat biotope. Part II. Plankton community with evidence for depletion. *Mar. Biol.* 22:1-24.

Glynn, P.W. 1976. Some physical and biological determinants of coral community structure in the Eastern Pacific. *Ecol. Monogr.* 46:431-456.

Grassle, J.F. and J.P. Grassle. 1977. Temporal adaptations in sibling species of *Capitella*. In *Ecology of marine benthos*, ed. B. C. Coull, pp. 177-190. Columbia, S.C.: Univ. of S. Carolina Press.

Grassle, J.F., G.P. Patil, W.K. Smith, and C. Taillio. 1979. *Ecological diversity in theory and practice.* Statistical Ecology Series. Vol. 6. Burtonsville, Md.: International Co-operative Publishing House.

Green, R.H. 1980. Multivariate approaches in ecology: the assessment of ecological similarity. *Ann. Rev. of Ecol. Sys.* 11:1-14.

Grice, G.D. and M.R. Reeve. eds. 1981. *Marine mesococesms: Biological and chemical research in experimental ecosystems.* New York: Springer-Verlag.

Grigg, R.W. and J.E. Maragos. 1974. Recolonization of hermatypic corals on submerged lava flows in Hawaii. *Ecology* 55:387-395.

Gulland, J.A. 1969. *Manual of methods for fish stock assessment.* Part 1. *Fish population analysis.* FAO Manuals in Fisheries Science. No. 4. Rome: FAO.

Harris, L.B. 1976. Field studies of benthic communities in the New England offshore mining environmental study (NOMES). Final Report. National Oceanic and Atmospheric Administration: Environmental Research Laboratories.

Hartzband, D.J. and D.F. Boesch. 1979. Benthic ecological studies: meobenthos. Chapter 7 in Special Report in Applied Marine Science and Ocean Engineering No 195. Prepared by Virginia Institute of Marine Science, Gloucester Point, Va. under contract No. AA550-CT6-62 with the Bureau of Land Management, U.S. Dept of Interior.

Henderson, H.F., R.A. Ryder, and A.W. Kudhongania. 1973. Assessing fisheries potential of lakes and reservoirs. *J. Fish. Res. Bd. Can.* 30:2000-2009.

Hess, W.N., ed. 1974. *Weather and climate modification.* New York: John Wiley and Sons.

Hiscock, K. and R. Mitchell. 1980. The description and classification of sublittoral epibenthic ecosystems. In *The shore environment*, Vol. 2, eds. J. H. Price, D.E.G. Irvine, and W.F. Farnham, pp. 323-370. London: Academic Press.

Hobson, E.S. and J.R. Chess. 1976. Trophic interactions among fishes and zooplankters near shore at Santa Catalina Is., Calif. *Fish. Bull.* 34:567-598.

Hobson, E.S. and J.R. Chess. 1978. Trophic relationships among fish and plankton in the lagoon at Entwetak Atoll, Marshall Islands. *Fish. Bull.* 76:133-153.

Hocutt, C.H. and J.R. Stauffer, Jr., eds. 1980. *Biological monitoring of fish.* Lexington, Mass: Lexington Books.

Horne, A. 1978. Nitrogen fixation in eutrophic lakes. In *Water pollution microbiology*, Vol. II, ed. R. Mitchell. New York: Wiley.

Howe, C. and F.P. Linaweaver, Jr. 1967. The impact of price on residential water demand and its relation to system design and price structure. *Water Resources Research* 3(1): 13-32.

Hudson, J.H. 1977. Long-term bioerosion rates in a Florida reef: A new method. In *Proceedings: Third Int. Coral Reef Symp.* Vol. 2, ed. D. L. Taylor, pp. 9-14. Miami, Fla.: University of Miami Rosensteil School of Marine and Atmos. Science.

Hydrologic Engineering Center. 1971-1977. *Hydrologic engineering methods for water resources development.* 12 vols. Davis, Calif.: U. S. Army Corps of Engineers.

International Oceanographic Tables. 1966. UNESCO, Paris and National Inst. of Oceanography, Wormley, England, Vol. 1., p. 128.

Jacobs, F. and G.C. Grant. 1978. *Guidelines for zooplankton sampling in quantitative baseline and monitoring programs.* EPA- 600/3-78-026. Corvallis, Oregon: Environmental Protection Agency.

Jackson, J.J. 1977a. *Climate, water and agriculture in the tropics.* London: Longman.

Jackson, J.B.C. 1977b. Habitat area, colonization and development of epibenthic community structure. In *Biology of bethnic organisms*, eds. B.S. Keegan, P.O. Ceidigh, and P.J.S. Boaden. Oxford: Pergammon Press.

Jansson, B.O. 1967. The availability of oxygen for interstitial fauna of sandy beaches. *J. Exp. Mar. Biol.* 1:123-143.

Johannes, R.E. 1978. Flux of zooplankton and benthic algal detritus. In *Coral reefs: Research Methods*, eds. D.R. Stoddard and R.E. Johannes. Paris: UNESCO.

John, D.M., D. Lieberman, M. Lieberman, and M.D. Swaine. 1980. Strategies of data collection and analysis of subtidal vegetation. In *The shore environment*, Vol. 1. *Methods*, eds, J.H. Price, D.E.G. Irvine and W.F. Farnham, pp. 265-284. London: Academic Press.

Jokiel, P.L. and J.E. Maragos. 1978. Reef corals of Canton Island II. Local distribution. *Atoll Res. Bull.* 221:71-98.

Jokiel, P.L., J.E. Maragos, and L. Franzisket. 1978. Coral growth: buoyant weight technique. In *Coral reef research methods*, eds. D. R. Stoddart and R. E. Johannes, pp. 529-542. Paris: UNESCO.

Jones, R. 1976. Growth of fishes. In *The ecology of the seas*, eds. D.H. Cushing and J.J. Walsh, pp. 251-279. London: Blackwell Scientific Publs.

Jones, W.E., S. Bennell, C. Beveridge, B. McConnell, S. Mack-Smith, J. Mitchell, and A. Fletcher. 1980. Methods of data collection and processing in rocky intertidal monitoring. In *The shore environment*, Vol. 1. *Methods*, eds. J. H. Price, D.E.G. Irvine and W.F. Farnham, pp. 137-170. London: Academic Press.

Kaufman, K.W. 1981. Fitting and using growth curves. *Oecologia* 49:293-299.

Kaufman, L.S., D.S. Becker and R.G. Otto. 1980. *Patterns of distribution and abundance of macrobenthos at Taylor's Island, Maryland with implications for monitoring programs.* Special Report No. 81. Maryland: Chesapeake Bay Inst.

Katzenelson, E., I. Buium and H.I. Shuval. 1976. Risk of communicable disease infection associated with wastewater irrigation in agricultural settlements. *Science* 194:944-946.

Kendeigh, S.C. 1974. *Ecology.* Englewood Cliffs, New Jersey: Prentice-Hall.

Kerfoot, W.C., ed. 1980. *Evolution and ecology of zooplankton communities.* Hanover, N. H.: Univ. Press of New England.

Kessell, S.R. and R.H. Whittaker. 1976. Comparison of three ordination techniques. *Vegetation* 32:21-29.

Knight, S.J.T. and R. Mitchell. 1980. The survey and nature conservation assessment of littoral areas. In *The shore environment*, Vol. 1. *Methods*, eds. J. H. Price, D.E.G. Irvine and W. F. Farnham, pp. 303-323. London: Academic Press.

LaBarbera, M. and S. Vogel. 1976. An inexpensive thermistor flowmeter for aquatic biology. *Limnol. Oceanogr.* 21:750-756.

Lamb, H.H. 1972. *Climate: Present, past and future.* Vol. 1. *Fundamentals and climate now.* London: Methuen.

Lamberts, A.E. 1978. Coral growth: Alizarin method. In *Coral reefs: Research methods*, eds. D.R. Stoddart and R.E. Johannes, pp. 523-528. Paris: UNESCO.

Levins, R. 1976. Evolution in communities near equilibrium. In *Ecology and evolution of communities*, eds. M.L. Cody and J.M. Diamond, pp. 16-50. Cambridge, Mass.: Belknap Press of Harvard Univ. Press.

Levinton, J.S. 1977. Ecology of shallow water deposit-feeding communities. Quisset Harbor, Massachusetts. In *Ecology of marine benthos*, ed. B.C. Coull, pp. 191-227. Columbia, S.C.: Univ. of S. Carolina Press.

Levinton, J.S. 1979. The effects of density upon deposit-feeding populations: movement, feeding, and floating of *Hydrobia ventrosa* Montagu (Gastropoda: Prosobranchia). *Oecologia* 42:27-42.

Lewis, J.R. 1976. Long-term ecological surveillance: Practical realities in the rocky littoral. *Oceanogr. Mar. Biol. Ann. Rev.* 14:371-390.

Lewis, J.R. 1978a. Benthic baselines - a case for international collaboration. *Mar. Pollution Bull.* 9:317-320.

Lewis, J.R. 1978b. The implications of community structure for benthic monitoring studies. *Mar. Pollution Bull.* 9:64-66.

Lewis, J.R. 1980. Options and problems in environmental management and evaluation. *Helgolander Meeresunters* 33:452-466.

Lind, O.T. 1979. *Handbook of common methods in limnology.* 2nd ed. St. Louis, Mo.: C. V. Mosby.

Lloyd, J.W. ed. 1981. *Case studies in groundwater resources evaluation.* Oxford: Carendon Press.

Lloyd, M. and R. Ghelardi. 1964. A table for calculating the equitability component of species diversity. *J. Anim. Ecol.* 33:217-225.

Lobel, Phillip S. 1981. Oceanographic Group, Harvard University. Pers. comm.

II. AQUATIC ECOSYSTEMS

Lof, George O. G., and A.V. Kneese. 1968. *The economics of water utilization in the beet sugar industry.* Baltimore, Md.: The Johns Hopkins Univ. Press for Resources for the Future.

Lorenzen, M. 1978. Phosphorus models and eutrophication. In *Water pollution microbiology*, Vol. II, ed. Ralph Mitchell. New York: Wiley.

Loya, Y. 1975. Possible effects of water pollution on the community structure of Red Sea corals. *Mar. Biol.* 29:177-185.

Loya, Y. 1978. Plotless and transect methods. In *Coral reefs: Research methods*, eds. D.R. Stoddart and R.E. Johannes, pp. 197-218. Paris: UNESCO.

Lubchenco, J. and B.A. Menge. 1978. Community development and persistence in a low rocky intertidal zone. *Ecol. Monogr.* 48:67-94.

Mackereth, F.J.H., J. Heron, and J.F. Talling. 1978. *Water analysis: Some revised methods for limnologists.* Ambleside: Freshwater Biological Association.

Mandel, S. and Z. Shiftan. 1981. *Groundwater resources: Investigation and development.* New York: Academic Press.

Mandell, G.L., R.G. Douglas Jr., and J.E. Bennett. 1979. *Principles and practice of infectious diseases.* New York: Wiley.

Maragos, J.E. 1978a. Coral growth: Geometrical relationships. In *Coral reefs: Research methods*, eds. D.R. Stoddart and R.E. Johannes, pp. 543-550. Paris: UNESCO.

Maragos, J.E. 1978b. Measurement of water volume transport for flow studies. In *Coral reefs: Research methods*, eds. D.R. Stoddart and R.E. Johannes, pp. 353-360. Paris: UNESCO.

McArthur, J.M. 1977. Determination of manganese in natural water by flameless atomic absorption spectrometry. *Analyt. Chim. Acta* 93:77-83.

Meadows, P.S. and J.I. Campbell. 1972. Habitat selection by aquatic invertebrates. In *Advances in marine biology*, eds. F.S. Russell and M. Yonge, pp. 271-382. London: Academic Press.

Megard, R.O., J.C. Settles, H.A. Boyer and W.S. Combs, Jr. 1980. Light, secchi disks and trophic states. *Limnol. and Oceanogr.* 25:373-77.

Menge, B.A. 1976. Organization of the New England rocky intertidal community: Role of predation, competition and environmental heterogeneity. *Ecol. Monogr.* 46:355-393.

Miles, D.L. and C. Espejo. 1977. Comparison between an ultraviolet procedure and the 2,4-xylenol method for ,he determination of nitrate in groundwaters of low salinity. *Analyst* 102:104-109.

Miller, A. and J.C. Thompson. 1970. *Elements of Meteorology.* Columbus, Ohio: A. Bell and Howell Company.

Morgan, E. 1970. The effects of environmental factors on the distribution of the amphipod *Pectenogammarus planierurus* with particular reference to grain size. *J. Mar. Biol. Ass. U. K.* 50:769-785.

Moriarty, D.J.W. 1978. Estimation of bacterial biomass in water and sediments using muramic acid. In *Microbial ecology*, eds. M.W. Loutit and J.A. Miles. Berlin: Springer-Verlag.

Morton, J. and M. Miller. 1973. *The New Zealand seashore.* 2nd ed. Glasgow: Wm. Collins & Sons.

Murphy, J. and J.P. Riley. 1962. A modified single solution method for the determination of phosphate in natural waters. *Analyt. Chim. Acta.* 27:31-36.

National Research Council. 1981. *Managing climatic resources and risks.* Washington, D.C.: National Academy Press.

Nedelec, C., ed. 1975. *Catalogue of small-scale fishing gear.* London: Fishing News (Books).

Neff, N.A. and L.F. Marcus. 1980. *A survey of multivariate methods for systematics.* New York: privately published (write to the authors at the American Museum of Natural History, Central Park West at 79th Street, New York, New York 10024.

Newell, R.C. 1970. *Biology of intertidal animals.* London: Logos Press.

NOAA Current Tables. Annually since 1853. Washington, DC: U.S. Dept. of Commerce, National Oceanic and Atmospheric Administration.

North, W.J. 1974. Effects of heated effluent on marine biota particularly in California. In *Modifications thermiques et equilibre biologues*, ed. J. Pearce, pp. 41-60. Amsterdam: North Holland Pub.

Noy-Meir, I. and R.H. Wittaker. 1977. Continuous multivariate methods in community analysis: Some problems and developments. *Vegetatio* 33:79-98.

Odum, H.T., B.J. Copeland and E.A. McMahan. 1974. *Coastal ecological systems of the U.S.* Washington, D.C.: The Conservation Foundation.

Osman, R.W. 1977. The establishment and development of a marine epifaunal community. *Ecol. Monogr.* 47:37-63.

Osman, R.W. 1978. The influence of seasonality and stability on the species equilibrium. *Ecology* 59:393-399.

Owen, O.S. 1980. *Natural resources conservation.* 3rd ed. New York: Macmillan.

Paine, R.T. 1966. Food web complexity and species diversity. *Amer. Nat.* 100:65-75.

Paine, R.T. 1969. The *Pisaster-Tegula* interaction: prey patches, predator food preference, and intertidal community structure. *Ecology* 50:950-961.

Paine, R.T. 1971. Energy flow in a natural population of the herbivorous gastropod *Tegula fenebralis*. *Limnol. and Oceanogr.* 16:86-98.

Paine, R.T. 1974. Intertidal community structure: Experimental studies on the relationship between a dominant competitor and its principal predator. *Oecologia*. 14:93-120.

Paine, R.T. 1977. Controlled manipulations in the marine intertidal zone and their contributions to ecological theory. *Proc. Phil. Acad. Nat. Sci.* Philadelphia: ANSP.

Paine, R.T. 1980. Food webs: Interaction strength and community infrastructure. *Anim. Ecol.* 49:667-685.

Palmen, E. and C.W. Newton. 1969. *Atmospheric circulation systems.* New York: Academic Press.

Pamatmat, M. 1968. Ecology and metabolism of a benthic community on an intertidal sandflat. *Int. Revue ges. Hydrobiol.* 53:211-298.

Pamatmat, M. 1975. *In situ* metabolism of benthic communities. *Cah. Biol. Mar.* 16: 613-633.

Pamatmat, M.M. 1977. Benthic community metabolism: A review and assessment of present status and outlook. In *Ecology of marine benthos*, ed. B.C. Coull, pp. 89-112. Columbia, S.C.: Univ. S. Carolina Press.

Pamatmat, M. and D. Fenton. 1968. An instrument for measuring subtidal benthic metabolism *in situ*. *Limnol. and Oceanogr.* 13:537-540.

Patil, G.P., E.C. Pielou, and W.E. Waters. 1971a. *Sampling and modeling biological populations and population dynamics.Statistical Ecology Series.* Vol. 2. Univ. Park, Penn.: Penn State Univ. Press.

Patil, G.P., E.C. Pielou, and W.E. Waters. 1971b. *Many species populations, ecosystems, and systems analysis.* Statistical Ecology Series. Vol. 3. Univ. Park, Penn.: Penn State Univ. Press.

Patil, G.P. and M.L. Rosenzweig. 1979. *Contemporary quantitative ecology and related ecometrics.* Statistical Ecology Series. Vol. 12. Burtonsville, Md.: International Co-operative Publishing House.

Patil, G.P. and C. Taillie. 1979. A study of diversity of profiles and orderings for a bird community in the vicinity of Colstrip, Montana. In *Contemporary quantitation ecology and ecometrics.* eds. G.P. Patil and M. Rosenzweig, pp. 23-48. Fairfax, Md.: Internat. Coop. Publ. House.

Patrick, R. 1973. Use of algae especially diatoms in the assessment of water quality. In *Biological methods for the assessment of water quality*, A.S.T.M/S.T.P. 528. Philadephia, Penn.: American Society for Testing and Materials.

Pearson. T.H. and R. Rosenberg. 1978. Macrobenthic succession in relation to organic enrichment and pollution of the marine environment. *Oceanogr. Mar. Biol. Ann. Rev.* 16:229-312.

Peet, R.K. 1974. The measurement of species diversity. *Ann. Rev. Ecol. Sys.* 5:285-307.

Peterson, C.H. and C.V. Andre. 1979. An experimental analysis of interspecific competition among marine filter feeders in a soft-sediment environment. *Ecology* 61:129-139.

Pichon, M. 1978. Quantitative benthic ecology of Tulear reefs. In *Coral reefs: Research methods*, eds. D. R. Stoddart and R. E. Johannes, pp. 163-174. Paris: UNESCO.

Pickard, G.L. 1979. *Descriptive physical oceanography.* Oxford: Pergamon.

Pielou, E.C. 1977. *Mathematical ecology.* New York: John Wiley and Sons.

Pocinki, L.S., R.S. Greeley, and L. Slater, eds. 1980. *Climate and risk.* 2 Vols. Proceedings of a Conference Sponsored by the Center for Advanced Engineering Study at MIT. Report. No. MTR 80W322. McLean, Va.: Mitre Corporation.

Poldermann, P.J.G. 1980. The permanent quadrat method, a means of investigating the dynamics of saltmarsh algal vegetation. In *The shore environment*, Vol. 1. *Methods*, eds. J.H. Price, D.E.G. Irvine and W.F. Farnham, pp. 193-212. London: Academic Press.

Poole, R.W. 1974. *An introduction to quantitative ecology.* New York: McGraw-Hill.

Pond, S. and G.L. Pickard. 1978. *Introductory dynamic oceanography.* Oxford: Pergamon.

Porter, J.W. 1972a. Ecology and species diversity of coral reefs on opposite sides of the Isthmus of Panama. *Bull. Biol. Soc. Wash.* 2:89-116.

Porter, J.W. 1972b. Patterns of species diversity in Caribbean reef corals. *Ecology* 53:745-748.

Porter, J.W., K.G. Porter, and Z. Batac-Catalan. 1977. Quantitative sampling of Indo-Pacific demersal reef plankton. In *Proccedings: Third Int. Coral Reef Symp.* Vol. I, ed. D. L. Taylor, pp. 105-112. Miami, Fla.: University of Miami Rosensteil School of Marine and Atmos. Science

Porter, K.G., J.W. Porter, and S.L. Ohlhorst. 1978. Resident reef plankton. In *Coral reefs: Research methods*, eds. D.R. Stoddart and R.E. Johannes, pp. 499-514. Paris: UNESCO.

Pugh, D.T. 1978. Techniques for the measurement of sea level around atolls. In *Coral reefs: Research methods,* eds. D.R. Stoddart and R.E. Johannes, pp. 93-108. Paris: UNESCO.

Quinn, T.J., II and V.F. Gallucci. 1980. Parametric models for line-transect estimations of abundance. *Ecology* 61:293-302.

Rainwater, F.H. and L.L. Thatcher. 1960. Methods for the collection and analysis of water samples. *Wat-Supply Irrig. Pap. Wash.* 1454.

Reckhow, K.H. 1979. *Quantitative techniques for the assessment of lake quality.* EPA-440/5-79-015. Environmental Protection Agency.

Reckhow, K.H., M.N. Beaulai and J.T. Simpson. 1980. *Modeling phosphorus loading and lake response under uncertainty: A manual and compilation of export coefficients.* EPA-440/5-8--011. Environmental Protection Agency.

Reid, G.K. and R.D. Wood. 1976. *Ecology of inland waters and estuaries.* 2nd ed. New York: D. Van Nostrand.

Rhoads, D.C. and D.K. Young. 1970. The influence of deposit feeding organisms on sediment stability and community trophic structure. *J. Mar. Res.* 28:150-178.

Rhoads, D.C., R.C. Aller, and M.B. Goldhaber. 1977. The influence of colonizing benthos on physical properties and chemical diagenesis of the estuarine seafloor. In *Ecology of marine benthos*, ed. B.C. Coull. Columbia, S.C.: Univ. of S. Carolina Press.

Riehl, H. 1954. *Tropical meteorology.* New York: McGraw Hill.

Riley, G.A., 1970. Particulate and organic material in sea water. *Adv. Mar. Biol.* 8:1-118.

Rohlf, F.J. 1972. An empirical comparison of three ordination techniques in numerical taxonomy. *Systematic Zoology.* 21:271-280.

Rosenfield, P. 1979. *The management of schistosomiasis.* Baltimore: The Johns Hopkins Univ. Press for Resources for the Future.

Roswall, T. ed. 1972. *Modern methods in microbial ecology.* Swedish Nat. Sci. Res. Council Bull. No. 17. Stockholm.

Russell, Clifford S., and W.T. Vaughan. 1976. *Steel production: processes, products, and residuals.* Baltimore, Md.: The Johns Hopkins Univ. Press for Resources for the Future.

Russell, G. 1980. Applications of simple numerical methods to the analysis of intertidal vegetation. In *The Shore environment*, Vol. 1. *Methods*, eds. J. H. Price, D.E.G. Irvine and W. F. Farnham, pp. 171-197. London: Academic Press.

Ruttan, V.W. 1965. *The economic demand for irrigated acreage: New methodology and some preliminary projects, 1954-1980.* Baltimore: Johns Hopkins Univ. Press for Resources for the Future.

Rutzler, K. 1978. Photogrammetry of reef environments by helium baloon. In *Coral reefs: Research methods*, eds. D.R. Stoddart and R.E. Johannes, pp. 45-52. Paris: UNESCO.

Rutzler, K., J.D. Ferraris and R.J. Larson. 1980. A new plankton sampler for coral reefs. *PSZN I.: Marine Ecology* 1:65-71.

Ryder, R.A., S.R. Keer, K.H. Loftus, and H.A. Regier. 1974. The morphoedaphic index, a fish yield estimator--review and evaluation. *J. Fish. Res. Bd. Can.* 31:663-688.

Saila, S.B. and P.M. Roedel, eds. 1979. *Stock assessment for tropical small-scale fisheries.* Kingston, R.I.: Internat. Center Marine Resource Development, Univ. Rhode Island.

Sale, P.F. 1977. Maintenance of high diversity in coral reef fish communities. *Amer. Nat.* 111:337-359.

Sanders, H.L. 1956. Oceanography of Long Island Sound, 1952-1954. *Oceanogr. Coll.* 15:345-414.

Sanders, H.L. 1958. Benthic studies in Buzzards Bay I. Animal-sediment relationships. *Limnol. and Oceanogr.* 111:245-258.

Sanders, H.L. 1960. Benthic studies in Buzzards Bay III. The structure of the soft-bottom community. *Limnol. and Oceanogr.* 113:138-153.

Saunders, R. and J.J. Warford. 1976. *Village water supply, economics and policy in the developing world.* Baltimore: The Johns Hopkins Univ. Press.

Sanders, H.L., J.F. Grassle, G.R. Hampson, L.S. Morse, S. Garner-Price, and C.C. Jones. 1980. Anatomy of an oil spill: long-term effects from the grounding of the barge *Florida* off West Falmouth, Massachusetts. *J. Mar. Res.* 38:256-380.

Sanders, H.L., R.R. Hessler, and G.R. Hampson. 1965. An introduction to the study of the deep sea faunal assemblages along the Gay Head-Bermuda transect. *Deep Sea Res.* 12:845-867.

Schaake, J.C. and Z. Kacamarek. 1979. Climate variability and the design and operation of water resource systems. In *Proceedings of the world climate conference*, pp. 290-311. Geneva: World Meteorological Organization.

Scheer, G. 1978. Application of phytosociologic methods. In *Coral reefs: Research methods*, eds. D. R. Stoddart and R. E. Johannes, pp. 175-196. Paris: UNESCO.

Scrimshaw, N.S. 1970. Synergism of malnutrition and infection: Evidence from field studies in Guatemala. *Journ. Amer. Med. Assoc.* 212:(10).

Sebens, K.P. 1981. Reproductive ecology of the intertidal sea anemones *Anthopleura xanthogrammica* (Brandt) and *A. elegantissima* (Brandt): Body size, habitat, and sexual reproduction. *J. Exp. Mar. Biol. Ecol.* 54:225-250.

Sebens, K.P. 1982a. The limits to indeterminate growth: An optimal size model applied to passive suspension feeders. *Ecology* 63:209-222.

Sebens, K.P. 1982b. Competition for space: Growth rate, reproductive output, and escape in size. *Amer. Nat.* 120:189-197.

Sebens, K.P. In press. Population dynamics and habitat suitability of the intertidal sea anemones *Anthopleura elegantissima* (Brandt) and *A. xanthogrammica* (Brandt). *Ecol. Monogr.*

Seed, R. 1969. The ecology of *Mytilus edulis* L. Lamellibrachiata on exposed rocky shores. I. Breeding and settlement. *Oecologia* 3:317-350.

Segar, D.A. and A.Y. Cantillo. 1975. Direct determination of trace metals in seawater by flameless atomic absorption spectrophotometry. *Adv. Chem. Ser.* 147:56-81.

Shinn, E.A. 1966. Coral growth-rate, an environmental indicator. I. *Palcont.* 40:233-240.

Simon, J.L. and D.M. Dauer. 1977. Reestablishment of a marine community following natural defaunation. In *Ecology of marine benthos*, ed. B.C. Coull, pp. 139-154. Columbia, S.C.: Univ. of S. Carolina Press.

Smith, K.L., Jr. 1973. Respiration of a sublittoral community. *Ecology* 54:1065-1075.

Smith, K.L., Jr. 1974. Oxygen demands of San Diego Trough sediments: An *in situ* study. *Limnol. Oceanogr.* 19:939-944.

Smith, K.L., G.T. Rowe, and J.A. Nichols. 1973. Benthic community repiration near the Woods Hole sewage outfall. *Estuar. Coast. Mar. Sci.* 1:65-70.

Smith, R.L. 1966. *Ecology and field biology.* New York: Harper and Row.

Smith, S.V. and R.S. Henderson, eds. 1978. Phoenix Island report I: An environmental survey of Canton atoll lagoon, 1973. *Atoll Res. Bull.* No. 221. Washington, D.C.: Smithsonian Inst.

Sneath, P.H.A. and R.R. Sokal. 1973. *Numerical taxonomy: The principles and practice of numerical classification.* San Francisco: W. H. Freeman.

Sokal, R.R. and F.J. Rohlf. 1969. *Biometry: The principles and practice of statistics in biological research.* San Francisco: W. H. Freeman.

Sousa, W.P. 1979a. Experimental investigation of disturbance and ecological succession in a rocky intertidal algal community. *Ecol. Monogr.* 49:227-254.

Sousa, W.P. 1979b. Disturbance in marine intertidal boulder fields: The nonequilibrium maintenance of species diversity. *Ecology* 60:1225-1239.

Stanier, R.Y., E.A. Adelberg and J. Ingraham. 1976. *The microbial world.* 4th ed. Englewood Cliffs, New Jersey: Prentice-Hall.

Steedman, H.F., ed. 1976. *Zooplankton fixation and preservation.* Paris: UNESCO.

Stephenson, W. and W.T. Williams. 1971. A study of the benthos of soft bottoms, Sek Harbour, New Guinea, using numerical analysis. *Aust. J. Mar. Freshwat. Res.* 22:11-34.

Stoddart, D.R. 1978a. Mapping reefs and islands. In *Coral reefs: Research methods*, eds. D.R. Stoddart and R.E. Johannes, pp. 17-22. Paris: UNESCO.

Stoddart, D.R. 1978b. Descriptive reef terminology. In *Coral reefs: Research methods*, eds. D.R. Stoddart and R.E. Johannes, pp. 5-16. Paris: UNESCO.

Stoddart, D.R. 1978c. Mechanical analysis of reef sediments. In *Coral reefs: Research methods*, eds. D.R. Stoddart and R.E. Johannes, pp. 53-66. Paris: UNESCO.

Strickland, J.D.H. and T.R. Parsons. 1974. *A practical handbook of seawater analysis.* 3rd ed. No. 167. Ottawa: Bull. Fish. Res. Bd. Can.

Stringer, E.T. 1972. *Foundations of climatology.* San Francisco: W. H. Freeman.

Strong, C.W. 1966. An improved method of obtaining density from line-transect data. *Ecology* 47:311-313.

Stumm, W. and E. Stumm-Zolinger. 1972. The role of phosphorus in eutrophication. In *Water pollution microbiology*, Vol. I, ed. R. Mitchell. New York: Wiley.

Sutherland, J.P. 1980. Dynamics of the epibenthic community on roots of the mangrove *Rhizophora mangle* at Bahia de Buche, Venezuela. *Mar. Biol.* 58:75-84.

Sutton, O.G. 1977. *Micrometeorology.* Huntington, New York: Robert E. Krieger Co.

Tables for Sea Water Density. 1952. U.S. Naval Oceanographic Office Publ. No. 615. Washington, D.C.

Tenore, K.R. 1977. Food chain pathways in detrital feeding benthic communities: A review, with new observations on sediment resuspension and detrital recycling. In *Ecology of marine benthos*, ed. B.C. Coull, pp. 37-54. Columbia, S.C.: Univ. of S. Carolina Press.

Testing For The Effects of Chemicals on Ecosystems. 1981. A report by the Committee to Review Methods for Ecotoxiocology. Washington, D.C.: Nat. Acad. Press.

Thomas, M.K. and G.A. McKay. 1978. *The need for a climatic information service.* Mimeographed report. Downsview, Ontario: Atmospheric Environment Service, Environmental Canada.

Thomassin, B.A. 1978. Soft-bottom communities. In *Coral reefs: Research methods*, eds D.R. Stoddart and R.E. Johannes, pp. 251-262. Paris: UNESCO.

Tillman, R.E. 1981. *Environmental guidelines for irrigation.* Washington, D.C.: U.S. Man and Biosphere Programme, Department of State and U.S. Agency for International Development.

Todd, D.K. 1980. *Groundwater hydrology.* 2nd ed. New York: John Wiley.

Tranter, D.J. and J.H. Fraser, eds. 1968. *Zooplankton sampling.* Geneva: UNESCO.

Tsuda, R.T., S.S. Amesbury, S.C. Moras, and P.P. Beeman. 1975. *Limited current and underwater biological survey at the Pt. Gabert. wastewater outfall on Moen, Truk.* Tech. Rep. 20. Agana, Guam: University of Guam Marine Laboratory

United States Water Resources Council. 1978. *The Nation's water resources, 1975-2000: Second national water assessment.* 4 vols. (052-045-00051-7). Washington, D.C.: Government Printing Office

Ure, A.M. and M.C. Mitchell. 1975. Lithium, sodium, potassium, rubidium and cesium. In *Flame emission and atomic absorption spectrometry*, Vol. 3. *Elements and matrices*, eds. J.A. Dean and T.C. Rains, pp. 1-32. New York: Marcel Dekker.

van der Leeden, F. 1975. *Water resources of the world.* Port Washington, New York: Water Information Center.

van Zijl, W.J. 1966. Studies of diarrheal diseases in seven countries by the WHO Diarrheal Diseases Advisory Team. *Bull WHO* 43.

Virnstein, R.W. 1977. The importance of predation by crabs and fish on benthic infauna in Chesapeake Bay. *Ecology* 58:1199-1217.

Vollenweider, R.A. 1969. *A manual on methods for measuring primary production in aquatic environments.* IBP Handbook No. 12. Oxford: Blackwell Scientific Publications.

Vollenweider, R.A. 1976. Advances in defining critical loading levels for phosphorus in lake eutrophication. *Mem. Ist. Ital. Idrobiol.* 33:53-83.

Watling, L., P.C. Kinner, and D. Maurer. 1979. The use of species abundance estimates in marine benthos studies. *J. Exp. Mar. Biol. Ecol.* 35:109-118.

Weber, J.N., E.W. White and P.H. Weber. 1975. Correlation of density banding with environmental parameters: The basis for interpretation of chronological records preserved in the corolla of corals. *Paleobiology* 1:137-149.

Wegman, E.J. and D.J. DePriest. 1980. *Statistical analysis of weather modification experiments.* New York: Marcel Dekker.

Weinberg, S. 1978. The minimal area problem in invertebrate communities of mediteranean rocky substrata. *Mar. Biol.* 49:33-40.

Welch, P.S. *Limnology.* New York: McGraw-Hill.

Westlake, D.F. 1969. Macrophytes. In *A manual on methods for measuring primary production in aquatic environments*, ed. R.A. Vollenweider, pp. 25-32. Oxford: Blackwell Scientific Publications.

Wetsclaar, R. 1981. Nitrogen inputs and outputs of an unfertilized paddy field. In *Terrestrial nitrogen cycles*, eds. F.E. Clark and T. Rasswaull. Ecological Bulletin (Stockholm) 33:573.

Wetzel, R.G. and G.L. Likens. 1979. *Limnological analyses.* Philadelphia: W.O. Saunders.

White, G.F., ed. 1978. *Environmental effects of arid land irrigation in developing countries.* MAB Technical Notes 8. Paris: UNESCO.

Whitlatch, R.B. 1977. Seasonal changes in the community structure of the macrobenthos inhabiting the intertidal sand and mud flats of Barnstable Harbor, Mass. *Biol. Bull.* 152:275-294.

Whittaker, R.H. and J.G. Gauch, Jr. 1973. Evaluation of ordination techniques. In *Handbook of vegetation science*, Part V. *Ordination and classification of vegetation*, ed. R.H. Whittaker, pp. 289-321 The Hague: Junk.

Whittaker, R.H., ed. 1978. *Ordination of plant communities.* The Hague: Junk.

Whyte, A.V.T. 1977. *Guidelines for field studies in environmental perception.* Man and Biosphere Technical Note 5. Paris: UNESCO.

WMO (World Meteorological Organization). 1974. *Guide to hydrological practices.* Geneva: WMO.

Wollman, N. and G.W. Bonem. 1971. *The outlook for water: Quality, quantity, and national growth.* Baltimore: The Johns Hopkins Univ. Press for Resources for the Future.

Wood, J. and R.E. Johannes. 1975. *Tropical marine pollution.* Amsterdam: Elsevier Publ.

Woodin, S.A. 1974. Polychaete abundance patterns in a marine soft-sediment environment: The importance of biological interactions. *Ecol. Monogr.* 44:171-187.

Woodin, S.A. 1978. Refuges, disturbance and community structure: A marine soft-bottom example. *Ecology* 59:274-284.

World Bank. 1978-1982. Appropriate Technology for Water Supply and Sanitation Series. Volumes are listed in numerical order.

    1. *Appropriate sanitation alternatives: A technical and economic appraisal.* 1982. (An expansion of *Technical and economic options*,

Vol. 1). J.M. Kalbermatten, D.S. Julius and C.G. Gunnerson. Baltimore: Johns Hopkins Univ. Press.

1a. *A summary of technical and economic options.* 1982. J.M. Kalbermatten, D.S. Julius and C.G. Gunnerson. Washington, D.C.: World Bank Publications Unit.

1b. *Appropriate sanitation alternatives: A planning and design manual.* 1982. (An expansion of *A planner's guide*, Vol. 2). J.M. Kalbermatten, D.S. Julius, C.G. Gunnerson and D.D. Mara. Baltimore, Md.: Johns Hopkins Univ. Press.

3. *Health aspects of excreta and sullage management--a state-of-the-art review.* 1981. R.G. Feachem, D.J. Bradley, H. Garelick and D.D. Mara. 318 pp. Washington, D.C.: World Bank Publications Unit.

4. *Low-cost technology options for sanitation: a state-of-the-art review and annotated bibliography.* 1978. W. Rybczynski, C. Polprasert and M. McGarry. 184 pp. Washington, D.C.: World Bank Publications Unit.

5. *Sociocultural aspects of water supply and excreta disposal.* M. Elmendorf and P.K. Buckles. 52 pp. Washington, D.C.: World Bank Publications Unit.

6. *Country studies in sanitation alternatives.* 1981. R.A. Kuhlthau, ed. Washington, D.C.: World Bank Publications Unit.

7. *Alternative sanitation technologies for urban areas in Africa.* 1981. R.G. Feachem, D.D. Mara and K.O. Iwugo. Washington, D.C.: World Bank Publications Unit.

8. *Seven case studies of rural and urban fringe areas in Latin America.* 1981. M. Elmendorf, ed. Washington, D.C.: World Bank Publications Unit.

9. *Design of low-cost water distribution systems.* 1981. Section 1 by D.T. Lauria, P.J. Kolsky and R.N. Middleton: Section 2 by K. Demke and D.T. Lauria; and Section 3 by P.B. Hebert.

10. *Night soil composting.* 1981. H.I. Shuval, C.G. Gunnerson and D.S. Julius. 81 pp. Washington, D.C.: World Bank Publications Unit.

11. *Sanitation field manual.* 1980. J.M. Kalbermatten, D.S. Julius, G.C. Gunnerson and D.D. Mara. 86 pp.

12. *Low-cost water distribution--a field manual.* C. Spangler. Washington, D.C.: World Bank Publications Unit.

World Bank. 1982a. *A model for the development of a self-help water supply program.* C. Glennie. Technology Advisory Group Working Paper, TAG/WP/01. World Bank Technical Paper No. 2. Washington, D.C.: World Bank.

World Bank. 1982b. *Ventilated improved pit latrines--recent developments in Zimbabwe.* P.R. Morgan and D.D. Mara. World Bank Technical Paper No. 3. Washington, D.C.: World Bank.

Wuhrmann, K. 1974. Some problems and perspectives in applied limnology. *Mitt. Internat. Verein. Limnol.* 20:324

Young, R.A. and S.L. Gray. 1972. *Economic value of water: Concepts and empirical estimates.* Final Report to the National Water Comm. Contract No. NWC 70-028. NTIS PB210 356. Springfield, Va.: NTIS.

Zar, J.H. 1974. *Biostatistical analysis.* Englewood Cliffs, N.J.: Prentice Hall.

Suggested Readings

WATER QUALITY, SURFACE WATER AND CHEMICAL ANALYSIS

American Public Health Association. 1980. *Standard methods for the examination of water and wastewater.* 15th ed., with supplement. Washington, D.C.: American Public Health Association.

Environmental Protection Agency. 1979a. *Methods for chemical analysis of water and wastes, 1978.* EPA-600/4-79-020. Environmental Protection Agency, Cincinnati, Ohio. Springfield, Va.: National Technical Information Service.

Strickland, J.D.H. and T.R. Parsons. 1974. *A practical handbook of seawater analysis.* 3rd ed. No. 167. Ottawa: Bull. Fish. Res. Bd. Can.

Wood, J. and R.E. Johannes. 1975. *Tropical marine pollution.* Amsterdam: Elsevier Publ.

Golterman, H.L. and R.S. Clymo. 1969. *Methods for chemical analysis of freshwaters.* IBP Handbook No. 8. Oxford: Blackwell Scientific Publ.

Environmental Protection Agency. 1973. *Biological field and laboratory methods for measuring the quality of surface waters and effluents.* EPA-670/4-73-001. Environmental Protection Agency. Springfield, Va.: National Technical Information Service.

Patrick, R. 1973. Use of algae especially diatoms in the assessment of water quality. In *Biological methods for the assessment of water quality*, A.S.T.M/S.T.P. 528. Philadephia: American Society for Testing and Materials.

Reckhow, K.H. 1979. *Quantitative techniques for the assessment of lake quality.* EPA-440/5-79-015. Environmental Protection Agency.

PRIMARY PRODUCTIVITY, MICROBIAL ECOLOGY, AND MONITORING ZOOPLANKTON

Cairns, J., Jr., ed. 1971. *The structure and function of freshwater microbial communities.* Research Division Monograph 3. American Microscopical Society Symposium. Blacksburg, Va.: Virginia Polytechnic

II. AQUATIC ECOSYSTEMS

Institute and State University. Distributed by Charlottesville, Va.: University of Virginia.

Environmental Protection Agency. 1978. *Microbiological methods for monitoring the environment, water and wastes.* Environmental Protection Agency, Cincinnati, Ohio. Springfield, Va.: National Technical Information Service.

Giger, W. and P.J. Roberts. 1978. Characterization of persistant organic carbon. In *Water Pollution Microbiology*, Vol. II, ed. R. Mitchell. New York: Wiley.

Jacobs, F. and G.C. Grant. 1978. *Guidelines for zooplankton sampling in quantitative baseline and monitoring programs.* EPA-600/3-78-026. Corvallis, Oregon: Environmental Protection Agency.

Roswall, T. ed. 1972. *Modern methods in microbial ecology.* Swedish Nat. Sci. Res. Council Bull. No. 17. Stockholm.

Vollenweider, R.A. 1969. *A manual on methods for measuring primary production in aquatic environments.* IBP Handbook No. 12. Oxford: Blackwell Scientific Publications.

FISH ASSESSMENT AND POPULATION ECOLOGY

Bagenal, T., ed. 1978. *Methods for assessment of fish production in fresh waters.* 3rd ed. I.B.P. Handbook No. 3. Oxford: Blackwell Scientific Publ.

Bazigos, G.P. 1974. The design of fisheries statistical surveys--inland waters. *FAO Fish. Tech. Paper 133.* Rome: FAO.

Gulland, J.A. 1969. *Manual of methods for fish stock assessment.* Part 1. *Fish population analysis.* F.A.O. Manuals in Fisheries Science. No. 4. Rome: FAO.

Grassle, J.F., G.P. Patil, W.K. Smith, and C. Taillio. 1979. *Ecological diversity in theory and practice.* Statistical Ecology Series. Vol. 6. Burtonsville, Md.: International Co-operative Publishing House.

Hocutt, C.H. and J.R. Stauffer, Jr., eds. 1980. *Biological monitoring of fish.* Lexington, Mass: Lexington Books.

Saila, S.B. and P.M. Roedel, eds. 1979. *Stock assessment for tropical small-scale fisheries.* Kingston, R.I.: Internat. Center Marine Resource Development, Univ. Rhode Island.

LIMNOLOGY

Lind, O.T. 1979. *Handbook of common methods in limnology.* 2nd ed. St. Louis, Mo.: C. V. Mosby.

Mackereth, F.J.H., J. Heron, and J.F. Talling. 1978. *Water analysis: Some revised methods for limnologists.* Ambleside: Freshwater Biological Association.

ESTUARIES AND COASTAL

Barnes, R.S.K. 1974. *Estuarine biology*. London: Edward Arnold.

George, J.D. 1980. Photography as a marine biological research tool. In *The shore environment*, Vol. 1. *Methods*, eds. J. H. Price, D. E. G. Irvine and W. F. Farnham, pp. 45-116. London: Academic Press.

Stoddart, D.R., and R.E. Johannes, eds. 1978. *Coral reefs: Research methods*. Paris: UNESCO.

GROUNDWATER

Freeze, R.A. and J.A. Cherry. 1979. *Groundwater*. Englewood Cliffs, New Jersey: Prentice-Hall.

Lloyd, J.W. ed. 1981. *Case studies in groundwater resources evaluation*. Oxford: Carendon Press.

Mandel, S. and Z. Shiftan. 1981. *Groundwater resources: Investigation and development*. New York: Academic Press.

ENVIRONMENTAL RESOURCE PLANNING AND ECONOMICS

Dunne, T. and L.B. Leopold. 1978. *Water in environmental planning*. San Francisco, Calif.: W.H. Freeman.

Erickson, P.S. 1979. *Environmental impact assessment. Principles and applications*. New York: Academic Press.

Owen, O.S. 1980. *Natural resources conservation*. 3rd ed. New York: Macmillan.

Ruttan, V.W. 1965. *The economic demand for irrigated acreage: New methodology and some preliminary projects, 1954-1980*. Baltimore, Md.: Johns Hopkins Univ. Press for Resources for the Future.

Saunders, R. and J.J. Warford. 1976. *Village water supply, economics and policy in the developing world*. Baltimore: The Johns Hopkins Univ. Press.

Wollman, N. and G.W. Bonem. 1971. *The outlook for water: Quality, quantity, and national growth*. Baltimore: The Johns Hopkins Univ. Press for Resources for the Future.

Young, R.A. and S.L. Gray. 1972. *Economic value of water: Concepts and empirical estimates*. Final Report to the National Water Comm. Contract No. NWC 70-028. NTIS PB210 356. Springfield, Va.: NTIS.

HEALTH ISSUES

Feacham, R., M. McGarry and D. Mara, eds. 1977. *Water wastes and health in hot climates*. New York: Wiley-Interscience.

II. AQUATIC ECOSYSTEMS

Mandell, G.L., R.G. Douglas Jr., and J.E. Bennett. 1979. *Principles and practice of infectious diseases.* New York: Wiley.

Rosenfield, P. 1979. *The management of schistosomiasis.* Baltimore: The Johns Hopkins Univ. Press for Resources for the Future.

CLIMATE

Jackson, I.J. 1977. *Climate, water and agriculture in the tropics.* London: Longman.

Sutton, O.G. 1977. *Micrometeorology.* Huntington, New York: Robert E. Krieger Co.

WMO (World Meteorological Organization). 1960. *Guide to climatological practices.* WHO No. 100 T.P.44. Geneva: World Meteorological Organization.

WMO (World Meteorological Organization). 1974. *Guide to hydrological practices.* Geneva: WMO.

STATISTICAL PLANNING

Poole, R.W. 1974. *An introduction to quantitative ecology.* New York: McGraw-Hill.

Sokal, R.R. and F.J. Rohlf. 1969. *Biometry: The principles and practice of statistics in biological research.* San Francisco: W. H. Freeman.

Zar, J.H. 1974. *Biostatistical analysis.* Englewood Cliffs, N.J.: Prentice Hall.

Part III -- Soils

Marion Baumgardner, Chair
Pierre R. Crosson
Harold Dregne
Matthew Drosdoff
Frederick C. Westin

III. SOILS

Introduction

In this report, methods for conducting soil resource inventories and the soils-related portions of environmental baseline studies in developing countries are reviewed. We focused on four topics: (i) the usefulness of a variety of soils information in development planning at national, regional, and local levels; (ii) the role of inventory and baseline studies in acquiring soils information; (iii) methods that have been evaluated under socioeconomic and environmental conditions characteristic of developing countries; and (iv) methods most appropriate to meeting development goals in a timely and cost-effective way. Our specfic objectives were: (i) to specify the kinds of soils information required by development planners, and (ii) to describe and assess methods for conducting inventories and baseline studies.

Soil resource *inventories* identify soils in specific geographic areas and by variety, quantity, distribution, and other factors. *Baseline studies* examine the status or condition of soil resources as components of ecosystems in dynamic interaction with climate, vegetation, animals, and man. They are concerned with the assessment of ecological processes that maintain soil quality and fertility and with the consequences of development on soil quality and the ecosystem as a whole.

Soils are an integral component of every terrestrial ecosystem, and their degradation by inappropriate exploitation and management may result in declining productivity. Soil productivity and stability can be improved as well as degraded by human intervention and thus intensification of human intervention has implications that may prove to be positive or negative depending the use of global soil resources.

The scale and scope of the intensification and expansion of cultivation required, as well as the rapidity with which it must be accomplished, dramatize the urgency and fundamental importance of improved inventories as a foundation for decision-making in agricultural development. These and other soils information needs are discussed in chapter 2. Methods for soil survey, classification, evaluation, and analysis are presented in chapters 3, 4, 5, and 6. Soils are considered primarily as a resource for agricultural and rangeland development.

1.1 Soils as the Foundation of Food Production: Resource Inventories

The importance of both soils inventories and baseline information has increased with the growth of populations and development pressures, especially those associated with increasing food production. Since the land area of the world is essentially finite, increasing demands on land and soil resources and the ecological stress accompanying them require significant improvements in soil management if desired food production gains are to be achieved and sustained without soil degradation and unacceptable decreases in environmental quality. Several recent studies give eloquent testimony to this fact and discuss courses of action in response: Food and Agriculture Organization's (FAO) *Agriculture Toward 2000* and *Land Resources for Populations of the Future* (FAO, 1981; Dudal, 1980a); the International Rice Research Institute's *Soil-Related Constraints to Food Production in the Tropics* (IRRI, 1980) and the report of Bonn Conference on Agricultural Production: *Research and Development Strategies for the 1980's* (Rockefeller Foundation, 1979).

Differences in soil quality have been recognized by human groups since the beginnings of agriculture and are still part of traditional knowledge used by many contemporary agriculturalists. Some of the indigenous soil classification systems are discussed in chapter 3. Today's population growth, technological change, and new economic influences as well as demands associated with modernization have disrupted former relationships.

Of the some 13 billion hectares (32 billion acres) of land on earth, little more than 10 percent is cultivated, almost a third is forest and woodland, and nearly a fourth is rangeland. The world's potentially cultivable land is estimated at just over 3 billion hectares or 22 percent of the total land area (Dudal et al., 1982) (see Table 1.1).

During the 20-year period 1957-1977, the amount of arable land increased an estimated 9 percent while world population grew more than 40 percent (Dudal, 1978). Food production was intensified on lands already under cultivation. One result was a 300 percent increase in fertilizer consumption, with a large part of the increase taking place in the technologically advanced countries. However, the percentage increase in fertilizer use was greater in the developing world.

With an estimated world population of 6 billion by 2000, an agricultural output 50 percent greater than that in 1960 will be required. Demand for food and agricultural products in developing countries will double. Increased production to meet this demand is expected to be achieved primarily through higher yields from already cultivated lands (60 percent) and increases in cropping intensity (14 percent) (FAO, 1979).

Because the required higher yields and cropping intensities exceed production capacities of most rain-fed agriculture, expanded irrigation is essential. Some 230 million hectares of cropland worldwide are under irrigation (FAO, 1978), but less than half the land with irrigation potential may be developed (Berry, 1980). Assistance agency and national investments in irrigation are already massive, but projections call for close to 50 percent expansion of irrigated areas in developing countries by 2000 (FAO, 1981).

Substantial increases in total production in developing countries must occur through the extension of agriculture into currently uncultivated lands (FAO, 1981). Among developing countries as a whole, potential cultivable lands cover 2154 million hectares, of which some 784 million are in use (Table 1.2). The largest reserves of potentially cultivable lands are in Latin America and Africa. Less than a tenth of the Far East's potential is unused, and little more than a third in the Near East, that primarily in the Sudan (Table 1.3). A 20 percent increase in cultivated area in the developing world as a whole is expected by the year 2000 and is likely to be required if per capita food production levels are to be main-

Table 1.1 Major categories of land in billions of hectares. [Source: Dudal (1982)]

Land in cultivation	1.3
Potentially cultivable land	3.0
Forest and Woodland	4.3
Rangeland	3.5
Other	1.3
World land area	13.4

tained. This means that as many as 200 million hectares should be opened and brought into production over the next two decades.

Already in many developing countries expanding human populations have forced cultivation onto marginal lands, such as steep slopes with severe water erosion hazard or semiarid areas subject to drought and wind erosion, resulting in even more accelerated soil degradation, accompanying diminished yields and further social impoverishment. And while large areas of land could still be brought under cultivation, essentially all *prime* agricultural land in the world is in use.

The use and management of soils for meeting the food and fiber needs of the future must be based on the principle of sustainable production, and this will require fundamental knowledge of the soil resources, their response to use and development, and the effects of intervention on the total ecosystem.

The agricultural orientation of this description is typical of most common meanings of soil but may be too narrow when it neglects other interests and values of soil resources. For example:

> The word 'soil'...has several meanings...even within soil science...Soil, in its traditional meaning, is the natural medium for the growth of land plants, whether or not it has discernible soil horizons...Although there are many uses of soil, the people of the world are more concerned with soil because it supports plants that supply food, drugs, fibers, and other wants of men than any other reason. In this sense, soil covers land as a continuum, except on bare rock, in areas of perpetual frost, or on the bare ice of glaciers. In this sense, soil has a thickness that is determined by the depth of rooting of plants." (Soil Survey Staff, 1975)

1.2 Carrying Capacity

Ever since Thomas Malthus initially published his essay in 1798 (see Malthus, 1826, for 6th edition) comparing the exponential rate of popula-

Table 1.2. Land use and population (areas in million hectares).
Source: Dudal, Higgins, and Kassam (1982)

	Developing countries	Developed countries	Total world
Land area	7619	5733	13392
Percentage of world's total	57	43	
Population (1979) (millions)	3177	1218	4335
Percentage of world's total	72	28	
Potentially cultivable	2154	877	3031
Percentage of land area	28	15	22
Percentage of world's potential	71	29	100
Presently cultivated	784	677	1461
Percentage of potential	36	77	48
Percentage of world's total	54	46	100
Persons per cultivated hectare	4.0	1.8	3.0

tion increase with the arithmetic increase in food production, concerned persons have pondered the issue of global carrying capacity. Carrying capacity has been used as a measure of the number of individuals of a species that a particular environment can support. Maximum or absolute carrying capacity for humans is the largest population that can be supported over the long term by the resources of an environment at a given level of subsistence and at a given level of technology. In the Malthusian sense of the term this refers to the number of human inhabitants that the resources of the earth can support through the provision of minimum daily nutritional requirements. Global carrying capacity must involve trade and is quite different from a site specific carrying capacity estimate. Ravenstein (1890) suggested that world resources could support a population of 6 billion, and Penck (1928) increased this number to 16 billion. Revelle (1976), enlightened by the technological improvements in food production, proposed a figure of 30 billion, and Buringh (1978) suggested that under optimal conditions these technologies could increase global food production by a factor of 30.

TABLE 1.3 Land Use and Population in Developing Countries (areas in million ha). [Source: Dudal, Higgins and Kassam (1982)]

	Africa	S.W. Asia	S.E. Asia	Central Asia	South America	Central America
Land area	2,886	677	897	1,116	1,770	272
Percentage of world's total	(21)	(5)	(6)	(8)	(13)	(2)
Population (1979) (millions)	427	153	1,232	947	239	119
Percentage of world's total	(10)	(3)	(28)	(22)	(6)	(3)
Potentially cultivable	789	48	297	127	819	75
Percentage of land area	(27)	(7)	(33)	(11)	(46)	(27)
Percentage of world's total	(26)	(2)	(10)	(4)	(27)	(3)
Presently cultivated	168	69	274	113	124	36
Percentage of potential	(21)	(144)	(92)	(89)	(15)	(49)
Percentage irrigated	(4)	(16)	(24)	(44)	(6)	(18)
Persons per presently cultivated hectare	2.5	2.2	4.5	8.4	1.9	3.3

None of the studies have addressed indepth problems related to the inequitable distribution of resources, the availability of technology, or the maintenance of desirable qualities of life given the environmental stress associated with attaining the estimated carrying capacities. Any ecosystem isn't only a "natural system"; it includes the existing technology and the social system which drives it, as well as "expatriate technology" (most externally introduced technology), and the social system that drives that. These factors are crucial in judging carrying capacity. Indeed, estimates that fail to give them serious consideration cannot be taken seriously.

Equally important is the fact that resources are interdependent. Development of soil for food production, for example, reduces its capacity to supply timber or firewood, and water resources diverted to irrigation may no longer be available for other uses. In a general sense, carrying capacity is related to ecosystem productivity. Carrying capacity is not fixed; it varies with natural factors such as fluctuation in weather and climate, and it is also continually modified by human action and level of technology.

For early man as a hunter-gatherer in tropical forests, the carrying capacity was on the order of one person per square mile. With the the domestication of plants, and animals, carrying capacity increased in areas suitable for cropping and grazing. In developed countries and in many developing ones, carrying capacity has been increased even more through importing resources and energy and by the use of fertilizers. Some technologies, however, may mask long-term deterioration of the resource base.

Ways to estimate the capacity of different types of indigenous agricultural systems have been the subject of various investigators (Allan, 1948; Milne, 1938; Richards, 1939; Gluckman, 1941 and 1943; Allan, 1965; Reining, 1979; and Fanale, 1982). Details are discussed in chapter 3. The results of carrying capacity research suggest that 0.5 hectare in crops per person per year is common under extensive cultivation systems and, with intensification, the area required can be reduced to 0.25 hectares per person per year (Allan, 1965; Netting, 1969 and 1979; Reining, 1979). Although many of these indigenous systems are now under stress, they have survived for centuries. Selective pressures eliminate those that cannot be sustained. In examining local carrying capacity, it is important to consider the longstanding agricultural practices that can be adapted to present-day use, as stated in the following example:

> Not only the land itself but the agricultural practices of its inhabitants must be carefully studied and clearly understood before any attempt can be made to estimate the land requirements of the traditional systems or to assess the way in which they have changed or are changing. We must first try to see the situations through the eyes of the African cultivators and put aside for the time being our own preconceived ideas, prejudices, and conceptions of good land use, which derive from very different societies and environments (Allan, 1965).

1.3 Soils and Man in the Ecosystem: Baseline Studies

Recognition is growing that soil characteristics of concern to agriculture are neither static, unchanging properties nor features independent of the broader environment in which they exist. A narrow focus on edaphic factors (that is, that set of soil properties directly affecting plant growth) is too limited. Soil is more than an anchoring place and nutrient reservoir for plants. More broadly defined, soil is seen as the biologically inhabited and active layer of the earth's crust and a product of the interaction between the plants and animals that inhabit it, time and climate, and the underlying rock or parent material. In this sense, soil is seen as a component of the larger ecosystem--a compartment in the network of interactions among soils, plants, animals, climate, and, increasingly human beings.

This perspective on soils emphasizes the continual exchange between living and non-living components in the system and the physical, chemical, and biological processes that link them. Of particular interest among the processes are those associated with the chemical elements or nutrients essential for living organisms. Under natural vegetation and over the course of ecological succession, the progressively more efficient capture of these elements--as they become available from substrate weathering, atmospheric inputs, and indigenous nitrogen fixation and their immobilization in soil colloids and the biota itself--results in the gradual increase of soil fertility. In the short term, however, the pool of available nutrient elements within an ecosystem is essentially fixed and they are constantly reused, moving from organic to inorganic states and back again in more or less circular routes known as biogeochemical or nutrient cycles.

Soil organisms play an important role in nutrient cycling and availability. In the sections on plants and wildlife in this volume, photosynthetic energy capture by green plants is discussed as the foundation of ecosystem productivity and the base of the grazing food chain. Most plant productivity, however, remains unused and is returned to the soil surface as litter where it supports another food chain of considerably more importance in most ecosystems. This pathway, known as the detrital or microbial food chain, is supported by the breakdown and decomposition of organic material and is responsible for the return of nutrients to forms in which they are again available for uptake by plants.

A vast and prolific array of specialized fauna and flora inhabiting the soil is associated with decomposition and remineralization processes (Alexander, 1977; Gray and Williams, 1975; Tansey, 1977). These include larger soil animals such as insects, mites, spiders, and worms, which assist in the mechanical breakdown of litter. By increasing soil aeration and moisture retention they also enhance soil respiration. But of major significance is the largely microscopic microbial mass consisting of bacteria, fungi, actinomycetes, and yeasts which through biochemical action liberate the nutrients bound up in plant and animal remains.

Among the vital soil processes associated with complex microbial activities are those resulting in the fixation of gaseous nitrogen and the remineralization or release of inorganic nitrogen from organic compounds. The nitrogen and phosphorous cycles are diagrammed in Fig. 1.1. Nutrients circulate continuously in the soil-plant ecosystem and the greatest stores tend to be in the organic matter, released by the activities of microorganisms. There is input of nitrogen by fixation and phosphate by release from soil minerals, and little is lost. It is estimated that nitrogen-fixing microorganisms presently supply close to 70 percent of global nitrogen, or some 175 million tons of nitrogen annually (Burns and Hardy, 1975).

Climate has perhaps the paramount influence on soil formation, and dynamics and the magnitude of the earth's climatic diversity have resulted in the evolution of a great variety of soils. The extreme heterogeniety of tropical environments is reflected in the variability of soils. Characteristic temperature and precipitation extremes, among other environmental factors, greatly affect the intensity of physical, chemical and biological processes in tropical soils. Rates of biological activity are much greater under high temperatures when moisture is not limited, with weathering often more advanced, and leaching more pronounced. Soils formed in these humid tropics are often greatly different in behavior from those formed in subhumid, temperate regions or the arid tropics.

Natural vegetation also affects the properties of soils. Soils formed under tall prairie grasses, for example, generally have higher inherent productive capacity than those in humid tropical forests where most nutrients have been leached or remain bound to the vegetation and cycling systems are more closed and direct. Figure 1.2 shows the general relationships among climate, vegetation, and soils.

During the first half of the this century, most research on soil characteristics and productivity was conducted in the temperate climatic regions and was geared to the needs of socioeconomic and agricultural systems which developed there. Attempts to transfer agricultural technologies and management approaches to the tropics without adaptation were of limited success. The research which did take place in the tropics was largely confined to the requirements of cash crops for the export market. During the past 20 years, greater attention has focused on the soils of the tropics. Research during this period has documented the differences between temperate and tropical soil resources and clarified environmental factors of critical significance to their use and management.

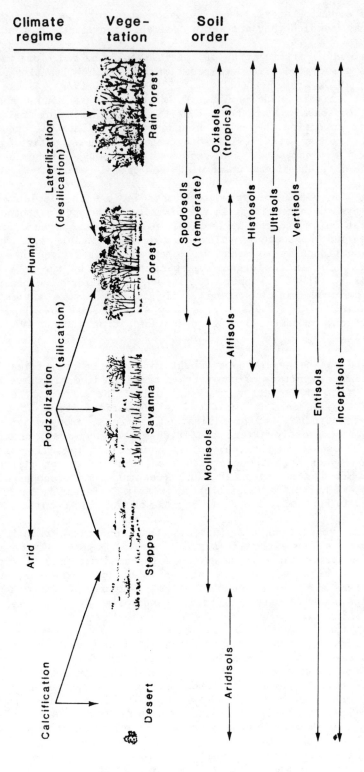

Figure 1.2 General relationships among climate, vegetation and soils. [Source: Erhlich et al. (1977)]

Introduction   197

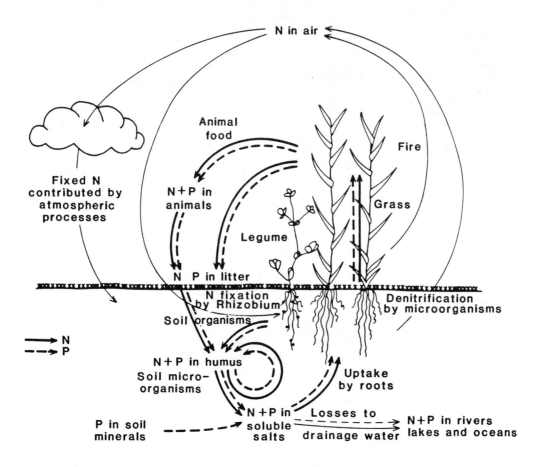

Figure 1.1 The nitrogen and phosphorus cycles and the soil. [Source: Bradshaw and Chadwick (1980)]

Because human activities have become such powerful agents in the disruption and acceleration of nutrient cycling, a better understanding of this phenomenon in tropical ecosystems is essential. Land clearing for agriculture, for example, results in the replacement of relatively large and stable nutrient pools (found in trees) by small pools with rapid turnover typical of annual crops. Unless care is taken, such conversions can result in considerable loss of the accumulated biological capital these nutrients represent, and, in consequence, diminished productivity (NRC, 1980). Figure 1.3 shows some of the relations between deforestation and potential crop failure that need careful management attention. Deterioration also occurs when nutrients removed by cropping exceed inputs or when removal of vegetative cover accelerates leaching and erosion. For these reasons, an understanding of nutrient cycles, the biological mechanisms that drive them, and the environmental factors that influence their flux is essential in maintaining or increasing the productivity of ecosystems, even those converted to intense human use.

The usefulness of an ecological perspective does not end when natural ecosystems are converted to human use. Even the most intensively managed cropland depends on the continuation of ecological processes and

198   III. SOILS

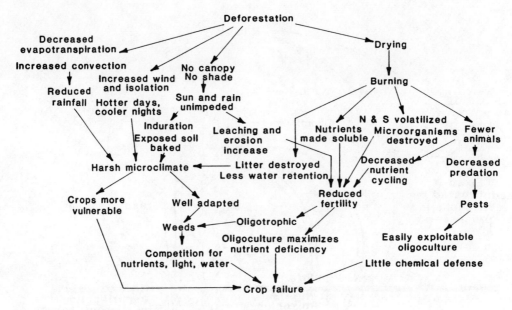

Figure 1.3 The relation between deforestation and potential crop failure.
[Source: Goodland and Irvin (1975)]

the essential functions they perform. Cultivated land can be seen as an agricultural ecosystem or agroecosystem. Human management inputs reduce competition and provide services which the natural system formerly paid for out of its own photosynthetic energy budget. With its reduced respiration or maintenance costs, a greater portion of a system's productivity can be channeled into yields.

Each agroecosystem has thresholds beyond which the deterioration of ecological services (for example, soil maintenance) exerts costs (for example, erosion) which can outweigh the incremental benefits in additional yields achieved by surpassing them. Because these thresholds are poorly understood in many cases, initial baseline studies and subsequent monitoring play an important role in guiding adaptive management responses. Aspects of soil degradation usually considered in baseline studies include water and wind erosion, loss of organic matter content, soil compaction, surface crusting, loss of fertility, waterlogging, and salinization. These and methods for their study are discussed in chapter 6.

Agroecosystems present complex problems in analysis. Basic ecological processes continue to operate although they are manipulated in favor of desired crop species. Major modifications occur in inputs, such as fertilizers, pesticides, and management energy, and in outputs, such as, harvest, leaching, and run-off. These modifications affect structure and rate relations within the system. In the most general sense, the natural system is shifted from self-maintenance to cultural control and the economic system governing inputs and outputs thus becomes an integral component of the agroecosystem to be reconciled with its overall sustainability.

The objectives of those who develop methods for estimating carrying capacity and for analyzing sustained yield agroecosystems are the same: how to derive enough information about the potential in soils, and plants, and water, and cropping of herding practice and management, to estimate how yields and capacity can be sustained (carried) over time. It is well understood that climate fluctuates, that insects invade and other factors intervene. Although the terms carrying capacity and sustained yield (or sustainable development) are independently used, they can mean the same thing. Analyses of fundamental ecological processes, such as those resulting from long-term field programs, are in the realm of basic research and potentially provide to more accurate projections of yields or capacity.

Much remains to be learned about the prerequisites of sustained yield agroecosystems in the tropics, and, in particular, specific features of the fundamental ecological processes underlying them (Janzen, 1973; NRC, 1980). Relatively more is known about the socioeconomic component of tropical agriculture and the ways in which farming systems in tropical countries, especially those of small holders, differ from those in temperate developed regions. To be of practical value in development, soils information must be responsive to the environment of the environment of the farmer as well as to that of plants. Baseline studies encompass both these dimensions of the environment.

1.4 Resource Inventories and Baseline Studies: The Basis of Soil Management

Baseline studies must account for the interaction of socioeconomic, biological, and physical processes in a dynamic system. Directed change to increase productivity is the basic objective of development. Establishing baseline information permits the monitoring of these changes, and the evaluation of progress, as well as points to the need for and corrective action when adverse trends are revealed.

Socioeconomic information derived from baseline studies of the man-soil-environment interface is especially needed in developing countries, where farming systems are profoundly different from those of the industrial countries. The differences affect the ways soils should be assessed.

A soil survey describes the location and extent of soils which are typed and systematically classified on the basis of a standard array of soil characteristics and properties, but these are open to wide interpretation. Classification is usually referenced to a natural or universal system such as soil taxonomy (Soil Survey Staff, 1975), but interpretation must be targeted to the specific needs of local farming systems, whether small holder cultivation or high-input, mechanized production of export crops. A soil characteristic that may be a benefit in rice production can be a liability to corn or cassava. Various technical soil classifications and related land evaluation approaches that integrate socioeconomic, soil, and broader environmental features in ways responsive to developing country needs are outlined in chapter 3.

1.5 Survey of the World's Soils

Soil resource inventories are generated through soil surveys, and the data, presented in the form of soil maps, become a basis for development and management decisions. The extent to which soils have been mapped varies greatly in different parts of the world, with wide differences in the survey methods used and the level of detail possible in the resulting maps.

There are hundreds of agencies and groups engaged in soil survey. One of the most ambitious efforts is the Soil Map of the World, recently completed by the Food and Agriculture Organization (FAO) and the United Nations Economic, Scientific and Cultural Organization (UNESCO). It is a compilation and extrapolation of existing soil survey information from many sources. Each map sheet provides the source of the information used, estimating the percent of the area covered by a soil survey and lists all the soil surveys used for making the map. Thus the Soil Map of the World provides a quick look at the soils of the world with a uniform legend but at a very small scale.

1.6 Soil Information Management

For more than half a century soil scientists in many countries have been collecting data related to specific soils and their properties. These inventories of soil resources have generated the maps, but a continuing challenge has been the development of a method to deliver the information to decisionmakers in a useful, accurate, and timely way. The growth in volume and complexity of potentially useful soils data prompted the International Society of Soil Science, more than a decade ago, to create a working group on soil information systems. The group has met several times in different countries to exchange and examine approaches to "viable, productive, self-supporting scientifically sound soil information systems usable by the rich and the poor" (Bie, 1980).

Perhaps a more appropriate term to use in describing these systems would be soil information management systems, since the purpose for such systems is to manage information about soils. When a decision-maker needs information about a single feature, such as a map of the drainage characteristics of a specific area, or a chart listing soil series, areal extent, potential productivity, and erosion hazard of each series, the information management system should be able to extract just that information from the mass of data stored and deliver it in a usable format. Although most of the development and use of these systems has occurred in industrialized countries, systems may soon extend to other areas of the world. Soil resource assessment calculations may have their steps burned into a chip or integrated circuit--special function boards (collections of chips and associated circuitry) representing several assessment techniques may be available as options, to be inserted into the system for specific resource management problems. Analyses involving the use of digital terrain models, soil erosion prediction models, certain types of multivariate statistics, and other quantitative methods may be adapted in similar ways.

Such information management systems are likely to be relatively inexpensive in the near future. The microprocessor and peripherals (screens, printer, disk storage) will constitute the main costs. Bie (1980) suggests that the "resource assessment microprocessor" (RAMP) is clearly within our technical capabilities to create. If the industrialized nations with their resource assessment specialists can make such a system available for general use, the developing countries may well gain the incentive to make significant investments in the collection of data for the assessment of their own natural resources. Such growth would, no doubt, give rise to methodological advances for the benefit of all nations.

Development for Sustainable Production: Needed Soils Information

A fundamental question is addressed in this section. What information about soils is needed to enable policy-makers and planners to develop cost-effective, sustainable programs and projects for agricultural development? Before we answer this question, some elementary comments on agricultural development may be helpful.

2.1 Perspective on Agriculture

The primary goal of agricultural development programs is to increase production per hectare and per person in ways which are sustainable over the long term. Development in this sense has both an economic and an ecological objective. The economic and ecological objectives are complementary, not competitive. Sustained economic growth requires attention to the integrity of the ecological system.

The economic objective is to achieve the most with the least. For the agricultural development planner, this means achieving the largest increases in yields at the smallest cost. In both economic theory and the real world of project planning, this involves complex choices. First, from a wide array of competing strategies, the one that will provide the greatest return with scarce means must be selected and at the same time benefits from the chosen course must be measured against the cost of other opportunities lost by the choice. Second, in deciding from among the chosen opportunities, the combinations that will be most efficient must be selected.

These choices force the planner to appraise carefully the resources at his disposal--land, labor, and the array of goods and services which investment funds can supply--and to calculate how best to deploy the resources.

In making these calculations it is essential that all costs and benefits be counted, including those that are external to the project under consideration. For agricultural projects, for example, the downstream disbenefits of fertilizer or pesticide pollution or of accelerated sedimentation of reservoirs must be counted as costs of the project. Too often in project planning these costs are ignored. Yet they are a drain on the nation's resources in precisely the same way as the on-site costs of the project. To be sure the off-site costs usually are harder to calculate, but this is not an argument for ignoring them.

The most effective way to achieve the most with the least is to take full advantage of the natural resources. For agriculture these include climate and weather, biological control of agricultural pests, and, of course, the soil resource itself. The soil as a storehouse of nutrients and a medium for plant growth is a form of natural capital accumulated over long periods. In a sense the ecological objective of agricultural development is to maximize the benefits, utility, and proportional contribution of the ecosystem to goods or services that must be imported or purchased.

With human use, the ecosystem is modified either to increase its overall productivity or that proportion of it available to human use. Two points require emphasis and suggest the context and kinds of soil information required by planners and the role of soil resource inventories and baseline studies in acquiring it.

First, agriculture has been, is, and agricultural development continues to be, a process of adapting ecosystems to human use, of converting them to agroecosystems which will supply growing needs for food and fiber. Adaptation is a two-way street, and development plans must be responsive to the specific features of each system. This requires improved understanding of soils, culture, and the other resource domains. Development approaches derived from one ecological region may require inappropriate modifications and costs in another region.

Second, because agricultural development may induce extreme changes, information is needed on the capacity of both the particular ecosystem and its soils to adapt to change. Experience in many types of agricultural development suggests that sustainability requires that the changes induced by development not violate the integrity of the essential ecological processes on which production depends. Of vital importance, for example, is the retention of habitats for beneficial animals and insects--pollinators and predators and parasites of crop pests. Protecting the processes of soil regeneration and nutrient cycling that maintain soil quality is equally essential.

2.2 Categories of Soils Information

With the goal of increased production in agricultural development and the accompanying economic and ecological considerations, three general categories of soils-related information can be distinguished for purposes of discussion. This discussion will introduce some principles to assist planners and policy-makers in determining the specific kinds of soil information needed to formulate and implement agricultural development programs. The three categories of soil-related information are:

1. Inventory information groups individual soils into units defined by their characteristics, properties, and evolution and describes their spatial distribution in the landscape. Such information also reveals conditions of use and opportunities for development.

2. Baseline information on man-soil-environment relations involves the status or condition of the soil resource in the context of the agroecosystem. This includes both socio-economic and farming-system information and information on critical ecological processes; it helps to define specific soil-related requirements and key environmental parameters for the system. It also provides a basis for monitoring subsequent changes and taking corrective action when necessary.

3. Management information is used to integrate resource inventory and baseline information, identify for particular soils and cropping systems specific constraints to development, and guide the formulation of plans or practices to overcome them and improve utilization.

Among these, basic inventory information on soils is of fundamental interest to the planner. At present little more than one-fifth of the world's soils have actually been surveyed, with the highest percentage in Europe and the lowest in Africa (Dudal, 1978). More large-scale surveys are needed, and remote-sensing techniques can reduce costs, increase the speed of survey at smaller scales, and identify, for more intense study, soils with greatest agricultural potential. Considerable effort is being expended in many developing countries on such soil resource inventories (SMSS, 1981), and they have been the subject of a major evaluation carried out by Cornell University and the Soils Management and Support Services (SMSS) program of the U.S. Soil Conservation Service supported by the U.S. Agency for International Development (AID). This comprehensive evaluation is a valuable reference for all phases of inventory design, implementation, and analysis (see chapter 7).

In developing countries the obstacles to success in soil resource inventories are great. Funds are often insufficient and administrative support and appreciation for their value is still limited. It should be emphasized that, although reliable information about soils themselves is necessary for cost-effective planning for agricultural development, it is in no sense sufficient. Information about markets, transport and other infrastructure needs, and sociocultural factors determining farmers' responses to government programs and incentives is at least as important, as information about soils. These nonsoil factors differ widely between countries and within them, but details about these information needs are for the most part beyond the scope of this discussion. Those which are most frequently associated with soil resource development and are important in determining specific soils-related information needs for planning are considered generally in this discussion; they pertain to baseline studies. Together with a soil-resource inventory, baseline information on these man-soil and soil-environment interfaces, as they occur within the agro-ecosystem, provide a foundation for soils management in the context of agricultural development.

The categories of inventory, baseline, and management information are not mutually exclusive. They interact and affect the kinds of information collected and its uses. Furthermore, they need to be considered together because information pertinent to each often has both a common source--the soil survey--and a common mode of presentation--the soil map.

Soil surveys are the principal mode by which soils information is acquired. Field data from a well-designed soil survey, backed by laboratory analysis not only provide basic inventory information, they can also frequently satisfy many baseline needs and indicate what, if any, supplementary studies may be required. Further, they can establish the basis for prediction and interpretation necessary for many soils mangement and development decisions.

It should be remembered, however, that even the most complete, well-collected, and potentially useful information is worthless until it is organized in a form which has practical value for planning decisions. Since more soil surveys fail on this criterion than any other, aspects of mapping, scale, and legend design are also extremely important to successful surveys. More discussion of soil survey, its methods, and related scale and mapping concerns can be found in chapter 4.

2.3 Determining Soils Information Needs

The information needed on soils will depend, of course, on the type, extent, and purpose of the agricultural development planned. Some stages of development planning require broad information about soils across extensive areas; others are concerned with particular sites. The information required to formulate programs and evaluate alternative uses of the soil, for example, is very different from that required for project locations, design, and implementation.

It is not possible here to consider the vast range of development objectives that might be under consideration. It is only practical to discuss the concerns that are likely to be common to most agricultural development projects. In the coming decades, higher yields and more valuable crops must be drawn from lands that are already farmed, and hundreds of thousands of hectares of new land must be brought into production. It can be assumed that the lands in use and their existing levels of production reflect limits of various kinds that have prevented higher yields, just as problems in marginal lands have presented obstacles to agricultural expansion. Many of these limits are socio-economic or institutional but many we know to be climate or soil-related.

A concern common to all agricultural development projects is the capacity of the soil and land. Disruption of the complex, interacting, chemical, biological, and physical properties of soil can be rapid when land is turned to new uses--for example, to row crops from forest or to mechanized production from hand labor. It is essential that information be collected to permit prediction of the effects of change. Information must also be collected periodically to monitor the changes.

The soil information needs of developing countries can focus on general themes: intensification and conservation. Agricultural intensification subsumes the range of development programs and projects associated with the efforts to increase food production described above. Soil resource should be an integral aspect of agricultural intensification and figures directly in the sustainability of increases in food projection. But conservation also subsumes nonagricultural programs and projects where information needs differ and will therefore be discussed separately.

2.3.1 Agricultural intensification.

Agricultural development to increase food production, as noted in section 2.1, has both economic and ecological dimensions, and both are associated with processes of intensification. In the ecological sense, agricultural intensification involves increasing the yield or rate at which an agroecosystem produces useful crops. In an economic sense, agricultural intensification means increasing the real value of agricultural production per hectare. This may be done either by shifting the land from less valuable to a more valuable use or by increasing the per-hectare output value of current land use. In either case, agricultural intensification requires an increase in the per-hectare quantity of nonland inputs (labor, fertilizer, machinery, number of animals, and so forth) more management skill, or, more typically, both.

This definition is consistent with an agricultural development policy that would reduce intensification on overexploited land such as steep hillsides planted in row crops within the catchment areas of impoundments for irrigation or hydroelectric power generation. Although many indigenous cropping systems in the Andes, East Africa, and South Asia were adapted to steep slopes, row-cropping for the market, as practiced by recent settlers, may not be as well suited to the conditions. The social costs of accelerated downslope erosion, altered runoff, and sedimentation may exceed the net value of the land's output. Shifting the land to a year-round cover of grass or production or protection forest, while possibly reducing the value of its per-hectare production, might add more to the social product than row crops. Thus, a policy to achieve more intensive use of the land in general may well mean that on some land intensity will decline.

Agricultural intensification, in its traditional sense, usually occurs on land that is already cultivated; however, colonization of unsettled areas to expand agricultural output is also a form of agricultural intensification, as is land reclamation through irrigation and drainage. In each of these cases, something is done to the land to increase the real value of agricultural output per hectare. In general, the forms of soils information needed to plan effectively for agricultural intensification are not different whether the focus is on already occupied land, colonization of new land, or land reclamation. Consequently, the soil information needs for these three modes and others can be introduced together, with the important differences in degree and emphasis of required information addressed in greater detail later in this discussion and in chapter 3.

Soil surveys will inform planners about what soils exist where so that they can plan how best to use them. In determining the value of a soil as a resource, differences among soil covers usually are addressed. Some of the principal differences examined are the arrangement of layers or horizons in the soil, color, texture, structure, depth, consistency, and reaction such as pH. As a whole these characteristics determine differences in its physical, chemical and biological properties. Figure 3.1, in the following chapter, shows a soil horizon in an idealized profile and the relation of the profile to the landscape.

Variations in climate, vegetation, and parent material are responsible for many of the differences, as are aspects of location such as landform, slope, and drainage. On the basis of these factors and others

TABLE 2.1 Soil and site characteristics, related land qualities and methods of assessment. [Source: McRae and Burnham (1981)]

Climate characteristics and related land qualities

Climatic characteristic	Main methods of assessment	Related land qualities
Temperature	M S DF	Frost risk; temperature regime, (length of growing season, etc.); moisture availability; evapotranspiration
Precipitation, including distribution and intensity	M S DF	Water erosion hazard; flood risk; moisture availability
Wind speed and direction	S DF	Evapotranspiration; exposure; climatic hazard (storms)
Net radiation	S DF	Evapotranspiration
Hail/snow	S DF	Climatic hazard
Evaporation	S DF	Evapotranspiration

Topographic characteristics and related land qualities

Topographic characteristic	Main methods of assessment	Related land qualities
Slope angle and length	R	Ease of cultivation; local access; water erosion hazard; civil engineering factors; irrigability
Altitude	R M DF	Climatic predictions (temperature, length of growing season, rainfall, exposure)
Landscape position including aspect	R M DF	Related to soil-mapping unit; climatic factors (temperature regime, exposure, frost risk); ease of cultivation; water erosion hazard; wind erosion hazard; salinity/nutrient availability; drainage; civil engineering factors; flood risk

Soil and site wetness characteristics and related land qualities

Soil and site wetness characteristic	Main methods of assessment	Related land qualities
Depth to water-table	M DF	Moisture availability; drainage and aeration; civil engineering factors
Presence of springs	R DF M	Ease of cultivation; civil engineering factors
Frequency of flooding	DF	Flood risk; civil engineering factors

TABLE 2.1 *(cont.)*

Soil characteristics and related land qualities

Soil characteristic	Main methods of assessment	Related land qualities
SOIL CHARACTERISTICS COMMONLY USED		
Soil texture and stoniness	M DL DF	Ease of cultivation; moisture availability; drainage and aeration; fertility; water erosion hazard; wind erosion hazard, soil permeability; irrigability; rootability
Visible boulders/rock outcrops	R M DF	Ease of cultivation; moisture availability
Soil depth	M DF	Moisture availability, ease of cultivation, rootability
SOIL CHARACTERISTICS SOMETIMES USED		
Soil structure, including pans, crusting, compaction	R M DF DL	Wind erosion hazard, water erosion hazard; rootability; moisture availability
Organic matter and root distribution	M DF DL	Moisture availability; wind erosion hazard; water erosion hazard; ease of cultivation
pH (reaction)/CaCO$_3$/gypsum	M DF DL	Soil fertility; soil alkalinity
Clay mineralogy	M DL	Water erosion hazard; ease of cultivation
Chemical analysis, e.g. extractable NPK, toxic constituents	DL	Fertility (i.e. nutrient availability); toxicities
Soil permeability	M DF DL	Drainage and aeration; moisture availability; irrigability
Available water capacity	DL	Moisture availability
Infiltration/run off	DF S	Water erosion hazard
Soil salinity	DL	
Soil colour and mottling	M DF R	Drainage and aeration
Soil parent material	M DF	Fertility (i.e. nutrient availability including deficiencies and toxicities)

R—remote sensing; M—maps; S—spacially located data; DF—direct observation in the field; DL—direct measurement in the laboratory.

that permit ever finer distinctions, each soil can be specifically described. Classification can vary in the amount of detail, from the farmer's field, to a watershed, to a region, to an entire country, depending on need and the resources available for collection of the data.

It is the combination of a soil's characteristics and location that determine its value and suitability of the land for specific objectives. Table 2.1 illustrates the relation of some soil and site characteristics commonly collected in soil surveys to assess land qualities and how these characteristics can be assessed.

If land quality is to be assessed for feasibility of colonization schemes, for example, it is vital to know how the productivity of the soil will be affected by replacing its existing cover with crops or livestock forage. Also essential is detailed information about soil properties as they affect nutrient and water-holding capacity in the target region and the response to repeated tillage or trampling by animals.

In making judgments about the potential benefits of irrigation over dryland farming, it would be useful to have information like that provided by the U.S Bureau of Reclamation's Irrigation Suitability Classification system (see section 3.3.5). The classification system shows the amount and location of land for which soil droughtiness or dryness of climate limits the utility of land for crop production. Although such information alone is not sufficient to define the relative merits of irrigation or dryland farming in development policy, it is a valuable part of the necessary information.

In general the context in which a determination of a soil's suitability can be made must also include: (i) an appraisal of other land attributes affecting use, (ii) information on soil and land dynamics under use in the ecosystem, (iii) the specific soil needs of the crop intended, (iv) other land and soil requirements for farming system operations, and (v) socioeconomic factors affecting process, availability of inputs, degree of management, and so forth.

Soil and land classifications and evaluation are discussed in detail in chapter 3. What should be stressed here is that planners need information on soil characteristics and properties which is targeted to the kinds of intensification contemplated. As development efforts are likely to be focused increasingly on marginal lands and problem soils, attention must be directed to those aspects of soils that exert the greatest limitations to expanded use or yield increases, namely, ease of land preparation and management operations, response to management, and susceptibility to degradation.

The suitability of various soils to the range of management activities associated with land preparation is a critical factor influencing the kinds and levels of development that can be supported. Slope, depth, stoniness, wetness, and field size, for example may determine the types of machinery that can be employed in clearing, leveling, diking, as well as in road construction and the overall ease with which production-related operations can be performed.

Prediction of productivity in response to management is most important in determining suitability for development, but estimating productivity is very difficult. Yield predictions can seldom be made on the basis of direct site observation but usually require extrapolations from other geographic areas or from other soils. Moreover, predictions are only useful when yields are related to the set of management techniques needed to achieve them and to the costs measured against the increase predicted. Because of the range of farming systems, it is often necessary to make yield predictions for several levels of management and combinations of input factors.

With information about the inherent productivity characteristics of a group of soils, planners can draw on experimental work and actual farm experience with similar soils in similar climates to infer, with reasonable accuracy, how much particular soils can produce on a sustained basis under alternative systems of technology and management. With additional information about markets, costs of alternative technologies, and availability of management skills specific decisions can be made about what to produce, which technology to use, and so on. The Benchmark Soils Project and the FAO Agroecological Zones Project discussed in chapter 3 provide examples.

Susceptibility of a soil to degradation should be considered early in planning agricultural intensification and is often a soil quality that can be predicted readily from soil characteristics measured during the soil survey. The most serious hazard is soil erosion, but other forms of deterioration are also important. Certain soils, for example, are prone to hardening after clearing, and others to the formation of acid sulfates. In considering irrigation, determination of which areas are highly likely to experience excessive calcium and sodium salt accumulation is needed. Projected soil deterioration, associated with loss of nutrients and declines in organic matter content, is an important consideration.

Together, information on these various factors will provide a basis for determining which soils are most suitable for various development objectives. It also helps to identify the key constraints to intensification and the measures necessary to overcome, compensate for, or eliminate them. These data will help in deciding whether the highest payoff would be in applying more intensive techniques to land already in production, in colonizing new land, or in investing in drainage or irrigation. Reliable soil information will indicate the consequences of these alternatives for maintaining the productivity of the land under the proposed new use. For example, because of high temperature and heavy rainfall, many soils in the tropics have low nutrient-holding capacity. Knowing this will help to prevent costly mistakes in deciding among alternative modes of agricultural intensification.

III. SOILS

Much attention is currently being given to the identification of soil-related constraints to agricultural intensification in developing countries. Present understanding of these constraints has been summarized in the recent *Priorities For Alleviating Soil-Related Constraints to Food Production in the Tropics* (IRRI, 1980). For planners this volume provides extremely valuable guidance on the occurrence of specific constraints by region and the major knowledge gaps which exist for each.

Dudal (1980) summarized many of these constraints and others by major soil group (Table 2.2). In this table, FAO soil categories are given with equivalents from Soil Taxonomy, the official soil classification system of the United States which is used extensively in other countries. The generalizations should be considered only as indicative of problems to be anticipated in intensification and suggestive of soils information that is likely to become important.

Table 2.2. Some major soil-related constraints to agricultural development: Soil classification by FAO legend; Soil Taxonomy equivalents given in parentheses. (*p.p.=pro parte, that is, the two terms correspond only in part) [Adapted from Dudal (1980)]

CONSTRAINTS BY SOIL GROUP

Ferralsols (Oxisols)

- low level of major nutrients
- low cation exchange capacity
- weak retention of bases applied as fertilizers or amendments
- strong phosphate fixation and deficiency on fine textured soils
- sulfur deficiency
- nitrogen leaching under high rainfall without recycling of plant cover
- acidity
- aluminum toxicity
- low calcium content resulting in limited rooting volume and moisture stress hazard
- frequent trace element deficiencies
- microelement toxicities in soils formed from ultrabasic rocks

Acrisols (Ultisols, pp)

- low nutrient levels
- exchangeable aluminum
- nitrogen loss through leaching under high rainfall
- maintenance of nutrient reserves depends on continuous recycling by vegetation unless fertilized
- high sensitivity to erosion in surface layer

CONSTRAINTS BY SOIL GROUP *(continued)*

Acrisols (continued)

- waterlogging during heavy rains
- compaction
- trace element deficiencies, e.g., boron and magnesium

Nitosols (Paleudults, Paleustults, Paleudalfs, Paleustalfs, pp)

- nutrient deficiencies similar but less acute than Ferralsols
- phosphate fixation in eutric types
- manganese toxicity in dystric types

Luvisols (Alfisols, pp)

- nutrient deficiencies related to low activity clay fraction on strongly weathered parent material in subhumid zone
- microelement deficiencies, e.g. zinc, in semiarid zones where saturation complex is dominated by calcium
- moisture stress
- sensitivity to erosion
- low aggregate stability and surface sealing common
- subsurface plinthite, ferruginous concretions, or hardpans in subhumid wooded savannah

Vertisols (Vertisols)

- physical properties hamper tillage
- low phosphorous availability
- texture and expanding-type clay minerals result in narrow range between moisture stress and water excess
- erosion during fallow
- seasonal waterlogging

Planosols (Albaquults, Albaqualfs)

- slow permeability of subsurface horizons and drainage problems
- waterlogging in wet season
- acute water stress in dry season
- strong leaching and low cation exchange capacity in surface horizons result in nutrient deficiencies
- in strongly developed Planosols aluminum toxicity
- cobalt and copper deficiencies for range use

CONSTRAINTS BY SOIL GROUP *(continued)*

Arenosols (Psamments, pp)

- low nutrients associated with coarse-textured quartzy material, low water retention, and low cation exchange capacity
- minor element deficiencies (zinc, manganese, copper, iron)
- low fertilizer efficiency due to high leaching
- sulfur and potassium deficiencies
- weak structure leading to compaction and erosion problems

Andosols (Andepts)

- high phosphate, borate, and molybdate fixing capacity
- nutrient imbalances in basic volcanic ashes in presence of ferro-magnesium minerals
- aluminum toxicity frequent in more acid Andosols
- manganese deficiencies

Podzols (Spodosols)

- acute nutrient deficiencies related to coarse-textured quartzy materials, excessive leaching, and complex organic matter-metallic compounds
- lack of nitrogen and potash
- phosphate availability reduced by presence of exchangeable aluminum
- low retention capacity results in rapid loss of applied nutrients
- copper and zinc deficiencies common
- many have waterlogging in wet season, water stress in dry
- low rate of regeneration of natural vegetation cover

Cambisols (Tropepts)

- similar to more developed soils they are associated with: Vertisols, Ferralsols, Luvisols, Acrisols, Gleysols; but less pronounced

Xerosols (Aridisols, pp)

- moisture stress
- when water is sufficient fertility problems result from high calcium carbonate content and reduced phosphorus availability
- salinity and alkalinity hazards
- iron and zinc deficiencies
- engineering problems in irrigation projects in high gypsum content Xerosols

---
CONSTRAINTS BY SOIL GROUP *(continued)*

Yermosols (Aridisols, pp)

- drought
- fertility problems and salinization hazards similar to Xerosols

Solonchaks (Salorthids and saline phases, pp)

- moisture stress and inhibition of ion uptake by plants as a result of high salt concentration in the soil solution
- excess of elements depending on nature of salt present

Fluvisols (Fluvents)

- flood hazard in areas with seasonal rainfall patterns
- in thionic Fluvisols waterlogging and extreme acidity with drainage
- low pH accompanied by aluminum, manganese, and iron toxicities

Gleysols (Aquepts, Aquents)

- water excess
- effectiveness of nitrogen application hampered by denitrification under wet conditions
- plinthite or concretionary layers common in humid and subhumid zones with danger of hardening or hard pan with drainage

Histosols (Histosols)

- waterlogging
- low bearing capacity
- weak foothold for plants
- subsidence with drainage
- microelement deficiencies frequent
- shrinking of organic material with drying
- acid peat low in major nutrients
- silica deficiencies in highest organic matter content Histosols

---

Swindale (1980) outlined some major constraints to intensification that have implications for the kinds of soil information planners are likely to need. Soil fertility and plant nutrition relations require more effective integration of the information collected. Intensification will require large increases in inputs of nitrogen, phosphorus, and potassium fertilizers and efficiency in their use will depend on a more careful appraisal of needs and responses. Requirements for soil amendments is another area in which planners need better and more detailed information especially with respect to the use of lime, manure, and silicates. Soil acidity is a severe constraint to the use of many nonmarginal lands, and estimation of lime requirements is of great significance.

Assessment of soil physical conditions and constraints to use also needs to assume greater importance as intensification proceeds. Soil moisture, the dynamics of water balance, and overall hydrology of the soil profile are factors that planners will need to consider, as are surface and subsurface impedance and tilth of the soil structure.

Research in tropical soils indicates the particular significance of minerological properties and surface chemistry (Uehara and Gillman, 1981). Variable electrostatic charge and surface phenomena have major effects on fertilizer use and efficiency, and improved information on these properties will be needed. More detailed information on soil biology will also become important, especially with regard to biological nitrogen fixation, mycorrhizal influences on phosphorus utilization, and the relation of soil organic matter and reactivity to fertilizers. Excellent overviews of soil-related constraints by region and suggestions for the kinds of soils information that will require additional emphasis in agricultural intensification efforts within them are included in Moormann and Greenland (1980) on humid tropical Africa, Dent (1980) on Southeast Asia, Sanchez and Cochrane (1980) on tropical America, and Kampen and Burford (1980) on the semiarid tropics.

2.3.2 Conservation.

Monitoring the condition of the soil resource once intensive use has begun is essential if planners are to collect the information they need to direct conservation measures. Conservation concerns not only the soil but the processes of land degradation and other fundamental ecological processes on which soil regeneration and continued productivity depend.

Much of the information useful to soil and land conservation can be aquired during soils survey. But special studies may also be required when more comprehensive data are needed. Because direct observation and measurement of soil and land degradation is often beyond the time and resources available in developing countries, organizations such as the FAO, in conjunction with United Nations Environment Program (UNEP) and UNESCO, are trying to develop methodologies for making indirect (and less costly) estimates of land and soil degradation.

Both soil and land are renewable resources, but "mining" them beyond their capacity to recover or renew inevitably leads to losses in productivity. Loss of land through degradation is of obvious importance to planners, but land and soil degradation is a more insidious process that can eventually affect productivity severely. Overall production potential declines almost imperceptibly as good lands are reduced in quality to average, and average to poor. Higher value crops must often be replaced by those of lower value that are less demanding of the soil. Baseline information on yields, organic matter, and soil depth, will provide a way of monitoring such changes.

Since the ecology of many regions where development can be expected to occur is still poorly understood, planners can expect that the more intensive use of agroecosystems may have unexpected effects on nutrient cycling and the maintenance of soil fertility (Janzen, 1973). Decisions on how to clear land, for example, can have a profound impact on the subsequent agricultural quality of the soil. Among the available guides to help planners determine what soils information should receive emphasis are, for the humid tropics, UNESCO (1978), IUCN (1975a and 1975b), and Farnsworth and Golley (1974); for irrigation, USMAB (1981); and for watershed development FAO (1979).

Conservation and watershed management projects are undertaken because of interdependencies in the way the land is used. For conservation the principal interdependency is between the present and subsequent generations. Watershed management involves interregional interdependence; upland land use, for instance, has immediate consequences for valleys and floodplains. These are not rigid distinctions. Upland erosion of the land may reduce its productivity for future generations, but it may also impose immediate damage downstream in clogged reservoirs and irrigation systems.

Since a principal objective of watershed management is control of runoff in upland areas, information about the water retention capacity and runoff characteristics of soils in the watershed under various cover and alternative technologies is critical (see also Aquatic Ecosystems, part II). Water moves both across the surface of the land and in subsurface flows; information about soil characteristics will help in predicting the relative importance of these two paths of movement under various conditions. Information on socioeconomic factors and farming practices is also required for effective conservation planning (Dudal, 1980).

2.3.3 Specifying information needs.

The last two sections have stressed the importance of targeting soils information to development objectives. A careful assessment of needs is essential in determining the kinds and levels of information that will be most effective. There is no easy guide to fit all circumstances, but there are some general rules to help the planner define the needs in specific cases and select appropriate soil survey methods.

First, soil surveys record the geographical distribution of soils and identify characteristic properties. The measureable attributes of soil, however, are so many, and the intricateness of the patterns of their distribution in a landscape so complex that the information obtained from these should be interpreted and used with care.

Second, it is important to remember that planning decisions are usually made at some remote distance from the soil resources to be developed--often in provincial or national capitals. Not infrequently, the decision-maker may never actually visit most of the areas involved, and the soil surveyor will be the only one to have traversed the area systematically. In summarizing the Cornell study on resource inventories in developing countries and some of its recommendations on soil information needs, Cline (1981) observed that what the planner, though at a distance, needs before him is the same information that would be necessary in an on-site appraisal.

The Cornell study identified five kinds of information needed to predict soil performance that should be gathered in an on-site appraisal. These form the basis of a set of principles that can guide the development planner in specifying what information he needs and may also serve as a series of steps to ensure maximum utility and relevance of the information collected. These steps should be considered both in evaluating existing information and in designing to collect resource inventory or baseline data:

1. The land use objective for which soil resources are to be evaluated must be explicitly established.

2. The level of detail of information that will be required to evaluate soil resources for this objective must be specified.

3. The soil properties that will be critical for the projected land use must be identified.

4. The degree of limitations that critical soil properties could impose on the projected land use must be ascertained.

5. The effects that the geographic distribution of these limiting soil conditions could have in the projected use must be determined.

Each statement specifies information about the soil resource needed by a planner to predict soil performance for a particular land use objective. Obviously, the first step is the most influential: identifying land use objectives establishes the requirements of the remaining four. In a field appraisal, the soil surveyor tests site and soil conditions against the limitations to use. Planners remote from the site, basing decisions on the data collected in soil surveys or resource inventories, require the same kinds of information. Its usefulness in decision-making can be evaluated, first, in relation to the adequacy with which it supplies the information needs determined in steps three through five and the detail of step two for the requirements of each land use objective contemplated; and, second, by the effectiveness with which this information is transformed into a map and accompanying text which permit it to be used. Key criteria here are the quality and ground control of the base map, its legibility, and the reliability of both map and associated text.

Classification and Evaluation of Soils and Land Resources

3.1 Introduction: Classification and Evaluation

This discussion of the methods by which information on soils is compiled and presented for resource management will include a review of national and international methods of soil classification (section 3.2), methods of soil and land evaluation (section 3.3), and indigenous systems of soil classification and methods designed to integrate the work of soil scientists, ecologists, agronomists and anthopologists (section 3.4). The soils information required for agricultural development are derived from two basic processes--classification and evaluation. The distinction between them is not rigid but the different processes indicate the need for an important shift in emphasis and orientation during the soil surveys. If the division between classification and evaluation is too pronounced, however, a gap that is difficult to bridge can open between the types of information collected and the types of information planners need. This is a problem common to soil surveys as conducted in many developing countries.

3.1.1 Soil units for classification.

Soil classification systems permit soils to be grouped into categories that are useful in understanding the natural system. The most generally accepted concept of a soil is that it is a natural body formed from interactions of parent or rock materials, climate, living matter, topography, and time. Its morphological, physical, and chemical properties can be described. To be classified individually, soil units must differ in one or more properties to such a degree that the combination of all the properties results in a different response to management treatments, to engineering manipulations, or other uses that require appreciable investment (Soil Survey Staff, 1975).

Soil units on a soil map may not necessarily be separated by a distinct boundary from adjoining units. The differences between soils in the transition or boundary areas may be almost indistinguishable because of the gradual changes in one or more soil boundaries. A boundary may be abrupt, as along a vertical scarp, or, as is more common, gradual over a distance of a number of meters, or even several kilometers.

There are two general kinds of soil classification, technical and natural. Planners are concerned with technical classifications because they provide answers to specific questions; however long-term planning should also be concerned about natural classifications because soils play a fundamental part in ecological processes, and that in turn in the long-term viability of soils for all purposes. A technical classification or, more properly, a technical grouping of soils defines them according to response for a specific use or management, highlighting one or more characteristics or conditions--for example, steeply sloping soils, poorly drained soils, soils requiring phosphorus or lime, and soils suitable for septic tanks or other uses. Because the soil properties identified are usually limited to those significant for specific objectives, the data are usually not useful for other groupings. Most land evaluations for agricultural development are of a combination of such technical groupings.

A natural or taxonomic classification, on the other hand, is designed to show relations among the greatest number and the most important soil properties, without a specific practical objective. It is a multiple-category system in which the highest category has a small number of general classes with a few differentiating characteristics and the lowest category has a large number of classes with many differentiating characteristics. The former can be said to have a high level of generalization, the latter a low level of generalization. For example, in the U.S. Soil Taxonomy (Soil Survey Staff, 1975), the highest category is the order, in which there are ten classes or taxa, and the soil series is the lowest category with about 12,000 classes.

Ideally, technical soil groupings should use the classes of the lowest category of a natural soil classification in order to have uniformity in the units needed for the technical groupings. Too often, however, this is not done. A complete discussion of the principles and bases for the different kinds of soil classification and interpretations can be found in Cline (1949); Bartelli (1978 and 1979); Orvedal and Edwards (1942); Buol, Hole, and McCracken (1980), and Beinroth (1978).

Natural soil classification systems are intended to be useful for a variety of purposes, and data collection for them is guided by the scientific standards of the soil system in use for the primary purpose of solving the problem of classification. Beek (1981), among others, has argued that the primary problems which soil surveys in developing countries should address are not those of soil classification but those of land use. This is not to say that a properly made soil profile description does not supply a great deal of valuable data about the internal and external characteristics and certain of the properties of the piece of the landscape in which it exists. In the land use problems posed by agricultural intensification, information on these characteristics is of obvious relevance.

3.1.2 Describing a soil profile or pedon.

The basic data for natural soil classification systems are descriptions of soil profiles, or *pedons*. Profiles are vertical cuts made into the soil to expose its layers or horizons. The composition and arrangement of soil horizons in the profile are the major determinant in classification; soil profile descriptions are based on comparisons of each horizon to descriptive standards (Fig. 3.1).

In soil taxonomy, for example, data are collected for each horizon on its depth, color, texture, structure, reaction, and boundary. Also, the site is described with regard to its climate, parent material, landform, relief, slope, aspect, erodibility, permeability, drainage, groundwater, soil moisture, root distribution, salinity or alkalinity, and stoniness. Samples are taken for laboratory tests on physical, chemical, and behavioral properties.

Soil surveys would make a greater contribution in developing countries if soils information pertinent to development problems were their direct objective rather than an incidental by-product of classification. For instance, the survey might emphasize soil properties critical to soil use, especially those hindering intensification and development. But because classification of soils tends to focus on permanent and unchanging factors, the uppermost layer of soil, down to the plow line, is not empha-

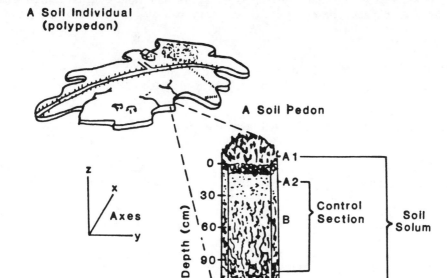

Figure 3.1 A soil individual is a natural unit in the landscape, characterized by position, size, slope, profile, and other features. [Source: Buol et al. (1973)]

sized since it is disturbed and does not reflect original conditions. Yet this zone is often the most important to crop performance.

It is information about the dynamic properties of soils and its response to use that is in shortest supply. Many of these properties can be deduced from the classification categories (Eswaran, 1977), but such interpretations are usually beyond the expertise of most decision-makers. Other soil attributes that limit performance, beyond those which are individually self-evident such as acidity or slope, can be inferred from the measured soil characteristics and properties in the profile description. The point to be made, however, is that greater attention to these factors during the survey can save time, increase the value of the data collected, and facilitate their use by the decision maker.

The soil evaluation process can provide the essential interpretative bridge between soil classification and use. Soil surveys oriented toward the solution of land use problems will demand complex information about the soils. Soil evaluation requires broad environmental information as well s an understanding of farming requirements. Soil and land evaluation is thus a more comprehensive process than soil classification because it looks at interactions in the agroecosystem that must be assessed in order to identify specific potentials for and limitations to

220   III. SOILS

development. Some of the approaches to soils and land evaluation that may be useful in developing countries are described in section 3.3.

3.2 Soil Classification Systems

3.2.1 National systems of soil classification.

The earliest efforts to classify soils were based either on physical properties important to agriculture, such as depth, texture, structure, or stoniness; or properties related to the geological origins of the soil. Early classification systems developed for soil survey and mapping used a descriptive hierarchy with series (soils derived from similar geological material), types (soils within a series with similar textures), and phases (soils within a type with similar depth and stoniness). These systems were applied to physiographic regions or soil provinces. The series has remained the basic unit of soil mapping in many countries at scales of 1:50,000 or greater although there are many national variations.

At the end of the 19th century Russian soil scientists developed the concept of soils as "natural bodies" and of zonal soils, with well-developed climatic-vegetation zones. Natural, genetic systems of soil classification based on soil types, were introduced in North America in the 1930's and became the basis of the official system adopted in the United States in 1938.

Most natural soil classification systems are primarily national systems that were developed to meet the conditions and needs of a particular country. In addition to U.S. Soil Taxonomy and the 1938 U.S. Department of Agriculture (USDA) classification some of the more established systems are shown in the accompanying listing, "Various National Systems of Soil Classification" from Butler (1980). Many of these systems were introduced during colonial rule and some continue to be used in various donor-country technical assistance programs. In many developing countries, soils information may be available in widely varied forms, resulting in a confusing array of terminology and names to describe similar processes, horizons, and profiles. There are other national classification systems such as the Brazilian, Belgian, and Hungarian to name a few, but those summarized above are representative of those in use.

Various National Systems of Soil Classification

Russian soil classification	Partial hierarchial system that emphasizes processes of soil formation in relation to environment.
Australian system	Initiated with a key; no conceptual basis for differentiating soil characteristics; simple, defined profile characteristics, most of which can be readily determined in the field.

| Kubiena's system for European soils | Has six categories: divisions, classes, types, subtypes, varieties, and subvarieties. The system is designed around keys which use diverse criteria, viz. morphological characteristics and broad environmental and soil-process attributes. |

| Soil classification for England and Wales | Has 10 conceptual classes (major groups), two lower categories formed by an elaboration of these, namely 43 groups and 109 subgroups. The subgroups are further subdivided into soils series which is the mapping unit in detailed soil survey. The classification is based on soil profile characteristics and excludes site and climatic factors. |

| Canadian system | Has three categories: orders (8), great groups (23), and subgroups (165). The subgroups are subdivided into families (800-1000) which are groupings of about 3000 soil series. The family groups are separated according to texture, texture contrast, mineralogy, depth, temperature class and moisture class. The series is a soil body in the field of limited defined variation in texture, color, and consistency. |

| South African soil classification | Has two categories, 41 soil forms and 504 soil series. Soil forms are defined by reference to five surface horizons and 15 subsoil diagnostic horizons. Differentiating characteristics of the surface horizons are percentage of organic matter, nutrient content structure, percentage of clay, etc.: of the subsoil horizons, they are color, structure, parent material, presence of a bleached or mottled horizon, etc. Soil series are defined as profiles belonging to a soil form which in addition have one or two specific differentiating characteristics such as percentage of clay, grade of sand fraction, acidity or alkalinity, color, etc. |

| French classification system | Widely used in French-speaking Africa and other tropical countries; has 11 classes in the highest category which are differentiated on several different bases: stage of development of the soil, characteristics of the entire profile, properties due to wetness of the soil, and degree of

movement of clay particles within the profile. Most of the classes in the highest category are subdivided into two or more subclasses on the basis of a variety of factors or properties such as climate, topography, iron content, parent materials, etc.

3.2.2 International systems of soil classification.

There are two international soil classification systems now in use in many developing countries: FAO-UNESCO classification system, the basis for the Soil Map of the World (FAO-UNESCO), and the U.S. Soil Taxonomy (Soil Survey Staff, 1975). The need for a universal taxonomic system has long been recognized by soil scientists in many countries.

The FAO-UNESCO classification system is, strictly speaking, not a classification or taxonomic system at all but a worldwide grouping of soils with some attributes of a classification system. The groupings were established primarily to prepare a soil map of the world with a uniform legend. To do this the diverse soil units of different countries were correlated with a set of newly defined soil classes--26 major classes and 106 subclasses. (To correlate similar soil groups from many countries into these classes and subclasses necessitated many compromises that broadened the variances of the classes). The 26 soil classes are shown below:

Fluvisols	Solonchaks	Luvisols
Gleysols	Solonetzs	Polzoluvisols
Regosols	Yermosols	Podzols
Lithosols	Xerosols	Planosols
Arenolsols	Kastanozems	Acrisols
Rendzinas	Chernozems	Nitosols
Rankers	Phaeozems	Ferralsols
Andosols	Greyzems	Histosols
Vertisols	Cambisols	

Many of these class names have been used traditionally for soil groups in a number of countries but were often not defined differently and with varying detail from country to country. An example of how differently named but similar soil groups in various countries were correlated with one of the classes and its subclasses in the the FAO-UNESCO classification is illustrated for Ferralsols (FAO-UNESCO, 1974).

Ferralsols: latosols (Brazil); ferralsols (Zaire); sols ferralitiques (French); oxisols (U.S.A.): lateritic soils, ferralitic soils (U.S.S.R.)

Orthic ferralsols: red-yellow latosols (Brazil): hygroferralsols; hygroxeroferralsols (Zaire); Sols ferralitiques moyennement a fortement desatures (French); red-yellow latosols (Indonesia, Vietnam); orthox, torrox, ustox (U.S.A)

Xanthic ferralsols: pale yellow latosols (Brazil); hygroferralsols (Zaire); sols ferralitiques janunes fortement desatures (French); orthox (U.S.A.)

Rhodic ferralsols: latosols roxo (Brazil); Hygroxeroferralsols (Zaire); sols ferralitiques faiblement a moyennement desatures (French); orthox, torrox, ustox (U.S.A.)

Humic ferrsalsols: humic latosols (Brazil); sols ferralitiques fortement desatures humiques (French); humox (U.S.A.)

Acric ferralsols: acrox (U.S.A.)

Plinthic ferralsols: plinthaquox (U.S.A)

Soil taxonomy (Soil Survey Staff, 1975) is a comprehensive soil classification system developed by soil scientists from many countries. The classification system is used in many countries, especially in developing countries. The aim was to construct a system with greater universal application, especially above the series level.

The system, which is hierarchical, has six categories. From the highest to the lowest level of generalization, they are: order (10), suborder (47), great group (about 230), subgroup (about 1300), family (about 5000 in the United States), and series (about 12,000 in the United States). The orders, differentiated on the basis of profile characteristics, which are the results of broadly differing genetic processes of soil formation, are: (i) Entisols, recently formed soils; (ii) Inceptisols, inception of soil formation with a small differentiation between soil layers; (iii) Alfisols, soils with clay subsoil relatively high in bases; (iv) Ultisols, old soils with clay subsoil and low in base; (v) Oxisols, highly weathered soils with much iron and aluminum oxides; (vi) Vertisols, soils high in shrinking and swelling clays; (vii) Mollisols, mineral soils high in organic matter; (viii) Spodosols, soils with light colored subsurface horizon overlying a subsoil with organic matter and iron accumulation; (ix) Aridisols, soils of arid regions; and (x) Histosols, organic soils such as peats.

The relation between the ten soil orders of the Taxonomy and the great soil groups of the 1938 USDA classification, which has been widely used in developing countries, is shown in the next box (from Buol et al., 1973).

Soil Taxonomy is the official taxonomic system used in the United States and taught in courses in most U.S. universities. This system is used in most technical journals and bulletins published in the U.S. as well. Many other countries use Soil Taxonomy, either as a primary system or as a secondary system. Data on the use of Soil Taxonomy in different countries were compiled by Cline (1980).

Reactions in the international literature to this soil taxonomy have ranged from absolute rejection to substantial endorsement (Cline, 1979). Russian soil scientists have tended to be negative about it; Australian, British, French, Belgian, Canadian, and New Zealand soil scientists were divided. The system has been criticized for its complexity; it is highly demanding and some of the field criteria are difficult to apply. Lack of

Entisols	Azonal soils, some low humic gley
Vertisols	Grumusols
Inceptisols	Ando, Soil Bruns Acides, some Brown Forest, Low Humic Gley, humic gley
Aridisols	Desert, Red Desert, Sierozems, Solonchak, some Brown and Reddish Brown soils, associated Solonetz
Mollisols	Chestnut, Chernozems, Brunizems, rendzina, some Brown, Brown Forest, associated Humic Gley, and Solonetz
Spodosols	Podzols, Brown Podzolic, Ground-Water Podzols
Alfisols	Gray-Brown Podzolic, Gray Wooded, Noncalcic Brown, Degraded Chernozems, associated Planosols and Half-Bog
Ultisols	Red-Yellow Podzolic, Reddish-Brown Lateritic, associated Planosols, and some Half-Bog
Oxisols	Laterite soils, Latosols
Histosols	Bog soils

laboratory support for the determination of some criteria and lack of data from which soil moisture and temperature regimes can be estimated are some obstacles to using the system. Many critics of the taxonomy, nevertheless, have found considerable merit in the concepts and nomenclature of the system. Although it is incomplete for soils of the tropics, several international committees are now considering ways to improve it for use in tropical regions.

3.3 Evaluation of Soil and Land Resources

3.3.1 Introduction.

The evaluation of soil and land is a more comprehensive process than soil classification, requiring broad environmental information which can be collected either directly or indirectly. Direct methods involve estimates of growing crops, collection of information on yields, and field trials (see chapter 5). Direct methods are important in predicting soil performance and can provide a basis for extrapolation to areas with similar soils. In developing countries, indirect methods that make it possible to evaluate the potential uses of extensive areas are usually more practical.

As with soils, many methods of land evaluation are founded on an initial classification of land according to its physical features. "Land classification" is often used synonymously with "soil classification," but the concept of land is much broader than that of soil. Land has been defined as the physical environment including soils, climate, relief, hydrology, and vegetation (FAO, 1974 and 1976). Land classification approaches are outlined in section 3.3.2. A reconnaissance of the broad physical features of the land can be derived from remote-sensing data before subsequent evaluations are made at more intense levels.

All indirect methods of land evaluation assume (i) that certain soils and site characteristics and properties will influence particular land use objectives in predictable ways and (ii) that evaluation can be performed indirectly by deduction when these characteristics and properties are observed. Basically, the intent of land evaluation is to identify what Kellogg (1961) has called "soil qualities" against which the appropriateness of specific land use objectives can be assessed. This concept has been broadened to include site characteristics other than soil that affect land use. Evaluation thus focuses on identifying "land characteristics" that will influence the success of a particular use. Complexes of interacting land characteristics are frequently summarized and referred to as "land qualities" (FAO, 1976).

In developing countries, soil and site information useful for delineating land characteristics and qualities can be collected during the soil survey and should be considered in the design of the survey. General discussion of soil and site attributes that are important in indirectly evaluating land for agricultural and other development can be found in Butler (1964), Mulcahy and Humphries (1967), Riquier, Bramao, and Cornet (1970), Beek and Bennema (1972), Bennema (1976), Beek (1977), Bartelli (1979), and Miller and Nichols (1979). These have been summarized by McRae and Burnham (1981). See also Table 2.1, which lists soil and site characteristics most frequently translated into land qualities influencing use. Tables 5.1 and 5.2 provide information on field methods for measurement.

> Interpretation of soil and site data makes it possible to evaluate land with respect to its capability and its suitability. Land capability is the more general concept and refers to the obvious *limitations* that certain land qualities such as stoniness or slope, impose on use. Evaluation of land capability emphasizes its versatility for a variety of intensive purposes and usually ranks land on this basis. Evalation of suitability emphasizes land qualities in relation to more narrowly defined *uses* such as irrigation or the production of a certain crop. Clearly both suitability and capability evaluations of land can provide information that is directly useful to development objectives and planning.

Factors that should be considered in the selection and implementation of land evaluation methods are discussed in section 3.3.3. In sections 3.3.4 and 3.3.5, specific methods of evaluating land capability and suitability are discussed. First, however, it would be useful to review the three principal approaches to land evaluation.

3.3.2 Approaches to land evaluation.

Mabbutt (1968) distinguishes three main approaches to classifying land for reconnaissance surveys: the genetic approach, the landscape approach, and the parametric approach. With the genetic approach, land is subdivided into natural regions on the basis of environmental factors, particularly climate and structure. With the landscape approach, the land units are components with similar landforms, soils, and vegetation; the units occur in groups that can be mapped from airphotos or images. The parametric approach seeks to establish land types by mapping key attributes in quantitative terms.

The genetic approach was an outgrowth of the development of physical geography in the 19th century under the influence of botanists and geologists concerned with genetic groupings of natural phenomena. This approach may be useful in providing a coordinating framework at a high level, such as the Oxford climatically determined land zones broken into divisions of the various morphostructural units (Brink et al., 1966). The genetic approach is also a way of taking a "first look" at an area before a detailed field sampling effort is initiated.

The landscape approach originated with Unstead (1933), who divided the landscape into distinctive geographic entities identified first by a unity of relief with characteristic structures, hydrology, plant cover, and land use. The concept proved useful as foresters and economic planners began to analyze terrain patterns from aerial photographs (see Bourne, 1931). More recently, satellite imagery interpretation has employed the landscape approach. Although the landscape approach is suited to reconnaissance surveys because recognizability is good and extrapolation from a limited sample base is required, few tests of the reliability of this approach have been carried out to date, and for intensive land use there must be a high level of reproducibility.

Mabbutt (1968) defined the parametric approach as a division of land on the basis of selected attributes such as elevation, with class limits at chosen contours. The parametric approach is claimed to achieve a more precise definition of land and to avoid the subjectivity inherent in the landscape approach. In the future new sensors may be developed to scan attributes formerly inferred from associated features, and the parametric approach favors the handling of electronic data. The choice of attributes and the delineation and definition of attribute classes are problematic as is the fact that soil character is largely hidden from the conventional surface scanning techniques used to define attributes.

In summary, with the landscape approach, the distribution of soil classes is inferred through the surface pattern of causal factors (parent material and landforms) and through their expression in the vegetation, whereas with the parametric approach, the same problem may require new sensing techniques, improvement of the sample base, and resolution of the soil complex into measurable component properties. At present, then the landscape approach offers the possibility of a rapid survey at a low cost.

Greater precision and reliability through improvements in data acquisition technology may make the parametic approach increasingly useful at the reconnaissance level.

3.3.3 Selection of methods for land evaluation.

The success of a land use objective can sometimes be evaluated indirectly if certain soil and site properties whose characteristics have been previously measured and described, are observed anew. Much of the following discussion has been drawn from McRae and Burnham (1981). Before evaluation methods are selected, those soil and site properties that will affect the success of the development enterprize must be identified. This done, the next step is to give the identified properties a range of values to define classes or categories within which any one observation can be placed or ranked. Although some existing classification systems are suitable for most objectives, planners do need to shop around for the best system for their purpose. Modifications will usually be required to adapt to particular needs, local conditions, and availability of data. A wide range of systems with examples of various modifications made to tailor them for use in developing countries is described in the next two sections, on land capability and suitability.

Whatever system is chosen, the more intensive the area being evaluated, the greater the system's dependence on extrapolated data and the greater the need for site checks to ensure precision (ground truth). It is advisable to test both the land evaluation system in its original form and after modifications have been made.

When extensive areas are being evaluated, the next step is to divide the area into uniform divisions of similar land qualities. When soil maps exist, it may be assumed that map units or simple division of them will be similar, but often map coverage is incomplete or legends are not uniform, and the maps can be used only for defining a broad sampling framework. Additional observations or assessments will be needed to define reasonable boundaries of units for evaluations.

Deciding what new information to collect to check or clarify the evaluation must be guided by cost and relevance. Direct acquisition of ground data is always desirable because it is the most accurate and reliable; however, specially commissioned surveys to acquire data for evaluation are expensive. Macroclimate, rainfall, slope, and vegetation-related land qualities can often be extrapolated from sampled sites, but soil-related land qualities are highly variable and difficult to observe from remotely sensed sources. For this reason land evaluations based on detailed soil surveys are the most satisfactory. General purpose soil surveys or resource inventories should be designed to facilitate this use when possible. Because, the major cost element in land evaluation of large areas is field work (see section 4.4.4), careful consideration must be given to a sampling design strategy as well as the intensity and density of information to be collected. As a general rule, it may be more effective in many developing countries to economize on the number of points sampled rather than on the amount of data collected at each. This is the principle behind the benchmark soil concept discussed in section 3.3.5 on suitability evaluation.

3.3.4 Land capability.

Land capability evaluations for agricultural development rank land into categories that reflect increasing limitations to use. Lands with the highest capability are those that will permit intensive use for the widest range of objectives. Land is assigned to categories by comparison of values of selected soil and site properties to the criteria established for each category.

The land capability evaluation system developed by the U. S. Department of Agriculture (Fig. 3.2) has been useful to agricultural management, and many countries have adopted or modified it (Beek, 1978). It should be emphasized, however, that this classification was developed for U.S. conditions almost 50 years ago and is not directly applicable to most developing countries without major changes.

The advantages of the USDA land capability classification are (adapted from McRea and Burnham, 1981):

1) Division of land into relatively small number of ranked categories is easily understood by planners.

2) Qualitative rather than quantitative which is a realistic approach given limited present knowledge of many soil/crop/environment interactions.

3) Versatility--limitations considered can be readily modified to suit local conditions.

4) Easily applied by both experienced and less highly trained staff.

5) Provides a general purpose classification which shows a clear division between lands capable of growing crops and those that are not.

6) Emphasizes, when correctly applied to local conditions, the adverse effects of unwise land practices and thereby promotes soil conservation.

7) Useful in relating soil, environmental, and technological information to practical farming.

8) Reflects current land capabilities at existing levels of management.

9) Results can be clearly and simply displayed on maps.

10) Gives reasonable and acceptable results which usually match local opinion.

The disadvantages of the USDA land capability classification are:

1) Subjectivity--when no limiting values for various criteria are set, allocation of land to classes is most often by opinion of evaluator.

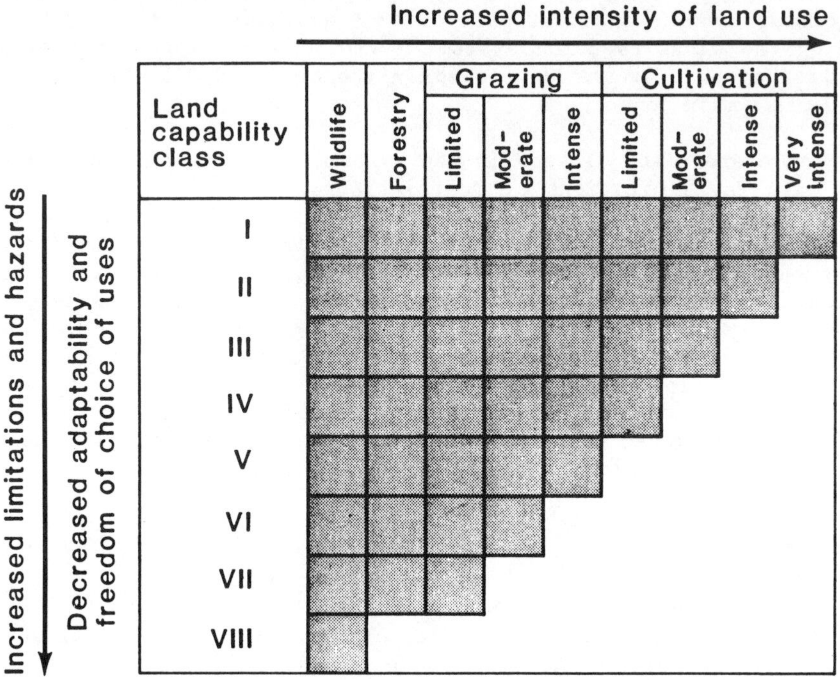

Figure 3.2 Relation between USDA land capability classification classes and the intensity with which classes can be used safely. [Source: Brady (1974)]

2) Interactions among limiting factors are difficult to take into account.

3) Division into only a few categories is too coarse and frequently does not permit more detailed consideration of relative merits of alternatives or comparisons between two pieces of land.

4) Implied rank order may give the wrong impression of the true value of the land. Class V and VI land, for example, may have high value for rice paddies or livestock, respectively.

5) Does not provide indication of the suitability of units for specific crops.

6) Neglect of socioeconomic factors means that classification does not include relative monetary value of the land or its profitability--information which may be vital to development planning.

7) Negative in orientation--emphasizing limitations rather than positive potentials of the land.

8) May be difficult to apply where soils information is lacking.

The land capability system is an interpretive soil classification that groups soil map units for detailed soil surveys "primarily on the basis of their capability to produce common cultivated crops and pasture plants without deterioration over a long period of time" (Klingebiel and Montgomery, 1966). The last part of that quotation is the key to understanding the classification system: "without deterioration over a long period of time." It was developed in the early 1930's when the erosion hazard was a main concern.

There are eight classes in the system, from class I, which includes soils with few limitations or restrictions on use, to class VIII, which includes soils and landforms with limitations that preclude their use for commercial plant production and restrict their use for recreation, water supply, and so forth. Class assignment is based on predictable deterioration from use over time; for example, if steeply sloping land (class IV) in an area of high rainfall is continuously cultivated, its productive potential will deteriorate very rapidly. Class descriptions emphasize the degree of limitation to the use of the land. Land capability subclasses indicate the kind of limitation. Subclasses are denoted by up to two suffixes appended to the class number. The classes and subclasses are briefly summarized below:

Class I soils have few limitations that restrict their use. Applications of fertilizer may be economically beneficial.

Class II soils have some limitations that reduce the choice of plants or require moderate conservation practices. Limitations may include effects of gentle slopes, moderate susceptibility to wind or water erosion, somewhat unfavorable soil structure and workability, some easily corrected salinity, and moderate wetness which can be easily improved by drainage.

Class III soils have severe limitations that reduce the choice of plants, require special conservation practices, or both. These limitations may result from the effects of moderately steep slopes, high susceptibility to water or wind erosion, frequent overflow causing some crop damage, slow drainage, shallow soil depth, moderate salinity, or low soil fertility which is not easily corrected. Conservation practices are usually more difficult to apply and maintain than for soils in class II.

Class IV soils have even more severe limitations that restrict the choice of plants and require very careful management. Use for cultivated crops is limited by the presence of one or more of such conditions as steep slope, severe susceptibility to wind or water erosion, shallow soils, poor drainage, severe salinity or alkalinity, and frequent overflows causing heavy crop damage.

Class V soils have severe limitations which are impractical to remove; use is largely restricted to pasture, range, woodland, or food and cover for wildlife. Examples of these limitations are stoniness, frequent flooding, climatic conditions such as a short growing season, and ponded areas that are not feasible to drain. Soils in this class are nearly level and have little or no erosion hazards.

Class VI soils have such severe limitations that their use is largely limited to pasture or range, woodland, or wildlife food and cover. They have continuing limitations that cannot be corrected such as steep slopes and severe erosion hazards--limitations which distinguish these soils from those in class V. Other limitations are excessive wetness or flooding, salinity or alkalinity and severe climate, making these soils generally unsuitable for cultivated crops.

Class VII soils have very severe limitations that cannot be corrected, such as very steep slopes, erosion, stones, shallowness, wetness, salts and alkalinity, and unfavorable climate. The restrictions are more severe than for class VI soils. They are not suited to any of the common cultivated crops. Under proper management, they may be used safely for grazing, woodland, wildlife food and cover, or for a combination of these.

Class VIII soils and landforms have limitations that cannot be corrected such as the effects of erosion or erosion hazards, severe climate, rocks, and salts and alkalinity. Examples are badlands, rock outcrops, sand beaches, areas with mine tailings, and barren or nearly barren lands. The use of these lands is restricted to recreation, wildlife, water supply or aesthetic purposes.

Subclass e (erosion) consists of soils for which susceptibility to erosion is the dominant limitation or hazard.

Subclass w (excess water) consists of soils where poor soil drainage, wetness, high water table, or overflow make excess water the major limitation.

Subclass s (soil limitations within the rooting zone) consists of soils which are limited by such factors as stones, salinity or high sodium, shallow depth, low moisture-holding capacity, or low fertility that is difficult to correct.

Subclass c (climatic limitations) consists of land for which climate is the only major limitation.

Efforts have been made to adapt this classification for use outside the United States, and these have been summarized by McRae and Burnham (1981). Some of the modifications include changes in the number of classes, use of different limiting factors, additional or other subdivisions of classes, quantification of limiting factors, and revisions in basic assumptions. Climatic subdivisions, for example, have frequently been dropped in regions with Mediterranean or tropical climates. In other cases, the number of subdivisions has been increased to incorporate topography gradient and to refine treatment of those subclasses in the original system. Efforts to quantify the criteria for class limits have been undertaken to reduce the subjectivity of the system. The USDA classification is intended to exclude socioeconomic factors but a high level of management is assumed; this may be difficult to assess in developing countries where management inputs vary markedly between traditional and more advanced and mechanized sectors. Steele (1967), suggested that separate capability assessments be made for each different farming system and respective level of management. Land capability classifications have also been adapted to evaluate land for forestry (USDA, 1967) and for other nonagricultural uses.

Another capability classification that is particularly noteworthy in light of the earlier discussion of soil-related constraints of agricultural intensification is the fertility capability soil classification

(Buol et al., 1975). Soil fertility is assessed from key soil properties in the uppermost layer of soil, and soil types are classed on the basis of texture. A number of modifiers can be appended to class types to indicate major chemical or physical limitations. It should be noted that subsoil characteristics may be equally or more important than surface ones in assessing the long-run productivity of the land. Where those characteristics are favorable, especially with respect to water-holding capacity, the soil can sustain much higher rates of erosion without reducing productivity than where the characteristics are less favorable.

3.3.5 Land suitability.

Evaluation of land suitability is extremely important in developing countries for identifying the areas with the particular potential for intensive development. Evaluation emphasizes opportunities rather than limitations and is performed for such purposes as choosing lands suitable for a specific crop or assessing the feasibility of irrigation or other engineering programs.

Because evaluations for suitability tend to be more specific than those for land capability, information needs must be clearly defined for the particular requirements of the land use proposed. Some requirements relate to the crop or land use itself; others will be a function of the overall farming systems and the agricultural practices associated with it; still others will be determined by socioeconomic and institutional environments. All these dimensions must be considered in establishing the criteria by which land suitabilities are established.

Over the last decade the FAO has coordinated a major international effort to develop an approach to land evaluation which encompasses most of these dimensions. The FAO approach is an effort to design an integrated methodology incorporating both soil- and social-related factors, that should be included if the effectiveness of land evaluation is to be increased. In the discussion that follows, the FAO effort is described first, and then other approaches of potential utility for developing countries, such as the U.S. Bureau of Reclamation irrigation suitability classification, are presented.

The *Framework for Land Evaluation* (FAO, 1976), the product of a number of international meetings and various pilot studies, was developed primarily for use in developing countries. It is not an evaluation system but a set of principles and concepts from which local, national, or regional evaluation systems can be constructed. The basic principles are summarized below (FAO, 1976):

1) *Land suitability is assessed and classified according to specified kinds of uses.* The concept of land suitability is only meaningful in relation to specific kinds of uses and their individual requirements. The qualities of each type of land must be compared to these requirements, and for this reason both the land and land use are fundamental to suitability evaluation.

2) *Evaluation requires a comparison of benefits obtained with inputs needed.* Suitability is assessed by comparing those inputs required,

such as fertilizers, labor, or road construction, with the goods produced or other benefits obtained.

3) *A multidisciplinary approach is required.* Evaluation requires contributions from many disciplines and always incorporates socioeconomic considerations. Evaluation teams require a range of specialists.

4) *Evaluation is made in terms relevant to the physical, economic and social context of the area concerned.* Factors such as regional climate, level of living and needs of the local population, market and land tenure conditions, and others form the context in which evaluation takes place. The assumptions underlying evaluation will vary between countries and often between regions.

5) *Suitability refers to use on a sustained basis.* The use must not bring about severe or progressive degradation. Consequences of land use changes need to be assessed as accurately as possible and taken into consideration in determining suitability.

6) *Evaluation involves a comparison of more than a single kind of use.* Evaluation is most reliable when benefits and inputs from any given use can be compared with alternatives. If only one use is considered in suitability appraisals other potentially beneficial uses may be neglected.

There are four levels of decreasing generalization in the FAO Framework for land evaluation as follows (FAO, 1976):

Land suitability orders reflect kinds of suitability.

Land suitability classes reflect degrees of suitability with orders.

Land suitability subclasses reflect kinds of limitations or kinds of improvements required within classes.

Land suitability units reflect no differences in management required within subclasses.

Land suitability orders denote whether the land is suitable (S) or not suitable (N) for the use considered. The designation "conditionally suitable" (Sc) is a phase of the order (s) and covers small areas that are unsuited at the present to the use considered but which would become suitable if certain conditions are fulfilled. Within each order, *land suitability classes* are numbered according to decreasing degrees of suitability. There is no limit to the number of classes within the order "suitable." Five is the recommended maximum and usually three are employed. There are normally two classes within the order "not suitable." An approximation of the amount of land in each class is taken from FAO (1979) and presented between parentheses. Typical suitable classes are:

1) Class S1, highly suitable--land with no significant limitation to sustained application of a given use or only minor limitations that will not significantly reduce productivity or benefits and will not raise inputs above an acceptable level. (788 million hectares)

2) Class S2, moderately suitable--land with limitations which in aggregate are moderately severe for sustained application of a given use; the limitations will reduce productivity or benefits and increase inputs to the extent that the overall advantage to be gained from the use, although still attractive, will be appreciably inferior to that expected on class S1 land. (362 million hectares)

3) Class S3, marginally suitable--land with limitations which in aggregate are severe for sustained applications of a given use and will so reduce productivity or benefits, or increase required inputs, that this expenditure will be only marginally justified. (893 million hectares)

Typical unsuitable classes are:

1) Class N1, currently not suitable--land with limitations which may be surmountable in time but which cannot be corrected with existing knowledge at currently acceptable cost; the limitations are so severe as to preclude successful sustained use of the land in the given manner. (375 million hectares)

2) Class N2, permanently not suitable--land having limitations which appear so severe as to preclude any possibilities of successful sustained use of land in the given manner.

All classes except S1 can be subdivided into land suitability subclasses depending on the nature of limitations. Subclasses are established to fit each situation; there is no formal list of subclasses. Land suitability units are used primarily in planning farm use and can be created as necessary. The units differ in production characteristics or in minor aspects of management requirements.

Central in implementing the FAO approach is the identification of *land utilization types* (LUT's) (Beek, 1974, 1975a, 1975b, 1978). LUT's are defined by subdividing major kinds of land use, such as rain-fed or irrigated fields and grassland or forest, into more detailed types to establish the different physical, economic, and social setting characteristics. Some of the attributes by which LUT's are specified include products such as crops or timber; services or other benefits; market orientation; capital and labor intensity; power sources; technical knowledge and attitudes of land users; technologies and inputs employed; infrastructure requirements; size and configuration of land holdings; land tenure; and income levels. Two examples of LUT's are (FAO, 1976):

1) Rainfed annual cropping of groundnuts and subsistence maize by small land holders with low capital resources, cattle-drawn farm implements, and high labor intensity on freehold farms of 5 to 10 hectares.

2) Extensive cattle ranching, with low levels of capital and labor intensity; land held and central services operated by a governmental agency.

Uses at different levels of management are recognized and multiple land uses can be accommodated in the LUT typology. A differentiation very important in developing countries exists between current and potential suitability. Current suitability refers to defined uses of land in its present condition without major improvements and thus corresponds to an estimate of indigenous carrying capacity. In contrast, potential land suitablity corresponds to what might be possible with new development initiatives. A distinction must be drawn between major and minor improvements on the basis of the capacity of individual farmers to effect them.

Success in the application of the FAO framework will depend on the skill and interdisciplinary communication which exists in the evaluation team. The extent to which land utilization types are correctly identified and corresponding land qualities made explicit is of paramount significance. A key element is the specification of soil-related needs of the type of agriculture for which land suitability is to be assessed.

One of the principal purposes of land evaluation, whatever the approach, is *crop suitability*, the establishment of classifications for specific crops. Soils information available in detailed soil maps or acquired in special surveys can be used to appraise suitability for the crops upon which development efforts are focused. Information on appropriate management practices for each can often also be derived. Such information is especially advantageous in revealing possibilities for specific crops on lands otherwise receiving a low-rating when classified on the basis of overall capability.

Thus, in planning soil surveys, the kinds of soil and site information needed to make suitability designations should receive careful consideration. In determining the kinds of soil information that will be most useful in guiding agricultural intensification and solving land use problems, planners should be aware that soil- and land-related requirements have been established in detail for a number of major crops. Those for sugar cane, for example, are discussed by Arens (1978), Yates (1978), and Thompson (1978). Other work describes requirements for rubber (Sys, 1978), wet rice (Brinkman, 1978), cocoa (Smythe, 1966), bananas (Arens, 1978), and oil palm (Wong, 1966; Ama, 1970). While these focus primarily on cash crops for export, ongoing research on the soil and land requirements for subsistence and food crop complexes typical of various developing regions is becoming available through the Consultative Group on International Agricultural Research's (CGIAR) network of research stations (see chapter 7 for more information). A number of sources outline soil-related factors in tropical agriculture which are suggestive of the kind of soil and site information that could be usefully included in developing country soil surveys, in addition to that required for soil classification. See Godin and Spensley (1971) for oilseeds, Kay (1973) for root crops, and Vink (1975), Young (1976), and Protz (1981). Requirements for paddy rice are discussed by Dent (1980), Tinsley (1981), Somasiri et al. (1981), and the Institute of Soil Science, Academica Sinica (1981). See also part III, Plants.

As discussed earlier, evaluation involves matching soil and land qualities with criteria established by crop needs, farming system operations, and other socioeconomic information. When suitability for specific crops, practices, or improvements is being considered, it is often these broader nonphysical aspects of land which become the dominant factors. Size of area, location, accessibility, distance to markets and transport become important as do the availability of management inputs and subsidies associated with the political system.

A good example of how soils information has been merged with other information to make land suitability evaluations is the FAO Agroecological Zones Project. Data on soils, soil limitations, climate and rain-fed production potential are examined at two levels. Climate data are overlain onto soils maps to delineate agroecological zones, and the extent of various soil units within major climatic and growing season zones is calculated. Biomass and crop yields by length of growing season are estimated by matching crop requirements with climatic data, and agroclimatic constraints to production for each area are determined as well as the probable increase in production to be expected at different levels of inputs. On the basis of predicted yields and the cost-benefit ratio of investments in required inputs, a suitability classification for each area is derived. This approach illustrates the kind of information that can help guide planners in determining the economic feasibility of implementing various improvements.

Soil suitability evaluations have been used in conjunction with soil classification as a basis for making extrapolation to other areas with similarly classified soils; for example, the Benchmark Soils Project. In this effort, carried out by the Universities of Puerto Rico and Hawaii, intense investigation of soil, crop, and management relationships were carried out on a variety of representative soils with the aim of transferring the findings to uninvestigated soils classified in other regions with similar climates.

Irrigation suitability is one example of suitability evaluations that can take into account the feasibility of land improvements and provide indications of how and at what effort development potentials could be realized. The kinds of soil and site information required for assessing land suitability for irrigation are summarized in Table 3.1. The capital expenditures in irrigation projects and the critical need for soil and water management make it essential to acquire much more information about soil and site characteristics than is obtained in a standard soil survey (Zimmerman, 1966; Hagan et al., 1967; Withers and Vipond, 1974; FAO, 1979; Stern, 1979; Hall et al., 1979).

Much of the information relevant to irrigation can be collected during standard soil surveys and supporting laboratory analyses. Measurements of water movement and retention will usually require coordination with hydrologists and drainage engineers (see Part II, Aquatic Ecosystems).

Irrigation is one of the most important ways of dramatically increasing the agricultural production in many developing countries, but it is very costly. AID, the World Bank, and other agencies plan to lend many billions of dollars in the coming decade for irrigation development, making assessment of suitable soils very important. The economic aspects of irrigation are a critical component of the evaluation. The FAO framework described earlier has also been applied to the selection of lands for irrigation (see Purnell, 1978 and 1979). The best example of an evaluation system which has integrated economic considerations is that of the U.S. Bureau of Reclamation.

There are six classes in this system, as follows:

Class 1 has the highest level of suitability, hence the highest payment capacity.

TABLE 3.1 Soil and site information required to assess land suitability for irrigation. [Source: McRae and Burnham (1981)]

Climate	Soil	Drainage and hydrology	Topography and vegetation
Rainfall	Field	Field	Macrorelief
Potential evapo-transpiration	Texture	Profile morphology	Microrelief
Frost risks	Structure	Depth to water-table	Erosion hazard
Storms	Stoniness	Infiltration rate	Existing vegetation (especially shrubs and trees)
Temperature	Arrangement of horizons	Hydraulic conductivity	Position and accessibility
Length of growing season	Depth of rooting zone	Surface drainage and outlets	Flood hazard
Seasonal variability	Consistence		
	Laboratory	Laboratory	
	Particle size distribution	Water quality	
	Bulk density and porosity	Hydraulic conductivity	
	Structural stability	Moisture retention and available moisture capacity	
	Organic matter content		
	CEC and base saturation		
	Electrical conductivity		
	pH		
	Soluble and exchangeable cations expecially sodium and magnesium		
	Nutrient levels including possible toxicities		
	Sulphate and carbonate		
	Effect of leaching		
	Clay mineralogy		

Class 2 has intermediate suitability and payment capacity.

Class 3 has the lowest suitability and payment capacity.

Class 4 designates classes of land for special use, such as orchards, or designates land with excessive deficiencies which special engineering and economic studies have shown to be irrigable.

Class 5 is used as a special designation for lands requiring special studies before a final class designation can be made.

Class 6 is for land not suitable for irrigation development.

In the Bureau of Reclamation system, appraisals of land resources provide information that is useful to economists, hydrologists, and engineers concerned with planning. These appraisals examine the interaction of productivity, cost of production, and land development costs, and the nature of the interactions determine the land class. Current land use is designated as an aid to subsequent economic analyses, and broad groupings are used; L, nonirrigated cultivated; P, nonirrigated permanent grassland; G, brush or timber; and C, irrigated cultivated. A farm water requirement appraisal is sometimes made with A, B, and C indicating low, medium, or high requirements, respectively. Drainage appraisals, relating to the 1.5- to 3-m depth of the soil and substrate are also made, with X, Y, and Z indicating good, restricted, or poor drainability conditions. Other designations made use g for slope, u for undulations, f for flooding, k for shallow depth to sand, gravel, cobble, and so on. These symbols are shown in the map unit designation in Fig. 3.3.

The Bureau of Reclamation's Irrigation Suitability Classification was designed to help select lands for irrigation in the western United

238   III. SOILS

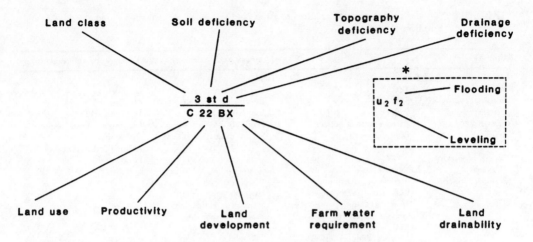

* Other use limitations which may be noted in mapping symbol.

Figure 3.3 Example of mapping symbols used in irrigation suitability classification. [Source: U.S. Bureau of Reclamation (1953)]

States, but with modification it may have application in developing countries (Maletic and Hutchings, 1967; U.S. Bureau of Reclamation, 1953). Land classes have similar physical and economic attributes that affect their suitability for irrigation; it is an expression of the relative level of payment capacity. Payment capacity is the capacity of land to provide a return greater than production costs under a given set of conditions. Economic and physical factors are correlated through the relation of soil, topography, and drainage factors to productive capacity, cost of production, and land development costs for a given project.

As with the land capability classification described earlier, the irrigation suitability classification of the Bureau of Reclamation, or some modification of it, is used in a number of countries. In some instances, however, the land classification will not meet the objectives of the intended project. As noted by Maletic and Hutchings (1967):

> Each potential project setting presents its particular land classification challenge. Land classification surveys should therefore be designed and land classes defined to meet development goals and economic requirements existing within the physical setting of the project.

Engineering suitability analysis. Classification of soils into simple groups may help determine their usefulness for some engineering purposes such as brick-making, foundations for structures, secondary roads, human waste disposal, embankments, aggregate soil material, and mining (gravel, bauxite, etc.). Extensive on-site engineering tests and deep borings may be needed for detailed assessment of the soil, but pedological soils data can help engineers and planners to make preliminary evaluations. Table 3.2 lists soil characteristics useful for assessing and classifying soils for engineering purposes.

Field investigations of many of the requisite soil properties are comparatively simple, but detailed investigations can be a sophisticated

TABLE 3.2 Soil Classification for engineering projects. [Source: FAO (1973b)]

Parameter	Uses[1]	Soil Rating		
		Good	Average	Poor
1. Slope (percent)	RFWP	0–8	8–15	>15
2. Depth to bedrock (cm)	RF	>100	50–100	<50
3. Texture, class[2]	R	GW, GP, SW, SP, GM, GC, SM, SC	ML, CL with plasticity index <15	CL with plasticity index >15; CH, MH
	FA	GW, GP, SW, SP	GM, GC, SM, SC, CL	ML, CH, OL
	P	Medium texture	Medium coarse to fine	Very coarse to fine
4. Shrink-swell potential	RF	Low	Moderate	High
5. Susceptibility to frost	RF	Low	Moderate	High
6. Stoniness class	R	0, 1, 2	3	4–5
	FW	0, 1	2	3–5
	P	0	1	2–5
7. Rockiness class	RFW	0	1	2–5
8. Soil drainage class	RF	Well drained	Moderately drained	Poorly drained
9. Permeability class	W	Rapid	Moderate	Poor
10. Hydraulic conductivity (cm/hr)		>2.5	1.5–2.5	<1.5
11. Percolation rate (mm/hr)	WP	>180	120–180	<120
12. Depth to water table (cm)	F	>150	71–150	<75
	W	>180	120–180	<120
13. Thickness of material (cm)	P	>40	2–40	<20
14. Coarse fragments (percent)	P	<3	3–15	>15
15. Soluble salts (mmhos/cm)	P	<4	4–8	>8

1. Uses: R, secondary roads on soils; F, building foundations in soils; W, human waste disposal in soils; P, transported plant material; A, aggregate soil material.

2. Texture classes: GW, well graded gravel; GP, poorly graded gravel; GM, silty gravel; GC, clayey gravel; SW, well graded sands; SP, poorly graded sands; SM, silty sands; SC, clayey sands; ML, inorganic silts; CL, inorganic clays; OL, organic silts; MH, inorganic elastic silts; CH, inorganic clays of high plasticity.

job requiring both expertise and costly equipment. Agencies engaged in civil engineering work, such as highway departments and construction firms in the public and private sector can be contacted for advice as can academic institutions with civil engineering departments and international organizations such as the FAO.

3.3.6 Analysis of indigenous systems.

As earlier defined, land capability denotes constraints that limit the uses of soils to certain agricultural, grazing, forest, or other purposes. Land suitability identifies the best crops or other uses for particular soils. Land carrying capacity provides a measure of the intensity of use or potential for use. Indigenous populations have developed a substantial understanding of both the capability and suitability of the soils available to them. These ethnic groups, occupying land areas with more than one soil type, must make practical discriminations about soil characteristics in order to use the land successfully. More complex multiethnic societies typically engage in resource partitioning which includes soil classification systems. Cultural ecologists have developed methods to identify indigenous soils classification systems and uses, and their work can provide a practical guide to agricultural potential in some developing countries. Other workers, especially Allan (1965) and his colleagues, have studied the relation of soil type, vegetation types and indigenous land use systems to establish critical population densities, and their work provides a carrying capacity methodology.

Methods for cultural ecology. Major environmental factors considered by cultural ecologists are climate, precipitation, topography, and soil types (Netting 1968 and 1974). Major sociocultural factors considered are (i) size, density, and aggregation of population; (ii) division of and composition of productive groups; (iii) rights to the means of production; and (iv) holding size. Table 3.3 shows the observed match between environmental factors (resource distribution) and sociocultural characteristics. Over time, there may be fine tuning between resources and the resource-using group, and Abruzzi (1982) has pointed out how ethnic group differentiation and resource partitioning are functionally related. However, socially mobile or migratory ethnic groups show social flexibility and comparatively rapid change or "resilience" (Vayda and McCay, 1975). Thus, the balance between population and resources is not always precisely tuned, even though population density in rural areas is an important indicator of resource concentration.

Specific methods. The field methods of cultural ecologists may provide an adequate substitute for soil surveys and classifications by soil scientists in certain areas, especially if continued or increased agricultural use is expected. Such methods may even be preferred in some instances since the classification developed by soil scientists is not directly useful to agriculturalists, and a sustantial amount of work is required to go from soil class type to agricultural application (Allan, 1965).

The field method of cultural ecology typically involves 12-36 months of direct observation to gather information on husbandry techniques, crops, animals, calendar of seasonal activities, settlement pattern and organization of labor and resource rights (Netting, 1968). Structured interviews yield useful data on household composition and relative crop yield, although the latter are notoriously unreliable as an absolute measure since farmers both lack quantitative measures and tend to conceal harvest results from outsiders. Field size can be measured directly (compass and chain or compass and pacing) (Netting, 1968; Reining, 1967) and current plantings or other vegetation noted. Interviews also supply information about soils types and planting rationales. Thus land capability and suitability can be determined.

Indigenous soil taxonomy studies. The work of Allan (1965) and the earlier work of Trapnell (1959) and Trapnell and Clothier (1959) demonstrate the striking pattern of relationships between vegetation, soils, and the African systems of land use. Netting (1968) describes such an indigenous classification system developed by the Kofyar farmers of Nigeria:

> The suitability of soil types and conditions for different crops is accurately judged by the Kofyar farmer. Sorghum should have brown soil whose color indicates a fair content of humus. Groundnuts will grow either on red volcanic or light-colored leached soils, doing best when there is not too much organic matter in the field. Kofyar

TABLE 3.3 Types of resource use and settlement characteristics. [Adapted from Netting (1968 and 1974)]

Resource Distribution (Climate, topography, soils*)	Settlement (Population size, density and aggregation)	Domestic and Societal division of labor	Rights to resources, land and water
Dispersed	Dispersed	Extended family	Shared
Intermediate (both, or some dispersed and some concentrated)	Clustered	Transitional	Lineage
		Unilineal descent	House property complex
Concentrated	Nucleated	Nuclear family	Impartible inheritance
Concentrated/very intensive		[Complex, heterogenous]	State

*The range of dispersion—concentration refers to the distribution of arable soils and seasonal constraints on their use.

farmers classify soils by color, texture, and moisture content. Dark brown or black earth (yilchip) is preferred for the demanding crops: sorghum, yams, and coco yams, but most homestead fields would be classed as baan, light reddish brown. Red volcanic soil (jing), often occurring with black pitted cinders in old lava flows, is considered inferior. It will grow sorghum, but it is less satisfactory for millet. It is also more difficult to work, clinging in a gluey mass to the hoe when wet. Waterlogged soils such as the swampy jagat and the alternately muddy and hard, cracked jak are used for moisture-loving rice and coco yams. Soils that are leached of organic matter are characterized as yilpiya, white land, or es, sand. No farming at all is attempted on seriously degraded plots or on areas of hard, lateritic material (sange) and earth used for building purposes (wanju).

As fertility declines, the Kofyar change crops to suit it. When a Bong homestead is deserted, the field may still be kept in coco yams for several years though no manure is applied. When production falls off, it is switched to the groundnut-acha or the late millet-acha rotation. Migrant bush farmers start with sesame on the deep, reddish brown virgin soil of Namu. Yams are planted next, but they grow best on fresh soils, so a new field is cleared for them annually. They are followed by the millet-sorghum-cowpea complex, which may be maintained for seven or eight years. By this time the soil has been visibly depleted, becoming lighter in color and increasingly granular in texture. A sandy field of this type can then support groundnuts or cotton for three or four years before being fallowed. (Netting, 1968:82-83)

A study which combines cultural ecology with history and linguistics adds a significant time dimension. Such a study of the Nahuatl of Texcoco, Mexico provides additional insights on the usefulness and accuracy of indigenous soil classification systems (Williams, 1977). Pre-Columbian Nahuatl codices supplied information on a variety of soil characteristics: texture, structure, consistency, organic and mineral content, color, drainage, topography, parent material, genesis, and fertility. Size of land holdings were detailed in the codic document, as were household size and data leading to measures of soil quality and variations in landholding and soil fertility patterns. Nahuatl-speaking informants provided contemporary soil nomenclature and taxonomies. Soil qualities as perceived by contemporary Nahuatl farmers were obtained by interview and supplemented with standard soils tests (phosphorus, potassium, magnesium, calcium, and nitrogen). Through aerial photographs historical field boundaries were reconstructed, and from these, the interviews, and the codices, four centuries of continuous use of soil resources were deduced. The study clearly demonstrates the Nahuatl's substantial understanding of soil types, and their efficient utilization of resources based on an indigenous soil classification system in existence for at least 400 years.

A number of studies have documented indigenous soil classification systems of other groups or describe methods and/or the theory necessary to do so. Most are cited in Conklin (1980). Listed by year of publication, they are the following:

Name of group or type of study	Location	Reference (in date order)
Nahuatl	Mexico	Penafiel, 1885
Glossary of terms	worldwide	Knox, 1904, reprinted 1968
"Races"	Borneo	Hose, 1905
Pomo	California	Barrett, 1908
Maps of primitive peoples	worldwide	Adler, 1911
Tewa	New Mexico	Harrington, 1916
Kwakiutl	British Columbia	Boas, 1934
Trobrianders	Trobriand Islands (off New Guinea)	Malinowski, 1935
Kiriwina	Trobriand Islands	Austen, 1939
Theory		Sorokin, 1943
Methods		Pough, 1953 Richards and Dobyns, 1957
Hanuoo	Philippines	Conklin, 1957
Methods		Berry, 1858
Hanuoo	Philippines	Conklin, 1969
Theory		Bunge, 1962
Methods		Chisholm, 1964

Theory		Orloci, 1966
Mountain people	Taiwan	Mabuchi, 1966
Methods		Rayner, 1966
Theory		Brookfield, 1969
Kekchi	Guatemala	Carter, 1969
Theory and Methods		Watson and Watson, 1969
Methods		Kasmar, 1970
Baruya of Wonenara	New Guinea	Ollier et al., 1971
Hanuoo	Philippines	Conklin, 1980

Carrying capacity studies. Carrying capacity (CC) determinations also provide a relatively direct and reliable basis for land utilization planning (Williams, 1977). Allan (1965) and his colleagues developed a methodology for determining critical population density (CPD) utilizing techniques from agronomy, soil science, and anthropology. They defined CPD as the maximum population that a given land area (at a given level of technology) will support permanently without damage to the land. As noted (section 3.3.6), the CPD for an area is equivalent to its CC at the existing level of technology.

The methodology invoves determining three major factors: (i) land capability classes, (ii) land use factors, and (iii) cultivation factors. As with the methodology discussed in the previous section, determination of these factors depends on knowledge supplied by indigenous farmers as well as information from field work, expert surveys, and aerial photographs and/or satellite images.

1) Land capability class (LCC). Soil and natural vegetation surveys; topographic, settlement, and other maps; and aerial photographs provide the data required to determine the total area of each soil-natural vegetation type (SVT) and the area presently or potentially under cultivation in that SVT. From these is computed the percentage of cultivable land (CL) for each SVT in the total area inhabited by an ethnic groups (or defined by some administrative boundary). For each SVT there is an equivalent LCC, and a typical measurement area might encompass three to eight LCC's.

2) Land use factors (LUF). The LUF is empirically derived (from observations, interviews, and/or literature sources) and is a function of the frequency and extent of cultivation in each SVT area. In Allan's scheme the LUF is given a numerial value by scaling from 1 to 20+ according to the following criteria: land which is cropped every year is assigned a value of 1--Permanent cultivation--or between 1 and 2, indicating occasional fallowing. Land in short-, medium-, or long-term recurrent cultivation (RC) would have LUF's of 3 to 5, 5 to 7, or 7 to 10. (That is, the area in crop in the current year represents 1/3 to 1/5 of the total land cultivated over time for short-term RC land, 1/5 to 1/7 for medium-term, and 1/7 to 1/10 for long-term RC land.) Shifting cultivation

land would have a LUF greater than 10, and terrain with only isolated pockets of arable land, or with no arable land at all, would be assigned still higher values.

3) Cultivation factors (CF). The CF is defined as the amount of land in use per person per year for each SVT under particular climatic conditions. It is derived from survey data obtained, typically, by stratified sampling of households of an ethnic group, with the data partitioned by SVT's. (Data collected for each household surveyed includes household rights, or tenure, in land, which yields the holding sites in each SVT; persons per household; and location of household relative to land holdings.) Total number of households and climatic conditions (principally yearly rainfall pattern and altitude) are also obtained for the survey area. The actual size of holdings in hectares is most easily obtained from aerial photographs (when the scale is precisely known).

Table 3.4 contains measured holding size per person for 27 major ethnic groups in sub-Saharan Africa. It can be seen that the hectares/head cropping season is roughly constant at a value of 0.2 to 0.3, with much of the remaining variation probably accounted for by soil quality and degree of technological development.

The critical population density, or the CC, is a function of (LUF x CF/CL) and is usually expressed in terms of population density per square mile of total area. The method allows calculation of each SVT and for each land use, so that the area required on a per capita basis is the sum of each land type held and worked. The method allows a precise determination of extensive or intensive land use, from permanent cultivation to waste land and can be modified to include aquatic ecosystems. Where good data already exist, the method also provides a measure of the effects of intensification.

Information from Landsat can be used for determining the incidence of major categories of soil types, certain vegetation classes, present and former cultivation sites and for locating nucleated settlements (contiguous houselots); however, the spatial resolution and other constraints of Allan's methodology preclude substitution of Landsat images for aerial photographs (Reining, 1980). Landsat also provides a new data source for regional as well as local approaches. Fanale (1982), for example, has carried out a sampling experiment on the regional scale necessary for an ecological study of an ethnic groups which could not have been done without Landsat. Fanale analyzed the "customary range usage" of the Navaho in a manner paralleling Allan's determination of customary land usage. (Note that "customary range usage" is not the same as the "range carrying capacity" of modern range science since the latter is determined without reference to indigenous systems or knowledge.) Satellite information provided regional environmental data, helped define SVT's and other regional-scale land uses, and aided in precise location of settlements, transportation routes and other significant features. Landsat has greatly improved construction of accurate area frame samples and has provided a new dimension in land use planning. Research continues on how best to combine remote sensing technology and field data methods appropriate to local sites. An important example is the ten year research project underway in Kordofan, in the western Sudan. Part of this project involves Large Area Data Sampling (LADS) for remote sensing applications and statistical analysis of

Table 3.4 Size of landholding by ethnic population or ethnolinguistic group in Africa. [Constructed from Allan (1965: 56-65)]

Group	Garden size (hectares/head)	Rainfall Pattern	Altitude (High/Low)
ZAMBIA			
Plateau Tonga	.5	one	H
Tonga	.4	one	H
Ngoni-Chewi	.4	one	H
Lamba	.4	one	H
Lunda-Ndembu	.4	one	H
Tumbuka Mzimba	.4	one	H
Swaka	.7(a)	one	H
Swaka	.4(b)	one	H
KENYA			
Kikuyu	.2	two	H
Nyeri	.2	two	H
Kamba (Machakos)	.2	one(c)	H
UGANDA			
Lango	.3	one(c)	H
Kasherengenye	.3	two	H
Alur	.2	two	H
Teso	.4	one(c)	H
Ganda	.2	two	H
Chiga-Kabale	.2	two	H
Soga-Mpita	.3	two	H
TANZANIA			
Nyakyusa	.2	two	H
NIGERIA			
Yoruba	.2	two	L
GHANA			
Akim	.2	two	L
Ga	.2	two	L
Ashanti	.2	two	L
Dogomba	.2	one	L
Mamprusi	.2	one	L
Kusasi	.4	one	L
FraFra	.4	one	L

(a) weak soils
(b) red loams
(c) secondary rainy season is weak

environment (Olsson and Stern, 1981). Another part involves an analysis of Landsat data for wood resources monitoring (Hellden and Olsson, 1982). Some of this research is centered on Umm Ruwaba as shown in Fig. 4.2. This work is ongoing and involves more than one discipline.

Soil Survey and Soil Maps

4.1 Concepts and Definitions

The term soil survey encompasses the procedures by which a wide variety of information on soils is collected and compiled. The survey is a basic mode of research in soil science and can contribute significantly to the success of development programs. General design considerations and aspects of survey structure and organization have been mentioned in preceding sections of this report.

Soil surveys serve a diversity of purposes. According to the Soil Survey Staff (1951):

> Soil survey includes those researches necessary (1) to determine the important characteristics of soils, (2) to classify soils into defined series and other classificational units, (3) to establish and to plot on maps the boundaries among kinds of soil, and (4) to correlate and to predict the adaptability of soils to various crops, grasses, and trees, their behavior and productivity under different management systems and the yields of adapted crops under defined sets of management practices.

Prediction is an important purpose of soil surveys in developing countries; it can guide investments to increase yields and adapt soils to more intense uses, as outlined above, and it is essential in engineering projects, such as pipeline, highway, and airport construction and in planning for waste disposal systems or recreation. In each case, the soil map and the accompanying description and interpretation of soil conditions and environment give the user information without a visit to the site.

Considerations related to the predictive power and utility of soil surveys have been raised in earlier discussions. This section summarizes these considerations and provides information that should be helpful in selecting appropriate survey methods. The level and reliability of the predictions that can be made are determined by the kind of field procedures employed; survey intensity and map scale are especially important. Minimum size delineations, mapping units, and map legends are also of major concern. Cost-benefit factors are presented, criteria for survey evaluation are outlined, and the role and application of remote sensing are discussed.

The evaluation of soil surveys involves a number of criteria: purity of the maps, reliability of the base map, information on the map units, and size of the delineations on the map in relation to its scale (intensity of the survey). These criteria are discussed by Forbes et al. (1982). Other characteristics used for evaluating soil maps are described by Eswaran et al. (1981). Some of these are summarized below:

> 1) *Map legibility* refers to the ease with which a user can read the information on a maps. Factors influencing the readibility of a map include the number and size of delineations, the choice of colors or patterns to represent the delineations, ground control, and quality of map presentation.

2) *Minimum size delineation* is the smallest delineation inside which a simple map unit symbol can be printed legibly or the smallest area that can be easily discerned by the map user. About 0.16 cm (1/6 inch) is considered a minimum size delineation.

3) *Least size delineations* are the smallest areas delineated on a specific map by a soil surveyor. Many of these very small delineations reduce map legibility.

Other criteria for soil map evaluation are map texture, intensity, minimum scale of reduction, and sizes of delineations in reference to boundary representation and map scale.

4.2 Intensities of Soil Survey and Map Scales

Soil surveys are made at several intensities. These have been designated in the United States as *orders*, and were developed to provide consistency among soil surveys made for different uses. Formerly, such terms as detailed, semidetailed, reconnaissance, and schematic were used to designate the different scales, but these terms are imprecise and are often confusing. There are five orders of soil surveys in the United States, described by Rourke (1981):

1st Order Soil Survey. The First Order survey is the most detailed, providing soil information for highly intensive land-use planning, such as developing agricultural experimental sites or other projects requiring complex, high-cost inputs (building sites, intensive irrigation systems). First Order surveys produce the largest scale maps, 1:12,000, (1 cm = 120 m, or approximately 5 inches=1 mile) or larger, with minimum size map unit of about 0.6 hectàres (1.5 acres) or smaller. Soils are classified in the most precise classes (which in Soil Taxonomy are phases of soil series) delineated by transecting and transversing the area at closely spaced intervals using direct observation. A sufficient number of field observations is required to locate and delineate areas of dissimilarities at the desired map unit size.

The soil series is the lowest and the most homogeneous category in the system. It is differentiated using all the differentia of the higher categories plus those additional, significant properties which separate one series from another. Some of the properties used to differentiate series are: the kind, thickness, and arrangement of the soil horizons (or layers of soil) down to a depth of one or two meters; structure, color, texture, consistency, acidity or alkalinity and humus content; kind and content of salts; content of rock fragments; and mineralogical composition. A significant difference in any one of these properties can be the basis for recognizing a different series.

Very rarely, however, do two soil series differ in just one of these characteristics since most characteristics are often genetically related and therefore change interactively. A phase of a series or higher category, while not an integral part of the classification, can be superimposed on a class for a utilitarian purpose, e.g. eroded phase, stony phase.

In the United States not many soil surveys qualify as Order 1; most have ranges too wide in some characteristics to meet the objectives for which Order 1 surveys are made and so are assigned to Order 2. For this reason, phases of soil series are generally specified to add sufficient specificity to meet the higher objectives.

2nd Order Soil Survey. Second Order surveys are detailed enough to provide soil information for areas of intensive land uses, such as high-value agricultural areas and urban or industrial locations. The map scale is from 1:12,000 (1 cm = 120 m; or 5.28 inches = 1 mile) to about 1:32,000, (1 cm = 320 m; or 2.7 inches = 1 mile). The most common map scales now used in the United States fall into this range: 1:20,000 (1 cm = 200m; or 1 inch = 0.316 mile) and 1:15,840 (1 cm = 158.4 m; or 4 inches = 1 mile). At these scales, areas as small as 1.0 to 1.6 hectares (2.5 to 4 acres) can be delineated on the map. Field mapping procedures are similar to those for Order 1. Map units are usually consociations of phases of soil series and complexes of phases of soil series: a *consociation* is a map unit in which only one identified class (series, in this case plus allowable inclusions) occurs in each delineation; a *soil complex* is a map unit in which different kinds of soils occur in a geographic pattern which is so intricate that the individual components cannot be delineated separately at scales of 1:20,000 or larger.

3rd Order Soil Survey. Third Order surveys provide information for extensive land uses, such as woodland management, wildlife management, and watershed management, and for general land use planning, for example, assessment of potential for cropland, pastureland, woodland, and urban development. Map scales range from 1:24,000 to 1:250,000, permitting delineations of dissimilar soils from tens to hundreds of hectares. The map scale chosen, however, must be large enough to accommodate delineations necessary for the purpose of the survey. Map unit boundaries are plotted by observation and by interpretation of remotely sensed data with some observations; the soils in each delineation are identified in the field by transecting and traversing. Field observations should be frequent enough to locate and plot areas with significant dissimilarities.

If soils in the survey area are known at the series level, mapping should be by associations of phases of soil series. (An association is a map unit in which two or more distinctive phases of soils occur in a repeating geographic pattern; the individual components can be delineated separately, at map scales of 1:20,000 or larger.) If the soils are not known at the series level, the map units could be consociations, associations, and some complexes of phases of the higher taxonomic categories, such as families, subgroups, and great groups.

4th Order Soil Survey. Fourth Order surveys provide information for broad land use planning at the multi-county, regional, state, or provincial level, for example, to identify areas having potential for more intensive development. Tract size could range from several hundred to several hundreds of thousands of hectares with map scales ranging from 1:100,000 to 1:300,000. General accessibility to the area for field soil survey may be limited.

If the soils of the survey are known, the map units could be associations of families of soil series: families of soil series are soils with similar physical and chemical properties that affect their response to use and management. Appropriate interpretations would be general rather than specific, e.g. for cropland or urban development rather than for wheat yields or limitations for septic tanks.

5th Order Soil Survey. Fifth Order surveys provide information for very broad, general land use planning at multicounty, regional, state, provincial or national levels, e.g., potential for cropland, pastureland, woodland, urban development, etc. This order of survey is also useful for identifying areas with potential for more intensive development. Once the potential of large tracts is identified, then detailed surveys are required for final selection of areas to be developed.

The field procedures for Fifth Order surveys are similar to those for Order 4: soils representative of landscapes are identified. Then the composition of map units is determined by mapping selected areas (35 to 50 square kilometers) using the field procedures of 1st and 2nd Order soil surveys or, alternatively, by transecting selected areas.

Map scales ranging from 1:250,000 to 1:1,000,000 are suggested for soil surveys intended for very broad land use planning. Map units are designed to separate landscapes or segments of landscapes that, because of location or physical factors, will be used in the same or similar ways, as determined by the purpose of the survey. Map units would rarely be composed of a single kind of soil, even if recognized at the higher categorical levels; they could be associations of phases of subgroups, great groups, suborders, and orders of the U.S. Soil Taxonomy system.

The kinds of soil surveys, ranges of scale, and relationships among survey terminology or nomenclature are illustrated in Table 4.1 which was prepared by the panel.

4.3 Soil Classification Units and Soil Map Units

The distinction between classification units and map units is a critical one and can considerably affect the kinds of predictions made from soil surveys. The term "soil classification unit" is often used erroneously as a synonym for "soil map unit." A soil map unit, with the name of a taxon or class, consists predominantly of that taxon or a phase of that taxon or class, but it may also include other components because of limitations imposed by the scale of the map and by the limited number of points in the landscape that can be examined.

When mapped limits of soil classes are superimposed on the pattern of soils in nature, areas of taxonomic classes rarely, if ever, coincide with mappable areas. There are several reasons for this. Some soil classification units are too small to be delineated on the map and so are included in delineations named for another soil; moreover, boundaries

TABLE 4.1 Orders and types of soil surveys, their characteristics, data sources and uses.

	Order of soil survey				
	5th order	4th order	3rd order	2nd order	1st order
Survey type	←——— Reconnaissance ———→		Semidetailed	Detailed	Intensive
Survey scale	1:300,000–1:1,000,000	1:125,000–1:300,000	1:32,000–1:125,000	1:12,000–1:32,000	1:1000–1:12,000
Size of mapping unit	35–50 km²	500–500,000 ha	10–1000 ha	1.0–1.6 ha	0.5 ha or smaller
Kind of mapping unit	Associations of phases of subgroups/great groups, suborders, orders	Associations of families of soil series	Associations of phases of soil series	Consociations of phases of soil series	Phases of soil series
Use in development planning	←——— Resource inventory ———→ ←——— Project location ———→		←——— Feasibility surveys ———→	←——— Management surveys ———→	
Common or potential remote sensing data sources	Landsat 1–4 MSS and TM (images) ———→ NOAA 6/7 ———→		Landsat 1–4 MSS + TM (digital) ———→ Aerial photography (high altitude) ———→	Landsat 4 TM (digital) ———→ Aerial photography ———→	Aerial photography (low altitude) ———→
Socioeconomic and land use features	Broad land use categories	Regional land use	Sets of villages	Village pastures, open fields	Village fields, village residential areas

between soil units are not everywhere so obvious that they can be precisely plotted on a map, and so part of one soil unit is often included in the delineation of an adjacent unit. Some soil classification units are so intermingled that mappable areas must be identified in terms of two or more classes; other soil units are not easily distinguished from similar adjacent ones and are included in delineations named for other soils. In these cases apparent differences in use and management are small.

The prevalence of such inclusions and the relative contrast with the surrounding taxa form a guide to the taxonomic purity of the map units. Purity of map units is in part the result of a conscious decision by the surveyor, based on the intended use of the soil survey. In defining map units, the surveyor exercises judgment about the effects of inclusions on management and the effort that is justified to keep inclusions to a minimum. If an inclusion does not restrict the use of entire areas or impose limitations on the feasibility of management practices, its impact on predictions for the map unit is small; however, if it has a significantly lower potential for use than the dominant component in the map unit or if it affects the feasibility of meeting management needs, then even a small amount in a map unit can affect predictions greatly. For example, a small area with a slope of 15 to 25 percent on a map unit dominated by slopes of 4 to 8 percent can seriously affect the use of the map for many purposes. Therefore, such irregularities are often excluded for practical reasons.

It is especially critical that delineations be made for soils that cannot be used for the same purposes as the surrounding soils. If the map scale permits it, the soils are separately delineated; areas too small to be delineated may be identified on the map by special symbols. Some inclusions may be limiting for one use and nonlimiting, or even beneficial, for another use. For example, when a soil with a high water table is included in a map unit dominated by a well drained soil, the area may be unsuitable for cultivation but well suited for excavated ponds (Soil Survey Staff, 1980b).

4.4 Costs and Benefits of Soil Surveys

The main components of survey costs have been listed by Bie and Beckett (1970) as follows:

1) Salaries and wages of soil surveyors and their auxiliaries in the field;

2) salaries and wages of the surveyors in the office and laboratory with their supporting staff there;

3) subsistence and other logistic and travel expenses in the field;

4) scientific equipment and materials for field work and for supporting laboratory, drafting, and administrative work;

5) maps and aerial photography;

TABLE 4.2 Major cost elements in consultant contract proposals for soil surveys. [Adapted from Western (1978)]

Publication scale	Survey Intensity	Country[b]	Year	Approximate cost ($/acre)[b]	Approximate extent (acres)	Cost element[a] (% of total)				
						A	B	C	D	E
	Exploratory	Yemen (P)[d]	1976	0.002	49,000,000	68.3	1.3	15.2	—	15.2
	Exploratory	Thailand (I)	1973	0.002	12,500,000	62.6	—	4.8	—	32.6
1:250,000	Exploratory	Brazil (U)	1973	0.001	60,750,000	67.6	—	7.6	—	24.8
	Low	Indonesia (I)	1973	0.13[d]	62,500	60.1	—	7.8	13.3	18.8
	Low	Zambia (U)[e]	1967	0.53	600,000	37.0	3.2	11.6	10.2	38.0
1:100,000	Low/medium	Thailand (I)	1971	0.37	332,500	44.4	3.2	9.3	10.9	32.2
	Medium	Ivory Coast (I)	1970	0.41	272,800	?	?	?	?	?
	Medium	Fiji (I)	1968	0.41[d]	50,000	56.1	—	7.3	18.3	18.3
	Medium	Ethiopia (I)	1973	0.89[d]	67,500	—[f]	?	?	?	?
1:25,000	Medium	Nigeria (I)	1972	0.93	92,500	48.3	1.2	5.8	27.2	17.5
	High	Somalia (P)	1975	5.23	37,500	38.4	—	26.9	15.9	18.8
1:10,000	High	Greece (I)	1972	6.64	2,500	35.3	—	7.5	34.6	22.6

[a] Cost elements: A, professional staff; B, aerial photography, etc.; C, reporting: typing, drafting, binding, printing, etc.; D, Laboratory analyses; E, mobilization, organization, equipment, overheads, fees, etc.
[b] Recalculated from English monetary units rising $2.5 = 1.
[c] (I), proposal implemented; (P), proposal pending at time of writing; (U), proposal unsuccessful.
[d] Minimum figure; real cost may be greater.
[e] The only self-contained soil survey; all others are part of an integrated study.
[f] Only professional staff cost is available.

6) report publication; and

7) administrative or headquarters overhead, including academic research and proposal preparation.

Survey costs are usually expressed in terms of cost per unit area, since the size, intensity, and scale of the survey determine the size of professional staff, by far the greatest part of the cost. The cost of a survey increases substantially as the map scale increases. Cost per unit area does not, however, take into account different types of surveys--exploratory or reconnaissance surveys covering large areas have economies of scale that intensive, large-scale surveys of a few hundred hectares do not. As shown in Table 4.2, the cost per unit area is much larger for the latter (Western, 1978). Use of data from remote sensing, when available, may dramatically reduce costs in some situations and should always be considered.

Costs of surveys by the Soil Conservation Service of U.S. Department of Agriculture (USDA) in 1965 were estimated at $0.30 to $0.70 (U.S.) per acre ($0.75 to $1.75 per hectare) for medium- to high-intensity surveys, with an average of $0.52 per acre ($1.25 per hectare). Assuming that the survey has a useful life of 25 years, the cost per year is only about $0.01 to 0.03 per acre ($0.025 to $0.074 per hectare). The costs of 14 surveys from various parts of the world are roughly equivalent for map scales of 1:20,000 to 1:50,000 (Western, 1978). It is estimated that 1981 surveys in Ohio and New York at a scale of 1:15,840 cost about $1.00 an acre ($2.25 per hectare) (K.R. Olson, personal communication).

Benefits from soil surveys have been demonstrated but are difficult to quantify. The literature yields a number of examples of urban and rural development projects that failed or incurred unnecessary costs because the soil was not taken into account--for example, buildings put on clay soils that swelled and cracked resulting in great damage, or water loss in unlined irrigation canals on deep sandy soils. But, as Beckett (1981) points out, the evidence is largely anecdotal and selective and difficult to evaluate. There is little information about how many projects were successful which did not use a soil map.

Beckett (1981) describes a procedure for relating the benefits to a farmer to the qualities of the soil map. The procedure, tested in a peach-growing area in New South Wales, Australia, showed that profits increased as the purity (quality) of the soil map increased from 20 to 90 percent. If relevant costs for inputs and outputs are available, similar calculations can be made for yield data in many survey reports.

4.5 Skills and Equipment Required for Soil Survey

The number of persons needed to conduct a soil survey depends on the objectives of the survey, size of the area, complexity of the soil patterns, accessibility of areas to be mapped, experience of the surveyors, and time available.

III. SOILS

The leader responsible for the survey normally should have a B.S. degree (or equivalent) in soil science and a minimum of 3 to 5 years as a field surveyor. The leader usually must train some of the field staff in mapping; an inexperienced person will require several months in the field, working with an experienced surveyor, before mapping alone. The leader will still need to check the work periodically.

Surveyors may have B.S. degrees in soil science or may have completed a 2-year college program. Subprofessional assistants trained in soil mapping can also do a satisfactory job. In many developing countries, vocational high school graduates with good training and motivation can be effective in field surveys and also in laboratory analysis. Unskilled labor is used in many surveys to make excavations, collect samples, cut lines through dense vegetation, and perform other essential tasks.

Vehicles for transportation, tools, mapping equipment, base maps, and reference maps are required for field mapping. A complete description of the equipment required is given in the revised USDA Soil Survey Manual (Soil Survey Staff, 1980b).

Nearly all base maps now used in soil surveys are derived directly or indirectly from aerial photographs or other remotely sensed data. For most surveys, it is more economical to purchase recent photographs and prepare field sheets at the dimension and scale that will be used for publication. Several kinds of aerial photographs are available: panchromatic, color, infrared, modified infrared and color infrared. Remote sensing with nonphotographic sensors can produce photograph-like images, which have been useful in smaller scale surveys and are being used increasingly in large-scale surveys.

4.6 Remote Sensing Methods for Soil Survey

"Remote sensing" in general describes the use of cameras or other sensors mounted in aircraft or spacecraft to view the surface of the earth. Remote sensing provides relatively rapid overviews of land and, for any but the smallest areas, is less time-consuming and expensive than field work. However, for accuracy of interpretation, it is best to use remotely sensed data in combination with field work or "ground truth." A discussion of remote sensing, including the potentials and limitations of various sensing systems, can be found in part I.

The amount and kind of remote sensing data acquired for soil surveys are largely determined by the map scale and the amount of detail desired in the survey. All orders of survey can benefit from the use of images acquired vertically from airborne or spaceborne sensors. During the past two decades remarkable advances have been made in data acquisition and analysis systems. This section gives a general description of data acquisition systems and an assessment of the utility of these systems for the five orders of soil surveys. Table 4.1 summarizes some of this information.

4.6.1 Aerial photography.

The general lack of maps and information on soils and the immediate data needs of planners have made aerial photograph interpretation and remote sensing techniques particularly significant in developing countries. Aerial photographs are a common source of base maps, and advances in techniques for extracting soil information have made them major sources of excellent and inexpensive information on soils for resource inventories and mapping at higher scales (Colwell, 1960; Reeves, 1975; Liang and Philipson, 1981; Paine, 1981). Their interpretation is primarily physiographic and has contributed most to the description of landform and other physical features. Chemical and biological information is difficult to extract, but topography (slope), drainage, soil erosion, vegetation, and land use characteristics can be assessed.

Since the 1960's, interpretation of color and color infrared photography has become an additional tool. In general, when crop and vegetation information is critical, the incremental increase in information they can provide justifies their additional expense and complexity. Interpretation of slope, drainage, and seepage is not significantly improved by color methods. Repetitive aerial photography in color and color infrared and analysis of soil-vegetation patterns can help in identifying and evaluating variation within soil map units (Milfred and Kiefer, 1976). Crop identification is possible (Philipson and Liang, 1975), but accuracy may be degraded because of the mixed cropping patterns characteristic of much of the agriculture in developing countries.

The quality of aerial photography depends on processing, atmospheric conditions, moisture, and ground cover and vegetation. Where sequential coverage is not available users must be sensitive to unrecognized influences and seasonal and successional ecosystem dynamics. Aerial photographs are usually an effective data source for information on soils in arid and semiarid regions and on exposed soils; greater cloud and especially vegetation cover makes soil and moisture interpretations more difficult in the humid tropics and other forested regions. Methods for aerial photograph interpretation are well established (Soil Survey Staff, 1966; Brenchley, 1974; Carroll et al., 1977).

More recently thermal infrared sensors and passive microwave and radar systems have been developed. Thermal infrared sensors have been used to measure surface temperatures; the low-resolution radar (which, like microwaves, has all-weather ability) has provided useful planning information at regional and national scales in areas of high cloud cover. An example is the Centro International de Agricultura Tropical evaluation of land resources at a scale of 1:1,000,000 (Cochrane et al., 1979; Projects RADAM-Brasil, 1972-1978).

4.6.2 Satellite data.

Since Landsat 1 was launched in 1972, earth and agricultural scientists in many nations have used satellite data to inventory and monitor earth surface features related to soils and agricultural production. Applications in developing countries are numerous (NAS, 1977). Landform units and features as well as settlement and land-use patterns can be differentiated with many applications to development planning (Reining, 1979).

Of special interest to land use planners is the Landsat system of repetitive coverage on an 18-day cycle with data collection at approximately the same local time. The Landsat 1-4 satellites follow a polar orbit at an altitude of 920 km, providing a synoptic view of the earth's surface in 185 by 185 km data collection frames. Temporal variations are particularly important for monitoring land-use changes. Repetitive passage of a sensor system over the same area on the ground provides a temporal record of important changes.

The remote sensors on Landsat include a four-wavelength-band multispectral scanner (MSS) and a three-wavelength-band return beam vidicon (RBV) system. The scanner gathers spectral reflectance in two bands in the visible portion of the spectrum (0.5 to 0.6 µm and 0.6 to 0.7 µm) and two in the near infrared (0.7-0.8 µm and 0.8-1.1 µm). The 80-m resolution of the MSS system on Landsats 1 to 4 means that 0.5 ha is the minimum area detectable on the ground. Landsat 4, launched in July 1982, also carries a new sensor system, the thematic mapper (TM). The TM has a spatial resolution of 30 m, giving a theoretical minimum areal coverage of .09 ha. Characteristics of the TM are discussed in part I. The TM's middle infrared and thermal bands provide information not previously available for delineating soils differences.

One of the most obvious agricultural applications is in identifying and measuring the area of cultivated crops (NASA, 1978). Equally important but somewhat more complex is the use of these techniques to predict crop yields. Research is being performed on the use of remotely sensed data for improved accuracy and updating of crop yield predictions, but applications remain limited in developing countries.

During the past 15 years, scientists have explored methods of using satellite-derived images for delineating soil boundaries and characterizing soil conditions. Recent results have shown promise of enhanced speed and accuracy in soil survey preparation (Weismiller and Kaminsky, 1978; Westin and Frazee, 1976; Barret and Curtis, 1976; Stoner and Baumgardner, 1979).

Landsat images are used to prepare preliminary soil legends and map general soil differences, often following work with aerial photographs for a more detailed investigation. A single satellite image provides a synoptic view of a contiguous area covering 34,000 km^2. Such a synoptic view is valuable to the soil surveyor in correlating soils across wide areas which may have been classified under a variety of systems by different surveyors at different times. For resource inventories in many developing countries, especially in arid regions such remotely sensed data are being used extensively (Liang and Philipson, 1981; Poulton, 1982; Westin 1982) and have been applied at the project level in some cases (Rib and Braida, 1982). Data from the four Landsat MSS bands can be subject to both visual interpretation with a color additive viewer and digital analysis from computer compatible tapes (CCT's). Digital analysis can provide a way of accurately delineating and quantifying soil map unit composition (Kirshner et al., 1978). Similarly, TM bands 5 and 6 provide soil moisture information, and data have been available to the research community since September 1982. This could be important in areas where more intensive development, such as resettlement projects, is being considered and more detailed soil information becomes necessary. Soil map units represent the dominant soil within the unit. In many developing regions in the tropics, soil variability is extreme and is not adequately represented by the dominant soil mapped at smaller scales.

Black-and-white and false-color images produced from Landsat data were visually interpreted by Westin and Frazee (1976) to produce general soil maps at scales of 1:250,000 or less with map units of 260 hectares or more. Brink et al. (1981) described a similar use for engineering information. Weismiller et al. (1979) visually interpreted black-and-white and false-color images produced from Landsat data to determine soil parent material boundaries in Jasper County, Indiana U.S.A. These boundaries were confirmed by field observations, and then digitized and overlaid on the four Landsat MSS, bands which had been adjusted to a scale of 1:15,840. Spectral maps showing soil differences were then produced at this scale with map units as small as 1 hectare. These maps, delineating more than 50 different spectral classes in six different parent material areas, are being used in more detailed soil surveys (Fig. 4.1). The digital format of Landsat MSS data permits rapid and easy merging and/or recombination of spectral classes to produce a wide array of smaller-scale base maps for soil surveys at different levels of generalization.

Landsat data can be useful for mapping and monitoring large-scale land degradation caused by wind erosion, water erosion, salinization and flooding. Here the sequential nature of Landsat coverage is especially helpful. A striking example of denudation and sand dune encroachment caused by wind action can be seen in an area along the Wadi Abu Habl in Western Sudan (Figs. 4.2 and 4.3). The three spectral classes of the cultivated sands seemed to be well correlated with areas of millet, peanuts, and fallow.

Severe erosion caused by rainfall has been detected and separated spectrally in the humid temperate region of northern and central Indiana (Fig. 4.4). In these cultivated soils the exposure of subsoil resulting from severe erosion gives reflectance values measured by the satellite sensors which are considerably different from those of the less eroded surrounding soils.

The ability to interpret MSS measurements is greatly dependent on understanding the spectral reflectance properties of earth surface features--soils, crops, trees, and water. Studies are being conducted to determine quantitative relationships between spectral and other properties of soils. Soils do not have a unique characteristic reflectance curve, but in general soil reflectance increases as wavelength increases from 0.3 to 2.5 µm. A variety of soil parameters and conditions individually and in association contribute to spectral reflectance. These include the physicochemical properties of organic matter, moisture, texture, iron oxide content, and other less well defined variables. Conditions that affect the reflectance characteristics of soils in their natural state are green vegetation, shadows, surface roughness, and non-soil residue, all of which vary with tillage operations, cropping systems, or naturally occurring plant communities. There is an extensive literature on the characteristic variations in visible and near-infrared reflectance of minerals and rocks. Considerable work remains in defining the spectral characteristics of earth surface features, especially in tropical environments, but progress has been made. An example of spectral

258   III. SOILS

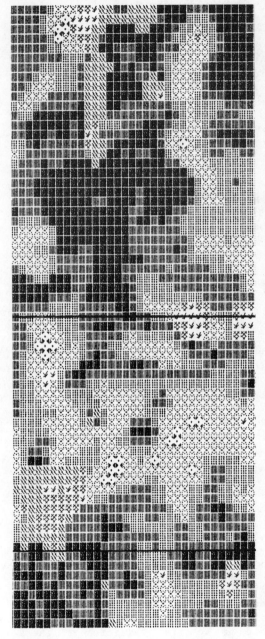

Class 1. Excessively to well drained soils with significant inclusions of moderately well drained soils.

Class 2. Excessively to well drained soils with significant inclusions of moderately well drained soils. Minor inclusions of somewhat poorly drained and very poorly drained soils occur.

Class 3. Excessively to well drained soils with a nearly equal amount of moderately well drained soils. Minor inclusions of somewhat poorly drained and very poorly drained soils occur.

Class 4. Very poorly drained soils with significant inclusions of excessively to well drained, moderately drained, and somewhat poorly drained soils.

Class 5. Very poorly drained soils with significant inclusions of somewhat poorly drained soils. Minor inclusions of excessively to well drained and moderately well drained soils occur.

Class 6. Very poorly drained soils with significant inclusions of excessively to well drained and moderately well drained soils.

Class 7. Very poorly drained soils. Minor inclusions of excessively to well drained, moderately well drained, and somewhat poorly drained soils occur.

Class 8 (Vegetation). Predominantly moderately well drained soils with significant inclusions of excessively to well drained and very poorly drained soils.

Figure 4.1   Spectral map derived from Landsat MSS data delineating soil characteristics in Jasper County, Indiana; original scale 1:15,840. [Source: Baumgardner (1982)]

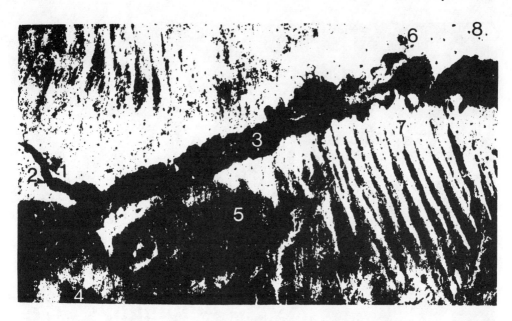

Figure 4.2 Landsat image showing denuded areas and dune encroachment along Wadi Abu Habl in Western Sudan; original scale, 1:250,000. [Source: Baumgardner (1979)] Key to prominent features in the landscape: 1, Er Rahad (town); 2, Er Rahad reservoir; 3, floodplain of Wadi Abu Habl; 4, Jebel Ed Dair (granite hill); 5, upland clay plain; 6, Umm Ruwaba (town); 7, sand dune; and 8, villages.

analysis is the comparison of three different soils described by soil surveyors in the field as dark red and assigned Munsell color notations of "2.5 YR 3/6" (Fig. 4.5).

In delineating differences between soils and describing the characteristics of a soil profile, color is one of the most obvious and useful attributes. Although the differences in visible reflectance (0.4 to 0.7 μm) are not great, the differences in the near and middle infrared (0.7 to 2.5 μm) are dramatic (Stoner et al., 1980). Similar spectral differences exist between severely eroded soils, non-eroded soils, and depositional soils among soils associated in an erosional sequence. Such spectral differences are important because they may indicate significant differences in texture, internal drainage, water holding capacity, degree of weathering, fertility status, and potential productivity.

The major limitations of Landsat data are similar to those in other remote sensing methods. Cloud cover is a problem in many tropical areas, as is the generally lower resolution of Landsat data (approximately 80 m). Lowe et al. (1974) provide an economic evaluation of remote sensing from satellites in developing countries and Allan (1980) summarizes some cautions.

260   III. SOILS

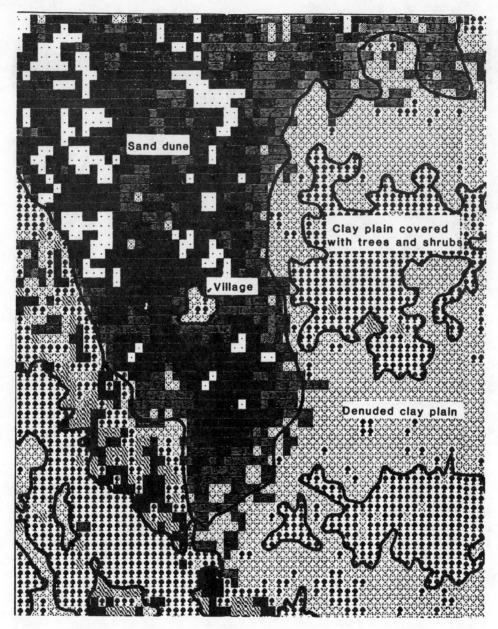

Figure 4.3 Detailed spectral map of dune area south of Wadi Abu Habl in Western Sudan; derived from Landsat MSS data. Original scale 1:25,000. [Source: Baumgardner (1982)]

Soil Survey and Soil Maps 261

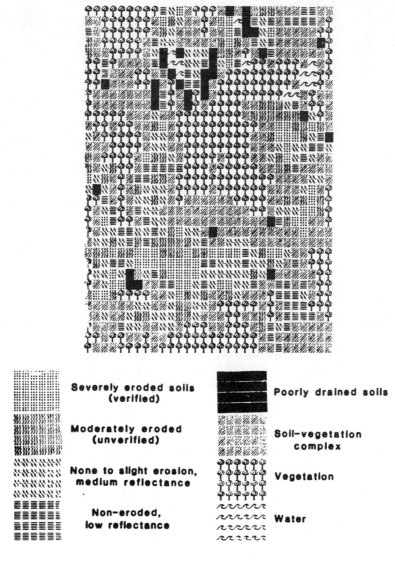

Figure 4.4. Spectral map units distinguishing severely eroded soils from other soil and vegetation classes; each symbol represents 0.45 ha. [Source: Baumgardner (1979)]

Access to remotely sensed data on the earth's surface is provided by the Earth Resources Observation System (EROS) Data Center in Sioux Falls, South Dakota administered by the U.S. Geological Survey. AID has provided assistance for remote sensing in developing countries, and data distribution and analysis facilities have been established in many developing regions. Imagery from a series of environmental weather satellites beginning with TIROS 1 in 1960 is available through the National Climatic Center of the National Oceanic and Atmospheric Administration (NOAA) in Asheville, North Carolina. Generally low in resolution, this imagery provides a more synoptic view which may be of value for many applications. Further information about these sources and the range of remote sensing information available is provided in the list of institutions at the end of this report.

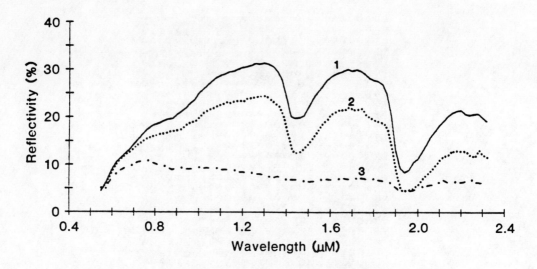

	Organic matter	Fe_2O_3
1 Udic Ustochrept (Okla., USA)	0.6%	0.87%
2 Typic Chromoxererts (Spain)	1.28%	2.00%
3 Typic Haplorthox (Brazil)	2.28%	25.60%

Figure 4.5. Reflectance curve for three dark red surface soils having Munsell color notations of 2.5 YR 3/6. [Source: Stoner et al. (1980)]

Field Methods for Soil Inventory, Classification, and Assessment

5.1 Introduction

Broadly speaking, field methods for inventorying soil resources are conducted for four purposes:

1) Soil classification,

2) Evaluation of soil suitability for agriculture,

3) Assessment of soil productivity and input needs.

4) Assessment of soil degradation and reclamation needs.

Several specific observations are needed for each type of information. Some important ones are listed in Table 5.1, along with comments on the relative ease with which they can be made, approximate equipment costs, and time required. Many of these observations are relevant for more than one of the purposes mentioned above.

Field Methods for Soil Inventory, Classification, and Assessment

The following sections briefly review the four purposes of field tests, summarizing rationale and procedures and making some recommendations about the suitability of methods, resources needed, and sources of assistance.

5.2 Soil Classification

The classification of a soil is based on a detailed examination of the typifying soils in a landscape. In the process, many soil profiles are examined. A *soil profile* is a vertical section through the various layers of the soil down to its parent material. Its depth may vary from a few centimeters to several meters. Morphological characteristics such as color, depth, texture, and structure, as well as physical and chemical properties, are used to characterize and classify the soil and to determine its value for agriculture and other purposes. Field observations, supported by laboratory studies, are essential for meaningful characterization and classification of the soils.

Field observations include soil color, texture, structure, stoniness, slope, and depth of horizons, as well as an assessment of climatic and geomorphologic characteristics of the area (Table 5.1). These are verified by laboratory studies of the chemical, mineralogical, and additional physical properties of the soil horizons. Ideally, these classifications should be further verified by aerial photographic or other remote sensing observations.

Soil classification is expensive and time-consuming, but it can provide the policy-maker with information needed for long-term planning. *Guidelines for Soil Profile Description*, prepared by the Food and Agricultural Organization (FAO) (undated), summarizes the procedures used in field observation. Hodgson (1978) reviews methods of soil and site description and of soil sampling throughout the world. Black et al. (1965) summarize specific methods available for individual tests. The Soil Survey Staff (1951) treats factors such as texture and structure that are important in soil classification and describes in detail the attributes of various factors that permit their identification.

Most countries have soil survey departments from which additional information can be obtained. The in-country offices of the USAID and FAO may also be helpful.

5.3 Evaluation of Soil Suitability For Agriculture

Field tests for soil suitability assessment include soil texture, structure, slope, stoniness, depth, permeability, erodability, depth to

TABLE 5.1 Field observations for inventorying and monitoring soil resources.

Observation	Unit of Measurement	Frequency of Measurement	Equipment Needed	Cost of Equipment (US$)	Time Needed For Test	Expertise Needed For Test
1. Locational						
1. Latitude & longitude	degrees	once	read from map	negligible	negligible	negligible
2. Elevation	meters	once	altimeter	100–500	½ hour	negligible
3. Temperature	degrees C	daily	hygrothermograph; also: thermometer	400–450 20–100	negligible	negligible
4. Relative humidity	percent	daily	hygrothermograph	50–450	negligible	negligible
5. Rainfall	mm	daily	rain gauge	150–200	negligible	negligible
6. Cloud cover	descriptive	daily	visual observation	—	negligible	negligible
7. Solar radiation	langleys/day	daily	radiometer	400–500	negligible	negligible
8. Evapotranspiration	mm	daily	evaporation pan set	600–1000	negligible	negligible
9. Wind run	km/hour	daily	anemometer; vane	500–1000	negligible	negligible
10. Geomorphology	descriptive	once	visual observation	—	large	large
11. Relief & slope	percent	once	abney level	50–100	½ hour	medium
12. Native vegetation	descriptive	seasonal	visual observation	—	large	large
13. Current land use	descriptive	once	visual observation and interview	—	large	large
14. Parent material	descriptive	once	visual observation and geologic map	—	large	large
15. Hazard of flood/wind storm/ rain storm/hail storm/fire	descriptive	once	visual observation and interviews	—	large	large
16. Depth to water table	meter	seasonal	boring/auger hole/pit	50–500	medium	negligible
17. Land degradation	descriptive	seasonal	visual observation; soil loss measurement	—	large	large

TABLE 5.1 (cont.)

Observation	Unit of Measurement	Frequency of Measurement	Equipment Needed	Cost of Equipment (US$)	Time Needed For Test	Expertise Needed For Test
II. Soil Related						
18. Soil color (by horizon)	hue, value, and chroma	once	Munsell color chart	40–50	<½ hour	medium
19. Soil depth (by horizon)	meter	once	measuring tape	negligible	<½ hour	medium
20. Texture (by horizon)	descriptive	once	by feel of dry and moist soil	—	½ hour	large
21. Structure (by horizon)	descriptive	seasonal	hand lens	negligible	½ hour	large
22. Consistency (by horizon): stickiness, plasticity, shrink-swell potential	descriptive	seasonal	by feel of dry, moist, and wet soil	—	½ hour	large
23. Soil pH (by horizon)	unit	seasonal	soil test kit	25–150	½ hour	medium
24. Free CaCO$_3$ (by horizon)	effervescence	seasonal	soil test kit	25–150	½ hour	medium
25. Hard pan	descriptive	once	pocket penetrometer; other pocket tools	50–100	½ hour	medium
26. Organic matter and fauna	descriptive	seasonal	visual inspection	—	½ hour	large
27. Soil temperature	degree C	daily	soil thermometer	negligible	½ hour	medium
28. Soil moisture	millibars	seasonal	tensiometer	30–200	constant	large
29. Soil salinity	mmhos/cm	seasonal	wheatstone bridge	100–200	2 hours	large
30. Fertilizer requirement	kg/ha	seasonal	soil test kit; field trials	25–150 100–1000	½ hour season	medium large
31. Gypsum/lime requirement	tons/ha	seasonal	soil test kit; field trials	25–150 100–1000	½ hour season	medium large
32. Soil erodibility	tons/ha	once	field trials	100–1000	season	large
III. Socio-economic						
33. Ease of communication, infrastructural and support; land tenure, etc.	descriptive	annual	field observations and interviews	100–10,000 plus	months	large

5.3 Evaluation of Soil Suitability For Agriculture

Field tests for soil suitability assessment include soil texture, structure, slope, stoniness, depth, permeability, erodability, depth to water table, salinity/alkalinity, and toxicity/acidity (Table 5.2). Seven key parameters that affect soil suitability assessment are discussed below:

Topography. The greater the slope, the less suitable the land for general agriculture. To be classified as "good," the slope should be less than 2 percent. However, some crops (for instance, tea, coffee, and rubber) grow best on slopes of 20 percent or more.

Soil texture. Soils with a coarse (sandy) texture are generally less fertile and less able to retain moisture; they need higher rates of fertilizer application and more frequent irrigation. Fine (clayey) soils may be more difficult to cultivate and may have poor drainage--particularly if they have a high content of montmorillonitic (swelling) minerals--but fertility and moisture retention are higher. Incorporation of organic matter usually helps to improve both types of soils. Medium-textured (loamy) soils are desirable for most crops. Soil texture, however, should also be considered in connection with the crop to be grown; for example peanuts grow better on lighter soils, while rice prefers heavier soils.

Soil structure. Granular structures are generally the best to work with. They also facilitate root growth, water movement, and air circulation. Incorporation of organic matter helps to improve soil structure. Dispersed soils due to alkali conditions (see below) may need gypsum as a soil amendment.

Soil depth. Soils that are 1 m or more deep provide ample room for root development and water penetration. Soils which are shallow have bedrock close to the surface, have an impervious layer at a shallow depth, or have severe erosion provide little room for root development. They may also have poor drainage.

Soil reaction. Acid soils (pH less than 7.0) may need to be limed, as the toxicity of some nutrient elements (such as aluminum and manganese) and deficiency of others (calcium and magnesium) may increase with acidity. Alkaline soils (pH more than 7.0) may need gypsum application. Gypsum is added to improve the physical condition of the soil, as the presence of alkali (when there is 15 percent or more exchangeable sodium) usually causes soils to become dispersed and impermeable (see chapter 6, on soil degradation). Availability of plant nutrients at various pH values is illustrated in Fig. 5.1. Soil pH can be determined in the field with a portable soil testing kit. The pH should also be considered in connection with the crop to be grown: apple, for example, is adapted to acid soils, while date palm is more suited to alkaline conditions.

Soil salinity. It is important, in developing countries, for this concept to be understood when large irrigation projects are being considered. A high soluble salt-content in the soil (electrical conductivity of the saturation extract greater than 4 millimhos per centimeter at 25 degree C) is toxic for most plants although some plants in arid regions, such as date palm, can tolerate levels as high as 16 mmho/cm (Salinity

TABLE 5.2 Indicative characteristics of good, medium, and poor soil classes.*

Parameter	Characteristics for soil suitability rating as		
	Good	Medium	Poor
1. Slope of land (%)	<2	2 to 5	>5
2. Soil depth (meter)	1.0 or more	0.25 to 0.9	<0.25
3. Soil texture	medium	moderately coarse or fine	very coarse or very fine (with 2:1 clays)
4. Soil structure	strong	moderate	weak
5. Soil pH	6.0 to 7.5	5.5 to 6.0; or 7.5 to 8.0	<5.5; or >8.0
6. Organic matter content (%)	more than 1.0	0.2 to 1.0	<0.2
7. Soil salinity (mmhos/cm)	less than 4.0	4.0 to 6.0	>6.0
8. Soil alkali (ESP, %)	less than 15	15 to 20	>20
9. Soil fertility/response to fertilizer use	high	medium	poor
10. Soil drainage	good	medium	poor
11. Hydraulic conductivity, (cm/hr)	>2.5	1.5 to 2.5	<1.5
12. Hard pan within soil	absent	present	present
13. Depth to watertable (meter)	>2	1 to 2	<1
14. Stoniness in soil	absent	some	high
15. Degradation hazard	absent	some	high

*These are for illustrative purposes only as crop requirements for various soil conditions vary significantly. The unsuitability of one parameter, even if all others are favorable for the crop/usage in question, may change soil rating from good to poor for that particular usage. Alternative usages would then have to be explored.

Stoniness. The more stones in the soil, the more difficult it is to cultivate. Stones also hinder root growth and water movement.

Like soil classification, soil suitability assessment provides the policy-maker with invaluable information for long-range development. However, it is less expensive and can be undertaken with simple field equipment used in soil surveys. With some training, high school graduates can learn this assessment technique.

5.4 Assessment of Soil Productivity

"The proof of the pudding lies in the eating." Similarly, the final test of the agricultural productivity of soil is its ability to support a high level of plant growth. The textural, structural, mineralogical, pedological, and chemical properties of soils; the prevailing climatic and geomorphological conditions; and the available technology and management systems all interact to produce the crop. Field tests, by simulating actual conditions, provide valuable information for planning. Human intervention, through improved farm management practices and land reclamation, helps to enhance soil productivity, but ecological and cost considerations are important in deciding on the desirable level of human intervention.

Broadly speaking, two types of field studies may be conducted:

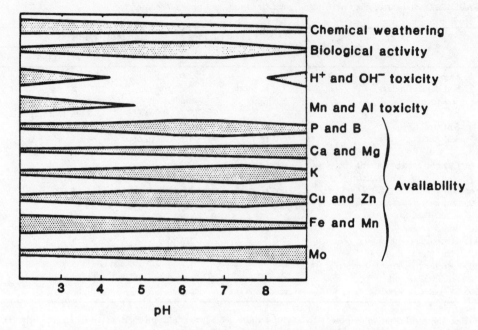

Figure 5.1 Soil pH and nutrient availability. (The wider the dark band, the greater the activity, availability, or toxicity.)

1) Those undertaken to assess the inherent productivity of a soil and/or the level of inputs (fertilizer, water, pesticides, seed) needed.

2) Those undertaken to assess the magnitude of land degradation and/or determine the conservation or reclamation procedures needed.

In this section we examine methods for the first type of studies; the second type is reviewed in section 5.5.

5.4.1 Examination of growing crop.

The simplest and quickest method for assessing the productivity of a soil is field inspection. A good stand of any crop generally indicates a good level of soil productivity and/or crop management (usually with the use of needed inputs) and the right type of climatic and geomorphologic conditions. A poor stand usually indicates poor productivity due to (i) poor land management (including inadequate use of needed inputs) and/or some problem in the soil (poor drainage or excess salt, alkalinity, acidity, or toxicity), or (ii) factors unrelated to soils, such as adverse climatic or geomorphologic conditions, poor seed germination, and/or pest attack. Spotty plant growth may indicate unevenness in field preparation, seed germination, fertilizer distribution, and/or pest attack. It may also indicate soil heterogeneity in fertility, salt, alkalinity, acidity, or toxicity levels.

Field Methods for Soil Inventory, Classification, and Assessment

It is always good practice to find out what the farmer considers to be the reasons for the condition of the crop, as often these reasons may be unrelated to the soils. Factors such as unfavorable geomorphologic conditions will also become apparent during field inspection.

It is difficult to judge soil productivity from field observations when the soil is covered with a dense forest, bare, or lying fallow. In such cases field productivity studies, backed up by laboratory examinations, must be undertaken.

5.4.2 Field productivity studies.

Field experiments help determine quantitative relations between measured values of individual components of an agricultural system and the yields obtained from that system (for instance, the effect of applied nitrogen on crop yield). If other components of the system (water, pests) also influence yield, it is necessary to: (i) hold these other components constant and develop a relation at this "standard" level of these components (such as, by providing sufficient water or adequate pest control to all experimental plots equally), or (ii) conduct more complex studies to determine the effects of the various components and their interactions (Hanway, 1967). When more than one factor influences crop yield and there are interactions among the variables (for instance, among levels of applied nitrogen, phosphorus, and potassium), and where the levels of these variables can be controlled, a factorial experimental design can be used (LeClerg et al., 1962). Where the levels of a particular factor cannot be controlled, studies may be conducted at the different measured levels of the factor as they occur naturally and the relation may be developed by use of regression techniques.

In all cases, the levels of different variables must be measured; the relations between the input(s) used and crop yields obtained may be described quantitatively. If results of two or more studies are to be combined to develop a relation, standardized methods and rates must be used at all sites.

Field experiments vary in size from small test plots with controlled and replicated input treatments to large field-scale demonstrations which may or may not be replicated and may or may not be controlled. Although the latter do not require a high level of precision, they must be performed with enough forethought and care that the results are reliable and meaningful.

Some cardinal rules for field productivity experiments, as summarized by the Fertilizer Institute (1972) and others, are given below:

1) Analyze the problem and plan a course of action accordingly. Vary only the elements that you wish to measure. For example, if you wish to test only nitrogen and phosphorus, do not vary potassium and other elements. Decide on the statistical method to be used--preferably in consultation with a statistician.

2) Use soil test results for background information. Soil samples should be collected in advance by standard procedures (FAO, 1970a;

Peek and Melsted, 1967). The nutrient status of the soil can be qualitatively assessed in the field with a portable soil testing kit. Note that the chemicals in these kits deteriorate over time; also, assessments of nutrient status by inexperienced individuals may be erroneous.

3) To ensure that field test results are not masked by deficiencies of other plant nutrients, apply adequate amounts of these as "standard" or "blanket" treatments uniformly across all plots. The treatment levels should conform with local recommendations and/or be based on soil test results.

4) For the nutrient to be tested, select treatment rates both above and below the probable optimum rate. Include the rate that farmers use for that crop in the area (if any).

5) Select test sites carefully, keeping in mind the following:

   a) Choose sites on soils that are fairly representative of those in the area.

   b) Use farmers' fields to the extent possible; avoid experiment stations as their soils and management practices are usually different from those of the farmers in the area.

   c) If testing for fertilizers or water, use leveled fields; unevenness of land may mask treatment effects.

   d) Do not use edges of fields as border effects may mask treatment effects. For the same reason, avoid areas of heavy livestock feeding, unique management situations (for instance, rice nursery areas or crop threshing or storage fields), unusual stoniness, and other unrepresentative characteristics.

   e) Do not use areas with heavy or unequal shading (such as wooded areas).

   f) In general, choose land, soil, and topography that are fairly representative of the area. Avoid fields with soil variability; if this is not possible, use larger plot sizes, as these tend to integrate the variations, and increase the number of replications. Other things being equal, use an approximately square experimental design, as this tends to reduce the effects of heterogeneity.

6) Stake or mark out the plot boundaries.

7) Make a plot diagram, showing all plots in the experiment.

8) Erect signs so that others may be able to identify individual plots.

9) Make and record observations several times during the crop growing season.

10) Follow all crop management practices (irrigation, interculturation, pesticide application) as recommended.

Field Methods for Soil Inventory, Classification, and Assessment

11) Calculate costs and return after the experimental crop is harvested.

12) To the extent possible, especially if the experiment is conducted on a farmer's fields, involve the farmer in day-to-day operations, from initial planning to postharvest care and benefit/cost calculations.

13) Inform others, particularly farmers in the areas, of your results.

Large-scale "demonstration" field studies should also be based on a quantitative assessment of needs. These provide an appropriate mechanism for comparing the recommended fertilizer and other input practices with those commonly used in the area. At an appropriate time before harvest when differences in yields corresponding to different practices are visible, field days should be organized to which farmers are invited to see differences. The crops should be harvested and data analyzed for statistical significance and benefit/cost considerations; visual inspections are at best only qualitative.

5.4.3 Conducting specific tests

Field trials are undertaken to assess the productivity of a soil as well as the crop's need for additional inputs (plant nutrients, pesticides, water, seed, mechanization) and the region's need for reclamation measures. Thus the crop's response to particular input to a soil can be evaluated by varying the input rate. Portable soil testing kits enable a qualitative assessment of the levels of various nutrients in a soil through the development of a color. Soil samples are usually collected before the field study (Peek and Melstead, 1967) and analyzed in the laboratory for physical and chemical properties. This is an important step and helps establish correlations between the soil physical and chemical properties and the response to the inputs being tested.

Field studies for productivity assessment are time-consuming and require long-term planning (at least over one season) and a commitment of resources but they do not require sophisticated equipment--land can be prepared by the farmer's methods, and the additional material needed is usually limited to strings and pegs, a weighing balance, measuring tape, fertilizer, seed, and pesticides. Techniques for plot layout and harvest are relatively simple and straightforward and can be easily learned with practice.

Detailed procedures for field tests for soil fertility assessment are given by the FAO (1970b); they are also applicable for assessing the needs for other inputs. Guidelines for conducting field experiments are summarized by the Fertilizer Institute (1972). Sampling procedures for soil testing are explained by Peek and Melsted (1967), and guidelines for calibrating soil tests for fertilizer recommendations are provided by the FAO (1973a). Agricultural experimentation is discussed in detail by Little and Hills (1978) and by Hanway (1967).

In addition to the agencies listed at the conclusion of the soil classification discussion (section 5.2), private and public groups engaged in research and development on fertilizers, pesticides, seed, machinery, and so on, may be contacted for additional in-country information.

5.5 Assessment of Reclamation Needs

Leaching water needed to reclaim saline soils, gypsum for alkaline soils, and lime for acidic soils can be assessed in the field in the same way as fertilizer requirements (section 5.4): through field trials with different amendments rates. Care should be taken to ensure reproducibility of results. Care in site selection and the other caveats mentioned in section 5.4 apply here also. Field studies should be backed up by laboratory tests to increase the reliability of the results.

Because of site specificity, there are no standard methods for quantitatively assessing reclamation needs due to wind and water erosion or for physical and biological degradation caused by increased bulk density. Erosion is usually controlled by providing suitable ground covers and adopting appropriate management practices; biological degradation is also controlled by measures to control erosion; physical degradation may be controlled by incorporating organic matter into the soil and by management practices.

In most cases, procedures for assessing the extent of degradation (salinization, sodification, acidification, toxicity, waterlogging, and for physical and biological degradation) and the reclamation measures needed are straightforward and relatively inexpensive. They require about the same level of expertise and resource allocation as for assessment of soil fertility. Soil erosion by wind and water, on the other hand, is assessed mostly through visual observation; its quantification is rather complex, involving detailed observations over a period of time, followed by calculations using the universal soil loss equation (Wischmeier and Smith, 1978), as discussed in chapter 6.

Methods for assessing soil degradation due to various processes have been summarized by the FAO (1977). Methods for assessing and managing soil erosion by water are reviewed by El-Swaify (1981). Methods for diagnosing and reclaiming saline and alkaline soils are described by the Salinity Laboratory Staff (1954) and the FAO (1967b). Acidity and toxicity and their correction are discussed by Brady (1980). Specific field and laboratory tests (for hydraulic conductivity, salinity, alkalinity, acidity, and so on) are described by Black et al. (1965). Land degradation and soil conservation methods are dealt with in other FAO documents (1971, 1976a, respectively).

In addition to the agencies mentioned in the preceding sections, agencies responsible for land reclamation, irrigation, land or soil conservation, and so on in various countries would be able to assist and respond to specific questions.

5.6 Laboratory Support for Field Studies

In characterizing soils, laboratory soil analysis is an essential complement to field study. It helps the soil surveyor define the soil units observed and described in the field and place the soil in the classification system. Furthermore, it enhances soil survey reports and provides information useful for soil management. It provides information on physical, chemical, and mineralogical properties required for technical and interpretive classification systems, as well as information on the levels of toxic or deficient elements that constrain the use of the soils. An understanding of the process of soil formation (soil genesis) is often increased through soil analysis; other uses of soil analyses are discussed by Eswaran (1981).

5.6.1 Determination of chemical properties.

The kinds of chemical analyses needed depend on how the information is to be used. The minimum analyses required to supplement field studies include organic matter; nitrogen; cation exchange capacity; exchangeable calcium, magnesium, sodium, and potassium; pH in water and in potassium chloride; and extractable aluminum. Other analyses may be required for specific purposes.

Laboratory methods and procedures for collecting soil samples have been published by the SCS (1972). Irrigation water, when used, should be analyzed for the kind and amount of dissolved salts and suspended solids, acidity, and toxic substances. In many countries these procedures are modified to suit local conditions, and confusion about the classification of the soils can result from inconsistencies in the analyses. A comprehensive volume on laboratory methods of soil analysis for chemical properties has been published by the American Society of Agronomy (Black et al., 1965).

For interpretative soil classification systems--for instance, soil fertility assessments (Buol and Cuoto, 1980)--simple soil tests are required for acidity and lime, available potassium, phosphorus, calcium, magnesium, and so on. Analytic methods have been published in journals, bulletins, and books on soil tests (e.g., Walsh and Beaton, 1973).

5.6.2 Determining physical properties.

Particle size distribution (sand, silt, clay) must be determined for soil classification, and a number of methods can be used. The pipet method is a standard procedure of the U.S. Soil Conservation Service (1972); it or some modification of it has been generally adopted. Other physical properties used in characterizing soils are infiltration rate, permeability, and water-holding capacity (the difference between the soil water held at field capacity and permanent wilting percentage). Standard procedures have been published in a number of technical journals and books; a widely used reference work is Black et al. (1965).

5.6.3 Determining mineralogical properties.

Data on mineralogical properties of soils are used primarily to gain an understanding of soil formation processes and, to a very limited degree, to characterize soils for classification and for land use. In the U.S. soil taxonomy, mineralogical data are helpful in differentiating the lowest categories (series and families); however, mineralogical properties can be largely inferred from other soil properties such as bulk density and cation exchange capacity, which are more easily determined. Methods for determining mineralogical properties are given in the publications cited above (Soil Conservation Service, 1972; Black et al., 1965) and in many other technical journals and books.

5.6.4 Determining microbiological properties.

Variations in soil texture, in particular the particle size of component minerals, and the widely variable aggregates associated with microbial activity make the soil environment extremely complex and heterogeneous. Even a small, apparently uniform plot presents to microbial organisms a world as diverse as our own and as rich with adaptive opportunities. Soils contain many discrete microhabitats, often in close proximity.

Microenvironmental physical, chemical, and biotic interactions in each of those habitats are widely varying and dynamic; in consequence, there are great fluctuations in microbial activity and composition. Detailed analysis of microlevel variation is beyond both the scope and needs of most research related to development. We therefore discuss only techniques for assessing microbial activity across microhabitats. For this reason, statistical considerations il.microhabitats associated with sampling design, inference, and interpretation are important.

In general, microbiological laboratory and analytical methodologies require large commitments of time, expertise, and equipment. The information obtained would generally not be useful in any but the most intensive resource inventories, as adequate assumptions about microbial activities can be made on the basis of other soil and soil-vegetation characteristics. However, soil microbial information can be valuable in baseline environmental studies and ecological monitoring, particulary when natural ecosystems are converted to agriculture, disrupting previous litter fall, decompositon, and nutrient cycling. The effects of inorganic fertilizers on indigenous nitrogen fixation and the microbial community generally are often of importance and interest in the development of certain tropical soils.

The application and appropriateness of particular methods are discussed by Alexander (1977), Litchfield and Seyfried (1979), Parkinson et al. (1971), and Rosswall (1973). Methods should be chosen with advice from a soil microbiologist.

Field Methods for Soil Inventory, Classification, and Assessment

5.6.5 Skills and equipment required.

The minimum skill required to direct a laboratory for soil analysis to supplement field studies is an M.S. degree with 1 1/2 to 2 years of chemistry, including analytical chemistry, two or three soil courses, including soil chemistry, and one or two courses in statistics. Experience as an apprentice or intern for 1 or 2 years in an analytical laboratory would be advantageous. For routine soil analyses to supplement field studies, the laboratory director can train technicians who have had only 2 years at a vocational college or have a good high school education.

The equipment required for a routine soil characterization laboratory is relatively modest. Balances, ovens, hot plates, pH meters, spectrophotometers, conductivity meters, furnaces, waterbaths, centrifuge glassware, and chemicals normally suffice. In addition, chemistry benches or tables with storage space, sinks, sources of gas or electricity, compressed air, and vacuum are required. Distilled water of good quality is essential. Equipment needs for soil laboratories are discussed by Golden et al. (1966). The capital investment and operating costs of such a laboratory are a small part of the total cost of making a soil survey, probably 5 to 10 percent. Unfortunately, in many soil surveys in developing countries, runs of laboratory data and laboratory manuals needed for the interpretation of field data are incomplete or absent. Although it is difficult to estimate the benefits from soil analyses, there is little doubt that supporting laboratory data are well worth the cost. Most laboratories in developing countries are well equipped to perform the analyses required for soil surveys (Brogan et al., 1965).

5.6.6 Evaluation of laboratory results.

Two important ways to evaluate soil analyses are in terms of the precision and accuracy of the data. A sufficient number of samples of a soil unit is required to estimate variability in the field and to replicate determinations in the laboratory to measure the precision of the methods. The accuracy of the methods is determined by checking the analyses with standard control samples. Failure to follow this procedure results in erroneous interpretations, which seriously limit the value of the data. Lack of precision and accuracy may be due to factors such as inadequate training of laboratory personnel, faulty equipment, or chemicals that have deteriorated.

Methods for Assessing Degradation

Soils are subject to degradation by water and wind erosion, salinization, compaction, and water logging. It is important to have methods for discriminating between human and natural causes of soil degradation and to be able to assess the impacts of degradational processes on productivity. The first part of this section is devoted to an evaluation of methods. Some models for estimating the impacts from erosion--generally the major cause of land degradation--are considered in the second part.

6.1 Methods of Assessing Soil or Land Degradation

Soil degradation is manifested by accelerated water and wind erosion, soil compaction, surface soil crusting, loss of fertility, waterlogging, and salinization. Salinization and waterlogging are usually associated with irrigated land but they also are problems in certain dryland farming areas of Australia, Canada, and the United States. The other degradational processes occur through a variety of exploitative land uses, including agriculture, mining, and recreation.

Attempts have been made to devise models of water erosion that would permit reasonably accurate predictions of soil loss. More recently, similar attempts have been made to characterize wind erosion as well as salinization, soil compaction, surface crusting, and loss of fertility. A degree of success has been achieved with water erosion, much less success with wind erosion, and practically no success with devising methods for predicting other kinds of degradation.

6.1.1 Water Erosion

The USDA has been in the forefront of efforts to develop equations for estimating accelerated water erosion. A wealth of data has been provided by soil erosion experiment stations established in the early 1930's and has been supplemented by research elsewhere. These data formed the basis for the equation developed by Zingg (1940) to predict soil losses from slope factors. The Zingg formula was expanded by Smith (1941) to include crop and conservation management factors and by Browning et al. (1947) to include soil erodibility factors. The fundamental elements of the present prediction equation were finally assembled when a method for calculating the rainfall factor was devised by Wischmeier (1959). Those elements formed the basis for the universal soil loss equation (USLE) published by Wischmeier and Smith (1965). A revised version of the equation and procedures for using it for the entire United States were published by Wischmeier and Smith in 1978. The equation was designed to predict long-time average soil losses from sheet and rill erosion under specified rainfall, slope, soil, and cropping conditions. It is an empirical model based on measured erosion rather than a theoretical model, although erosion theory is being developed from the analyses made to test the USLE.

The USLE is expressed as $A=RKLSCP$, where A is the predicted amount of soil loss (tons per acre per year), R is the rainfall factor, K is the soil erodibility factor, L is the length-of-slope factor, S is the slope gradient, C is the crop management factor, and P is the conservation practice factor. Graphs and standard tables provide values for R, K, and C for the particular sample area, local soil series, and cropping pattern. Values of L and S are measured in the field, and P depends on whether the land is cultivated on the contour. Terracing also affects the value of the P factor. Metric equivalents for use in the equation have been used in many countries.

The USLE has been used in other countries as well. The U.N. Food and Agricultural Organization used the equation as the basis for the water erosion portion of its soil degradation assessment methodology (FAO, 1979). Detailed analyses of the suitability of the USLE for humid tropical regions have been made in Nigeria by the International Institute of Tropical Agriculture (Aina et al., 1977) and in West Africa by Roose (1977). The West African tests involved annual measurements of erosion on 50 experimental plots in five French-speaking countries and provide one of the most extensive collections of erosion data in the world. Australian soil conservationists have begun testing the USLE and believe that it could provide the basis for an Australian approach to erosion prediction (Edwards and Charman, 1980).

Calculation of the rainfall factor is the most daunting problem in countries outside the United States. Adequate data for calculating R usually are not available. Furthermore, in regions where the rainfall distribution pattern differs markedly from those in the United States (e.g., wet summers and dry winters), there is uncertainty about the effect of rainstorm intensity on soil erosion. In addition, where crops are different from those grown in the eastern United States or intercropping is practiced, a new crop management factor must be established.

Attempts have been made to use the USLE to estimate soil loss for rangelands, forest lands, mined land, construction sites, and recreational areas, with varying degrees of success. Erosion models for these noncrop lands are incomplete or nonexistent, so that the USLE is the only potentially suitable model, however imperfect it might be. Problems that have become apparent include the effect of snow on runoff, the value of the crop management factor for rangelands where shrubs are dominant, the exclusion of gully and streambank erosion factors from the equation, and the inaccuracies that arise when average values of the factors are used to characterize conditions over an entire watershed. The USLE has even been used to estimate sediment loads in streams and lakes, although the equation was not intended to be used in that way.

The factors in the USLE are continually being refined. The equation is widely used in the United States for a variety of purposes and is routinely used by the Soil Conservation Service to assist in planning for erosion control. A calculator has been devised to facilitate solution of the equation.

A simplified procedure for estimating annual soil losses from sheet erosion on cultivated lands in southern Africa was developed in Zimbabwe (Elwell, 1978). The prediction model is called the soil-loss estimation model for southern Africa (SLEMSA). It provides a framework for devising specific models for localities with similar environmental conditions. Such a model has been used, with reasonable success, in the Zimbabwe Highveld (Elwell and Stocking, 1982).

The model is based on minimal data on field erosion. Established theory, expert opinion, laboratory test data, and field plot experiments are used to determine numerical values for five factors: seasonal rainfall energy, soil erodability, rainfall energy intercepted by a crop, slope steepness, and slope length. The authors of the Zimbabwe Highveld prediction model believe that modifications must be made in the SLEMSA to adapt it to different localities; the model is not intended to be universal.

6.1.2 Wind erosion.

A wind erosion prediction equation has also been developed by the USDA, principally at the Wind Erosion Center at Kansas State University (USDA, 1961). W.S. Chepil was primarily responsible for developing the original equation, which has been modified by Woodruff and Siddoway (1965) and is still being improved. Portable and stationary wind tunnels were used to obtain the data for developing the equation. Virtually all of the wind tunnel tests in the field were conducted in the Great Plains, beginning in the 1950's. The theoretical basis for the equation came from Bagnold's (1943) treatise on sand.

The wind erosion equation is $E=f(I', K', C', L', V)$, where E is the amount of soil loss (tons per acre per year), I' is the soil erodibility factor, K' is the soil ridge roughness factor, C' is the climatic factor, L' is the unsheltered field length along the prevailing wind direction, and V is the vegetation factor. These five variables are called equivalent variables because they are obtained by grouping ten primary variables. For example, V is calculated from information on three primary variables: quantity, kind, and orientation of vegetative cover. Whereas the USLE is calculated by straightforward multiplication of the numbers assigned to each factor, the wind erosion equation must be solved by use of graphs, tables, nomograms, and charts. It is a tool for determining the potential erosion from a field and for determining what field conditions (soil cloddiness, roughness, vegetative cover, wind barriers, and width and orientation of fields) are necessary to reduce potential erosion to a tolerable amount.

Little testing of the wind erosion equation has been done, because data for some factors are difficult or impossible to obtain, the values assigned to some factors are dubious, and the equation is cumbersome to solve. According to a report on the 1980 appraisal of the Soil and Water Resources Conservation Act, an attempt by the Soil Conservation Service to use the equation outside the Great Plains states was unsuccessful (RCA, 1981). Information is needed on the prevailing wind erosion direction for areas outside the Great Plains, the vegetation factor for crops outside the Great Plains, and seasonal and annual soil erodibility variations. In addition, an improved method for calculating surface moisture is needed.

The FAO (1979) attempted to use the wind erosion equation in preparing a soil degradation map of Africa north of the Sahara, but ended up developing a different simplified method in which the variables are multiplied as in the USLE. The FAO method has not been field tested.

Attractive though the wind erosion equation is to those concerned with wind erosion and its control in developing countries, there have been few attempts to improve the equation after the first disappointment in the anomalous results that are usually obtained. Wind erosion is not as ubiquitous and attention-getting as water erosion. A recent book entitled *Assessment of Erosion* does not mention wind erosion (De Boodt and Gabriels, 1980), nor does an earlier book with the comprehensive title *Soil Erosion: Prediction and Control* (Foster, 1977). Yet wind erosion is a major problem to cultivators and pastoralists in the arid and semiarid regions of the world, in both developed and developing countries. In the drylands of the world, wind erosion is much more difficult to control than water erosion.

6.1.3 Salinization.

Land degradation due to salinization is primarily associated with irrigated land and a few dryland cropping areas in North America and Australia. It is considered the major threat to long-continued canal irrigation. Sodic soil formation in irrigated land is one aspect of salinization.

Despite an extensive literature on soil salinization and attention to the problem throughout the world, no numerical prediction model of salinization or sodic soil formation has been formulated. Descriptive models abound, particularly for the salt component of irrigation return flows. Numerical models of certain phases of salinization, such as movement of salt during leaching or upward movement of salt and water from a groundwater table, have been constructed (Dregne, 1977). They generally have only modest practical value because soil and water conditions are too dynamic to permit more than gross predictions of effects on soils. One method for predicting soil salinity (FAO, 1979b) is actually a risk assessment and does not appear to be particularly useful. The same FAO publication gives a method for predicting sodic soil formation, but, again, it is of dubious value.

Most commonly, soil salinization is predicted by relating some combination of irrigation water quality, soil properties, and depth of water table to the likelihood that a salinity or sodicity problem will arise. The prediction is qualitative because of the variability of field conditions.

6.1.4 Waterlogging.

As a land degradation problem, waterlogging is largely due to irrigation. It is often stated that irrigation and drainage go hand-in-hand, yet irrigation systems continue to be constructed without considering the drainage problems that nearly inevitably arise.

As with salinization, numerous descriptive models of water movement in soils and geologic strata have been constructed and used to estimate the likelihood of waterlogging. Once a likelihood has been established, design formulas can be used to calculate drain device spacing, depth, slope, and so on. In practice, little careful attention is given to predicting the occurrence of waterlogging, so that prediction is inexact. Much attention has been given to the design of drainage systems (Luthin, 1957).

6.1.5 Loss of soil fertility.

Soil fertility losses are associated with soil erosion, removal of nutrients in harvested plants, and leaching. There is no numerical model for predicting these soil fertility losses. The principal problem is the difference in availability to plants of nutrients held in organic and inorganic forms in soils. Nitrogen availability, for example, is a function of the amounts of nitrogen in solution, adsorbed on the exchange complex, and held in various organic forms. These vary throughout the year in a way that is, at best, only grossly predictable. Progress is being made in predicting postharvest soil nitrogen, phosphorus, and potassium levels when crop yields are known, and soil testing has been done over a period of years to establish baseline conditions. No comprehensive studies are under way to elucidate the relation between fertility loss and either erosion or leaching. A method for assessing the risk of chemical degradation (leaching of exchangeable bases) of soils has been proposed but has not been tested (FAO, 1979b).

6.1.6 Soil compaction.

Soil compaction, in the context of land degradation, refers to compression of soil particles due to tillage, machinery pressure, and livestock trampling. Tillage operations are the most important cause of soil compaction in cultivated fields.

Soil engineers have developed reliable equations for predicting soil compression in the field from laboratory tests of one-dimensional consolidation (Yong and Warkentin, 1966). These equations can be applied to agricultural as well as engineering operations, with modifications to accommodate shear effects in tillage operations. The prediction problem in agriculture is principally one of specifying soil moisture conditions, which usually are variable at the time when tillage is performed or when livestock are walking over the land. In practice, livestock effects on cultivated and rangeland soil compaction are significant in some cases but not important enough to be studied carefully. Similarly, the effects of tillage operations can be and are predicted, but usually nothing is done until a compaction problem must be resolved. Emphasis has been placed on compaction principles, measurement of compaction, and manipulating the soil to increase or reduce compaction (ASAE, 1971).

Compaction is a widespread problem affecting plant growth, especially in cultivated soils. The area affected has only been assessed in California, where about 1,200,000 ha. of acricultural land are said to be affected by anthropogenic soil compaction (Department of Conservation, 1979).

6.1.7 Surface soil crusting.

Although surface soil crusts are found in nondegraded soils, the problem is generally associated with cultivated land where humans have adversely influenced the soil structure and with rangelands where the vegetative cover has deteriorated. Direct impact of raindrops on exposed soil is probably the principal immediate cause of surface crusting but puddling of soils by livestock and tillage instruments are important contributors. The cause of crusting is dispersion of soil aggregates by an externally applied force (Baver et al., 1972). No attempt appears to have been made to develop an equation for predicting surface soil crusting.

6.2 Methods of Assessing Effects of Degradation on Productivity

Erosion by wind and water, soil compaction, waterlogging and salting reduce the productivity of land. The effects of water erosion have been studied most and are the focus of this discussion.

The distinction often made between data and information is particularly relevant to the analysis of erosion-productivity relations. It is relatively easy to design experiments that show the effect on crop yields of removing various amounts of topsoil from small plots of land. It is not easy to transform these experimental data to information that will be useful to policy-makers and planners in designing large-scale projects of agricultural intensification, conservation, and watershed management.

Here we briefly describe the experimental methods that have been used in the United States to obtain data about erosion-yield relations, but give most of our attention to models now under development that would put these data in a form useful for policy. Most of the models discussed predict the effects of erosion on crop yields, because that is the principal focus of research in this area. However, damage to productivity caused by sedimentation of eroded soil after it leaves the farm is also important and is discussed briefly.

6.2.1 Experimental data.

In the United States, systematic experimental research on erosion-productivity relation began in the 1920's and extended into the 1950's. Research in this area was begun again in the 1970's, when rapid growth of U.S. exports of grains and soybeans resulted in much additional land in crops and a renewal of concern about erosion. This type of research is illustrated by the studies by Latham (1940), Hays et al. (1948), Stallings (1950), Odell (1950), Whitney et al. (1950), Englestad et al. (1961), Reuss and Campbell (1961), Peterson (1964), Lyles (1975, 1977), Fryrear (1977), Olson (1977), and Power et al. (1979). A more general treatment is given by K. Young (1978), and three excellent recent reviews of the literature are by the National Soil Erosion-Soil Productivity Research Planning Committee (1981), Shrader (1982), and Langdale and Shrader (1982). In 1981, the USDAs Agricultural Research Service (ARS) established a program of research on erosion-productivity relations. A workshop was held at Purdue University in September 1981 to discuss research needs and methodologies. The program is viewed by the ARS as a long-term commitment. For information about the program contact H.L. Barrows, USDA-SWAS, SEA-AR, Beltsville Agricultural Research Center, Beltsville, Maryland.

The methodologies used in the experimental work involve a comparison of crop yields before and after removal of topsoil. In some cases the topsoil was removed as part of the experiment; in others, areas where topsoil had been removed by railroad or highway cuts, land leveling, or surface mining were studied. Generally, the studies were conducted over a period of 2 to 5 years. They all attempted to hold the effects of weather and technology constant, with varying degrees of success. It was found that removal of topsoil significantly reduced yields, but the reduction varied widely, depending on the nature of the subsoil, the crop, and other local conditions.

6.2.2 Predictive models.

In designing agricultural development projects, policy-makers and planners should ask how much the projects will cost in erosion-induced loss of productivity. Such productivity losses are costs in the same sense as those incurred in building dams, clearing land, and using machinery and fertilizers. A proper cost-benefit analysis of projects must include these erosion costs.

Experimental data on erosion and yield relations are useful for this purpose, but are not enough. What is needed is a function showing the annual loss of yield resulting from given amounts of erosion continued indefinitely or for a specific period of time. The experimental data typically provide only two points: yields before and after a given amount of erosion. Moreover, the data are too site-specific to be used with confidence in planning projects in areas for which experimental data are lacking.

Development of erosion-productivity models is in its early stages in the United States. The first attempt was in connection with the Resource Conservation Act Appraisal (RCA), published by the USDA in 1981. A so-called yield soil-loss simulator was developed which, when combined with estimates of erosion from the USDA National Resources Inventory (NRI), gave regional estimates of erosion-induced losses of yields for ten crops over a 50 year period. The simulator makes crop yield on various soil types a function of soil depth. (Data on yields and soil depth were taken from county level soil surveys.)

A problem with the simulator is that it was never adequately documented. The best appraisal is in an unpublished memorandum by Benbrook (1980), an economist with the Council for Environmental Quality and a member of the RCA Coordinating Committee. A major point was that soil texture, water holding capacity, and other soil characteristics change as soil depth is reduced by erosion, and these changes and their effects on yield are not adequately represented simply by changes in depth.

In work under way in connection with the USDA's 1985 RCA appraisal, a new and much more sophisticated simulator, called EPIC, is being developed. EPIC is being developed under the direction of P.T. Dyke and J.R. Williams (1982) at the regional office of the USDA at Temple, Texas. EPIC consists of six major submodels: (i) soil, water, and nutrient budgets; (ii) weather simulation; (iii) biomass-crop growth simulation; (iv) tillage and residue management simulation; (v) soil erosion simulation; and (vi) production cost budgets. The first submodel incorporates information on soil depth, porosity, water holding capacity, bulk density, particle size distribution, pH, and nutrient content. The initial state of these variables is specified, and changes in them are traced over time in response to erosion, weather, cropping patterns and other factors. With weather, cropping patterns and other factors held constant, the model predicts the effect of erosion on soil characteristics and hence on yields.

EPIC represents the most ambitious research under way in the United States to predict the effects of erosion on crop yields. Its data requirements are great. If it works, it will make a major contribution to soil conservation planning in the United States. Its applicability in other countries, where the necessary data are less readily available, is an open question. Simplification of the model would be necessary in most if not all cases.

A less ambitious but promising modeling effort was begun in 1980 under the direction of W. Larson of the University of Minnesota, Saint Paul. Drawing on work by L.L. Neill, C.L. Scrivner, and M.E. Keener at the University of Missouri, Larson developed a methodology for predicting the effects of sustained erosion on crop yields. There are two principal steps in the methodology.

The first step relies on the crop rooting model developed by Neill et al. (1981), which relates yield to five soil characteristics: bulk density, potential available water, pH, electrical conductivity, and aeration. Nutrient content of the soil is not specified. Nutrient supply in the form of fertilizer is treated as a management function. The combined effect of these characteristics on yield varies with soil layer, becoming less favorable at lower layers. In his adaptation of the model of Neill et al., Larson distinguishes three soil layers to a depth of 1 meter. The yield from 1 hectare of soil 1 meter deep is the weighted average of the contribution of each of the three layers, where the weight is the relative rooting concentration in each layer. In an uneroded or slightly eroded soil, rooting concentration is greatest in the top layer, and this layer has the greatest weight; the weights of the second and third layers are successively less. As erosion occurs, the top layer declines in weight relative to the second and third layers. Since the soil characteristics of the lower layers are generally less favorable, the shift in weights causes yields to decline.

In the second step, data from soil surveys are used to calculate soil productivity indexes for all soils covered by the surveys. Erosion data for these soils, taken from the USDA National Resources Inventory, can be used in the crop rooting model to predict the effects of erosion or yield.

The Larson model and the EPIC model predict the effects of erosion on yields with technology held constant. This can be misleading if the effects of technology are not subsequently taken into account. In the United States technological advances have increased crop yields substantially, even in areas of high erosion. In considering how best to promote productivity, planners should consider investments in both erosion control and technological advances. Planners can go astray if they are so concerned with the negative effects of erosion on yields that they neglect the positive effects of technology.

The models discussed deal only with the effects of erosion on the productivity of the eroded land. They do not address off-farm productivity damages resulting from such effects as accelerated siltation of reservoirs. Under some circumstances, these off-farm damages may substantially exceed the on-farm productivity loss (Swanson, 1978). In analyzing the erosion effects of agricultural development projects, planners should incorporate these off-farm damages. The models developed at the University of Illinois should prove useful for this purpose.

Institutions

Hundreds of agencies and groups have been or are engaged in making soil surveys and analyzing and publishing the results. A few are cited below.

1) Land and Water Development Division of the U.N. Food and Agricultural Organization. R. Dudal is the director. The address is Via delle Termi di Caracalla 00100, Rome, Italy.

2) International Soil Museum, Wageningen, The Netherlands. W. G. Sombroek is the director. The museum is a repository of soil samples from many parts of the world. Its staff is reviewing the status of soil surveys by country in order to make accessible any larger scale maps that can be used for management at the national and local levels.

3) Consultant Group on International Agricultural Research, 1818 H Street, N.W., Washington, D.C. 20433.

Other sources of information about soil surveys for particular countries are the ministry of agriculture or a similar national agency of the country and the International Division of the USDA Soil Conservation Service (USDA/SCS), which is currently engaged in soil surveys in other countries. Information about these surveys is maintained in a card file in the national office of the Soil Classification and Mapping Branch of the SCS, Washington, D.C.

FAO-UNESCO is continuing its work on soil surveys, as are agencies in other countries that provide technical assistance to developing nations. The Soil Management Support Services program of the SCS, funded by the U.S. Agency for International Development (U.S.AID), provides international technical assistance in soil survey and classification and in soil conservation, use, and management. It is a valuable source of information and information on soils in many developing countries. (Addresses of the above are provided in the list of institutions at the end of this report).

Soil survey work in the United States is being carried on by several agencies. The SCS and the Forest Service, which are both part of the USDA, along with the Bureau of Land Management (BLM) and the Bureau of Indian Affairs (BIA), both in the U.S. Department of Interior, are responsible to Congress for mapping all of the lands under their jurisdiction and are actively engaged in soil surveys. As of 1982, approximately 50 percent of the United States had been mapped in detail. The USDA/SCS coordinates mapping among these agencies and annually prints a list of published soil surveys, grouped by state and county name.

References Cited

AASHO. 1961. *Standard specification for highway materials and methods of sampling and testing.* 8th ed. Washington, DC: American Association of State Highway Officials.

Adler, B. 1911. Maps of primitive peoples. Translation from Russian and resume by H. de Hutorowicz. *Bulletin of the American Geographical Society* 43(9): 669-679.

Aina, P.O., R. Lal, and G.S. Taylor. 1977. Soil and crop management in relation to soil erosion in the rainforest of western Nigeria. In *Soil erosion: Prediction and control*, ed. G. R. Foster, pp. 75-82. Ankeny, Iowa: Soil Conservation Society of America.

Alexander, M. 1977. *Introduction to soil microbiology.* 2nd ed. New York: John Wiley and Sons.

Allan, J.A. 1980. Remote sensing in land and land use studies. *Geography* 65:35-43.

Allan, W. 1965. *The African husbandman.* Edinburgh: Oliver and Boyd.

Allan, W., M. Gluckman and others. 1948. Landholding and land usage among the Plateau Tonga. *Rhodes Livingstone Papers*, No. 14. Manchester: Manchester University Press.

Ama, J.T. 1970. Report on the detailed soil survey of the proposed oil palm research centre., Kusi, Tech. Rep. *Soil Res. Inst. Ghauna Acad. Sci.*, No. 82.

Arens, P.S. 1978. Edaphic criteria in land evaluation. *Wild Soil Resour.* Rep. 49:24-31.

ASAE. 1971. Compaction of agricultural soils. *ASAE*, p. 471. St. Joseph, Mich.: American Society of Agricultural Engineers.

Austen, L. 1939. The seasonal gardening calendar of Kiriwina, Trobriand Islands. *Oceania* 9(3): 237-253.

Bagnold, R.A. 1943. *The physics of blown sand and desert dunes.* New York: Wm. Morrow.

Barrett, S.A. 1908. *The ethnogeography of the Pomo and the neighboring Indians.* Univ. of California Publications in American Archaeology and Ethnology, 6. Berkeley.

Bartelli, L.J. 1978. Technical classification system for soil survey interpretation. *Adv. Agron.* 30:247-89.

Bartelli, L.J. 1979. Interpreting soil data. In *Planning the uses and management of land*, eds. M. T. Beatty et al., pp. 91-116. Madison, Wisc.: ASA, CSSA, SSSA.

Baumgardner, M.F. 1982. Remote sensing for resource management: Today and tomorrow. In *Remote sensing for resource management*, eds. C.J. Johannsen and J.L. Sanders, pp. 16-29. Ankeny, Iowa: Soil Conservation Soc. Amer.

Baumgardner, M.F. 1979. Assessment of arable land. In *Fertilizer raw material resources, needs and commerce in Asia and the Pacific*, eds. R. Sheldon et al. Honolulu: East West Center.

III. SOILS

Baver, L.D., W.H. Gardner, and W.R. Gardner. 1972. *Soil physics*. 4th ed. New York: John Wiley.

Beatty, M.T., G.W. Peterson, and L.D. Swindale, eds. 1979. *Planning the uses of management of land.* Madison, Wisc.: American Society of Agronomy.

Beckett, P. 1981. The cost-benefit relationships of soil surveys. *Soil resource inventories and development planning.* Technical Monograph No. 1. Soil Management Support Services, Soil Conservation Service. Washington, D.C.: U.S. Department of Agriculture.

Beek, K.J. 1974. The concept of land utilization types. *FAO Soil Bull.* 22:203-220.

Beek, K.J. 1975a. Identification of land utilization types. *Wild Soil Resour. Rep.* 45:89-102.

Beek, K.J. 1975b. Land utilization types in land evaluation. *FAO Soils Bull.* 29:87-106.

Beek, K.J. 1977. The selection of soil properties and land qualities relevant to specific land uses in developing countries. In *Soil resource inventories.* Agronomy Mimeo 77-23, pp. 143-62. Ithaca: Cornell University.

Beek, K.J. 1978. *Land evaluation for agricultural development*. Publication No. 23. Wageningen, The Netherlands: Intl. Inst. for Land Reclamation and Improvement.

Beek, K.J. 1977. The selection of soil properties and land qualities relevant to specific land uses in developing countries. In *Proceedings, Soil resource inventories workshop at Cornell University*, April 4-7, 1977. Agronomy Mimeo. 77-23, pp. 143-162. Ithaca: Agronomy Department, Cornell University.

Beek, K.J. 1981. The selection of soil properties and land qualities relevant to specific land uses in developing countries. In *Soil resource inventories and development planning, Proceedings of Workshops at Cornell University 1977-1978.* Technical Monograph No. 1. Washington, DC: SCS, USDA.

Beek, K.J. and J. Bennema. 1972. *Land evaluation for agricultural land use planning: An ecological methodology.* Wageningen, The Netherlands: Agricultural State University of Wageningen.

Beinroth, F.H., 1978. Some fundamentals of soil classification. In *Soil resource data for agricultural development*, ed. L.D. Swindale. Honolulu: Hawaii Agricultural Experiment Station, University of Hawaii.

Bennema, J. 1976. *Land evaluation for agricultural use planning.* Benchmark Soils Project. Honolulu: University of Hawaii.

Berry, B.J.L. 1958. A note concerning methods of classification. *Annals of the Association of American Geographers.* 48(3): 300-303.

Berry, L. 1980. *The impact of irrigation on development: Issues for a comprehensive evaluation study.* Paper prepared for USAID. Program for International Development. Worcester, Ma.: Clark University.

Bie, S. 1980. A survey of soil information systems. Presented at the Sixth Annual Symposium on Machine Processing of Remotely Sensed Data. Purdue University, West Lafayette, Indiana USA (unpublished).

Bie, S.W. and P.H.T. Beckett. 1970. The costs of soil survey. *Soils and Fertilizers* 33:203-16.

Black, C.A., D.D. Evans, J.L. White, L.E. Ensminger, and F.E. Clark, eds. 1965. *Methods of soil analysis.* Parts I & II. Madison, Wisc.: American Society of Agronomy.

Boas, F. 1934. Geographical names of the Kwakiutl Indians. Columbia University Contribution to Anthropology, no. 20. New York: Columbia University.

Bourne, R. 1931. Regional survey and its relation to stockholding of the agricultural resources of the British Empire. *Ox. For. Mem.* 13:16-18.

Brady, N.C. 1974. *The nature and properties of soils.* 8th edition. New York: Macmillan.

Brady, N.C. 1980. *The nature and properties of soils.* 8th ed. New York: Macmillan.

Bradshaw, A.D. and M.J. Chadwick. 1980. *The restoration of land.* Berkeley: Univ. of California Press.

Brenchley, G.H. 1974. Aerial photography for the study of soil conditions and crop diseases. *Soil Surv Tech. Mongr.* 4:99-106.

Brink, A.B., J.A. Mabbutt, R. Webster and P.H.T. Beckett. 1966. Report 940. Christchurch, England: Military Engng. Exp. Establ.

Brink, A.B.A., T.C. Partiridge, and A.A.B. Williams. 1981. *Soil Survey for engineering.* Oxford: Clarendon Press.

Brinkman, R. 1978. Land suitability evaluation for wetland rice: criteria and required standards. *Wld Soil Resour. Rep.* 49:32-8.

Browning, G.M., C.L. Parish, and J.A. Glass. 1947. A method for determining the use and limitations of soil erosion in Iowa. *Journal of the American Society of Agronomy.* 39:65-73.

Bunge, W.W. 1962. *Theoretical geography.* Lund studies in Geography, Series C, General and Mathematical, No. 1. Lund: Royal University of Lund, Department of Geography.

Buol, S.W., F.D. Hole and R.J. McCracken. 1973. *Soil genesis and classification.* Ames: The Iowa State Univ. Press.

Buol, S.W. and W. Conto. 1980. Fertility capability soil classification system for use in the humid tropics. In *Characterization of soils in relation to their classification and management for crop production: Examples from some areas of the humid tropic*, ed. D.J. Greenland. Oxford: Oxford Univ. Press.

Buol, S.W., F.D. Hole and R.J. McCracken. 1980. *Soil genesis and classification.* 2nd ed. Ames: The Iowa State Univ. Press.

Buol, S.W., P.A. Sanchez, R.B. Cate, and M.A. Granger. 1975. Soil fertility capability classification: a technical soil classification for fertility management. In *Soil management in tropical America*, eds. E. Bornemisza and A. Alvarado, pp. 126-141. Raleigh: North Carolina State University.

Bureau of Reclamation. 1953. *Land classification handbook.* Publ. V., part 2. Washington: U.S. Dept. Interior, Bureau of Reclamation.

Buringh, P. 1978. Limits to the productive capacity of the biosphere. World Conference on Future Sources of Organic Raw Materials. Toronto, Canada (Mimeographed).

Burns, R.C. and R.N.F. Hardy. 1975. *Nitrogen fixation in bacteria and higher plants.* New York: Springer-Verlag.

Butler, B.D. 1964. Assessing the soil factor in agricultural production. *J. Aust. Inst. Agric. Sci.* 30:232-240.

Butler, B.E. 1980. *Soil classification for soil survey.* Oxford: Clarendon Press.

Carroll, D.M., R. Evans, and V.C. Bendelow. 1977. Air photo-interpretation for soil mapping. *Soil Surv. Tech. Monogr.* No. 8.

Carter, W. 1969. New lands and old traditions. Second Latin American Monograph no. 6. Gainsville: University of Florida.

Chan, H.Y. 1978. Soil-survey interpretations for improved rubber production in pennisular Malaysia. In *Soil-resources data for agricultural development*, ed. L.D. Swindale, pp. 41-66. Honolulu: Agricultural Experiment Station.

Chisholm, M. 1964. Problems in the classification and use of the farming type region. *Transactions of the Institute of British Geographers.* 35:91-103.

Cline, M.G. 1949. Basic principles of soil classification. *Soil Science* 67:81-91.

Cline, M.J.. 1979. *Soil classification in the United States.* Agronomy Mimeo no. 79-12. Ithaca, NY: Department of Agronomy, Cornell University.

Cline, M.J. 1980. Experience with soil taxonomy of the United States. *Advances in Agronomy.* 33:193-302.

Cline, M.J. 1981. Thoughts about appraising the utility of soil maps. In *Soil resource inventories and development planning.* Technical monograph no. 1. Soil Management Support Services, Soil Conservation Service, pp. 37-46. Washington, DC: U.S. Dept. of Agriculture.

Cochrane, T.T. 1979. An on-going appraisal of the savanna ecosystems of tropical America for beef cattle production. In *Pasture production in acid soils of the tropics*, eds. P.A. Sanchez and L.E. Tergas, pp. 1-12. Cali, Colombia: Centro International de Agricultura Tropical.

Cochrane, T.T., J.A. Porras, J. Azevedo, P.G. Jones and L.F. Sanchez. 1979. *An exploratory manual for CIAT's computerized land resource study of tropical America.* Cali, Columbia: Centro International de Agricultura Tropical.

Colwell, R.N., ed. 1960. *Manual of photographic interpretation.* Falls Church, Va.: American Society of Photogrammetry.

Conklin, H.C. 1980. *Folk classification: A topically arranged bibliography of contemporary and background references through 1971.* New Haven, Conn.: Yale University Department of Anthropology.

Conklin, H.C. 1967a. Ifugao ethnobotany 1905-1965; the 1911 Beyer-Merrill report in perspective. In *Studies in Philippine anthropology, in honor of H. Otley Beyer*, ed. M.D. Zamor, pp. 204-262. Quezon City: Alemar Phoenix.

Conklin, H.C. 1967b. Some aspects of ethnographic research in Ifugao. *Transactions of the New York Academy of Sciences*, Series 2, 30(1): 99-121.

Conklin, H.C. 1960. The cultural significance of land resources among the Hanunoo. *Bulletin of the Philadelphia Anthropological Society*. 13(2): 38-42.

Conklin, H.C. 1957. Hanunoo agriculture: A report on an integral system of shifting cultivation in the Philippines. FAO Forestry Development Paper No. 12. xii, 209 pp. Rome: FAO.

De Boodt, M., and D. Gabriels. 1980. *Assessment of erosion.* New York: Wiley.

Dent, F.J. 1980. Major production systems and soil-related constraints in Southeast Asia. In *Soil related constraints to food production in the tropics.* Los Banos, Philippines: IRRI.

Department of Conservation. 1979. California soils: An assessment. Draft report, California Department of Conservation, Sacramento, California.

Dregne, H.E., ed. 1977. *Managing saline water for irrigation.* ICASALS. Lubbock, Tex.: Texas Tech. University.

Drosdoff, M., ed. 1978. *Diversity of soils in the tropics.* ASA Spec. Publi. 34. Madison, Wisc.: American Society of Agronomy.

Dudal, R., 1978. Land resources for agricultural development. In *Plenary papers of 11th Congress*, vol. 2, pp. 341-359. Edmonton, Canada: International Soil Science Society.

III. SOILS

Dudal, R. 1978. *Adequacy of soil surveys and soil classification for practical applications in developing countries.* Land and Water Development Division. Rome: FAO.

Dudal, R. 1978. Adequacy of soil surveys and soil classification for practical applications in developing countries. *Proceedings, 2nd International Soil Classification Workshop*, Vol.1. Soil Survey Division. Bangkok: Land and Water Development Department.

Dudal, R. 1980. *Land resources for populations of the future.* Land and Water Development Division. Rome: FAO.

Dudal, R. 1980. Soil-related constraints to agricultural development in the tropics. In *Soil-related constraints to food production in the tropics.* Los Banos, Philippines: IRRI.

Dudal, R. 1980. An evaluation of conservation needs. In *Soil conservation, problems and prospects,* ed. R.P.C. Morgan. Chichester: John Wiley & Sons.

Dudal R. 1982. Land degradation in a world perspective. *Journal of Soil and Water Conservation* 37(5): 245-249.

Dudal, R., G.M. Higgins and A.H. Kassam. 1982. *Land the world's food production.* Proceedings 12th International Congress of Soil Science, New Delhi. Wageningen, The Netherlands: International Soil Science Society.

Dyke, P.T., and J.R. Williams. 1982. *Erosion productivity impact calculator (EPIC): A demonstration of process modeling for policy analysis.* Temple, Tex.: U.S. Department of Agriculture.

Edwards, K., and P.E.V. Charman. 1980. The future of soil loss prediction in Australia. *Journal of the Soil Conservation Service of N.S.W.* 36:211-218.

El-Swaify, S.A. 1981. Soil erosion by water as a concern in development program planning. In *Natural systems information for planners,* ed. R.A. Carpenter. Honolulu: East-West Center, Environmental & Policy Institute.

Elwell, H.A. 1978. *Soil loss estimation.* Compiled works of the Rhodesian Multidisciplinary Team on Soil Loss Estimation. Department of Conservation and Extension. Salisbury: Government of Rhodesia.

Elwell, H.A. and M.A. Stocking. 1982. Developing a simple yet practical method of soil-loss estimation. *Tropical Agriculture* 59:4347.

Engelstad, O.P., W.D. Shrader, and L.C. Dumenil. 1961. The effect of Surface Soil Thickness on Corn Yields: I. As determined by a series of field experiments in farmer-operated fields. *Proceedings of the Soil Science Society of America* 25:494-499.

Ehrlich, P.R., A.H. Ehrlich and J.P.V. Holdren. 1977. *Ecoscience.* San Francisco: W.H. Freeman.

Eswaran, H. 1977. An evaluation of soil limitations from soil names. In *Soil resource inventories: proceedings of a workshop held at Cornell University April 4-7, 1977*, pp. 289-313. Agronomy Mimeo 77-23. Ithaca, New York: Agronomy Department, Cornell University.

Eswaran, H. 1981. Soil analysis for soil surveys. In *Soil resource inventories and development planning.* Technical monograph no. 1. Soil Management Support Services, Soil Conservation Service. Washington, DC: U.S. Department of Agriculture.

Eswaran, H., T.R. Forbes, and M.C. Laker. 1981. Soil map parameters and classification. In *Soil resource inventories and development planning.* Technical monograph no. 1. Soil Management Support Services, Soil Conservation Service. Washington, D.C.: U.S. Department of Agriculture.

Fanale, R. 1982. Navajo land and land management: A century of change. Unpub. Ph.D. Dissertation. Washington, DC: Catholic University of America.

FAO. Undated. *Guidelines for soil profile description.* Rome: FAO.

FAO. Undated. *Agriculture toward 2000.* Rome: FAO.

FAO. 1970a. *Methods for physical and chemical analysis of soils.* Soils Bulletin No. 10. Rome: FAO:

FAO. 1970b. *Soil fertility investigations on farmers' fields.* By G. F. Hauser. Soils Bulletin No. 11. Rome: FAO.

FAO. 1971. *Land degradation.* By R. Rauschkolb. Soils Bulletin No. 13, p. 105. Rome: FAO.

FAO. 1973a. *Guide to the calibration of soil tests for fertilizer recommendations.* Soils Bulletin No. 18. Rome: FAO.

FAO. 1973b. *Soil survey interpretations for engineering purpose.* Soils Bulletin No. 19. Rome: FAO.

FAO. 1974. *Irrigation suitability classification.* FAO Soils Bulletin No. 22. Rome: FAO.

FAO. 1974. *Approaches to land classification.* Soils Bulletin No. 22. Rome: FAO.

FAO. 1976a. *Soil conservation for developing countries.* By I. Constantinesco. Soils Bulletin No. 30. Rome: FAO.

FAO. 1976b. *Prognosis of salinity and alkalinity.* Soils Bulletin No. 31. Rome: FAO.

FAO. 1976c. *A framework for land evaluation.* Soils Bulletin No. 32. Rome: FAO.

FAO. 1977. *Assessing soil degradation.* Soils Bulletin No. 34. Rome: FAO.

FAO. 1978. *Report on the agro-ecological zones project.* Vol. 1. *Methodology and results for Africa.* World Soil Resourc. Rep. 48/1. Rome: FAO.

FAO. 1979a. *Soil survey investigations in irrigation.* Soils Bulletin No. 42. Rome: FAO.

FAO. 1979b. *A provisional methodology for soil degradation assessment.* Joint FAO/UNEP/UNESCO publication. Rome: FAO.

FAO. 1979c. Land evaluation criteria for irrigation. Report of an expert consultant. *World Soil Resourc.* Rep. 50. Rome: FAO.

FAO. 1979d. *Agriculture: Toward 2000.* Rome: FAO.

FAO-UNESCO. 1970-1980. *Soil map of the world.* Prepared by the Food and Agricultural Organization of the United Nations. 10 volumes, 18 soils map sheets, scale 1:5,000,000. Paris: UNESCO.

FAO-UNESCO. 1974. *Soil map of the world.* Vol. 1. *Legend.* Paris: UNESCO.

Farnsworth, E. and F. Golley, eds. 1974. *Fragile ecosystems. Evaluation of research and applications in the neotropics.* New York: Springer-Verlag.

Fertilizer Institute. 1972. *The fertilizer handbook.* Washington, D.C.: The Fertilizer Institute.

Forbes, T.R., D.G. Rossiter, and A. Van Wambeke. 1982. *Guidelines for soils resource inventory evaluation.* Ithaca: Cornell University. Department of Agronomy.

Foster, G.R., ed. 1977. *Soil erosion: Prediction and control.* Ankeny, Iowa: Soil Conservation Society of America.

Fryrear, D.W. 1977. *Long term effect of erosion and cropping on soil productivity.* Paper given at the Feb. 1977 meeting of the AAAS, Denver (was to have been published in the Proceedings).

Godin, V.J. and P.C. Spensley. 1971. *Oils and oilseeds.* London: Tropical Products Institute.

Golden, J.D., P. Lemos, R.E. Carlyle, and C.R. Rreitas. 1966. *Guide for general and specialized equipment for soils laboratories.* FAO Soils Bull. No. 3. Rome: FAO.

Goodland, R.J.A. and H.S. Irwin. 1975. *Amazon jungle: Green hell or red desert?* Amsterdam: Elsevier.

Goodman, L.J. and R.H. Karol. 1968. *Theory and practice of foundation engineering.* New York: Macmillan.

Gluckman, M. 1941. *The economy of the Central Barotse plain.* Rhodes-Livingstone Papers, No. 7. Manchester: Manchester Univ. Press.

Gluckman, M. 1943. Essays on Lozi land and royal property. Rhodes Livingstone Papers, No. 10.

Gray, T.P. and S.T. Williams. 1975. *Soil microorganisms.* New York: Longman.

Hagan, R.M., H.W. Haise, T.W. Edminster, eds. 1967. *Irrigation of agricultural lands.* Madison, Wisconsin: ASA.

Hall, W.A., G.W. Hargreaves, G.M. Cannel. 1979. Planning large-scale agricultural systems with integrated water management. In *Planning the uses and management of land*, ed. M. T. Beatty et al., pp. 273-289. Madison, Wisc.: ASA, SCCA, SSA.

Hanway, J.J. 1967. Field experiments for soil test correlations and calibration. In *Soil testing and plant analysis.* Part I. pp. 103-114. Special Publication No. 2. Madison, Wisc.: Soil Science Society of America.

Harrington, J.P. 1916. The ethnogeography of the Tewa Indians. *Annual Reports of the Bureau of American Ethnology.* 29:29-618.

Hays, O.E., C.E. Bay, and H.H. Hull. 1948. Increasing production on an eroded loss-derived soil. *Journal of the American Society of Agronomy.* 40:1061-1069.

Hodgson, J.M., ed. 1974. *Soil survey field handbook.* Soil Surv. Tech. Monogr. No. 5.

Hodgson, J.M. 1978. *Soil sampling and soil description*, eds. P.H.J. Beckett, V.C. Robertson, and R. Webster, p. 241. Oxford, England: Clarendon Press.

Hose, C. 1905. Various methods of computing the time for planting among the races of Borneo. *Journal of the Royal Asiatic Society* Straits Branch, 42:1-5, 209-210.

Institute of Soil Science, Academia Sinica, ed. 1981. *Proceedings of symposium on paddy soil.* New York: Springer-Verlag.

IRRI. 1980. *Soil-related constraints to food production in the tropics.* Los Banos, Philippines: IRRI.

IUCN (International Union for the Conservation of Nature). 1975a. *The use of ecological guidelines for development in the American humid tropics.* IUNC New Series No. 31. Morges, Switzerland: IUCN.

IUCN. 1975b. *The use of ecological guidelines for development in tropical forest areas of south-east Asia.* IUCN New Series No. 32. Morges, Switzerland: IUCN.

Janick, J., R.W. Schery, F.W. Woods, and V.W. Ruttan. 1974. *Plant science.* 2nd ed. San Fransisco: W. H. Freeman.

Janzen, D.H. 1973. Tropical agroecosystems. *Science* 182:1212-1219.

Kampen, J. and J. Burford. 1980. Production systems, soil-related constraints, and potentials in the semi-arid tropics, with special reference to India. In *Soil-related constraints to food production in the tropics*. Los Banos, Philippines: IRRI.

Kasmar, J.V. 1970. The development of a usable lexicon of environmental descriptors. *Environment and Behavior.* 2(2): 153-169.

Kay, D.E. 1973. *Root crops.* London: Tropical Products Institute.

Kellogg, C.E. 1961. *Soil interpretation in the soil survey.* Washington, D.C.: USDA Soil Conservation Service.

Klingebiel, A.A. and P.H. Montgomery. 1960. *Land capability classification.* USDA Agricultural Handbook No. 210. Washington, D.C.: USDA.

Knox, A. 1968. *Glossary of geographical and topographical terms, and of words of frequent occurrence in the composition of such terms and of place names.* Detroit: Gale Research Co.

Langdale, G.W., and W.D. Shrader. 1982. Soil erosion effects on soil productivity of cultivated cropland. In *Determinants of soil loss tolerance.* Madison, Wisc.: Soil Science Society of America.

Latham, E.E. 1940. Relative productivity of the A horizon of Cecil Sandy Loam and the B and C horizons exposed by erosion. *Journal of the American Society of Agronomy* 32:950-954.

LeClerg, E.L., W.H. Leonard, and A.G. Clark. 1962. *Field plot techniques.* Minneapolis, Minn.: Burgess Publishing Co.

Liang, T. and R. Philipson. 1981. The use of airphoto interpretation and remote sensing in soil resource inventories with special reference to less-developed countries. In *Soil resource inventories and development planning.* Technical Monograph No. 1. Soil Management Support Services, SCS. Washington, DC: USDA.

Litchfield, C.D. and P.L. Seyfried. 1979. *Methodology for biomass determinations and microbial activities in sediments.* ASTM Special Publication 673. Philadelphia: American Society for Testing and Materials.

Little, T.M. and F.J. Hills. 1978. *Agricultural experimentation.* New York: Wiley.

Lowe, D.S., R.A. Summer, and E.J. Greenblat. 1974. *An economic evaluation of the utility of ERTS data for developing countries.* 2 Vols. Contract AID/CM/ta-73-38. Ann Arbor, Mich.: ERIM for USAID/Office of Science & Technology.

Luthin, J.N., ed. 1957. *Drainage of agricultural lands.* Agronomy Monograph No. 7. Madision, Wisc.: American Society of Agronomy.

Lyles, L. 1975. Possible effects of wind erosion on soil productivity. *Journal of Soil and Water Conservation* 30(6): 279-283.

Lyles, L. 1977. Wind erosion: Processes and effect on soil productivity. *Transactions of the American Society of Agricultural Engineers* 20(5): 880-884.

Malthus, T.R. 1826. *An essay on the principle of population.* 6th ed. 2 vols. London: John Murray, Albemarle Street.

Mason, H.L. and J.H. Langenheim. 1957. Language analysis and concept environment. *Ecology* 38.2:325-340.

Mabbutt, J.A. 1968. Review of concepts of land classification. In *Land evaluation*, ed. G.A. Stewart, pp. 11-28. London: Macmillan.

Mabuchi, T. 1966. Sphere of geographical knowledge and sociopolitical organization among the mountain peoples of Formosa. In *Folk culture of Japan and East Asia*, Monumenta Nipponica Monographs, No. 25, pp.101-146. Tokyo: Sophia Univ. Press.

Maletic, J.T. and T.B. Hutchings. 1967. Selection and classification of irrigable land. In *Irrigation of agricultural lands*, eds. Hagan, R.M., R. Howard and T.W. Edminster. Agr. Monogrpah no. 11. Madison, Wisc.: Amer. Soc. of Agron.

Malinowski, B. 1935. The language of magic and gardening. In *Coral gardens and their magic.* Vol. 2. London: Allen and Unwin.

McRae, S.G. and C.P. Burnham. 1981. *Land evaluation.* Oxford: Clarendon Press.

Milfred, C.J. and R.W. Kiefer. 1976. Analysis of soil variability with repetitive aerial photography. *Soil Sci. Am. J.* 40:553-557.

Miller, F.T. and J.D. Nichols. 1979. Soils data. In *Planning the uses and management of land*, eds. M.T. Beatty et al., pp. 67-89. Madison, Wisc.: ASA, CSSA, SSSA.

Milne, G. 1938. Essays in applied pedology, I Bukoba--High and low fertility in a laterised soil. *East African Agricultural Journal*.

Moorman, F.R. and D.J. Greenland. 1980. Major production systems related to soil properties in humic tropical Africa. In *Soil related constraints to food production in the tropics.* Los Banos, Philippines: International Rice Research Institute.

Mulcahy, M.J. and A.W. Humphries. 1967. Soil classification, soil surveys and land use. *Soils Fertil.* 30:1-8.

National Academy of Sciences. 1977. *Resource sensing from space: Prospects for developing countries.* Prepared for USA/Office of Science & Technology. Washington, D.C.: NAS

NASA (National Aeronautics and Space Administration). 1978. *Soil moisture workshop, a conference held at the U.S. Department of Agriculture, Beltsville, Maryland, January 17-19, 1978.* NASA Conference Publ. 2073. Wash., DC: Scientific and Technical Information Office, NASA.

Netting, R.M. 1968. *Hill farmers of Nigeria: Cultural ecology of the Kofyar of the Jos Plateau*. Seattle: Univ. Washington Press.

Netting, R.M. 1969. Ecosystems in process: A comparative study of change in two West African societies. In *Ecological essays*, ed. D. Damas, pp. 102-112. National Mus. Can Bull 230.

Netting, R.M. 1974. Agrarian ecology. *Annual Review of Anthropology*. 3:21-56.

Netting, R.M., D. Cleveland and F. Stier. 1979. The conditions of agricultural intensification in the West African savannah. In *Sahelian social development*, ed. Steven, pp. 187-506. Abidjan: REDSO/WA USAID.

NRC (National Research Council). 1980a. *Conversion of tropical moist forests*. Washington, DC: National Academy of Sciences.

NRC (National Research Council). 1980b. *Research priorities in tropical biology*. Washington, DC: National Academy of Sciences.

National Soil Erosion-Soil Productivity Research Planning Committee. 1981. Soil Erosion Effects on Soil Productivity: a Research Perspective. *Journal of Soil and Water Conservation* 36(2): 82-90.

Neill, L.L., C.L. Scrivner and M.E. Keener. 1981. *Evaluating soil productivity based on root growth and water depletion*. A contribution from the Missouri Agricultural Experiment Station, Missouri Department of Natural Resources, and USDA-Economics, Statistics, and Cooperatives Service, Columbia, Mo.

Odell, R.T. 1950. Measurement of the productivity of soils under various environmental conditions. *Agronomy Journal* 42(6): 282-292.

Ollier, C.D., D.P. Drover, and M. Godelier. 1971. Soil knowledge amongst the Baruya of Wonenara, New Guinea. *Oceania* 42(1): 33-41.

Olson, T. C. 1977. Restoring the Productivity of a Glacial Till Soil After Topsoil Removal. *Journal of Soil and Water Conservation* 32(3): 130-132.

Orloci, L. 1966. Geometric models in ecology. 1. The theory and application of some ordination methods. *Journal of Ecology* 54(1): 193-215.

Orvedal, A.C. and M.J. Edwards. 1942. General principles of technical grouping of soils. Soil Sci. Soc. Amer. Proc. (1941). Madison, Wisc.: Soil Science Soc. of America.

Paine, D.P. 1981. *Aerial photography and image interpretation for resource management*. New York: John Wiley and Sons.

Parkinson, D., T.R.G. Gray and S.T. Williams. 1971. *Methods for studying the ecology of soil microorganisms*. IBP Handbook 19. Oxford: Blackwell Scientific Publications.

PCA. 1962. *PCA soil primer*. Skokie, Ill.: Portland Cement Association.

Peek, T.R. and S.W. Melsted. 1967. Field sampling for soil testing. In *Soil testing and plant analysis*, pp. 25-35. SSSA Special Publication No. 2. Madison, Wisc.: Soil Science Society of America.

Penck, A. 1928. Das Hauptproblem der Physischen Anthropogeographie. *Proc. and Papers, 1st Int. Congress Soil Sci.* 2:98:116.

Peterson, J.B. 1964. The relation of soil fertility to soil erosion. *Journal of Soil and Water Conservation* 19(1): 15-19.

Philipson, W.R. and T. Liang. 1975. Airphoto analysis in the tropics: crop identification. In *Proceedings Tenth Int. Symposium on Remote Sensing of Environment*, pp. 1079-1092. Ann Arbor, Mich.: Environmental Research Institute of Michigan.

Pough, F.H. 1953. *A field guide to rocks and minerals.* (Peterson field guide 7.) Boston: Houghton Mifflin.

Poulton, C.E. 1982. *Effective resource inventory in the Sahel: Planning, implementation and Utilization.* Prepared for SDPT/Bamako through ST/FNR Forestry Support Program Contract No. 53-319R-2-94. Chevy Chase, Md: Earth Satellite Corporation.

Power, J.F., F.M. Sandeval, and R.E. Ries. 1979. Topsoil-subsoil requirements to restore North Dakota mined land to original productivity. *Mining Engineering* December.

Protz, R. 1981. Soil properties important for various tropical crops: Pahang Tenggara Master Planning Study. In *Soil resource inventories and development planning, proceedings of workshops at Cornell University 1977-1978.* Technical Monograph No. 1. Washington, DC: SCS, USDA.

Purnell, M.F. 1978. Progress and problems in the application of land valuation in FAO projects in different countries. *Wld Soil Resour. Rep.* 49:81-8.

Purnell, M.F. 1979. The FAO approach to land evaluation and its application to land classification for irrigation. *Wld Soil Resour. Rep.* 50:4-8.

RADAM-Brazil, Projecto. 1972-1978. Levantamento de Regiao Amazonica. Vols. 1-12. Ministerio das Minas e Energia, Departmento Nacional da Producao Mineral, Reio de Janeiro, Brasil.

Ravenstein, E.G. 1890. Lands of the globe still available for European settlement. *Proc. Royal Geographical Society.* 13:27.

Rayner, J.H. 1966. Classification of soil by numerical methods. *Journal of Soil Science* 17(1): 79-92.

RCA. 1981. *1980 Appraisal.* Part I. *Soil, water, and related resources in the United States: Status, condition, and trends.* Washington, D.C.: USDA.

Reeves, R.G., ed. 1975. *Manual of remote sensing.* Falls Church, Va: Am. Soc. of Photogrammetry.

III. SOILS

Reining, P. 1967. The Haya: The agrarian system of a sedentary people. Unpublished dissertation. Chicago, The University of Chicago.

Reining, P. 1979. *Challenging desertification in West Africa - Insights from LANDSAT into carrying capacity, cultivation and settlement sites in Upper Volta and Niger.* Africa Series No. 39. Athens, Ohio: Ohio Univ. Center for International Studies.

Reining, P. 1980. *Challenging desertification in West Africa. Insights from Landsat into carrying capacity and settlement sites in Upper Volta and Niger.* Papers in International Studies, Africa Series No. 39. Athens, Ohio: Ohio University Center for International Studies.

Reuss, J.O., and R.E. Campbell. 1961. Restoring productivity to eroded land. *Proceedings of the Soil Science Society of America*, 25(4): 302-304.

Revelle, R. 1976. The resources available for agriculture. *Scientific American.* 235(3): 164-178.

Rib, H.T. and R.L. Braida. 1982. Remote sensing applications in a road development project in Mauritania, Africa. A valuable tool for projects in arid and semi-arid environments. Paper presented at ERIM, International Symposium on Remote Sensing of Arid and Semi-Arid Lands. January 1982. Cairo.

Richards, A.I. 1939. *Land, labour and diet in Northern Rhodesia.* Oxford: Oxford Univ. Press for the International Institute of African Languages and Culture.

Richards, C.G. and H.F. Dobyns. 1957. Topography and culture: The case of the changing cage. *Human Organization.* 16(1): 16-20.

Riquier, J., D.L. Bramao and J.P. Cornet. 1970. *A new system of soil appraisal in terms of actual and potential producitivity.* AGL/TESR/70/6. Rome: FAO.

Rockefeller Foundation. 1979. *Agricultural production: Research and development strategies for the 1980's.* New York: Rockefeller Foundation.

Roose, E.J. 1977. Use of the Universal Soil Loss Equation to predict erosion in West Africa. In *Soil erosion: Prediction and control*, ed. G.R. Foster, pp. 60-74. Ankeny, Iowa: Soil Conservation Society of America.

Rosswall, T. 1973. *Modern methods in the study of microbial activity ecology.* IBP Bulletin from the Ecological Research Committee 17. Stockholm: NFR.

Rourke, J.D. 1981. Legend design for various kinds of soil surveys. In *Soil resource inventories and development planning.* Technical monograph no. 1. Soil Managment Support Services, Soil Conservation Service. Washington, D.C.: U.S. Department of Agriculture.

Salinity Laboratory Staff. 1954. *Diagnosis and improvement of saline and alkali soils.* USDA Handbook 60. Washington, D.C.: U.S. Department of Agriculture.

Sanchez, P.A. and T. T. Cochrane. 1980. Soil constraints in relation to major farming systems in tropical America. In *Priorities for alleviating soil-related constraints to food production in the tropics*, pp. 107-139. Los Banos, Philippines: International Rice Research Institute.

Shrader, W.D. 1982. *Effect of erosion and other physical processes on productivity of U.S. croplands and rangelands.* National Technical Information Service. Springfield, Va.: U.S. Department of Commerce.

Smith, D.D. 1941. Interpretation of soil conservation data for field use. *Agricultural Engineering* 22:173-175.

Smith, R.L, ed. 1980. *Ecology and field biology.* 3rd ed. New York: Harper and Row.

Smyth, A.J. 1966. *The selection of soils for cocoa.* FAO Soils Bull. No. 5. Rome: FAO.

Soil Conservation Service. 1972. *Training outline and notebook for soil mechanics classification systems for engineering use.* Fort Worth, Tex.: SCS, USDA.

Soil Conservation Service. 1972. *Soil survey laboratory methods and procedures for collecting soil samples.* USDA SSIR 1. Washington: U.S. Government Printing Office.

Soil Management Support Services (SMSS). 1981. *Soil resource inventories and development planning.* Technical monograph No. 1. Soil Conservation Service. Washington: USDA.

Soil Survey Staff. 1951. *Soil survey manual.* USDA Agriculture Handbook No. 18. Washington: Superintendent of Documents, Government Printing Office.

Soil Survey Staff. 1966. *Aerial-photo interpretation in classifying and mapping soils.* USDA Agriculture Handbook No. 294. Washington: USDA.

Soil Survey Staff. 1975. *Soil taxonomy.* USDA Handbook No. 436. Washington: U.S. Government Printing Office.

Soil Survey Staff. 1980a. National Soils Handbook Notice 60. Appendix 1. Soil Survey Manual. Chapter 3. Preparing for mapping. Unpublished mimeo.

Soil Survey Staff. 1980b. National Soils Handbook Notice 63. Appendix 1. Soil Survey Manual. Chapter 6. *Map Units.* Unpublished mimeo.

Somasiri, S., R.L. Tinsley, C.R. Panabokke and F.R. Moorman. 1981. Evaluation of rice lands in mid-country Kandy District, Sri Lanka. In *Soil resource inventories and development planning.* Technical monograph no. 1, pp. 89-115. Soil Management Support Services, Soil Conservation Services. Washington, DC: U.S. Department of Agriculture.

Sorokin, P.A. 1943. *Sociocultural causality, space, time.* Durham: Duke Univ. Press.

Stallings, J.H. 1950. *Erosion of topsoil reduces productivity.* USDA-Soil Conservation Service, SCS-TP-98 (August). Washington, D.C.: USDA.

Steele, J.G. 1967. Soil survey interpretation and its use. *FAO Soils Bull.* No. 8. Rome: FAO.

Stern, P. 1979. *Small scale irrigation.* London: Intermediate Technology Publications.

Stoner, E.R. and M.F. Baumgardner. 1979. Data acquisition through remote sensing. In *Planning the uses and management of land*, eds. M.T. Beatty et al., pp. 159-85. Madison, Wisc.: ASA, CSSA, SSSA.

Stoner, E.R., M.F. Baumgardner, R.A. Weismiller, L.L. Biehl, and B.F. Robinson. 1980. Extension of laboratory-measured soil spectra to field conditions. *Soil Sci. Soc. Am. J.* 44:572-574.

Swanson, E.R. 1978. *Economic evaluation of soil erosion: Productivity losses and off-site damages.* Publication No. 86. Lincoln, Neb.: Great Plains Agricultural Council.

Swindale, L.D. 1980. Toward and internationally coordinated program for research on soil factors constraining food production in the tropics. In *Soil-related constraints to food production in the tropics.* Los Banos, Philippines: International Rice Research Institute.

Sys, C. 1978. The outlook for the practical application of land evaluation in developed countries. *Wld Soil Resour. Rep.* 49:97-111.

Tansey, M.R. 1977. Microbial faciliatation of plant mineral nutrition. In *Microorganisms and minerals*, ed. E.D. Weinberg, pp. 343-385. New York: Marcel Dekker.

Thompson, H.A. 1978. Land evaluation for sugar cane production. *Wld Soil Resour. Rep.* 49:73-80

Tice, J.A. 1979. Soil considerations in highway design and construction. In *Planning the uses and management of land*, M.T. Beatty et al. ed., pp. 555-579. Madison, Wisc.: ASA, CSSA, SSSA.

Tinsley, R.L. 1981. Special considerations in evaluating soil resoruce information. In *Soil resource inventories and development planning.* Technical Monograph No. 1. Soil Management Support Services. Soil Conservation Service. Washington, DC: U.S. Department of Agriculture.

Trapnell, G.G. 1959. *The soils, vegetation and agricultural systems of North Easterm Rhodesia.* Lusaka: Government Printer.

Trapnell, G.G. and J.N. Clothier. 1959. *The soils vegetation and agricultural systems of North-Western Rhodesia.* Lusaka: Government Printer.

Uehara, G. and G. Gillman. 1981. *The mineralogy, chemistry, and physics of tropical soils with variable charge clays.* Boulder, Co.: Westview Press.

UNESCO (United Nations Educational, Scientific, and Cultural Organization). 1978. *Tropical forest ecosystems.* Natural Resources Research XIV. Paris: UNESCO.

Unstead, J.F. 1933. A system of regional geography. *Geog.* 18:185-187.

USBR. 1953. *Bureau of reclamation manual.* Vol. V. *Irrigated land use*, Part 2, *Land classification.* Washington, D.C.: U.S. Dept. Interior

U.S. Department of Defense. 1968. *Unified soil classification system for roads, airfields, embankments and foundations.* (MIL-STD-619B). Washington, DC: U.S. Department of Defense.

USDA. 1961. *A universal equation for measuring wind erosion.* ARS 22-69. Washington, DC.: Agricultural Research Service, U.S. Department of Agriculture,

USDA. 1967. *Developing soil-woodland interpretations.* Soils Memorandum No. 26 (Revision 2). Washington, DC: USDA.

USDA. 1981. *RCA 1980 appraisal*, Part II. Washington, D.C.: USDA.

U.S. Man the the Biosphere Program (MAB). 1981. *Environmental guidelines for irrigation.* New York: Cary Arboretum.

Vink, A.P.A. 1975. *Land use in advancing agriculture.* New York: Springer-Verlag.

Walsh, L.M., and J.D. Beaton, eds. 1973. *Soil testing and plant analysis.* Madison, Wisc.: Soil Science Society of America.

Watson, R.A. and P.J. Watson. 1969. Man and nature: An anthropological essay in human ecology. New York: Harcourt, Brace and World.

Western, S. 1978. *Soil survey contracts and quality control.* Oxford: Clarendon Press.

Weismiller, R.A. and S.A. Kaminsky. 1978. Application of remote sensing technology to soil survey research. *J. Soil Water Conserv.* 33:287-289.

Weismiller, R.A., F.R. Kirschner, S.A. Kaminsky and E.J. Hinzel. 1979. *Spectral classification of soil characteristics to aid the soil survey of Jasper County, Indiana.* W. Lafayette, Ind.: Laboratory for Applications of Remote Sensing, Purdue University.

Westin, F. 1982. *Resource inventories of arid and semi-arid lands using LANDSAT.* Paper presented at ERIM, International Symposium on Remote Sensing of Arid and Semi-Arid Lands. Cairo. January 1982. Brookings: South Dakota State University.

Westin, F.C. and C.J. Frazee. 1976. Landsat data, its use in a soil soil survey program. *Soil Sci. Soc. Am. J.* 40:81-89.

Whitney, R.S., R. Gardner, and D.W. Robertson. 1950. The effectiveness of manure and commercial fertilizer in restoring the productivity of subsoils exposed by leveling. *Agronomy Journal* 42:239-245.

Williams, B.J. 1977. Ethno-pedology and social ecology of soils in the 16th century Valley of Mexico. Unpublished manuscript.

Wischmeier, W.H. 1959. A rainfall-erosion index for a universal soil-loss equation. *Soil Science Society of America Proceedings* 23:246-249.

Wischmeier, W.H. and D.D. Smith. 1965. *Predicting rainfall-erosion losses from cropland east of the Rocky Mountains.* USDA Agriculture Handbook No. 282. Washington, DC: USDA.

Wischmeier, W.H. and D.D. Smith. 1978. *Predicting rainfall erosion losses - A guide to conservation planning.* USDA Agriculture Handbook No. 537. Washington, DC: USDA.

Withers, B. and S. Vipond. 1974. *Irrigation: Design and practice.* London: Batsford.

Woodruff, N.P. and F.H. Siddoway. 1965. A wind erosion equation. *Soil Science Society of America Preceedings* 29:602-608

Wong, I.F.T. 1966. Soil suitability classification for dryland crops in Malaya. *Proc. 2nd Malays. Soil Conf.*, pp. 154-156. Kuala Lumpur.

Yates, R.A. 1978. The environemnt for sugar cane. *Wld. Soil Resource Rep.* 49:58-72.

Yong, R.N. and B.P. Warkentin. 1966. *Introduction to soil behavior.* New York: Macmillan.

Young, A. 1976. *Tropical soils and soil survey.* Cambridge: Cambridge Univ. Press.

Young, K.K. 1978. The impact of erosion on the productivity of soils in the United States. A paper given at the Workshop on Assessment of Erosion in the United States and Europe, Ghent, Belgium, Feb. 27 - March 3.

Zimmerman, J.D. 1966. *Irrigation.* New York: Wiley.

Zingg, A.W. 1940. Degree and length of land slope as it affects soil loss in runoff. *Agricultural Engineering* 21:59-64.

Suggested Readings

Beinroth, F.H., G. Uehara, J.A. Silva, R.W. Arnold, and F. B. Cady. 1980. Agrotechnology transfer in the tropics based on soil taxonomy. *Advances in Agronomy.* 33:303-339.

Bene, J.G., H.W. Beall, and A. Cote. 1977. *Trees, food and people: Land management in the tropics.* Publication No. 084e. Ottawa, Canada: International Development Research Centre.

Bentley, C.F., H. Holowaychuk, L. Leskiw, and J.A. Toogood. 1979. *Soils Report prepared for the conference agricultural production: Research and development strategies for the 1980's.* New York: The Rockefeller Foundation.

Bie, S.W. and J. Schelling. 1978. An integrated information system for soil survey. In *Factual data banks in agriculture*, pp. 45-53. Wageningen, Netherlands: Pudoc.

Boserup, E. 1965. *The conditions of agricultural growth: The economics of agrarian changes in population pressure.* Chicago: Aldine.

Brogen, J.C., P. Lemos and R.E. Carlyle. 1965. *A survey of soils laboratories in sixty-four FAO member countries.* FAO Bull. No. 2. Rome: FAO.

Buringh, P., H.J.D. Van Heemst and G. J. Staring. 1975. *Computation of the absolute maximum food production and the world.* Wageningen, The Netherlands: Agric. University.

Carpenter, R. A., ed. 1982. *Assessing tropical forest lands: Their suitability for sustainable use.* Wicklow, Ireland: Bray Co.

Cochrane, T.T., J.A. Porras, J. Azevedo, P.G. Jones and L.R. Gardner, and D.W. Robertson. 1950. The effectiveness of manure and commercial fertilizer in restoring the productivity of subsoils exposed by leveling. *Agronomy J.* 42:239-245.

Collins, J.B. 1981. Soil resource inventory for the small farmer. In *Soils resource inventories and development planning.* Technical Monograph No. 1, pp. 37-46. Soil Management Support Services, Soil Conservation Service. Washington, DC: U.S. Department of Agriculture.

Committee on Tropical Soils (CTS). 1972. *Soils of the humid tropics.* Washington, D.C.: Natl. Acad. Sci.

Crosson, P., ed. 1982. *The cropland crisis, myth or reality?* Baltimore: Published for Resources for the Future by The Johns Hopkins Univ. Press.

Echelberger, H.E. and Wagar, J.A. 1979. Noncommodity values of forests and woodlands. In *Planning the uses and management of land*, eds. M.T. Beatty et al., pp. 429-443. Madison, Wisc.: ASA, CSSA, SSSA.

Greenland, D.J., ed. 1979. *Characterization of soils in relation to their classification and management: Examples from some areas in the humid tropics.* U.K.: Oxford Univ. Press.

Haantjens, H.A. 1969. Agricultural land classification for New Guinea land resources surveys. *Tech. Mem. Div. Ld Res. CSIRO Aust.* No. 69/4.

Hauser, G.F. 1970. *A standard guide to soil fertility investigations on farmers' fields.* FAO Soil Bull. No. 11. Rome: FAO.

Higgins, G.M. 1978. A framework for land evaluation and its application. *Wld Soil Resour. Rep.* 49:3-12.

International Food Policy Research Institute (IFPRI). 1977. *Food needs of developing countries: Projections of production and consumption to 1990.* Research Report 3. Washington, D.C.: IFPRI.

ISFEIP (International Soil Fertility Evaluation and Improvement Program). 1974. *The evaluation and improvement of soil fertility in Latin America.* 1974 annual report and project summary. Raleigh: North Carolina State University.

Johnson, W.M. 1980. Soil-related constraints, soil properties and soil taxonomy: A terminology for exchanges of scientific information. In *Soil-related constraints to food production in the tropics.* Los Banos, Philippines: IRRI.

Kamarck, A.M. 1976. *The tropics and economic development.* Baltimore, Md: Johns Hopkins Univ. Press.

Know, E.G. 1981. The role of soil surveys in the decision-making process for development planning. In *Soils resource inventories and development planning.* Technical Monograph No. 1, pp. 37-46. Soil Management Support Services, Soil Conservation Service. Washington, DC: U.S. Dept. of Agriculture.

Krantz, B.A., J. Kampen, and S.M. Virmani. 1978. Soil and water conservation for increased food production in the semi-arid tropics. *ICRISAT* Journal Article 30, Hyderabad.

Lang, D.M. 1973. *Handbook of the land capability classification method for open-field (including mechanized) agriculture in northern Zambia.* Supplry Rep. No. 9, Ld Resour. Div. ODM, London.

Likens, G.E., F.H. Borman and N.M. Johnson. 1981. Interactions between major biogeochemical cycles in terrestrial ecosystems. In *Some perspectives of the major biogeochemical cycles, SCOPE 17*, ed. G.E. Likens. New York: John Wiley and Sons.

Mitchell, A.J.B. 1977. Wind erosion: Processes and effect on soil productivity. *Transactions of the American Society of Agricultural Engineers* 20(5): 880-884.

Moorman, F.R. and A. Van Wambeke. 1978. The soils of the lowland rainy tropical climates: Their inherent limitations for food production and related climatic restraints. In *Proceedings, 11th Congress of the International Soil Science Society II*, pp. 272-291. Edmonton, Canada.

Nye, P.H. and D.J. Greenland. 1960. *The soil under shifting cultivation.* Tech. Com. No. 51. Harpenden, U.K.: Commonwealth Agricultural Bureau.

Olson, G.W. 1964. Application of Soil Survey Problems to Health, Sanitation and Engineering. Memoir 387, p. 77. Ithaca, New York: Cornell University Agricultural Experiment Station.

Pimentel, D., E.C. Terhune, R. Dyson-Hudson, S. Rochereau, R. Samis, E.A. Smith, D. Denman, D. Reifschneider, and M. Shepard. 1976. Land degradation: Effects on food and energy resources. *Science* 194:149-155.

Protz, R. 1977. Soil properties important for various tropical crops: Pahang, Tenggara master planning study. In *Soil resource inventories.* Agronomy Mimeo 7-23, pp. 177-88. Ithaca, New York: Cornell University.

Rhoades, J.D. 1976. Measuring, mapping, and monitoring field salinity and water table depths with soil resistence measurements. In *Prognosis of salinity and alkalinity.* FAO Soils Bulletin No. 31, p. 159-186. Rome: FAO.

Ruthenberg, H. 1971. *Farming systems of the tropics.* 2nd ed. London: Oxford Univ. Press.

Sanchez, P.A. 1976. *Properties and management of soils in the tropics.* New York: J. Wiley and Sons.

Smyth, A.J. 1981. The objectives of soil surveys of various intensities. In *Soil resource inventories and development planning,* Technical Monograph no. 1, pp. 27-36. Soil Management Support Services, Soil Conservation Service. Wash., DC: U.S. Dept. of Agriculture.

Soil Conservation Society of America. 1977. *Soil erosion: Prediction and control.* The Proceedings of a National Conference on Soil Erosion. Special publication no. 21. Ankeny, Iowa: SCSA.

Sombroek, W.G. 1979. Soils of the Amazon Region. In *The characteristics of soils of the humid tropics and their potential for intensified crop production.* London: British Society of Soil Science and Royal Geographical Society.

Thompson, T.R.E. 1979. Soil surveys and wildlife conservation in agricultural landscapes. *Soil Surv. Tech. Monogr.* 13:184-92.

U.S. National Research Council. 1972. *Soils of the humid tropics.* Washington, D.C.: National Academy of Sciences.

Part IV -- Plants

Cyrus McKell, Chair
Charles D. Bonham
J.R. Goodin
Daniel R. Gross
Charles E. Poulton
Samuel Snedaker

IV. PLANTS

Introduction

Plants provide the foundation for organic life on this planet through their ability to transform carbon dioxide and water into a simple carbohydrate in the process of photosynthesis. Varieties of plant species have evolved over millions of years in the diverse environments of the polar, temperate, and tropical regions. They have evolved adaptations that enable them to survive or flourish under the varying conditions of climatic zones, landscapes, soil types, animals, and competitor plants. The structure and function of living systems and the processes of interaction between the living and non-living environment constitute the ecosystem.

From an ecological point of view, humans are omnivores who make extensive use of plants for food, shelter, fiber, fuel, and other necessities. Early human strategies of adaptation did not have massive environmental effects, perhaps because technology had not evolved to a point where large amounts of energy were applied in the struggle for subsistence. Within the past 10,000 years, the invention of agriculture, the rapid increase of human populations, and many technological innovations gave humans far more power over the environment. The spread of technological innovations has not been a uniform process. Some areas of the world (for instance, North America and Europe) tend to adopt virtually every appropriate technological innovation in energy and agriculture after a short lag period. In other areas, there is little or no adoption of new technologies. The potential for environmental destruction is particularly high in such areas when innovations are introduced because the people and biota have not been gradually introduced to new technologies.

In recent years, conservationists have begun to express concern over the destruction of particularly "fragile" ecosystems such as the Amazonian rain forest (Goodland and Irwin, 1975). The concern is due to the increasing human impact on such environments and the evidence for loss of species diversity in many regions. The costs in terms of human suffering have dramatized events such as the Sahelian drought of the 1970's. As public awareness of these events grew, so did international resource development programs. While some of these programs have brought better management procedures, others have accelerated the process of environmental degradation. Some of the reasons for this are the following:

1) Modern industrial ideas and high technologies are not always directly applicable to the problems of the less developed countries.

2) Indigenous knowledge, practices, and customs were often ignored during the design and implementation of development assistance projects.

3) The dynamics and interrelations in many ecosystems were not well enough understood to allow development of ecologically sound management decisions and program designs.

This report describes appropriate methodologies for obtaining the information needed to mitigate many of the problems associated with reason 3 above. In addition, methods are described for determining the extent of indigenous knowledge, the local populations's perceptions of their environment, and their customs and practices regarding its utilization. It is assumed that this increased knowledge will be incorporated into project planning and design and that the local population will be brought into the mainstream of development planning (especially as they are often targeted as the population to benefit from the intervention). This would tend to minimize conflicts related to reason 1 above.

Before presenting a brief overview of this report we must begin, as with any project, by clearly defining its terms of reference and intended scope:

Resource inventories identify and usually map the resources within specific geographic areas. They are concerned with variety, quantity, quality, distribution, and other parameters that help determine what the resources are and where they are located. Inventories provide the basis for allocation of resources and for planning resource development, improvement, and management programs.

Baseline studies are concerned with the present status or condition of the resources and their interactions with other components of the ecosystem at a given moment in time (that is, climate, landforms and soils, and animals, including humans). Ecological processes are emphasized, with the ultimate goal of predicting the consequences of development programs for both the individual resources and the ecosystem as a whole. Environmental baseline studies establish a level from which departures can be measured to document changes. This latter activity is referred to as *monitoring*. Changes are monitored in order to determine the direct and indirect effects of development and management programs and to use this information to improve subsequent programs, or to better understand the impacts of certain natural phenomena.

Plant resources are any aspect of plants that serve a useful purpose in the functioning of an ecosystem, especially in providing for human needs for food, fuel, fiber, fodder, shelter, medicines, and so on.

Plant life can be divided into four types: soil microflora, nonvascular plants (mosses, lichens, and fungi), herbaceous vascular plants (grasses and forbs), and woody vascular plants (trees and shrubs). This report includes methodologies for measuring structural characteristics (such as biomass, density, and cover) and physiological-chemical feaures of vascular plants. This is not to minimize the importance of the nonvascular plants in the function of ecosystems or as site indicators.

Vegetation is a term that should be restricted to the grouping of plant species into assemblages of plants (that is, phytosociology) and connotes the total plant cover of an area.

Natural resources are materials or forms of energy that occur in nature and are useful to human cultures. A *natural area* is a portion of the landscape in which native or indigenous species predominate and the terrain or vegetation has not been significantly shaped or modified primarily by human actions. Although natural resources are ubiquitous, natural areas are rare, because local human populations and in-place subsistence systems have existed for centuries over the inhabitable parts of the world.

In many resource development projects, plants are either the subject of direct concern (for instance, increasing the growth rate or yield of forest trees for timber) or the means by which the major goal is achieved (such as developing additional forage for increased livestock production). In either case, the development activity is concerned with the unique capability of plants for primary production through photosynthesis. Thus, interventions involving plant resources may be designed to increase or maintain plant production by modifying land use patterns or environmental limitations, or to redirect plant production by substituting more useful plants for those found in the area. In many instances these activities might be combined.

One reason for attempting to improve the efficiency of production is to increase the sustainable yield of the biota for human use without damage to the environment. Another common goal is to increase the production of marketable products such as timber, grain, and beef. Such actions can actually lower the local carrying capacity for the human population. Certainly, any well-conceived development program should also be directed toward minimizing overall environmental degradation.

Before any of these goals can be realized, it is necessary to have enough information to make intelligent decisions about resource allocation and to formulate rational and ecologically sound development projects and management plans. Resource inventories and environmental baseline studies are necessary not only for determining the potential for increased production, but also for predicting the probabilities of undesirable effects of the intervention. One must never lose sight of the fact that component ecosystems are intricately interrelated. Plant-supported processes such as exchange of carbon dioxide for oxygen, uptake and release of soil nutrients, and protection of the soil surface by leaves and other organic litter significantly influence the stability of the ecosystem and its ability to continue to function optimally. The plant component of any ecosystem has a definite threshold of biomass that can be removed without an adverse impact.

Chapter 2 sketches the history of human resource use and introduces basic concepts of ecological management.

Chapter 3 begins by stressing the importance of defining exactly what problem is being addressed and what information is needed to make decisions concerning its resolution. Also reviewed are some general considerations relevant to project planning, which includes the possible presence of endangered species. The third part of chapter 3 describes five different levels of intensity for plant resource surveys, including the purpose and goals of each level; sources of information used; appropriate classification and mapping scales; vegetation information on plant species, growth forms, production, density, cover, and successional status; statistical reliability of data collected; and cultural issues associated with each level.

Chapter 4 describes an extremely useful tool for vegetation inventory and monitoring: remote sensing. The remote sensing systems most appropriate for plant resource investigations in developing countries are discussed and three methods for maximizing the usefulness of this technology are presented.

Chapter 5 describes field methods for plant resource inventories and baseline studies. The chapter begins with a discussion of the importance of collecting statistically valid data through the process of random sampling. Next, three analytical methods useful for designing vegetation sampling programs are discussed. The remainder of the chapter is a review of proven methods for collecting data on plant cover, density, and biomass.

Chapter 6 is a very brief overview of a highly specialized field in plant resource analysis: laboratory methods. It includes a summary of commonly used laboratory equipment and techniques, some special sampling considerations, and the handling and storage of samples.

Chapter 7 covers methods for determining human uses of plant resources and human response to environmental change. Included are guidelines for conducting interviews (the basic methodology of cultural anthropology and sociology) and the kinds of information required to assess plant resource utilization. Two standard methods for determining the importance of different utilization strategies are outlined, and several quantitative indicators of human responses to environmental changes are described.

Chapter 8 discusses the importance of classification systems for vegetation inventories. These systems provide the framework for communication and comparison of information between different workers and different areas. Types of classification systems are reviewed.

Chapter 9 gives sources of information pertinent to plant resource inventories and baseline studies. Chapter 10 contains References Cited and Chapter 11 contains Suggested Readings.

The Resource Base, Its Utilization and Management: A Summary

2.1 The Resource Base

Organisms that have the capacity to perform photosynthesis and produce sugars directly from carbon dioxide and water are called *autotrophic*. *Heterotrophic* organisms are those which are unable to fix energy themselves and therefore rely on autotrophs for sustenance. The autotrophs are referred to as the *primary producers*, while the heterotrophs are the *consumers*.

In ecological terms, the organic matter resulting from the primary productivity of green plants becomes the food for all the consumers of the world. Organisms that feed directly on the green plants are called *herbivores* or *primary consumers*. Those that feed on the herbivores are called *carnivores* and are the meat eaters or *secondary* and *tertiary consumers*. Some higher level consumers, including humans, eat both plants and meat and are referred to as *omnivores*.

Primary productivity is utilized by the biological community in a complex series of food and feeding relations known as food chains or more complex food webs. Each step in these feeding series is referred to as a *trophic level*. As a rule, the biomass of organisms at any trophic level is reduced by at least 90 percent compared to the biomass of organisms form the preceding level, on which it depends. This is because most of the total available energy is dispersed from the system principally through respiration, growth, reproduction, and excretion. Since no more than 10 percent of the total energy is actually incorporated into organic molecules at the next level, food chains tend to be very short. The concept of trophic levels is illustrated in Fig. 2.1.

In addition to carbon dioxide and water, plants require a suitable environment including such abiotic components as climate, landforms, soils, and a balanced set of mineral nutrients as building blocks of plant tissue, enzymes, and so on. The complex relations between these abiotic components and the biotic components (plants and animals), including the flow of energy originating from the sun, are organized into natural ecological systems, or ecosystems. Aspects of these systems related to climate and water are considered in the aquatic ecosystems part of this report, and soils are considered in another part. Within this total environmental framework, plant resources seek to achieve an ecological balance.

2.2 Resource Use

Humans came upon the ecological scene as relatively generalized omnivores, foraging for their food and other resources like many other animals. But humans with their advanced brains could make far more powerful tools and with symbolic language could coordinate their activities more effectively than other species. As the curve of technological innovation grew steeper, the human impact on environments grew as well. Accumulated experience, reflection, and very recently science gave humans the opportunity to correct particularly damaging strategies. It remains to be seen whether such knowledge can be acquired and applied at the same rate as environmental modification.

The primary reasons for making an inventory of the plant resources of an ecosystem are to determine the variety, quantity, quality, distribution, and productivity of the major autotrophic organisms in a geographic area of interest. Monitoring adds a temporal dimension to this information base. Baseline data give us the ability to gauge the effects that human activites have on the environment and possibly to apply knowledge of ecological processes to more effective management.

2.2.1 Origins of plant resource use and domestication.

Early hominids were probably omnivores, as are all the large primates, and probably used plants first as food and then for fuel, shelter, fiber, medicines, and other uses. Virtually all human groups known today or through archeological studies have used plants. Studies of the

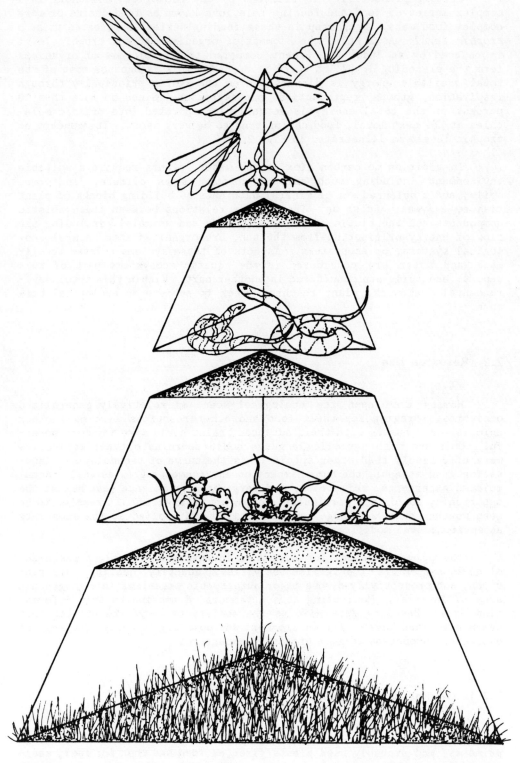

Figure 2.1 The concept of trophic levels. [From Northington and Goodin (in press)]

!Kung San hunter-gatherers of Botswana showed that 80 percent of their calories came from vegetable foods and only 20 percent from animals (Lee, 1969). Data on other hunter-gatherer groups also support the idea of a primary dependence on wild vegetable foods, except in the polar regions.

Plant and animal domestication occurred relatively recently in human history. Archeologists generally agree that (i) domestication was developed repeatedly and independently at widely separated times and places, (ii) the capability for domesticating plants and animals was available long before it was used, and (iii) domestication began to contribute significantly to human food supplies during the late Mesolithic, approximately 20,000 years ago.

Domestication can be defined as the manipulation by humans of a plant or animal population in order to increase the productivity and quality of that population. At its simplest level, domestication may consist only of controlling the natural predators or competitors of a desired species. All domestication involves concentrating energy within the environment, whether through human or animal muscle power, burning vegetation, machine power, organic or chemical fertilizers, herbicides, pesticides, or even agricultural research and development (McKell et al., 1972).

All present-day societies are dependent on plant products. The knowledge acquired over the centuries regarding various uses of wild plant species and the process of domestication constitutes a valuable data base for resource management. Some of these uses have not always been compatible with stability and sustained productivity of the environment and the development planner can benefit by understanding both the successess and the failures of human use of plant resources. For instance, domesticated plants cultivated for centuries in some underdeveloped regions may represent a valuable pool of well-adapted varieties, familiar to local cultivators and resistant to climatic fluctuations, diseases, and parasites. It seems wise to consider their advantages before replacing them wholesale with imported hybrid varieties. If there is a potential for loss due to a development project, care should be taken to preserve seed stocks or maintain natural areas where the range of genetic diversity may be sustained. Similarly, one should be congnizant of special dependences on certain indigenous plants which may provide medicines or reliable food crops during periods of droughts, lest they be destroyed as an inadvertent consequence of development actions.

2.2.2 Adaptation and modernization: Plant resource utilization.

There are many causes of human impacts on plant resources today. These occur within the matrix of institutions, goals, and methods and at different levels of magnitude. The impetus for a change in resource use strategies may emerge either from highly centralized planning or spontaneously within a population. Many centralized societies have what may be termed a planning elite which sets goals and directs implementation. Highly decentralized, spontaneous changes are more difficult to deflect or modify because they stem from parallel decisions made by many different people, perhaps for different reasons. One example of spontaneous change is the widespread adoption of the plantain as a subsistence crop among the native peoples of the Amazon Basin over the past 200 years.

Highly centralized decision-makers may be unresponsive to information flowing from dispersed sources. They may rely instead on geographically or socially biased sources. An example is the colonization program that accompanied construction of the Transamazon Highway in Brazil: some results fell far short of projected outcomes because colonists were assigned plots from maps with a scale of 1:1,000,000 and inappropriate seed varieties were distributed and squatters moved in (Moran, 1980; Furley, 1980). In other instances, development projects may be less than successful because farmers fail to adopt new crops and techniques due to previous bad experiences, fear, ignorance, or loss of local control (Gladwin, 1976; Brokensha et al., 1980). In some cases, when local farmers adopted new techniques, they found that their cash income increased but their ability to feed themselves and provide for their own needs actually declined (Gross and Underwood, 1971).

Modernization refers to the adoption of new technological, organizational, and economic modes by so-called traditional peoples. It can occur either spontaneously or as part of planned economic change. When planned, it is usually aimed at achieving higher productivity and economic growth. Planned modernization schemes frequently do not take traditional land use patterns into account, nor do they always make provisions for environmental or social impact. In some areas, the human population is virtually ignored. Several projects under way in Amazonia and in the Sahel have been framed as if the lands to be utilized were empty, even though indigenous populations not only are subsisting but often are producing an exchangeable surplus. Modernization can affect plant communities in a variety of ways. Two of the most significant modernization activities have been the "green revolution" and the exploitation of fossil fuels energy.

Development of high-yielding crops. The green revolution is one of the more significant and innovative developments since World War II. It consists basically of the introduction of new high-yielding varieties of rice, wheat, and other crops grown with improved techniques to replace the less productive varieties currently grown in many parts of the developing world. High-yielding wheat and rice have been adopted in many places, including the Philippines, Indonesia, and Mexico. Productivity per unit area has been dramatically increased, sometimes more than twofold (Plucknett and Smith, 1982).

Contrary to expectation, however, traditional cropping techniques have not been replaced due to the high cost of the cultivation practices necessary for the green revolution crops. The higher capital requirements of agricultural modernization resulted in concentration of land holdings and income and displacement of weaker farmers in many parts of the world. Many social scientists now consider this a typical process.

Development specialists have realized that the green revolution must be implemented as a "package" in developing countries. The package includes the new seed varieties, new cultural practices, new inputs of fertilizer and water, and new credit and commercial arrangements. Not all farmers in developing areas are able to absorb and deal with this package. Small farmers may lack sufficient technical ability. Even where technical assistance and financing are available, some farmers may not be able to take advantage of them because they lack title to their land.

Another problem with the green revolution is that it tends to make agriculture more energy-dependent. As world energy prices rise, many observers have suggested that high yields should not be the sole criterion for adoption of technological innovations in agriculture. Some have urged that less intensive, relatively extensive, agricultural practices requiring smaller increments of energy would be more appropriate for much of the developing world. A final problem is that green revolution varieties are highly selected and often exogenous to the regions where they are introduced. In use they may become more vulnerable to crop diseases and other hazards. Although it is too early to assess the impact of green revolution varieties on world food production and on the economies of the areas where they are produced, plans involving their introduction should (i) take into account the potential negative effects, and (ii) carefully consider the available locally adapted alternatives to green revolution varieties.

Significance of fossil fuel energy. A break with the past occurred when fossil fuels began to be applied in agricultural processes. With the advent of chemical fertilizers, farm machinery, herbicides, pesticides, and canning, vacuum-packing, refrigeration, and other processes, agriculture entered the industrial era. One outcome of this industrialization of agriculture was that remote regions were integrated in the world economic system, resulting in specialization and in some regions losing the self-sufficiency in food and other agricultural products. The productivity, income, health, and welfare of entire regions became dependent on international markets.

The rising dependence of agriculture on fossil fuels made many areas even more vulnerable to calamities such as the oil price increases of 1973 and 1979. Until about 1973, most of the developed countries treated energy from oil as essentially a free good, and the dependence on fossil-fuel based technology for transportation, agriculture, food preparation, and manufacturing spread rapidly to the developing world. More recently, petroleum importers have been severely stressed by price increases, and many countries have had to adopt rationing schemes and to cut back or abandon ambitious development programs. Even oil-rich countries such as Mexico and Nigeria have been severely stressed because of a temporary slump in world oil demand. In several cases, the development programs involved resources that would have been devoted to food production, particularly land and agricultural inputs in the case of biomass energy schemes. Although the issue has yet to be thoroughly assessed, it appears that intensification of agriculture, with the consequences noted above, is still going on.

2.2.3 Impacts of modernization on plant resources.

Although there are different ways to consider the impacts of modernization on plant resources, we discuss them below in terms of population growth, commercialization, transportation and land use for urbanization.

Population growth. Human population growth is an ingredient of change in many parts of the world, although it is rarely the single cause of change in ecosystems. Where growth results in increased population density, there is frequently an accompanying increase in the intensity of resource use (Boserup, 1965). For example, fallow cycles may be shortened or rangeland may be grazed more frequently or more closely. Where populations rise precipitously, as in parts of India, Indonesia, and Latin America, governments frequently intervene to spur the process of agricultural intensificaton. Dense human populations are likely to become centralized both administratively and politically. There are, however, societies that have no centralized political controls, even with population densities over 250 per square kilometer (highland New Guinea, for instance).

Commercialization. Commercialization nearly always accompanies modernization. It has far-reaching effects on social and ecological systems because of the monetization of exchange. As monetary exchange diffuses, virtually everything becomes exchangeable as a commodity, including land, labor, and all sorts of products. This vastly accelerates the circulation of goods and has multiple effects on behavior. In some cases, goods used mainly for subsistence come into demand as markets commodities, as happened with the hardwood species in tropical forests of Sumatra, Brazil, and Indonesia. Commercialization often involves the introduction of new plant or animal species, such as the introduction of sisal (*Agave sisalana*) to Brazil and East Africa (Gross, 1970; Brockway, 1979). Some plantation crops like sugar cane and coffee have resulted in complete modification of ecosystems in Brazil (Stein, 1957; Margolis, 1973) and Indonesia (Geertz, 1963).

Transportation. Changes in transportation systems frequently accompany modernization and commercialization. Perhaps the most dramatic changes are those brought about by penetration roads or new railroads in a previously unserved region. The construction of riverine and maritime port facilities, canals, dredged channels, and airstrips can have similar effects. Major penetration roads built in Kenya (Kudat, Bates and Conant, 1981), Brazil (Moran, 1980), and elsewhere can induce a population already present to intensify agriculture (Kenya) or they can bring in new settlers (Brazil). Construction of roadways, rights-of-way, feeder roads, quarries, culverts, cuts, and embankments involves clearing existing vegetation, often with no possibility for regrowth. Transportation facilities encourage agricultural and livestock operations on a larger scale, while often restricting movements required for traditional management strategies. Changes in transportation systems can have a great indirect impact on ecosystems. An interesting example is the wood-burning river steamer trade that plies the Sao Francisco River in northeast Brazil and consumes large amounts of firewood cut in the already denuded hills of central Minas Gerais.

Land use for urbanization. Urbanization and industrialization annually remove hundreds of thousands of hectares of land used for growing plants as the ground is converted to roadbeds, buildings, and the like. These activities also increase the local demand for construction materials (such as lumber) and fuel (firewood and charcoal). Cities and industries produce various types of waste, the disposal of which as landfill or in dump sites may have important effects on ecosystems. Air- and water-borne waste emissions can also affect biotic communities hundreds of miles from the source.

2.3 Management of Plant Resources

Plant resource management involves the manipulation of autotrophic production to increase, decrease, or maintain the flow of output energy or products. Management tools include a wide array of practices and materials designed to stimulate plant growth processes, increase the favorability of the environment, reduce the detrimental effects of pests and diseases, or increase the efficiency of harvesting. Many specific practices or inputs have been developed in the advanced agricultural countries for these purposes. However, not all methods or materials are suitable for every situation because of the diverse nature of the world's ecosystems, social organization, and human customs, and the economic returns that may be expected for each increment of management intensity. In comparing the intensity of management appropriate to rangelands with a low production potential with that appropriate to high-potential croplands, Love (1961) suggested a relationship that recognized high outputs for *intensive* agricultural management of croplands and low outputs for *extensive* ecological management appropriate for rangelands (Fig. 2.2). In planning resource development projects, it is essential to recognize the limitations of the environment. Limitations may consists of an unfavorable temperature regime, inadequate precipitation, infertile or poorly developed soil, or topographic unsuitability. Resource managers must develop appropriate practices to overcome such limitations within the constraints of the expected economic returns, or restrict development to a level that the ecosystem can tolerate.

2.3.1 Extensive management.

Extensive management is practiced by many people in areas of marginal productivity or low population density, where low labor effort per unit is expended. Extensive management involves the use of relatively small quantities of resources scattered over wide areas of land over long periods of time. Intensive management techniques are sometimes indiscriminately practiced in areas more suited to extensive management. This may place demands on people and resources that cannot be met without serious dislocation or degradation. For example, the use of deep plowing may be inimical to achieving sustained yields in human tropical forests, where the humus is typically very thin. Development projects in such areas must make appropiate adjustments with regard to time, space, and numbers when transferring technologies. A more appropriate approach is the high-intensity, low-frequency grazing system proposed by Corbett (1978) and long used by grazing cultures. McKell and Norton (1981) discuss resource-compatible management practices that make it possible to use the land within the constraints of desert ecosystems. Whether management intensity can be increased with a resulting increase in productivity depends on a careful appraisal of the ecosystem.

320   IV. PLANTS

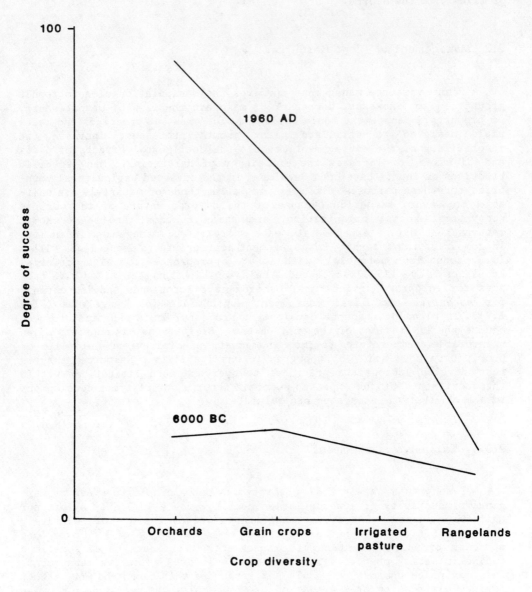

Figure 2.2. Productivity increased by amount of management and cultural manipulation.   [Redrawn from Love (1961)]

2.3.2 Intensive management.

Areas with high potential productivity can be managed on an intensive basis by using the high-value inputs of fertilizer, irrigation, improved plant and crop varieties, pest control, and mechanization of cultivation and harvesting. Whereas extensive management generally deals with multispecies plant communities, intensive management generally deals with crops grown in a monoculture or sequential crop or species rotation. A considerably higher yield is needed to pay for the cost of intensive management. As a result of the emphasis on maximum investment returns, ecological or external costs are often overlooked. Soil erosion, loss of beneficial insects, reduced genetic diversity, and dependence on a limited agricultural crop base (single-crop economy) are some of the undesirable side effects of some forms of intensive management. These side effects need not occur if they are considered in the overall resource management plan.

Determining the appropriate level of management intensity according to the limitations of a given agricultural system is a major problem. The extent of control desired over the risks of production involves corresponding levels of inputs and managerial costs. A technology or level of management appropriate to a plant production system from a developed country cannot be transferred effectively to a developing country without an assessment of potential impact, adequate baseline information on natural resources, a clear determination of the appropriate level of management, and acceptability by the local people. One must recognize that intensity of management may be low, moderate or high as appropriate to the soils and economic and sociological limitations of the project area. Management intensity is a relative concept. Intensities are not just low and high, nor is high intensity management limited to cropland areas. Even in the management and use of natural ecosystems such as grazing lands and forests, the concept of low, moderate, and high levels of technology, capital, and labor inputs must be clearly recognized, and development projects should strive to match an appropriate level of input with the ecological, social, and economic constraints or limitations of the area.

2.3.3 Use of energy in management.

The amount of energy applied per unit area over a given period can be used as an index of the intensiveness of management. Extensive management systems such as open-range grazing and shifting agriculture, involve relatively low expenditures of energy per unit area. Yields per unit of energy expended tend to be high, up to 20 to 1 when food yields are converted into calories. As management intensity rises (for instance, in irrigation agriculture, modern feedlot operations, and highly mechanized, chemical-dependent systems) energy efficiency tends to decline. Food energy yields per unit of energy applied are generally less than 1.0 in modern mechanized agricultural systems and less than 0.2 when the inputs involved in processing, packaging, and delivering food to consumers are taken into account.

Most of the energy expended in agriculture is used to channel energy and nutrients to desired species at the expense of competing species. However, agriculture still depends on natural ecological processes such as biogeochemical cycles, succession, and mutualism. In shifting or swidden agriculture, for example, soil fertility is managed by allowing forest succession to capture and store nutrients over several years (generally exceeding the duration of a cultivation cycle in annual crops). The vegetation is then clear-cut, and the slash is allowed to dry and then burned, releasing nutrients into the soil for uptake by crop plants. While this form of agriculture was once practiced throughout Europe and other temperate zones, it is now practiced mainly in the humid tropics. It may be well-adapted to these areas because nutrients tend to be more safely stored in living forms than in the soil. Swidden agriculture is highly productive per unit of labor, although productivity per unit area is low when both fallowed and cultivated areas are considered. Thus, swidden agriculture may be highly adaptive when practiced by low-density populations, which can move away when the land is exhausted, or when a single family or farmer can be allocated enough land to accommodate long term rotations.

There are many systems of more intensive nonmechanized agriculture involving use of the hoe, plow, manure, crop rotation, and simultaneous polyculture. Perhaps the most intensive nonmechanized agricultural systems involve the use of terraces, canals, reservoirs, and hydraulic devices. Such systems are found in China and the Philippines, and were used in ancient Peru and elsewhere. In these systems, the flow of water carries nutrients directly to the growing plants and reduces the dependence on rainfall. The upper limit on production in irrigated areas often seems to be set by labor availability during peak periods, such as transplanting and harvesting (Geertz, 1963). Thus, unlike the swidden agriculture system, these systems are amenable to rising population density.

2.4 Ecosystems: The Basis for Rational Management

The ecosystem is a useful concept for resource management because it includes both the living and nonliving parts of the environment and their interrelations. However, while much is known about ecosystem interrelations in temperate zones, the extent of our knowledge of tropical, semi-desert, and desert ecosystems is more limited. Thus ecologically oriented resource assessment is not a panacea for all possible environmental problems associated with economic development. The term ecosystem can be used in either the concrete or the abstract sense. In the concrete sense it refers to real or individual landscapes, and the in abstract sense to groups of landscapes having similar biotic components, that is, to classification units in the array of possible ecosystem types. According to Daubenmire (1968),

> "...the ecosystem concept may be applied to communities as small as a decaying blade of grass with its complement of fungi, protozoa, etc., or as large as the earth with its atmosphere and all known biologic organisms, and to units which appear stable as well as to others that are evidently in the process of change. It may be used in connection with a specific kind of landscape unit or to express the abstract concept of relatedness."

Thus, the ecosystem is a multilevel concept. It may be defined in a broad, highly generalized way or in a very specific way. The level at which an ecosystem should be defined and characterized is determined by the intended application of the knowledge that results from its characterization, classification, and mapping. In any given project one must, therefore, define the working ecosystem concept so that it is compatible with the problem being addressed and with the level of information needed as determined by the problem statement. Similarly, ecosystem is a multiscale concept. The information needed and nature of the problem also determine the mapping scale and the intensity or level of detail at which ecosystems are defined for practical inventory applications. This subject is elaborated on in chapter 3.

The ecosystem concept suggests that a change in any aspect of the environment will have effects on many other aspects. Considerations of food chains, energy flows, nutrient cycles, successions, and so on lead us to ask questions about these multiple effects and to formulate testable hypotheses about the possible outcomes of different resource management strategies.

Keeping in mind that any investigation must be defined in terms of the specific problem to be solved, we can now briefly consider component parts of ecosystems and consider how they are related to plant resource surveys. The complexity of such undertakings often make it mandatory to involve a multidisciplinary team, although this does not mean that all aspects of a unit must be investigated at the same time or intensity. Attention can be restricted to a portion of the ecosystem or a single aspect of its function, for example, forest undergrowth, the soil, microclimate, or grazing-induced succession. It is only necessary that these limited observations be carried out within a unit that is specifically delimited in space and time. As long as less comprehensive studies are defined according to an appropriately identified unit, the new information can be generalized to the full geographic extent of the unit. The idea of "integrated resource inventories" originated in a realization of the underlying importance of resource analyses within a unit framework. This topic is discussed in greater detail in part I of this report.

2.4.1 Abiotic components of terrestrial ecosystems.

The abiotic or nonliving components of terrestrial ecosystems, also referred to as physical environmental factors, include climate, physiography (landforms), and soils (edaphic factors). Water is another component of the physical environment. While it is easy to understand how precipitation and temperature affect the growth and distribution of plants, one must also consider the more subtle influences of relative humidity, wind, and light (intensity, duration, and quality). Edaphic factors such as soil texture, depth, pH, organic matter content, water holding capacity, fertility and salinity all combine in various ways to influence plant growth and the distribution of vegetation types. Soil characteristics also affect water-plant relations. In fact, soils are often considered the interface between the biotic and abiotic components of ecosystems. These abiotic components of ecosystems are discussed in more detail in the reports on to the soils and aquatic ecosystems and in the "data compendia" in chapter 2 of the report on wildlife.

2.4.2 Biotic components.

Although it is an important concept that all the biological organisms, including soil microorganisms and ground-dwelling and avian fauna, are part of the ecosystem, complications arise in defining the geographic boundaries of an ecosystem for plant resource analyses when the more mobile animal species are considered. Clearly defined boundaries on maps are essential for such resource management activities as determining the stocking rate (number of animals per unit area) or the volume of timber available in a forest. In practice, ecosystem boundaries are generally based on plant communities, vegetation and soil units, or other physical features, and the animals are "incorporated" as secondary users. The more mobile animals are considered to range over more than one ecosystem.

Plants and vegetation constitute the other biotic component of ecosystems. For environmental analysis, the characteristics of plant resources which are important are (i) the individual plant species (floristics); (ii) the characteristics of individual species, such as chemical composition; (iii) the groupings of individual species into plant communities; (iv) vegetation-soil-terrain relations; and (v) functional relations, especially the stability (or conversely the direction and rate of change of those between the biotic and abiotic components) of the ecosystem. When changes are identified, one should always attempt to determine the probable causes and whether they are intrinsic or extrinsic.

Stratification of observations according to vegetation-environment systems may be required. This involves recognition of the significant criteria (such as indicator species) that enable the managed resource to characterize these systems. Stratification can be facilitated by the multilevel nature of the ecosystem concept. At the most highly generalized levels, appropriate for national or regional inventories, the biotic component of the ecosystem is characterized by relatively homogeneous life forms and appearance (physiognomy) rather than by plant taxonomic criteria. The corresponding level for the abiotic component is represented by broad macrorelief features such as plains, hills, and mountains and by regional climates. At an intermediate level, floristics become a criterion for the biotic component, while homologous landscapes, characterized by similar landforms and topography, and more localized climatic influences distinguish the abiotic components. To use the ecosystem concept for local management and project level decisions, it is necessary to define it in the context of specific communities of plants and/or vegetation-landform-soil systems. Additional information on stratifying ecosystems within a hierarchical context can be found in chapter 3 (levels of plant resource assessments) and chapter 8 (plant resource classification systems).

2.4.3 Ecosystems dynamics: Vegetation succession and climax communities.

The environment is constantly changing. The changes are sometimes gradual, as in the weathering of rocks to produce new soil material, and at other times much more rapid, as in landslides or wildfires caused by lightning. Nature responds to these changes through ecosystem processes.

The response may be a return to the original ecosystem after a disturbance, or the development of a different biotic community better adapted to the new environmental conditions. It is within this well-established, and sometimes delicate, ecological framework that humans seek to manage nature. Accordingly, the development planner must understand the dynamics of ecosystems.

The process of change in ecosystems that results from constant interactions between organisms and their environment is called ecological succession. Krebs (1972) characterizes succession as follows:

> Communities may change with time either in a directional way (succession) or in a nondirectional way (cyclic)....Succession proceeds through a series of stages from the pioneer stage to the climax stage. The *monoclimax hypothesis* suggested that there was only a single predictable end point for whole regions, and that given time all communities would converge to the climatic climax. This hypothesis has been superseded by the *polyclimax hypothesis* which suggests that many different climaxes could occur in an areas...controlled by soil moisture, nutrients, fires, or other factors....Succession does not always involve progressive changes from simple to complex communities...because of short-term changes in climate and the cyclic changes of growth and decay in the community. For most communities, we observe changes over time but do not know the factors causing the changes.

The transitory communities that arise during this process are called seral stages (singular: sere), while the terminal stage is referred to as the climax. The climax community is in a dynamic equilibrium with the physical environment. An equilibrium state that is limited only by climate (that is, one that occurs on undulating or nonlimiting relief and normally on zonal soils) is referred to as a "climatic climax". Often, other limiting factors may make it impossible to attain climatic equilibrium. A divergent pathway of succession will lead to "edaphic climaxes," where soil development is abnormal for the climatic zone and the self-perpetuating vegetation that develops differ from the climatic climax for the area. "Topographic climaxes" develop a still different kind of equilibrium and self-perpetuating vegetation, with topography limiting the effect of the prevailing zonal climate on soil development and plant succession. Recurrent burning of vegetation can also result in divergent equilibrium conditions called a "fire climax." Fire climaxes are common throughout West Africa, where repeated burning favors herbaceous vegetation over woody shrubs or trees. Areas converted from perennial to annual grasses by abusive grazing or destructive wood gathering could be termed "zootic" or "biotic climaxes". Stable communities that are not the true climatic or other natural climax types because they are maintained by humans or domestic animals may be designated as disclimaxes (disturbed climaxes) or anthropogenic climaxes.

In some cases management may opt for seral conditions because of higher productivity of certain seral species over the climax species. For example, where a commercial timber species is dominant in a seral stage but does not reproduce under the full canopy of the climax community, silvicultural practice will favor heavy cutting and maintenance of a relatively open canopy. In certain shrub grassland communities, the proportion of shrubs is manipulated to stimulate greater grass production. Similarly, many forests and woodlands are managed for a more open canopy to stimulate greater growth of herbaceous plants for grazing. If the area is judiciously grazed, tree regeneration is not inhibited and may even be favored.

2.4.4 Application of ecosystem dynamics to resource inventories.

Knowledge of successional and cyclic processes in a particular environment is invaluable to a planner contemplating the introduction of major changes in resource use. This knowledge can be acquired only through painstaking, and often time-consuming, naturalistic observation. In areas where successional change is orderly and convergent, the probable outcome of certain modifications can often be predicted. Even where change is divergent, it is most useful to be aware of the range of possible environmental responses to perturbations.

Although a knowledgeable and astute plant ecologist can often draw reasonable inferences about ecological relations, it is difficult if not impossible to develop a classification, even by the conclusion of a normal survey, that accurately reflects the climax community that would occur in the project area. If one attempts to map these communities on the basis of inadequate information, the result may be mapping of erroneous ecological interpretations that will defy subsequent correction. Therefore, the safest course, especially in the developing nations, is to characterize and map *existing* vegetation together with landform and soils. Classifications and written descriptions should characterize the vegetation that is on the land at the time of the survey and clarify to the fullest extent possible the vegetation-landform-soil relations. If this is done with the concepts and procedures of synecological (plant community level) investigation in mind, it may be possible, by the conclusion of a survey, to make a map overlay of a first approximation to the climax community based on what is known. As knowledge grows, these interpretations can be brought up to date if the initial descriptive classification and field data are adequate.

2.4.5 Carrying capacity: A management concept.

Carrying capacity is the ability of an area (regional ecosystem, management unit, or local ecosystem) to produce the surplus necessary for its own maintenance. Because climate and other governing factors fluctuate over time, the carrying capacity also changes. Further, since the usurping population is supported by the surplus production of the ecosystem rather than its maintenance requisites (removal of which would result in degradation and nonsustainability) it must either be kept low enough to exist during the least favorable years or be adjusted (reduced) for low-production periods. Alternatively, certain limits to production can be overcome with the necessary inputs of management or resources, for instance, fertilizer as supplemental protein. Each ecosystem has a characteristic biomass, some of which is surplus productivity. Thus, an analysis of ecosystem carrying capacities that identifies the amounts of these surplus "yields" provides the management concept for sustained renewable resource exploitation.

The following discussion illustrates the application of the carrying capacity concept to rangelands and animal stocking rates. However, the principles also apply to human numbers in a finite area. Methods for estimating human carrying capacity which involve both a knowledge of vegetation and of soils is discussed in soils, part III.

Vegetation is the critical determinant of the capacity of an area to sustain herbivorous animals. Vegetation provides both food and cover (protection). The aboveground biomass of palatable herbage and browse, the uniformity of consumption over the area, and the period in the plant growth cycle at which removal takes place, strongly affect carrying capacity. In addition, different kinds of animals have different preferences for the same kinds of plants, and these preferences also vary with season of the year.

The concept of carrying capacity also implies sustained productivity (Stoddart et al., 1975). Failure to hold herbage removal within the tolerances of key forage species causes a reduction of plant vigor, production, and competitive ability. This can result in undesirable changes in plant composition over time and can promote range deterioration. Minimal plant cover may result in soil erosion and permanent environmental degradation.

The quality of vegetation as cover for animals is related to the growth form of the plant species and their stature or size, density, and distribution. As these features relate to the nonforage needs of animals for shade and protection, the carrying capacity may be further modified by the way cover affects the distribution of animals and the uniformity of grazing of the forage-producing areas.

Finally, major variation in forage production from year to year is influenced by climatic variations, including seasonal fluctuations in precipitation, temperature, and wind. Forage production can vary more than 50 percent from year to year in some vegetation types. Therefore, measurement in a single year without adjustment for effective precipitation is little better than a guess. The average or long-term production from which carrying capacity might be estimated is very elusive. The best way to approach the determination of production is therefore as a part of monitoring after the initial baseline studies have determined the kinds and patterns of vegetation in an area.

For all of the reasons above, determination of carrying capacity as a definitive and meaningful figure is difficult. Furthermore, it is not necessary in practical management because a known level of use is set as an initial stocking rate. Monitoring techniques are used to make adjustments in stocking rates.

When additional effort can be justified for management, monitoring can include such additional determinations as (i) the proportion and uniformity of forage and browse removal, (ii) vegetation composition as an index of ecological conditions, (iii) evidence of the direction of change in ecological condition, and (iv) year-to-year variation in production. Under intensive management, careful sampling and gathering of quantitative data are easily justified. An intensive inventory that maps specified vegetation-soil systems, plant communities, or habitat types will establish a data baseline for this kind of monitoring (Heady, 1975).

In a specific management area, it is also possible to determine the optimum stocking rates under different patterns of management by experimental grazing trials. In this case, optimum stocking rate is estimated from data on animal gain in relation to forage production, where both gain per animal and gain per unit area are considered (Bement, 1969). The Food and Agricultural Organization/United Nations Development Programme (FAO/UNDP) and the U.S. Agency for International Development (AID) have funded successful projects of this nature.

It is difficult to realize an ecologically satisfactory stocking rate in the developing world because of the custom, tradition, and attitudes of people regarding their dairy herds. Indigenous herders are sometimes difficult to convince of the benefits of new management techniques since they often involve a radical change from dairy to beef. Knowledge of the direction in which to make stocking rate changes can be made, however, by an ecologist. The urgency of the need to change stocking or to institute practices to grow more fodder are easily assessed by a qualified range ecologist. Such ecologists would use the data derived from an intensive baseline study plus their field observations.

If regional or national estimates of animal production potential and of current and potential range productivity are needed, they can be obtained by multistage sampling methods, often starting from a reconnaissance survey. Data can be derived for broad policy, regional planning, and goal setting by government agencies. Given such estimates and a reliable census of animal population, an evaluation can be made of the current balance between animal demand and fodder availability on a national or regional basis. This would give a clearer idea of the benefits to be derived from a program to prescribe management strategies, improve ranges, increase cropland fodder and feed grain production, and improve animal breeding and management (Pratt and Gwynne, 1977).

Many Requests for Proposals (RFP's) to consulting groups from international assistance agencies call for determination of carrying capacity as though it were a simple, easily derived statistic. What should be asked for, if it is necessary, is an estimate of plant production and associated factors or conditions.

Techniques for accomplishing this analysis are described in the methods chapters to follow, but the concepts of ecosystems, sustained yields, and carrying capacities as they are related to ecologically sound management are perhaps the most important ideas of this report.

2.4.6 Conclusion.

Landscapes, natural or altered, are inventoried, measured, or monitored in terms of their plant and environmental elements. With this information, users of these landscapes can better understand potential influences on the environment and make sound decisions in the areas of renewable natural resources use, development, improvement, and management. The ecosystem concept is a powerful tool for characterizing landscapes in terms of similarity. Whereas different areas representing the same ecosystem may vary considerably in individual site factors (soil nutrients, depth, texture, moisture, temperature, land slope and aspect, elevation, and so on), the biological summation, as a result of factor interaction, is essentially the same throughout. Thus all areas comprising the ecosystem have very high environmental similarity. Among such similar areas, comparable responses to comparable management actions can therefore be reasonably expected.

Since the collective areas that are occupied by an ecosystem are similar, or homologous, in terms of the ecological sum or integrated effect of all the separate environmental factors, the ecosystem concept provides a basis for:

1) Determining homologous areas,

2) Determining and extrapolating site potential,

3) Cross-sectional analysis and the assessment of external impacts on the system,

4) Stratifying landscapes as a preliminary to quantitative evaluation, and

5) Monitoring improvement, management, and development.

Review of Problem Analysis and Levels of Plant Resource Surveys

Plant resources are dynamic. This variability is expressed on a year-to-year basis and even within the annual cycle. Such attributes as productivity, phenology (the timing of plant processes such as flowering), and species composition reflect this variability. In some cases, especially involving annual species or ephemerals, the variability ranges from abundance to complete absence. For example, if floristic studies are envisioned, one must plan surveys related to the seasons when the appropriate diagnostic plant parts, usually flowers and fruits (the basis for taxonomic distinctions), are present. One of the major factors controlling this variablity is climate.

Because of this variability, one must be cognizant of departures from the norm, or long-term average, during the time when inventory or monitoring data are collected. Measurements and observations made in a single year can be misleading unless the year happens to represent the normal condition. In some cases, data can be normalized by using an adjustment factor that represents the degree of departure for the norm. When production data are critical, it is best to measure the parameter annually for a minimum of 5 years, longer in areas of extreme climate variation. The resource planner should keep in mind, however, that the carrying capacity of the land also varies, and even normalized data may not be adequate for ecologically sound management decisions if a major drought period commences in the following year.

Furthermore, we should realize that information on plant resources affects perhaps the majority of subjects of interest in development programs. Data on plant resources are a direct factor in forestry and rangeland management, a virtual controlling factor for most wildlife species, often an indicator of agricultural suitability, vital for soil stabilization and development, and a key mitigating factor in stream and reservoir sedimentation. The list can easily be expanded.

Because of the importance of information on plant resources and because of the dynamic nature of these resources, the planning of plant resource inventories and baseline studies and the selection of appropriate methodologies are complex subjects. In this chapter we wish to:

- emphasize the need for a thorough problem analysis that results in a concise statement of the problem conditions surrounding the problem, and its causes;

- review some general planning considerations, including the need for quantitative data, the recognition of intersystem dependences, the mandate to consider endangered species preservation, and the opportunities for using expert advice in the planning process;

- consider the different objectives of five levels of plant resource assessment, including the sources of information, mapping scales, attributes of vegetation measured, and statistical reliablity.

Given a clear statement of the problem, a comprehensive program design, and recognition of the proper level of plant resource assessment and amount of detail required to resolve the problem, the methods of data collection can be selected as outlined in chapters 4, 5, and 6.

3.1 Problem Analysis and Statement

Resource inventories and environmental baseline studies are not ends in themselves. They are implemented for one primary purpose: to provide the information essential for rational decisions that will lead to solution of clearly stated problems.

In this context, a *problem* is defined as a deviation from a desired or essential condition. The *problem statement* involves a clear expression of the nature and magnitude of the deviation from a norm or expectation. The *problem analysis* is the process by which the normal condition or goal is clearly defined, the parameters/characteristics of the deviation are concisely expressed, and the cause or causes that meet or explain the parameters and characteristics of the deviation are clearly identified. When this is done, one knows precisely where to attack the problem (causes, not symptom) and the areas in which decisions are required.

The era of great, generalized exploratory surveys undertaken primarily "to see what is out there" is past. The severity of the problems facing the developed as well as the less developed countries, the urgency of solving the problems in terms of human and environmental continuity, and the high cost of project interventions all dictate that resource inventories and environmental baseline studies provide the specific information needed to make national decisions about clearly identified problems.

Comprehensive problem analysis is no mean task. The biological components above may be formidable, but one must also consider the socioeconomic impacts. For example, a major livestock development project in Upper Volta succeeded in obtaining long-term land tenure rights from the government which induced a group of nomadic Peul herders to adopt a sedentary village life style. Grazing areas were established; new wells were dug; a stocking rate was determined and maintained by convincing the herders to sell substantial numbers of stock on a regular basis; zebu cattle were successfully raised in an area heavily infested with ticks and tsetse fly through a comprehensive veterinary program. But the problem analysis failed to recognize the major socioeconomic need for the women to be able to market the surplus milk. There were no major towns within walking distance of the settlements. At this writing, this problem is still unresolved.

3.2 General Planning Considerations

In addition to addressing the problem and its causes, five considerations are of paramount importance in specifying information needs and planning a resource inventory or monitoring project. These are:

- The resource development, improvement, and management objectives to be attained;

- The administrative-management or executive level at which the information will be used and decisions made;

- The anticipated intensity of management, that is, the inputs of labor and capital to bring about the desired changes;

- The types of errors that are likely and their consequences (whether they are correctable); and

- The flexibility available in the decision and management process (whether decisions can be modified).

Each of these five considerations must be examined as one decides what the data output from an inventory must be and how the data are to be derived. They determine the scale and legend requirements, the sampling designs, the accuracy and confidence limits allowable in the output data, and the kinds of information produced--whether qualitative or quantitiative and whether productivity factors are omitted or estimated.

When the nature of a problem and its causes are clearly perceived, the level of decision-making is identified along with the consequences of error and the feasibility of downstream correction, and the objectives to be attained are clearly stated, the specification of information needed is virually self-evident. If this is not systematically done as a prelude to inventory project design, the information output will generally not meet the needs of the decision process. Decisions will be subject to ill-advised and costly downstream corrections, and the project will fail to achieve desired goals or result in irretrievable damage to the resource.

Following from the problem analysis, resource inventory and monitoring projects should always begin with an exact statement of (i) the kinds of information needed, (ii) why they are needed, (iii) the scale or intensity of examination appropriate to this needs, and (iv) how the information will be used in the decision process.

3.2.1 Need for quantitative data.

Experience and sound judgment are the basic requisites for recognizing the potential to improve the productivity of a plant resource. However, communicating this experience and judgment to others not familiar with the resource base, comparing the potential of different projects, and justifying certain elements of costs requires quantified data and the framework for communication. One of the most useful ways to organize plant resource information for communication is a classification system, and this is discussed in detail chapter 8. The importance of numerical and statistically valid data is discussed here because of its overriding importance in project planning of selection methods. Lack of quantified data has been a major roadblock to environmentally sound decision-making. The introductory comments to this chapter emphasize the variability of plant resources. Without some idea of the range of this variability and how accurately the information collected reflects the variability data can be misleading. Potential adverse impacts described only in qualitative terms may not be perceived as serious, and overly dramatic descriptions of potential impacts may be too convincing. Accordingly, a special effort is required to quantify or at least rank certain environmental influences, particularly those used to predict long-term effects.

3.2.2 Recognizing intersystem dependences.

Another consideration in project planning is intersystem dependences. There are two aspects that have direct relevance to plant resource development. The first is species-area relations. The populations of many species cannot be maintained in relatively small areas even if the species are highly protected. Hence assistance projects must be evaluated in the context of all activities within a region. A small project by itself may have a minor environmental impact, but if it is one of numerous small projects in a local area, the impact may be synergistically magnified. The second aspect concerns the influence of activities within a watershed on downstream systems. Intensive development of watersheds can greatly affect coastal estuaries and near-shore fisheries through sedimentation, pollution, and altered salinity regimes. A great deal of skill and experience is required to visualize the problems that might occur hundreds of kilometers from a project area. Lack of this type of evaluation has created many regional-size problems in both developed and developing countries. A generic methodology for considering these intersystem relations is presented in the Wildlife Panel report (part V) in chapter 2, on strategic assessment.

3.2.3 Endangered species.

Endangered plant species are of definite concern in the project planning process, as mandated by part I, sections 119 and 506, of the Foreign Assistance Act of 1961, as amended. This act specificially charges the Agency for International Development, the Environmental Protection Agency, and the Council on Environmental Quality to develop plans to protect biological diversity, particularly in tropical environments. It further specificies the identification, study, and cataloguing of animal and plant species. A practical justification for the legislation is the need to ensure genetic conservation of primative cultivars and closely related wild and weedy relatives, which form the gene pool for future crop breeding. The fact that in vitro culture techniques have not overcome the problems of maintaining genetic integrity indicates importance of in situ culturing, especially for species that still occur in natural ecosystems. This practice is one of the objectives of the Man in the Biosphere Programme. Unfortunately, adequate inventories do not exist for established reserves, much less unregulated wildlands. Thus, within the framework of resource inventores, there is a clear need to consider endangered plants for their potential in maintaining genetic varieties. The greatest threat to the survival of these species is posed by habitat alteration and removal. Since isolated preserves may not be able to include all the genotypes of certain species (geographic or clinal genetic diversity), project planning related to endangered species preservation must consider the regional perspective. This type of conservation and the legal mandate are discussed by Williams (1982) and Prescott-Allen and Prescott-Allen (1982).

3.2.4 Opportunities for outside expertise.

The entire plannning processs provides opportunities for recognizing problems and impacts. Unfortunately, the proper responses do not always jump out of the available documents, because (i) environmental questions may require different interpretations of raw data by experienced natural scientists and (ii) perception of environmental problems may be lacking in the experience or understanding of project planning personnel. Fortunately, solutions and assistance are available. One source of extremely useful information is the country environmental profiles; these brief summary documents stress the environmental perspective and highlight the prevalent problems. Another source of information is the consulting technical personnel involved in writing project identification papers and other documents leading to development actions. There are usually several steps in the planning process, each of which offers opportunities for substantive input from experts. If serious and demonstrable adverse environmental impacts are identified, it is environmentally sound to consider modifications, delays, or even cancellation of proposed projects. However, it may be unrealistic to do so in terms of the management process. A critical evaluation can be obtained by allowing contractors to recommend program modifications based on their experience and knowledge of particular environmental problems. Encouraging well-presented and well-documented environmentally advantageous changes (within the general scope of the project directive) during the contract award process would be an excellent source of free environmental expertise. A final opportunity to obtain valid suggestions is provided by the inception reports prepared by the contractor, once he is actually on site. Faced with the contractual obligation to perform, and unanticipated problems such as topographic maps being out of print or the season being wrong for sampling vegetation (particlary if there was a long delay in the contract award), he is likely to have salient comments regarding the feasibility of project implementation. Such comments should be carefully evaluated by project officers. The overriding concern should be to do the project right, not just to spend the money or complete the project as initially planned.

The next step is to consider exactly what level of detail is required.

3.3 Plant Resource Assessment

In general, the level of plant resource assessment required will be determined by the intensity of the technical intervention selected to resolve the specified problem or problems. Different types of development actions affect plant resources in different ways, and appropriate levels of survey detail, parameters to be included, data collection methodologies, and sampling designs must take this into account. Other factors include the geographic scope of the action and the environmental risks involved.

3.3.1 Levels of assessment.

Levels of plant resource assessments and the within the framework of the ecosystem. As pointed out in section 2.4, this is the only logical way to address the various ecological processes that will be affected by development. Some inventories are synonymous with the average perception of an ecosystem; an example would be the regional resource inventories made to identify the status of various resources for multiple-use development. Specific inventories--such as a determination of the distribution and quantities of jojoba (*Simmondisia chinensis*) for commercial exploitation of the seeds for oil extraction--can lose sight of the relationship of the intended crop to the rest of the ecological system. It is particularly in the context of the latter type of inventories that the ecosystem concept must be incorporated. Individual species are intricately related to soil types and climatic regimes, nutrient cycles, and a host of synergistic relations, so that a systems approach is required if a complete assessment is to be performed.

Ecosystems are presented here as a multilevel concept. The level at which the "working ecosystem" should be defined and characterized and the level of plant resources assessment are both determined by the intended use of the information derived from the investigation, which in turn is defined by the specific problem being addressed. Five sequential levels of plant resource assessment are defined, each with an increase in the amount of detail acquired. These five levels with their principal purposes and major goals are introduced in Table 3.1 which compares these levels under eight categories designed to assist the development planner in selecting the most appropriate level. These categories are (i) goals, (ii) information sources, (iii) classification, (iv) mapping scales, (v) vegetation descriptions, (vi) statistical reliability of data, (vii) cultural/societal factors, and (viii) watershed roles.

Given the tremendous diversity of geographic and cultural areas, the complexity of ecosystem processes, myriads of plant resources, resource management options, and resource development programs, it can be difficult to make clear distinctions between all the parameters considered under the five levels. In some cases there is probably not a clear separation. Depending on the scope and emphasis of the development program, multistage sampling designs, or other geographic stratifications, some projects will incorporate aspects of several levels of assessment.

Certainly, each successive level will build on the data base that begins in previous, more general levels. Thus, information is accumulated and progressively refined over time as the result of various surveys. Once the data base has been assembled, it can be used repeatedly for decisions and evaluations of subsequent projects, with data added as necessary. Progressive implementation of these levels will result in a true ecological data base having continuity for vegetation resource management.

Table 3.1. Levels of plant resource assessment for inventory or baseline studies.

LEVEL I: PRELIMINARY NATIONAL RECONNAISANCE

Purpose: Compile general information for the establishment of countrywide policies, priorities, and major objectives related to plant resources.

1. Goals.

   Identify major problems or questions to be addressed prior to design of field activities and provide information for policy deliberations.

2. Information sources.

   Summary documents (country profiles, regional synopsis reports, etc.) and generalized maps.

   Visual observations during reconnaissance overflights.

   Knowledgeable informants and specialists.

   Satellite imagery.

   Note: this level does not generally include on-ground visitations except for certain localized development actions.

3. Ecological classification.

Biome, province, formation or other generalized levels based on physiognomic distinctions without regard to family, genus, or species taxonomy. Regional climatic zones and broad physiographic characteristics (plains, hills, mountains, etc.) are other criteria.

4. Mapping scales--1:1,000,000 to 1:500,000

   Average map units: 750 square mile (1942 km^2) at 1:1,000,000

   200 square mile (518 km^2) at 1:500,000

   Minimum map units. 15.5 square mile (40 km^2) at 1:1,000,000

   4 square mile (10 km^2) at 1:500,000.

5. Vegetation descriptions.

   a. Physiognomy (including growth form and strata).

   Addressed only as major physiognomic types (grasslands, savannas, deserts, forests, etc.). Growth form and vegetation strata addressed only as attributes of physiognomic types.

   b. Plant species.

   Generally not addressed at this level, except as related to anticipated development of certain economic species (e.g., timber trees) or essentially monotypic plant communities (e.g., mangrove swamps).

   c. Productivity.

   Adressesed only in the literature reviews.

   Cover.

   Density.

   d. Successional status.

   Generally not addressed, except as regional agents if known (i.e., frequent burning of savanna lands).

6. Statistical reliability.

   None, except as excerpted from literature review.

7. Cultural/societal.

   Addressed according to stated national priorities, recognizing ethnic groups.

8. Watershed roles.

   Identify major areas of concern or recognized problems.

LEVEL II: NATIONAL OR REGIONAL PLANT RESOURCES SURVEY

Purpose: Compile and interpret information based on national or regional resource inventories to be used to set specific priorities and goals, necessary legislation, and development program direction.

1. Goals.

    Provide a generalized information base useful for evaluating resource distribution, status, and potential, for identifying specific areas for future development activities, and for eliminating unsuitable areas from further consideration.

    Assist with prioritization of project areas.

    Identify potential problems that might be aggravated by project actions.

    Determine previously unidentified actions that will be required for successful development of localized resources.

    Address intercommunity relations.

2. Sources of information.

    All available documents (published literature, unpublished reports, maps, archival records, and raw data such as climatic record; see chapter 9).

    Knowledgeable informants (local and international agencies, specialists, etc.).

    Remotely sensed data (Landsat satellite images, small-scale aerial photography, large-scale air-photo transects).

    Ground data collection (low to moderate intensity).

3. Ecological classification.

    Physiognomic criteria predominate; land forms are used as a criterion to segregate homogeneous landscape units; floristics used only at the family or generic level to identify important dominants or economically valuable plants.

4. Mapping scales-1:250,000 to 1:100,000

    Average map units   50 square mile (130 km^2) at 1:250,000

    7.5 square mile (19 km^2) at 1:100,000

    Minimum map units   1 square mile (259 ha) at 1:250,000

    100 acre (40 ha) at 1:100,000

5. Vegetation descriptions.

a. Physiognomy.

Recognized for major growth forms, such as trees, shrubs, and grasses (noting predominance of annual or perennial species). Vegetation strata are recognized only as determined by the growth forms of dominant species.

b. Plant species.

Plant species identified only for dominant plants in each classification category; other diagnostic/indicator species; plants of major economic importance.

c. Productivity (a qualitative description).

Indicated with general modifying comments such as open or closed woodland, dense grassland, sparse vegetative cover, i.e., barren lands, etc.

d. Successional status.

Addressed at general levels related to area potential, e.g., abandoned agricultural fields will eventually return to shrub lands, forests, and so on; major controlling factors should be identified (e.g., fire, ground water extraction, subsistence exploitation).

6. Statistical reliability.

Emphasis on accurate information with only a minor concern for precision; semiquantitative estimation and ordination techniques employed.

7. Cultural/societal.

Effort directed at collating information regarding cultural usage from literature and selected informants; all dominant and indicator plant species should be evaluated for provision of food, fuel, fiber, fodder or medicines.

8. Watershed roles.

Assess potential development programs in terms of water source areas, floodwater reservoirs or receiving areas.

LEVEL III: INTRAREGIONAL PLANT RESOURCES SURVEYS

Purpose: Establish baseline data and guide development program design for priority areas within a regional framework.

1. Goals.

Evaluate plant resources within a geographic region either as the focal point of a development project or to mitigate an anticipated effect of a project.

Clarify vegetation-physical environment relations.

Determine general productivity levels.

Address socioeconomic issues at a local level.

Consider special problems (disease or pest distributions).

2. Sources of information.

   Incorporate document review compiled for level I survey.

   Remote sensing information (Landsat imagery, intermediate-scale aerial photography, large-scale airphoto transects).

   Field work (moderate to high intensity).

3. Ecological classification.

   Species composition used as a classification criteria.

4. Mapping scales-1:80,000 to 1:40,000.

   Average map units 5 square mile (13 km^2) at 1:80,000

   1200 acre (486 ha) at 1:50,000

   750 acre (304 ha) at 1:40,000

   Minimum map units 64 acre (26 ha) at 1:80,000

   25 acre (10 ha) at 1:50,000

   15 acre (6 ha) at 1:40,000

5. Vegetation descriptions.

   a. Physiognomy.

      Indicate the predominant growth form for each species recognized at the generic or specific level; attention should be given to variations in growth forms as related to site conditions. Each plant community recognized should be described in terms of each stratum of vegetation, particularly as it relates to wildlife habitats.

   b. Plant species.

      All dominants, codominants, diagnostic or indicator plants should be identified at the species level.

   c. Productivity.

      Qualitative and semiquantitative techniques employed often within a multistage sampling design. Cover is either estimated or measured, usually from aerial photos, but for groups rather than for individual species.

   d. Successional status.

Emphasis is on recognizing seral stages (range land condition/trend, forest regrowth after harvesting, fallow agricultural fields, and so on).

6. Statistical reliability.

Semi-quantitative to quantitative techniques are used and data should have defined limits of statistical reliability, although actual sampling may be inadequate to represent the more variable resources.

7. Cultural/societal.

Complete information regarding cultural uses and values of plant resources is a major aspect of baseline studies at this level of plant resource assessment; field interviews are necessary.

8. Watershed roles.

Consider project interventions in terms of effects of activities on water resource processes, especially runoff, erosion, and siltation.

LEVEL IV: MANAGEMENT UNIT PLANT RESOURCES SURVEY

Purpose: Provide information necessary for planning and implementation of development/management projects directed at local plant resources, and for initiation of monitoring programs. Focus is on the specific project area.

1. Goals.

Determine productivity of specific vegetation types, and define plant physiological responses and ecological processes that may affect, or be affected by development actions; consider site potentials; and identify plant sociology relations.

2. Sources of information.

Field work (high intensity), organized if possible with both soil and plant ecologists.

Appropriate laboratory analyses.

Subdiscipline specialists (microbiologists, plant physiologists, geneticists, agronomists) consulted as necessary.

Remotely sensed data (some Landsat applications; medium- to large-scale aerial photography).

3. Ecological classification.

The most detailed level of classification is addressed at this level; specific vegetation types are classified with both plant species (and subspecies if necessary) and internal community structure (layering or strata) as criteria. At this level, individual plant community entities are mapped, rather than the vegetation complexes more typically mapped at the higher levels. Microclimate and/or habitat criteria can be used if necessary.

4. Mapping scales-1:30,000 to 1:10,000

   Average map units 450 acre (182 ha) at 1:30,000

   200 acre (81 ha) at 1:20,000

   50 acre (20 ha) at 1:10,000

   Minimum map units 10 acre (4.0 ha) at 1:30,000

   4 acre (1.6 ha) at 1:20,000

   1 acre (0.4 ha) at 1:10,000

5. Vegetation descriptions.

   A complete description of plant communities should be made at this level.

   a. Physiognomy.

      Recognizing variations in growth form and defining species composition of different vegetation strata are intricate aspects of level IV surveys; height, if relevant to survey objectives or management activities (forest resource evaluations or timber harvest planning) should be addressed with statistically valid confidence limits.

   b. Plant species.

      Complete floristic inventories.

   c. Productivity.

      A major activity at this level is required for management decisions; methods should be quantitative with adequate sampling intensity for statistically reliable results; cover and density are measured as necessary.

   d. Successional status.

      Emphasis on defining successional progressions and understanding the influences of at least the basic controlling factors.

6. Statistical reliability.

   Quantitative techniques, adequate sampling to reflect the variability of the resources; statistically reliable (narrow confidence intervals) and precise data.

7. Cultural/societal.

Expanding the data base compiled in levels II and III if necessary with additional interviews and research related to specific areas or problems.

8. Watershed roles.

Evaluate local watershed management potential.

LEVEL V: SITE SPECIFIC INVESTIGATIONS

Purpose: Address specific problems at the project site level found to be essential for development and management implementations, including intensive monitoring and necessary research activities.

1. Goals.

Within specific ecosystems, determine physiological characteristics of individual plant species as necessary, develop or refine management techniques, and define critical ecosystem processes.

2. Sources of information.

Intensive on-site investigations.

Additional laboratory analyses.

Very large scale aerial photographs (for monitoring).

3. Ecological classification.

Use representative sites of types classified at level IV.

4. Mapping scales-1:10,000 to 1:5000 (if mapping is required); often the only need is to locate the study site on previously made maps, or locate individuals or sample sites for measurements.

   Average map units   12 acre (5 ha) at 1:5,000

                          7.5 acre (3.0 ha) at 1:1,000

                          .12 acre (.05 ha) at 1:500

   Minimum map units   0.25 acre (0.1 ha) at 1:5,000

                          .01 acre (.004 ha) at 1:1,000

5. Vegetation descriptions.

   All of these categories can be quite variable depending on the focus of the particular investigation.

   a. Physiogomy.

      Very detailed measurements can be conducted, even on individual plants.

b. Plant species.

  Subspecies, varieties, hybrid biotypes.

 c. Productivity.

  Density and cover determinations are often made at this level; if repeated over time, accurate measurements of productivity can be obtained to characterize this parameter for individual ecosystems or seral stages.

 d. Successional status.

  Refinements of knowledge on successional progressions, particularly in complex situations; research on succession controlling processes, including crop rotation or cropping practices.

6. Statistical reliability.

 Highly quantitative, statistically reliable data.

7. Cultural/societal.

 Addressed only within a specific or unique resource utilization context; extension-education continued if necessary to expand local knowledge of then detailed intervention programs.

8. Watershed roles.

 Watershed research and monitoring of parameters related to watershed quality.

3.3.2 Background information on multilevel assessment.

Below we elaborate on several of the categories of information compiled for each of the five levels of plants resources (see Table 3.1).

1) and 2) *Goals and sources of information* are self-explanatory. Note, however, the decreasing reliance on available documents and the increase in new resource analysis and supporting field work intensity as one moves from reconnaissance to site-specific assessments.

3) *Classification of plant resources* indicates both the general levels of the classification hierarchy addressed and the criteria used. The hierarchical levels of vegetation classification systems and the five levels of plant resource assessment should not be confused. Because vegetation classifications are used to facilitate communications about defined types of plant communities. They are usually hierarchical so that the level of detail of interest can be better matched to the information needs identified in the problem analysis. The levels of plant resources assessments reflects the intensity of resource examination. Vegetation classification is a parallel concept useful for organizing the information for resource development planning and management. The level of

assessment determines which level of hierarchical classification is appropriate. Chapter 8 elaborates on vegetation classification systems. One point that should be emphasized is that if classification systems already exist, they should be used as a starting point. This enhances both host country acceptance and access to the regional literature. If it is necessary to develop or adapt another system, the new classification should be cross-referenced to any existing local system.

4) *Mapping scales and sizes of map units* are included to indicate the amount of detail addressed at each plant resource assessment level, because mapping usually plays a significant role in resource inventories. *Average* map unit sizes are based on the number of acres included in a 3-square inch (1x3 inches) polygon at the map scale indicated. *Minimum* sizes of map units are based on map delineations approximately 1/4 x 1/4 inch. Such small map units are reserved for very important, contrasting features only.

The 1x3 inch polygon "average size" selected for illustration is meant to be the actual size of the unit on the map, at whichever scale is indicated. Further, it does not imply that the area must be rectangular. The minimum size (1/4 x 1/4 inch) represents the smallest area which can be depicted on any map. The formulae used to derive the corresponding average numbers are:

(1) acres per square inch = $\frac{(scale) \times (scale)}{43560 \times 144}$

(2) (acres per square inch) x 3 = average mapping unit size (acres)

(3) $\frac{(acres\ per\ square\ inch)}{16}$ = minimum mapping unit size (acres) (1/4 x 1/4 inch)

Other relevant themes such as topography, geology and soils, land use and even political/administrative regions which certainly have direct influences on plant resources and their utilization can also be mapped. An ecological assessment of an area is not complete without the incorporation of landforms and soil types into the analysis of the vegetation components of the ecosystem. The most useful map of an ecosystem for land use planning is one which expresses combined landform/vegetation/soil mapping units. This map can be most efficiently produced from the collaborative efforts of plant ecologists and soil scientists. This collaborative approach is vastly superior to performing individual surveys, and will provide the development planner with the most useful information base.

5) *Vegetation descriptions* include physiognomic traits such as growth form, vegetation strata, and height; the classification of plant communities and productivity; and successional status. Each of these is discussed below as it is related to levels of plant resource assessment.

5a) *Physiognomy* or growth form is used primarily in vegetation classification systems. A similar characteristic of plant communities useful for both classification and sampling designs is layering or vertical strata (canopy, subcanopy, shrub layer, and ground cover). Strata are important features of plant communities and are therefore addressed at levels III and IV. It is important to recognize that strata can vary depending on successional status (see below). Certain physiognomic characteristics such as height conditions (especially with respect to availability of water: more water is usally associated with greater stature, or standing volume in timber surveys). Such characteristics are addressed only at survey levels, which permit on-ground measurements, or when they can be determined photogrammetrically from larger scale aerial photographs.

5b) *Plant taxonomy* or floristics is also a criterion of hierarchical plant resource classification but is generally introduced at more refined levels of detail, beginning with genera, then species, and finally subspecies or varieties if necessary. Taxonomy is used to reflect species composition of ecological types and not to express phylogenetic relations.

5c) *Productivity*. The major goal of many development programs is to increase productivity. A common error in designing vegetation inventories is that of addressing productivity or biomass measurements at too general a level. They become useful parameters only when they can be related to specific vegetation types or mapping units, which can sometimes be accomplished with estimates at level III and is a major objective of level IV surveys. Plant cover (percent of unit of area covered by overhead plant foliage) and density (the number of stems per unit area) are included in the tables not only as attributes of vegetation biomass, but also as indicators of the potential importance of the plant resource and of the growth-promoting (water, soil fertility) or growth-inhibiting (soil hardpans, windy exposures) characterisics of a site. Density and cover tend to be maximized under ideal growth conditions. Cover can generally be addressed at a more general level of plant resource survey than can density. Both density and productivity are best estimated onsite, while plant cover can be often be reliably estimated from aerial photography. Methods for determining plant cover, density, and biomass are discussed in chapter 4.

5d) *Successional status* of a plant community is important for resource management decisions and for monitoring the effectiveness of management activities. Succession as a process of ecosystem dynamics was discussed in section 2.4. Determining the seral stage of a succesional progression, the direction of change or trend, and the inferred climax type is a major aspect of plant resource assessment. Although this can be done in general terms in the level I and level II surveys, particularly with regard to the major causative factors operating within the region, the complexities of succession usually require considerable on-ground information acquired at levels III and IV and even from the site-specific investigations of level V. Almost invariably, this cannot be accomplished during rapid resource inventories, and in heavily disturbed areas it may be be impossible. The best approach is to characterize seral vegetation types according to standard phytosociological techniques (Daubenmire, 1968), being sure to locate observation sites accurately. More detailed measurements such as production, density, and height should be acquired as time and funding permit. As data are slowly accumulated, successional patterns and causes will begin to emerge. Nevertheless, knowledge of the factors controlling successional status is useful in determining what can be implemented in a region and what are the probable consequences of new development interventions. Knowledge can indicate whether a given pro-

6) *Statistical reliability* of the information collected during plant resource assessments should be given attention. Basing major development programs such as improved rangeland livestock production on statistically unsubstantiated forage productivity can result not only in project failure but also in major ecological and local economic problems. Although reliablity increases as more detailed information is collected, the characterization of level II in terms of "semiquantitative to quantitative techniques with defined limits of statistical reliablity" reflects the potential of multistage, stratified sampling designs to provide adequate statistical confidence even at the more general levels of assessment. Statistical considerations and methods are presented in chapter 5.

7) *Cultural/societal perspectives* should be recognized during the various levels of plant resource assessment. The fundamental reason for international development program should be to improve the quality of life of people in developing countries. Accordingly, we should be attentive to the values and adaptation of the people within areas of development. Similarly, ecologically sound range utilization programs should also consider the significance of indigenous herd size. In terms of levels of assessment, there are two basic sources of information on cultural perspectives. The first, addressed briefly in level I and more thoroughly in level II surveys, is the available literature. Level III begins plant resource assessment at the region level and directly affects local groups of people. Thus, a specific data base should be compiled which is relevant to these localized situations, and this requires a program of interviews to ascertain values and attitudes toward the development action. Further, since interviewing can be a substantial effort, it should be incorporated as a major activity at level III.

8) *Watershed roles* are included to emphasize the "resource systems" nature of plant resources. Terrestrial landscape units can be defined relative to water and its entrained materials as source areas, flood water reservoirs, or receiving areas. These areas should be identified in level II surveys. The watershed captures precipitation in source areas and permits its downward escape (via gravity) to occur over time. Natural floodwater reservoirs (for instance, floodplains along rivers) are a topographically defined feature that can accept large quantities of watershed runoff; downstream transport occurs relatively slowly. Receiving areas are usually the coastal plain and its estuaries although floodwater reservoirs can also act as receiving areas at times. Any alteration of the basic characteristics of a watershed can affect each of these areas in an ecologically and economically detrimental manner. For example, exploitative deforestation of source areas usually results in erosion and downstream flooding. Floodwater reservoirs become silted and their floodwater holding capacity is reduced. Receiving areas may also experience silting and sedimentation and pulsed inputs of fresh water instead of a steady seasonal discharge. Each affected area loses its ability to function in a natural state and significant economic losses can result in many areas. In addition to deforestation, water diversion projects and groundwater development schemes affect downstream plant communities and soils, usually through altered erosion and deposition patterns and surface and groundwater salinity intrusion. Level II and Level III surveys should have as a primary objective the determination of the

watershed roles of the plant communities in a project area and analysis of how the assistance project will affect those roles. Even minor disturbances in an upland watershed can have severe effects downstream by increasing downstream erosion. For example, grazing by just a few animals, such as cattle or goats, in an area of steep terrain and unstable soils can accelerate erosion in the upland watershed and sedimentation in downstream receiving areas. Frequently, resource development projects are designed without a good understanding of these problems in spite of some widely publicized examples such as the Aswan dam. In general, it is recommended that every project subjected to level III and level IV surveys be evaluated in terms of its potential effects on runoff, erosion, and siltation.

At intensive levels of assessment, the nature of the specific problem under consideration can be quite variable. Accordingly, once the program planner has decided that the task requires a level IV or even a site-specific level V investigation, the following parameters should be considered on a per species basis during intensive assessments:

1. Spatial distribution of a species within project area and within region. Do species have a continuous or discontinuous distribution? If discontinuous, is the distribution associated with climatic, atmospheric, or edaphic conditions?

2. Spatial variation in density or abundance. Is the species uniform in density or abundance throughout the area of distribution or is there a distinct spatial pattern?

3. Size characteristics and life form. How is a species described in terms of relative size and life form?

4. Size and age relations. What is the relation between the population size class distribution and the age class distribution? Is it a successional species or a species typical of a vegetation climax?

5. Phenology of key life-cycle events: What is the calendar correlated sequence of flowering, fruiting, seeding, germination, shoot initiation and elongation, leaf shedding, and so on?

6. Reproduction biology. How does the plant reproduce? Does it require an external factor such as a pollinator, fire, change in water availability, and so on?

7. Herbivory: What insects or animals consume all or parts of the plants, in what quantity, and at what frequency?

8. Habitat requirements: What are the nutritional, water, soil type, light, temperature, humidity, microbial, etc. requirements of the species?

Because of the sophisticated scientific nature of ecological monitoring of plant populations, it is necessary to use qualified specialists, particularly those experienced in the region of interest.

In conclusion, it is emphasized that the successful completion of any level of plant resource assessment depends entirely on the skill and effort that goes into the evaluation of the data collected. The existence of tabulated data does not in itself identify the proper actions to take, or the solutions to the problems addressed. An effort should be made to ensure that those doing the evaluation have a holistic viewpoint and a first-hand knowledge of the area.

Remote Sensing: A Tool for Vegetation Inventory and Monitoring

The utility and value of remote sensing technology for plant resources inventory assessment, and monitoring are well documented (Langley, 1978; Aldrich, 1981; Poulton, 1979; Rhode, 1978; Mouat et al., 1981; Bonner and Morgart, 1981; NAS, 1977; Nash, 1979; Treadwell and Buursink, 1981; Sayn-Wittgenstein, 1979; Reeves, 1975).

The application of remote sensing technology to resource inventories is not a highly esoteric and complicated discipline. An analogy can be made with the use of computer technology: one does not have to know about silicon chips, decision algorithms, or progamming to make use of the technologies. In fact, most remote sensing specialists who are involved in international resource development have academic backgrounds in the resource disciplines themselves (range management, forestry, soil science, etc.) and have since incorporated another tool into their repertoire.

It is important, however, for program design officers and project managers to have a basic understanding of the technology in order to interact with the remote sensing specialist. Since this basic material is of interest in relation to all of the panel reports, characteristics of the various remote sensing systems and technologies are summarized part I. Other useful references include Reeves (1975), Estes and Senger (1974), Sabins (1978), and the newly revised *Manual of Remote Sensing* (Colwell et al., 1983).

In this chapter, the three remote sensing systems most appropriate for plant resource studies in developing countries are described. These are aerial mapping camera photography, the Landsat multispectral scanner (MSS) and small-format camera aerial photography. Methods for maximizing the utility of remote sensing technology are then presented. These methods include interpretation, multistage sampling, and ground data collection.

4.1 Remote Sensing Systems for Plant Resource Studies

Three remote sensing systems have clearly demonstrated their appropriateness for most of the plant resource information needs of developing countries. These are, in order of familiarity, aerial mapping camera photography, the Landsat MSS, and small-format aerial photography.

4.1.1 Aerial mapping camera photography.

Considerable coverage from aerial mapping cameras (9x9 inch format) is available for most developing counties. Most of these photographs have been acquired over the past 30 years for a variety of reasons (including topographic map production) and by numerous donor countries and private aerial survey companies. These photographs are often housed in host country agencies, and the negatives (or first-generation positives) may be stored in the libraries of private companies or foreign government agencies.

Although it may sometimes be a formidable task to obtain copies of these photographs, they often prove applicable for certain projects and should be reviewed for availability and suitability. Scales of these photographs typically range from 1:20,000 for local areas to 1:80,000 for country-wide coverage; 1:50,000 and 1:40,000 are very common scales. Virtually all the earlier photographs were taken with black-and-white panchromatic film, but a fair amount of more recent coverage is in either natural color or color infrared. An appropriate use of the older aerial photos is to gain a historical perspective as a starting point for monitoring programs. The Tri-Metrigon aerial photographs taken by the United States over much of the world just after World War II are useful for this purpose; information about their availability can be obtained from the Central Intelligence Agency, Attention: R.T. S-3B, Washington, D.C. 20301. Inquiries should include the longitude and latitude coordinates of the area and a statement of intended use. Computer analysis of Landsat MSS data by competent ecologists is also appropriate at assessment level III.

For plant resource surveys at level II, IV, or V (see chapter 3), aerial photograpy is the most appropriate type of remote sensing information. It is also very useful as a subsample aid for the interpretation of Landsat imagery. This subject is elaborated in section 4.2.2.

Color infrared film (originally developed for camouflage detection during World War II) is uniquely suited for plant resource studies because green, growing vegetation is highly reflective in this region of the electromagnetic spectrum (see part I). When aerial photography is selected for vegetation analysis projects, natural color or color infrared is generally preferred for larger scales of coverage, while color infrared (or sometimes black-and-white infrared) is recommended for the smaller scale, more regional coverage. This is because scattering of light in the blue portion of the spectrum at the higher altitudes (from which small-scale photos are usually taken) can severely attenuate the quality of natural color film. Since color infrared films are always exposed through a minus blue filter, this haze does not present a problem. The timing of color infrared photography is critical for optimal discrimination of vegetation. Normally, immediately after the rainy season, before drying out or leaf drop, is the best time. If such seasonal coverage cannot be oabtained, it is probably better to use black-and-white panchromatic or infrared film.

4.1.2 Landsat multispectral data.

Beside aerial photography, Landsat satellite imaging is the source of remote sensing data most commonly known to resource managers, development planners, and host country scientists. However, there are many misconceptions about a "picture" that includes nearly 33,000 km^2 and has "red vegetation." The reader is referred part I for a review of the basic characteristics of the MSS and a comparison with the new-generation earth resources sensing system, the thematic mapper.

Landsat data are particularly appropriate for level I and level II inventories, where the final results will be presented at scales ranging from 1:1,000,000 to 1:100,000. If computer analyses of the digital MSS data are to be used as a means of extracting information, one can look at each 0.4 hectare in a scene and analyze features at an effective or comparable working scale of 1:24,000. However, the smallest ground resolution unit is 0.4 ha. This means that 0.4 ha is the smallest surface area that has a unique multispectral response. When working with natural vegetation or vegetation-soil systems, it is important to recognize that even with this coarse ground resolution there is much more spectral variation in the response of most homogeneous plant communities (ecosystems) than one might expect. This variable spectral response comes out as "noise" in the system which requires compensatory analysis. Thus success in converting four multispectral numbers for each 0.4-ha pixel (the basic spatial unit in remote sensing analysis) into valid and useful information requires perception and ecological awareness about the individual ecosystem. The MSS system is often sufficiently powerful to correctly classify landscapes at the specific plant community or habitat level, but doing so demands diligent work and large amounts of ground knowledge of the characteristic of each ecosystem and how they are likely to affect the spectral response. The fact that few analysts can provide this kind of ecological interpretation is one reason why most computer analyses of Landsat data on natural landscapes separate only gross features such as conifers, hardwoods, brush fields, grasslands, water, rock, and saline areas. Another reason is that some vegetation categories important to the resource ecologist and manager simply do not have unique and separable multispectral responses. Inclusion of temporal data in the analysis might solve many of these problems, but temporal data are harder to work with and Landsat, outside the range of receiving stations, is not reliable for obtaining such data at times that would enhance the discrimination of difficult vegetation subjects.

For natural resource analysis programs in developing countries, the preferred format for MSS data is computer-enhanced data photographically printed to a false-color composite image for visual interpretations. Up to now, the complexities of digital classifications and the need for real-time and continuous human-computer interaction have precluded the successful application of MSS data except under ideal circumstances. However, with the increasing use of micro-processors, new horizons for using computer-compatible data will be realized, providing a reliable power supply can be maintained. In many developing countries this might require use of an auxiliary power generator.

Old Landsat imagery can be used for inventories of plant and soil resources, but ground verification is important, especially if the weather at the time of image acquisition was markedly different from the weather when the interpretations are being made. Serious errors can be made if this is not done. Similarly, old aerial photography can be useful, particularly in support of satellite data interpretations. However, if the primary interest is land use change, forest harvesting, or deterioration of rangelands, imagery more than 2 to 4 years old may not be useful (depending on the rate of change).

4.1.3 Small-format camera aerial photography systems.

Small-format (35- and 70-mm) cameras have a demonstrated utility for plant resource assessments in developing countries (Rhody, 1981; Treadwell and Buursink, 1981; Panzer and Rhody, 1981). Although 70-mm photography can be used for complete coverage of substantial areas (thus competing with 9 by 9-inch aerial mapping cameras), the major contribution of small-format aerial photography is in subsampling. It is especially well suited for the next-to-last stage (the last being ground data collection) in a multistage remote sensing sampling program, typically at level IV. Small format aerial photography can be used to:

1) Provide supplemental large scale phototransects to facilitate extrapolation of ground truth information to Landsat imagery. Strips of 35-mm photos at a scale of 1:20,000 (1700 feet above ground level, 24-mm wide-angle lens) can be readily located on 1:200,000 Landsat images. Very large scale photos (1:2500) exposed simultaneously with a second camera equipped with a 200-mm telephoto lens can provide detailed information for the identification of certain diagnostic plants, percent ground cover, degree of surface erosion, and so on. Such photographs can be used as a surrogate for on-site observations where inaccessible terrain or lack of time precludes actual visitation, provided the interpreter has an adequate knowledge of the area. Transect locations are selected to represent mapping units delineated on Landsat imagery.

2) Provide sample photography for quantitative determination of productivity of forest, rangeland and agricultural sites. Scales, camera configurations, and data similar to those described above can be obtained. Transect locations are usually randomized in these quantitative sampling designs.

3) Provide repetitive coverage of selected sites for monitoring range trend, forest utilization, disease and insect infestation, and so on. Repeated transects can also be used to provide information at different times if adequate coverage is not available with Landsat imagery.

4) Provide 70-mm stereoscopic photos from fast-cycling cameras on a scale of 1:600. Such photography is suitable for photogrammetric measurements of vegetation. This technique has been used successfully for studies of range utilization, conditions, and trends.

5) Provide an inexpensive, permanent record of vegetation and surface soil conditions. Such photos can be used as evidence of land resource degradation or improvement. This kind of information on trends may be the most important single input for sound decision-making and resource allocation.

There are distinct advantages of small-format photography that make it particularly appropriate for applications in developing countries:

1) Cameras and film are inexpensive (compared to metric mapping cameras) and easily operated.

2) All basic film types (black-and-white, natural color, and color infrared) are available in 35- and 70-mm formats.

3) Most small aircraft available in developing countries can be adapted for small-format aerial photography. Camera mounts for windows, wing struts, and bellyports are inexpensive and easily fabricated.

4) The availability of small planes and cameras and the quality of the photography for sampling and monitoring preclude a need for expensive contracts with commercial firms. Films can often be processed in the country for rapid turnaround.

5) Recurrent costs of monitoring programs can be substantially reduced by training operators and leaving a small-format camera capability in the country, thereby eliminating the need for importing teams of experts.

6) Multiple camera configurations can be readily designed to obtain coverage with more than one film type, or through simultaneous multiscale photography (another way of reducing costs).

Additional information on small format aerial photography can be found in the *Operating Manual for 35 mm Aerial Photography* by Meyer and Grumstrup (1978). An annotated bibliography available through the EROS Data Center (Anderson and Wallner, 1978), includes more than 20 references dealing with the operations of various camera systems, the design and construction of camera mounts, and applications.

4.2 Methods for Maximizing the Utility of Remote Sensing Technology

Problems associated with interpretation, multistage sampling methods, and ground data collection are common to remote sensing programs. Lack of understanding of how information is extracted from images and photography (interpretation), lack of routine incorporation of multistage sampling into program design, and lack of understanding of the basic methods of ground data collection (verifying interpretations based on remotely sensed data) all suggest that an overview of these subjects can increase the utility of remote sensing technology for vegetation analysis.

4.2.1 Interpretation.

Interpretation is the process by which useful information is obtained from the images or photographs. Due to the aerial perspective, the scale, and such enhancements as false-color renditions, the image appearance of a forest is quite unlike the normal ground level view. Thus, image interpreters must use the characteristics of images and their experience and knowledge of the subject matter to develop an understanding of the object or condition in situ. A brief overview of this process can help the project design officer to understand the capabilities and limitations of remote sensing.

In remote sensing terminology, the characteristics of an image are referred to as image elements. They are the parameters, or attributes, of the picture that can be observed, measured, or used by the interpreter to identify subjects of interest. Image elements are the data inputs into the interpreter's inductive reasoning process which define the subject-image relations. Several basic texts on remote sensing, such as Avery (1977), Reeves (1975), and Paine (1981), describe these elements in detail, and we will only list the elements here. They are spatial resolution, spectral signature or resolution, tone (relative brightness), color, size, texture, shape and stereo-perception shadows, pattern, and convergent and associated evidence.

The experienced interpreter incorporates as many of these elements as possible into the analysis of image-subject relations. Quite often, the addition of a single new element may provide the necessary clue for identification. It is important to recognize that what an interpreter can derive from an image or photograph depends on how well the interpreter understands what is to be expected on the actual landscape.

One method of image interpretation that is useful for vegetation mapping relies heavily on terrain feature-vegetation correlates. Even with very large scale photography (such as 1:1000) the spatial resolution is not adequate to identify the species of most plants. However, the plant ecologist knows that many species have strong affinities for certain physical situations in the landscape. Elevation gradients are probably the most familiar example. Others include associations with particular soils and the influence of different slope angles and aspects of microclimate (primarily temperature and moisture).

Many physical characteristics of the landscape are readily identifiable on remotely sensed imagery, even at the small scales of satellite images. Accordingly, defining correlations between vegetation types or plant species and interpretable physical features of the environment greatly enhances capabilities for vegetation mapping. Although this may sound like a formidable task, many of these correlations are already well established in plant ecology (for example, coniferous species tend to be encountered at higher elevations or on cool, protected slopes; marshes occur in wet bottomlands). The fact that vegetation mapping is substantially based on these terrain feature correlates makes old Landsat imagery useful.

The main reason for establishing strong terrain feature-vegetation correlates and for carefully defining image-subject relations is to facilitate one of the primary advantages of remote sensing: the capability to extrapolate information from a site for which ground truth is available to a remote area through image interpretation. This is the major cost-saving contribution of the technology to resource inventories.

4.2.2 Multi-stage sampling methods.

Most resource inventories can benefit from the use of more than one scale of imagery or photography. Smaller scales provide a synoptic overview of the entire area of interest, while larger scales provide the detail required for specific interpretations. Remote sensing systems are extremely versatile in providing such multi-scale coverage.

In the context of remote sensing, multistage sampling refers to the combination of multiscale coverage with "nested" sampling designs (see Fig. 4.3). The purpose of each stage of examination is to minimize the cost of the successive, more expensive stages, of which the final and always the most expensive is field work.

Basically, these methods take advantage of the increasingly finer resolution (provision of more detail as the scale becomes larger) that can be obtained by using multiple sensor platforms (spacecraft, high-altitude and low-altitude aircraft) or by including multiple-focal-length lenses on a single platform. As the scale increases, the resultant smaller area coverage provides subsamples with more detail than in the previous stage. Within this sampling framework, the final stage is always ground data collection. Each scale and type of imagery is used only because it can provide information more efficiently than another stage.

In the first stage a countrywide base map was created as a mosaic from false-color composite Landsat image prints at a scale of 1:1,000,000. This map provided a synoptic view of the entire project area and was used to stratify the country into broad ecological provinces. It was later used for geographic summation and presentation of data. The second stage consisted of 1:500,000 Landsat images selected at random and interpreted for major resource systems (geobotanical units) and land use patterns (barren lands, urban and agricultural areas, and so on). In the third stage, black-and-white 1:50,000 aerial photographs were used to obtain a random subsample of each 1:500,000 Landsat scene. These photos provided the primary stereo interpretive base for vegetation cover types and soil/landform conditions. The fourth stage consisted of 1:3000 color aerial photo strips (also with stereo overlap) randomly selected to represent each 1:50,000 black and white photo. These photostrips were interpreted to verify vegetation types and to estimate range productivity, a primary objective of the project. In the fifth and final stage, a helicopter was flown along the route represented by each of the 1:3000 photo strips and a crew was landed at preselected random points for each vegetation type. At each stop ground data were collected to determine the species composition of the plant community and to obtain a soil profile description and visual estimates of forage production for each of ten randomly selected plots. In addition, one of the ten plots was clipped and weighed to provide a double-sample regression correction for adjusting the production estimates (see sections 5.2.4 and 5.7). Since all data were randomly collected or interpreted, proportional probability sampling theory could be used to make statistically valid extrapolations of these data to the entire country.

The use of weighted sample selection, based on total area of the vegetation type and relative importance for the objectives of the project, is particularly appropriate in multistage sampling schemes. Weighted samples allow emphasis to be given to the more important types, while statistical validity is maintained through randomization. For example, if forest production was the primary project objective, selection would be weighted in favor of woody vegetation over barren lands, agricultural areas, or herbaceous vegetation types.

This example illustrates the advantages of multistage remote sensing programs. These include increased information content, acquisition economics at each scale, and statistical analysis capability. The complexity and number of stages should always be determined by the problem addressed and the information required. The least number of stages required should always be used. Sometimes existing mosaics of aerial photographs can be incorporated to further reduce cost, although less than complete photographic coverage might jeopardize randomization (if it is required). Any remote sensing program consists of at least two stages: interpretation of imagery or photography, and ground truth. A three-stage system consisting of Landsat image interpretation, subsample aerial photography interpretation, and ground work is very common and often appropriate for use in developing nations.

4.2.3 Ground data collection (ground truth).

Ground data collection, or ground verification, is an essential component of resource inventories based on remotely sensed data. Remote sensing can *reduce* the amount of field work (the most expensive activity of an inventory) but should not be used to replace it. The only exception to this statement occurs when the ground is totally inaccessible. Under these circumstances, the best substitute for ground data collecion is low-level aerial observations if the project design and budget include this option. If adequate ground data are not acquired, the resulting maps and information will be proportionately inaccurate.

It is important to understand that any survey that does not actually examine every hectare (an economic impossibility) cannot be 100 percent correct in every boundary delineation, feature identification, or measurement of a specific parameter. Remote sensing survey does allow the examination of virtually every hectare, but rarely provides as much on-ground observation as traditional field surveys, in which ground observation sites can be more strategically selected.

Accuracy of remotely sensed data depends on the experience and skill of the image interpreter. Precision levels of 70 to 90 percent (and even 100 percent for certain unique features) are not uncommon. Unfortunately, some data users overreact to the occasional, but readily apparent, errors or misnomers on the final maps, and blame the tool, that is, remote sensing technology. It would be more constructive to reevaluate the program design, which might not have included adequate funding for ground verification (an independent activity usually omitted from contracts), or to consider the possibility of inadequate performance control, which may have resulted in individuals not performing interpretations or field work properly.

The intensity of ground verification required for a particular project is a function of (i) the complexity of the resources being addressed, (ii) the quality of the imagery (spatial and spectral resolution, dates of acquisition, and so on), (iii) the experience and knowledge of the interpreter, and (iv) the availability of corollary information. Additional information on these variables can be found in most texts on remote sensing.

The fundamental methods used for ground data collection are listed below. This summary is included so that the project planner will be able to adequately budget for them.

1) Once imagery is in hand, a reconnaissance field trip is conducted in order to become generally familiar with the project area, to collect preliminary data on plant communities for the development of an initial classification, and to determine the relations between the image characteristics and the features of interest. The last activity helps to define which parameters of interest can (and cannot) be directly identified with confidence by image interpretation. If subsequent overflights for additional aerial photography are planned, survey markers can be placed if needed. Aerial reconnaissance flights can also be used to gain a general perspective on the project area, identify complex areas that may require extra field time, and ascertain access routes.

2) The production phase of plant resource inventory projects based on remotely sensed data includes two interrelated and often cyclic activities. The first is image interpretation and initial designation of types, or mapping units. This involves separating the image into areas that look similar (based on image characteristics such as color, texture, and pattern), incorporating corollary information from existing maps and reports, and conferring with colleagues about feature identifications. Preliminary mapping results in increased familiarity with the area, which in turn allows for more efficient use of field time. The second activity is ground truth or verification of the delineations and identifications. In practice, these activities are often combined. A field verification trip might reveal several errors in mapping types of vegetation and necessitate a redefinition of image-subject relations. Maps, images, and even portable light tables and stereoscopes are often taken into the field to continue the interpretation process on-site.

3) Special recording forms are used in the field to ensure more complete and consistent records of field observations in a permanent and readily available format. Corollary data can also be included.

4) Evaluation of personnel is important for maintaining consistency in inventories where many interpreters and field observers are involved. This is generally accomplished by having all personnel map and verify the same test area and then comparing the results. When superior interpreters are identified, they should be assigned most of the interpretation and mapping work (with periodic field trips), while other experienced individuals do most of the ground verification and data compilation.

5) Selection of travel routes and ground observation sites is a critical aspect of ground verification methodologies related to nonrandomized sample designs. Travel routes are planned to include the greatest possible diversity of pretyped mapping units in the most efficient manner. It is important that the sites selected for observation and the data col-

lected be representative of the mapping unit and classification category. Some sites are selected because interpretation alone is inadequate for satisfactory feature identification and subsequent field work must "fill in the blanks." In the more typical verification and characterization work, large, homogeneous-looking units should be checked in several locations, while smaller units may require visits to several separate areas until the field worker is satisfied that the inherent variations in the type have been sampled. A useful technique for field verification is to anticipate changes on the basis of the imagery as one proceeds along a travel route. By following the image using the vehicle odometer or features such as drainageways, the observer attempts to predict when a boundary will be crossed and what the next type will be. A field form is filled out shortly after a new type is entered, and then the variations can be added as one continues through it.

Accurate site location and the use of field forms provide an indispensable record of observations made at a site. Even if the site is not representative of the mapping unit, or the extrapolation of remotely sensed data is incorrect, the field verification is still a valid record for that site. This allows future interpretations without additional field work. Also, it provides well-defined sites for future comparisons in monitoring studies. Thus, the benefits of ground data records are contingent on accurate location of the site. An easy way to accomplish this is to make a pinhole representing the site the photo, using large trees, drainageways, or road intersections for reference points. A number is assigned to this pinhole on the back of the photo and on the corresponding field form.

4.3 Conclusions

Remote sensing technology is quite variable in terms of data-gathering options and quite flexible in terms of applications. Substantial benefits can accrue from more careful examination of these options during the program design. Selection of the appropriate system or combination of systems from the wide array of meteorological satellites, radar and thermal scanning instruments, Landsat satellite sensing systems, and aerial photography capabilities (high or low altitude, large or small scale, large- or small-format cameras) should be determined by the nature of the problem to be addressed, the specific information needs, the conditions under which the system must operate, and the cost in relation to efficiency of data acquisition.

The primary advantage of remote sensing is that it can reduce the costs of the more expensive elements of inventory and monitoring: the time and cost of ground data collection. In addition, remote sensing provides a comprehensive view of resources that cannot be acquired on the ground and a permanent record of the conditions on the ground at the time of acquisition. With proper use of remote sensing technology, ground data collection is used only to obtain information which cannot be extracted from the imagery and to provide necessary training and verification.

Field Methods for Plant Resources Inventories

5.1 Introduction

The preceding chapter stated that it is cost-efficient to acquire as much information as possible from remotely sensed data and resort to field work only when necessary to verify interpretations or to collect data that the photography or imagery could not provide. Some general guidelines for verification, or "ground truthing", during remote sensing programs were also provided. The purpose of this chapter is to outline field methods for obtaining detailed plant resource information such as species composition, more reliable data on plant cover and density, and statistically valid data on plant biomass productivity. This is also the phase of the inventory or baseline study in which specimens/samples are collected for laboratory analyses, if required; this topic is covered in chapter 6.

5.2 Definitions of Plant Resource Parameters

Most of terms used in studies of plant resources have been used for decades, often with various meanings. While it is not possible to correct the sometimes contradictory usages found in the literature, we can clearly define the parameters discussed in the remainder of this chapter.

Percent species composition. Density, frequency, cover, or biomass of one species divided by the corresponding parameter summed for all species, expressed as a percentage.

Plant cover percentage. Plant cover is the percent of the ground surface area covered (as represented by the sample) by the downward vertical projection of the crown or stem portions of the plant. Several kinds of plant cover are recognized and are defined below.

Canopy or crown cover. Percent of the ground surface covered by the vertical downward projection of the outermost perimeter of the natural spread of plant foliage; small openings within the canopy are included. When determined on a per-species or life form basis, the total value may exceed 100%.

Ground cover. Percent of material, other than bare ground, covering the land surface. Materials included can be living or standing dead vegetation, litter (organic debris), cobble, stones and gravel (more than 5 mm in diameter), and bed rock. Cover estimates can be made for individual species and/or categories (total ground cover). Vegetation cover has been established as both canopy cover (which includes inner spaces) and absolute cover. Absolute cover is estimated by visually compressing all vegetative components together before assigning a cover value.

Relative ground cover. Proportion of the total ground cover that is contributed by a given plant species or category. The sum of relative ground cover values for all species is always 100% and relative cover measurements are useful for comparing areas with unequal total cover.

Basal area. Cross-sectional area of the stem or stems of plants, expressed as a percent of the area covered, for either individual species or total vegetation. Trees are usually measured at breast or other designated height [the diameter at breast height (dbh) of timber surveys], while herbaceous and small woody shrubs are usually measured at or near ground level. Rangeland survey measures of basal area for grasses and forbs have the advantage of being less dependent on the time of year of the measurement or the degree of current utilization.

Plant biomass. Total amount of living plant material both above- and belowground in an area at a given time.

Production. Plant material yield above ground level on a unit of land at any given time, typically expressed as weight per land area. Synonymous with standing crop and aboveground biomass.

Productivity. Rate at which radiant energy is converted into aboveground plant material on a unit of land. Productivity is expressed as weight per unit of land area *per unit of time*.

Current annual growth. Amount of vegetation production on a unit of land during the annual growing season for the area, expressed in weight units per unit of land area.

5.3 Magnitude and Diversity of Plant Resources

Obviously, one cannot make measurements or counts for all the plants within even local areas of concern. Instead, data must be recorded for a sample of the population of interest and extrapolated to the entire group. Plant communities, as definable ecological entities (see chapter 8.0, on classification), constitute the initial unit for sampling vegetation parameters because such characteristics as species composition, number of individuals per unit area, cover and production are relatively constant (Bonham et al., 1981). Characteristics within the community unit will not be as variable as those between communities, and a smaller number of measurements will be required to represent the entire population.

In addition to the horizontal distribution of vegetation characters expressed by different plant communities, there is vertical distribution. The vertical distribution expressed by stratifying vegetation by life form. For example, a forest may have tree, shrub, and herbaceous strata. Vertical or life form statification can be useful for addressing particular groups of plants such as trees for timber surveys or herbaceous species for rangeland surveys. Any means of separating vegetation into specific groups or units of interest (whether by horizontal or vertical stratification) is beneficial, because it reduces the total variability and therefore the number of samples required. This is important because field methods are time-consuming and therefore expensive. The fewer samples required, the less expensive is the procedure.

5.3.1 The goal of sampling: Statistical reliability.

The role of statistics in plant resource inventories is to assist in the collection, analysis, and interpretation of data in the most efficient and reliable manner. More important, inferences can be drawn from statistically valid data even though the data (samples) represent only a small portion of the total vegetation. Statistical methodologies are unique in providing these benefits and in assigning a level of probability to the estimated values--that is their statistical reliability.

5.3.2 Randomization of sampling for statistical validity.

Plant resource sampling procedures should be randomized. Random sampling makes it possible to obtain an unbiased, statistically valid estimate of the true mean and variation for a given parameter because each sample or observation has an equal opportunity of being selected. Statistical analyses of unbiased data yield the interval of confidence (reliability) around the estimate, and thus allow probabilities or risks to be evaluated for decisions based on those data.

Random sampling in the field is usually achieved by superimposing a coordinate grid network over the area (using measuring tapes, compass bearings, aerial photographs, and so on) and selecting sample sites on the basis of paired numbers in a table of random numbers, with one number for each axis of the grid. The spacing of the grid can be any distance necessary for practical field work within each vegetation stratum.

Non-random or systematic sampling usually does not allow unbiased estimates and the use of statistical analysis. Systematic location of sample sites (for instance, placement of quadrats at 10m intervals along a 100m transect) results in a predetermined (i.e., biased) pattern for sample observations which precludes randomization after the first sample site is selected. In some situations, systematic sampling can be used to reduce the variability of the data, and statistical analyses can still be applied if the systematic observations are averaged and treated as a single sample in the analysis. This approach, however, usually requires a greater expenditure of time to collect a sufficient number of independent samples. Accordingly, *systematic sampling is not recommended* for plant resource analysis unless the project design requires it for other than statistical reasons.

5.3.3 Accuracy versus precision.

Accuracy is how close the *estimated* value is to the true value. In statistical terms, accuracy involves the amount of deviation between sample observations and the true mean of the population (that is, the variance). *Precision* is defined as the size of the deviation from the mean obtained with repetitions of the same sampling procedure. Thus, a method can be precise but not accurate. Precision is the concept addressed in most sampling methodologies. Although high precision generally suggests a high degree of accuracy, the natural variability in most inventory or monitoring samples usually makes high precision extremely costly. As a result, both accuracy and precision must be considered in the design or selection of sampling methodologies. Some examples will illustrate these concepts.

Daubenmire (1968) presented a method for making visual estimates of cover using cover classes 0-5 percent=1, 6-25 percent=2, and so on. He reasoned that observers would not differ significantly in their estimates if the midpoints of each class were used in the analysis. While this proved to be true, the wide range of cover values in class 2 (6-25 percent) would always be assigned the midpoint value of 15 percent, regardless of actual value. Thus, precision was attained with respect to the midpoint, but accuracy remained undetermined. Many of the methods used to obtain estimates of vegetation characteristics do not meet accuracy requirements. If cover classes having 10 percent intervals are used, for instance, the estimates will not be more accurate than 10 percent. Thus, any plant with less than 10 percent cover (which is typical for most plants of arid regions) cannot be represented accurately.

Care must also be taken when defining absolute precision criteria; the objectives of the study and the eventual use of the data must be considered. For example, production estimates derived from clipping and weighing the herbage within 1-m^2 quadrats should not be measured to the nearest 0.1 g if livestock stocking rates will be decided in terms of kilograms per hectare. The cost of obtaining these very precise estimates cannot be justified in view of the management goals.

5.3.4 Other considerations in statistical sampling.

Sample size, which is the number of observations made for a particular measurement, often affects the precision of the estimate obtained. Although any random measurement can be assumed to provide an unbiased estimate of the mean, a single estimate may not be the best estimate of the mean. On the other hand, a large number of observations (that is, several hundred) does not necessarily provide a more precise estimate of the mean than 25 to 30 observations if the sampling is random. If the mean is the parameter needed for decision-making, then in order to obtain more representative samples it may be necessary to increase the size of the sampling units and make fewer observations. However, the quality of the data collected should never be compromised due to the large sample sizes needed for selected confidence and precision levels of probabilities. That is, methods that yield large quantities of data because they are rapid do not necessarily provide the most reliable estimates of vegetation characteristics. Although a sample size of 30 may be considered large from the statistical viewpoint, it may be very small from the ecological viewpoint because it may not include all species-habitat relations within the vegetation type. Conversely, a 10 percent sample (that is, 1 ha observed for every 10 ha of the vegetation type) would probably yield an excellent description of ecological characteristics, but would undoubtedly include more samples than necessary from a statistical viewpoint. An adequate sample is defined as the number of observations needed to obtain a specified level of precision and statistical confidence.

When plot or quadrat sampling designs are used, the size (dimensions) of the plot can influence the data collected. If the plot is too small, there may be a bias toward the dominant species and underrepresentation of more infrequent species. If it is too large, there may be an excessive expenditure of time to make all the necessary observations, with a relatively minor return of additional information. This is one situation where vertical stratification can be useful in selecting the appropriate size of the sampling unit. There are trade-offs between quadrat sizes and costs of sampling for different vegetation types, and the final selection will often be based on the experience of an ecologist within that particular type. As a rule, quadrat size may be determined by plotting the number of species that occur in a quadrat against different quadrat sizes. The point at which the curve (referred to as the species-area curve) flattens out suggests the trade-off between the expense of estimation and the gain in number of species encountered.

Vegetation data may be collected for individual species, categories of species, or some other group. There is value in obtaining estimates on a per-species basis since these estimates can later be combined to provide a desired level of information for decision-making (for instance, production per species summed to give total production). Field estimates for a category or an aggregate of species characteristics cannot be subdivided to the species level if that should ever be needed. These decisions need to be made before field sampling begins.

Double sampling methods can reduce the amount of time actually spent measuring certain parameters. In these methods, regression equations are used to estimate a particular parameter from other measurements that are more easily obtained and that can be correlated with the parameter being estimated. A well-known example of double sampling in vegetation studies is the use of ocular estimates of production and periodically clipped plots to provide a correction factor for estimating the productivity of herbaceous vegetation. The effectiveness of a double sampling method depends on the relation that is assumed to exist between the value being predicted and the variables being measured. Measurements should be made in a consistent and precise manner; otherwise, the variability associated with the data collection method is increased and cannot be separated from the natural variability. Any inconsistency in measurements of vegetation characteristics will reduce the effectiveness of the double sampling procedure.

There are many sources of error in the measurement of vegetation characteristics, and the magnitudes of the errors can vary with different observers as well as different techniques. Errors due to a method or technique can be easily compensated for, while others (such as inconsistency in the double sampling process) may remain uncorrected. It is much easier to change aspects of a method (for instance, the size of a quadrat) than to change the personal characteristics of an observer, which are involved in technique. A particular observer might tend to overestimate the cover of bunchgrasses while underestimating the cover of single-stemmed species, for example. If these errors are consistent and are recognized, they may be corrected. If not, they add to the overall variability of the data and decrease both accuracy and precision.

In conclusion, statistically valid sampling of plant communities may involve time-consuming methods and often requires special expertise to design the appropriate procedures. However, statistical methods are only a means with which to gather data with determinable probabilities of accuracy. Such data are only a starting point (albeit a sound one) for a complete characterization and interpretation of plant resources. Statistics cannot dictate the principles used in interpreting ecological data, but they are useful for determining when a particular attribute has been adequately described. Interpretation of ecological data usually requires other special expertise. That is, vegetation data often indicate some important biological relation other than the vegetation itself (such as good wildlife habitat or low soil erosion potential).

5.4 Beyond Single Parameter Sampling: Analytical Methods

In vegetation analysis for baseline studies, one can compare data from homologous vegetation units in different regions (comparative analysis), from units in the same region that have been exposed to different treatments (cross-sectional analysis), or from particular units as they change over time in a monitoring program (longitudinal analysis). It is within these frameworks that individual parameters such as species composition or cover can be used to analyze vegetation types as entities. Each of these methods is briefly summarized below.

5.4.1 Comparative analysis.

This method involves the selection of homologous units of vegetation that are widely separated in space but are believed to be similar in certain aspects of structure and function. The units may be chosen on the basis of presence of comparable genera, species, or other taxonomic units; common functional features; similar soils or climates; or a combination of such factors. Comparative analysis can yield generalizations about causal or developmental processes, but it is most revealing when correlations can be found between variables in systems that differ in one or more key respects.

5.4.2 Cross-sectional analysis.

Cross-sectional analysis is usually conducted within a single area. Its purpose is to detect differences between otherwise similar segments of a plant community which have been exposed to known treatments. It can also detect regularities in plant processes when only a short time has been allotted for the investigation. For this reason, cross-sectional analysis may be particularly important when recommendations must be made quickly. In practice, this method may involve examining different quadrats within a relatively homogeneous plant community which have been exposed to known influences at different times in the past.

A basic assumption in cross-sectional analysis is that the quadrats differ only with regard to specific, known influences. If the influence is from a recent disturbance, it is assumed that all quadrats were previously indistinguishable from each other. In other words, one attempts to find comparable quadrats in order to hold all variables but one constant. Measurements can often be standardized and an adequate size obtained in cross-sectional analysis.

The assumption that all variables but one are held constant may be difficult to justify. Although costly, increasing the sample size also increases confidence that exotic or unknown influences have not affected the results.

A example of cross-sectional analysis is a study of shifting agriculture among native groups of central Brazil (Gross et al., 1979). This study provided valuable data about the response of soils, vegetation, and cultivated crops under various conditions and permitted conclusions to be drawn about a period greatly exceeding the duration of the study period.

5.4.3 Longitudinal analysis.

Longitudinal analysis is possible when an adequate baseline has been established and when there is sufficient time to directly observe changes in a plant community at various points in time. Because of the rates at which succession and other processes occur, longitudinal analysis may require many years. However, there is no more reliable method for monitoring changes in a plant community. In longitudinal analysis, one can directly observe processes unfolding in time while recording data on the factors thought to be responsible for them. A basic assumption in longitudinal analysis is that the time span of the observations is typical in terms of time dependent variables such as rainfall and herbivory. With this method there usually is no question of comparability of sites, since the same area may be revisited repeatedly. Occasionally, problems arise in "controlling" extraneous influences in a study area, and legal and/or physical means of protection may be needed. One extraneous influence that is usually recognized is that of traffic involved in repeated sampling at study sites.

Longitudinal analysis is perhaps the ultimate monitoring method in terms of control and accuracy. It is most feasible if associated with field stations where there is a constant presence. Its greatest drawback is the time required to conduct the studies; planners often cannot wait long enough for such a study to be completed. Good planning, however, involves making provisions for ongoing monitoring and laying a basis for future longitudinal studies.

These three analytical methods are not mutually exclusive; they can be combined in supportive ways to provide innovative analyses. One of the authors, Snedaker, combined comparative and longitudinal analyses in his study of the effect of a dam on a tropical mangrove swamp in Bangladesh. Early in the study, using comparative analysis, he predicted that the damming of the Ganges River would deprive the downstream mangroves of essential nutrients, cause hypersalinization, and result in the deterioration and eventual destruction of the swamp. Subsequent longitudinal analysis confirmed this prediction and demonstrated the steps in the process and the effects on the economy of adjacent areas.

5.5 Plant Cover Estimation Methods

Plant cover measurements are often used to describe the structure of plant communities. Species composition and the distribution of individuals within the plant community determine the structure of the vegetation type. The amount of plant cover provided by a species is directly related to that species' ability to compete for and convert various resources (nutrients, water, sunlight) into above- and belowground biomass.

Methods commonly used to make plant cover estimates include ocular estimation within two-dimensional plots (belt transects, quadrats, and loops), line intercepts, variable plots, and point samples. One or more of these methods has demonstrated applicability for virtually any grassland, pasture, shrubland or forest vegetation type (see Table 5.1). More important than vegetation type, however, in selecting the most suitable method, is the type of cover data (ground, canopy, or basal) required to meet the objectives of the project. Another important consideration is how the data will be recorded, that is, for plant species, groups of particular species, lifeforms, and so on.

5.5.1 Ocular estimates with two dimensional plots.

Cover is considered the single most important parameter of a species in relation to other species of the vegetation type. Basal cover is perhaps the easiest type of cover to measure and fluctuates less than some of the other types with climate, season, or condition of the plant (overgrazed, for example). However, since canopy or crown measurements are easier to relate to, they tend to be the most commonly used.

Ocular (visual) estimation of plant cover from two-dimensional plots (quadrats or belt transects) is often more accurate than obtaining cover data by linear (line transects) or plotless techniques. A variety of quadrat shapes and sizes have been used successfully, including circular plots (ranging from 0.12 square inch to 100 square feet), rectangular ones (0.25 by 0.50 inch to belt transects), and squares. When sampling vegetation types with multiple strata, the concept of "nested" plots is often utilized. For example, a 10 by 10 m plot might be used for estimating tree cover, 4 by 4 m for shrubs, and 1 by 1 m for herbaceous plants. When appropriate, the size of the plot can be selected to represent 1/100 or 1/1000 ha, which facilitates certain calculations. However, the quadrat shape or size does not significantly influence the accuracy of ocular estimation of plant cover, and an adequate number of samples can be acquired more rapidly by increasing the number of observations rather than the size of the plot.

Experience and training are needed if quality data are to be obtained. Inaccuracies in plant cover estimation usually result from poor visual techniques and not the method used. Percent cover classes (10 percent intervals, for instance) can be used to reduce subjective error in cover estimation, but are not appropriate for sparse vegetation types because species contributing only small amounts of cover tend to be overestimated. In order to estimate cover visually by a quadrat method and

Table 5.1 Appropriate use of cover estimation methodologies in four vegetation categories.
Revised from Bonham et al. (1981)

Method	Used for
Ocular estimation (proportion of area)	Grassland, shrubland, forest
Belt transects (widths and lengths)	shrubland
Line intercepts (length only)	Grassland, pasture, forest
Point methods (usually 10 pin frames; number of hits)	grassland, pasture, forest*
Loop methods (3/4 inch; occurrence only)	grassland, pasture, shrubland, forest**

*Using cross-hair sighting techniques.
**Consider understory strata only.

obtain repeatable results, known areas of the quadrat (such as, 1 or 5 percent) must be used as references. One must also realize that estimates can never be more precise than the smallest area used for reference. For example, if a 5 percent area is used as the cover percent interval, then averaging cover estimates for the different quadrats will be accurate within 5 percent. Observers should also be trained for plant cover estimations based on quadrat sampling. Very accurate estimates of presence or absence of plant cover in shrub or grassland communities can be obtained by vertically lowering a small rod or pin at 25 to 30 random locations within the plot. The data obtained can be used for comparison with visual estimates, and such trials should be continued until the observer is consistently making accurate estimations. Although the pin or point method is more accurate (see section 5.5.3), ocular methods are more rapid and, with adequate training, are usually preferred.

5.5.2 Line-intercept (linear transect) methods for cover (estimation).

The line-intercept technique for plant cover measurements was developed to eliminate the subjectivity of ocular methods. However, total cover tends to be overestimated because interstices in the foliage are included in the intercept measurement, and cover of individual species can be underestimated because multiple layers of vegetation are easily overlooked. Data collection can be simplified by using the entire plant, rather than separate parts as the unit of measure.

In practice, a length of wire, string, or measuring tape is placed randomly within the vegetation type, stretched, and held tightly along the ground or at some selected height above the ground. The distance along this line occupied or overlapped by each plant (the intercept) is recorded as the cover for that plant. Individual cover measurements are summed to estimate the total cover.

Transect lengths reported in the literature range from 4 to 1800 m. Since the length of the line can affect the results, short lines placed at random may be more effective than a few long lines. When vegetative cover is sparse, however, longer lines reduce sampling variation. The degree of vegetation heterogeneity also influences the length of the line. Transects that cross heterogeneous areas, even within a type, may bias the cover estimate for certain species, even though the estimate of total plant cover for the type remains unaffected.

5.5.3 Point method for sampling plant cover.

In the point method, a long, narrow pin (or frame of pins) is lowered vertically through the vegetation until it makes contact with a plant part or the ground surface. Any contact between the pin point and a plant is recorded as a "hit" for that species. Cover is estimated by summing the number of hits per species and dividing the sum by the total number of pins lowered. For short-statured vegetation types (pastures, grasslands, and low shrub communities), this method yields precise and accurate estimates of plant cover. As with any quantitative sampling procedure, sample sites must be randomly located. Use of individual pins, rather than frames, is more efficient for obtaining sufficient independent samples, because successive pin drops from the same frame are no longer random once the first pin is lowered.

There are several sources of potential error in this method. The only inherent bias is that broad-leaved plants (typically exhibiting horizontal leaf orientation) are overestimated, while narrow-leaved plants (especially grasses with erect leaves) are underestimated. If the vegetation is too tall, or it is windy, there can be errors associated with plant movements. The vegetation must remain undisturbed for accurate estimates to be obtained. The size of the pins can also affect the results. Larger pins tend to overestimate percent cover since border area error increases with pin diameter. Commonly used pins range from 3.0 to 4.1 mm in diameter.

The point method of estimating plant cover, if carefully used with the proper vegetation types, will yield the most consistent estimates of absolute ground cover.

5.6 Measurement of Plant Density

Density, or number of individual plant per unit area, is probably the most easily understood parameter used to describe the ecological characteristics of a species or a plant community. Density measurements have been used to determine responses of plant species to various management treatments or environmental fluctuations. For other purposes, such as evaluating seedling emergence and survival on reforested lands, density may also be the most appropriate method.

There are two basic approaches for determining density. One involves simply counting the number of individuals within carefully delineated plots. Both the size and the shape of these plots can influence the results. The other approach includes several variations on plotless, or distance-measure, techniques. In general, quadrat or plot methods can address variations in plant distribution patterns better than plotless methods, especially for herbaceous species and small shrub and trees. Commonly used plot sizes range from 0.1 to 1.0 square meter for herbaceous plants and 4, 10, or 16 square meter quadrats for woody species. Tree species in forested communities have been sampled with 100 square meter quadrats, but are most frequently sampled by plotless methods (see section 5.6.2), which require less time, are more efficient, and may even be more accurate for some vegetation types. These two approaches to measuring plant density are described in greater detail below. First, however, we must address a fundamental question for any determination of density: What constitutes an individual plant?

Discrimination of individual plants is a basic requisite for making accurate estimates of plant density, yet it can sometimes be difficult and subjective. In general, an individual plant can be defined as the sum of the aerial parts that correspond to a single root system. Since most trees and many shrubs are single-stemmed, as are some grasses and most forbs, these plants pose no particular problem. However, some old and large bunchgrasses appear to break into smaller somewhat concentric "individuals" because the centers eventually die. Perennial forbs and grasses that spread vegetatively are difficult to count as individuals, as are many species of creeping or spreading shrubs. Multistem shrubs also require difficult decisions.

The only consistent definition of a counting unit may be the individual stalk or culm, but in many situations this will undoubtedly result in greatly increased densities of individual plants. Thus, this procedure must be evaluated in terms of the plants being studied and the use to which the density data will be put. In some cases, other vegetative measures such as cover or biomass may be more meaningful. Whichever definition of an individual is used, consistent decision criteria must be established and should be reported with the results.

5.6.1 Quadrat or plot methods for density determination.

Counts of individual plants within sample plots or quadrats can be used to estimate plant density. Counting can be a tedious and time-consuming method, depending on the ease of recognition of individual plants, and their degree of dispersion. Two critical considerations when using plot methods to determine density are: 1) the size of the quadrat, and 2) the number of plots required to obtain an adequate sample.

Of particular concern relative to the size of plots, is the opportunity for errors in determining whether plants along the boundaries are within or excluded from the plot. The smaller the plot, the greater the length of boundary per unit area and therefore the greater the chance of error. On the other hand, counting numerous individuals cannot be done accurately within large quadrats unless these plots are subdivided or the individual plants marked in some fashion after each is counted. Obviously, larger plots require more time per sample, too.

Thus, the choice of quadrat size depends on the characteristics of the plants being counted, the intended use of the density data, the effects of boundary, and convenience of sampling. Quadrat sizes are usually proportional to the size of the plants being counted (i.e., life forms). For example, trees are sampled with larger plots than grasses. A 10 by 10 m quadrat size for sampling trees is the most popular, but sizes have ranged from 0.25 milacre to 0.1 acre or more. A 10 by 10 m quadrat used in tropical forests had the lowest required sample size for the majority of species tested. Larger quadrats, of 10 by 140 m, have been used for density measurements in some hardwood forests. Strip or belt transects are essentially very long rectangular plots. Belt transects 2 to 10 m wide have been used in timber surveys and found to yield less biased estimates of density than quadrat measures alone. Herbaceous vegetation is most frequently sampled with 1 by 1 m plots. Smaller quadrats, such as 20 by 50 cm have been used for denser vegetation types such as meadows, pastures, annual grasslands, and other alpine-arctic, tundra and subtropical plant communities where the individuals are in close proximity. A general rule based on mathematical relations is that the most efficient quadrat size for estimating density of herbaceous species corresponds to not less than 20 percent absence of the species or a product of quadrat size and plant density of about 1.6. This relation is maintained in plots ranging from 0.7 to 3.0 m^2.

Whatever size or shape of plot is selected, it is critical to locate the boundary accurately. In the case of nonrigid boundary materials (string, rope, chain) this can be a significant source of error.

There are two general ways to determine the proper number of samples to take. The first is a function of the total size of the area to be sampled, and the second is a function of the variability of numbers occurring within successive sample plots.

A common practice in timber volume surveys is to make the sample size a function of the study area. One may set a standard of 5 or 10 percent sampling intensity; for example, if a particular vegetation type covers 10 h, a 5% sample would extend over an area of 0.05 x 100,000 square meters = 5000 square meters. This area could be sampled by counting individual trees in each of fifty 10 by 10 m plots.

A practical guide for estimating the adequacy of the sampling size is to calculate successive means. After an initial large variation, the curve flattens out, indicating that less variation is occurring between accumulated quadrats. Sampling should be terminated when additional quadrats do not significantly affect the mean of the more important species.

5.6.2 Plotless methods for determining density.

Plotless or distance-measure sampling methods have been used for the last 30 years to determine plant densities. These methods are based on the concept that the number of plants per unit area can be calculated from the average distance between individual plants. The major advantage of plotless methods is that density may be determined without first establishing plot boundaries. This may save considerable time and improve the accuracy of the density estimate because no plot boundary error exists.

The number of plants may be calculated by dividing the unit reference area (such as 1 ha=10,000 square meters) by the mean area of an individual plant. Mean area is defined as the reciprocal of density. Thus, density is estimated from the mean area of an individual, which is the average of distance measures. There are four commonly used methods for determining distances; based on the nearest neighbor, random pairs, point-centered quarter, and closest individual stem. Each of these methods can provide mean values of stem density (all species combined) close to the actual values if an adequate sample size is obtained. These four methods are defined below.

Closest individual. Distance is measured from a random point to the closest individual plant. This method is very simple but produces variable results, has high coefficients of variation, requires a large number of samples, and estimates are subject to some bias.

Nearest neighbor. The individual closest to a random point is determined, and the distance from this individual to its nearest plant neighbor is measured. This method is less variable than the preceding one, but also requires large sample sizes.

Random pairs. The plant closest to a random point is selected, and the distance from this plant to the nearest individual plant outside the 180 degrees inclusion angle is measured (see Bonham et al., 1981). This method is less variable than either of the preceding two.

Point-centered quarter. From a random point, the distance to the nearest plant in each quarter is measured. This is the preferred plotless method because it yields less variation in mean distances, provides more data for individual species per sampling point, and is least susceptible to bias.

5.7 Methods for Determining Vegetation Biomass

Vegetation biomass is a measure of plant material produced on a unit of land. Biomass is often expressed as weight per unit land area (for instance, grams per square meter or kilograms per hectare), although other measures such as volume (timber board-feet) are also used. Aboveground biomass is of great concern in plant resource management, as evidenced by the interest in agricultural yields, rangeland productivity, and timber volume measurements. Plant biomass production is often correlated with cover and used to determine dominant species. The production of plant biomass is an expression of the plant's ability to compete with other plants in its utilization of water, nutrients, space and other ecosystem resources.

Plant ecologists estimate aboveground biomass by a variety of harvesting methods, by indirect (regression) methods or by a combination of the two--that is, double sampling methods. Data can be summarized on the basis of species, aggregates of species, or lifeforms (that is, grass or shrub production). As always, the type of data provided, degree of confidence, and intended use of the information should be decided before data collection according to the objectives of the study. Sample site selection should, of course, be randomized to allow for valid statistical analysis of the data.

Harvesting methods for grass and shrublands entail randomly superimposing a sample plot of known area on the landscape, removing all the plant biomass within the three dimensional volume determined (area x height), and weighing this harvest.

Plot shapes can be circular, square or rectangular. Circular plots (hoops) are easily carried and put in place, but long, narrow rectangular quadrats generally give more accurate results. The size of the plot must be a function of the size and distribution pattern of the plants to be sampled.

5.7.1 Direct harvest methods.

Harvesting methods include hand plucking, clipping or cutting with shears, axes, machetes, sickles, or mowers. Harvesting can be done at different levels (heights) depending on the program objectives. For instance, it can be restricted to the current year's growth, some amount considered to be a proper utilization level for sustained yield (for instance, 50 percent), or, as in the case of shrubs, the level to which herbivores can reach (this level may be quite surprising with goats). When determining annual available production for shrubs, one should consider harvesting not only the vegetative growth (twigs) but the production of seedpods as well. Although these may develop out of the reach of many animals, on maturity they usually fall to the ground providing an often abundant and nutritious forage supply. A common source of error in production estimates for herbaceous plants derived from clipping is the variable "stubble" height left unclipped. This error can be eliminated by consistently clipping to ground level.

Another important error may occur in determining exactly what vegetation is within the volume bounded by the length and width (or radius) of the plot and the height of the vegetation. Any plant biomass within this three-dimensional area, even if the plant is rooted outside the boundaries, is harvested with the sample. Conversely, parts of plants rooted within the area that hang outside the vertical plane of the boundary are not included.

Harvested material should be oven-dried before weighing because moisture content can dramatically affect the weight. The material is placed in an oven at 70°C until a constant weight is obtained. When large numbers of samples are taken, weights can be measured in the field and a

10 percent sample brought back for oven-drying. This 10 percent sample can be used to determine a correction factor between oven-dry and field (moist) weights in order to convert the other 90 percent samples to their oven-dry equivalents. Determining weights to the nearest 1 gram, or even 5 grams, is typical for animal stocking rate calculations.

5.7.2 Double sampling methods.

Because plot clipping is time-consuming and many samples are typically required for statistical adequacy, double sampling techniques are commonly used for plant biomass estimates. Double sampling involves a combination of visual estimates of biomass with clipped and weighed samples. The visual estimates can rapidly produce an adequate number of samples, while the clipped and weighed plots are used to train the observer and to provide a mathematical correction factor. In practice, all sample plots are visually estimated, but only a certain proportion (ranging from 1:4 to 1:11) is clipped and weighed.

For precision with this method, the estimator must begin work with each new vegetation type by visually estimating a plot, then clipping it to determine the degree of error in the estimate. This should be repeated until the results of both methods are close. The worker must be consistent, and continuous training is often needed in order to maintain linear relations between clipped and visual estimates of production.

A linear relation between estimated values (visual and clipped) usually holds between certain actual weight limits. An upper bound for linearity may be in the range 100 to 120 grams per unit area of quadrat. Visual estimates are usually obtained by mentally adding units of 5 to 10 grams until the biomass for the quadrat is totaled. As more mental additions are made, errors accumulate more rapidly. To mitigate this, smaller quadrats should be used which limit the total biomass amount possible.

5.7.3 Indirect or regression methods.

Indirect methods are used to avoid destroying plant materials unneccesarily. They involve easily measured vegetation characteristics which are statistically correlated with production estimates. Production is usually calculated from regression equations of the form $y=a+bf(x)$ where y is the production estimate for each species, a is a constant, b is the slope of the line, and $f(x)$ is a function of the measured plant characteristic. The function may be log x or another relation that is dependent on the species growth form. These equations are developed for individual species and data can be combined for each species from different vegetation types in the same locale.

The following production parameters and correlated plant characteristics have been used in vegetation production sampling:

Production Parameter	Plant Characteristic
Leaf weight (shrub)	Leaf area Crown diameter Percent crown diameter Crown circumference x height
Twig production	Crown volume (basal diameter squared) x branch length Twig length Twig diameter
Current annual production (shrub)	Crown Logarithm of crown diameter Crown volume Natural logarithm of basal area
Herbaceous production	Leaf length Foliage cover Stem length (various measures) Number of stems (culm and flowering)

Methods for Laboratory Analysis of Plant Materials

In the preceding section, methods for sampling and measuring characteristics of vegetation by a variety of field techniques were discussed. Some projects also require information on specific attributes of particular plant species, or even individual plants. Common examples are analysis of the nutrient quality of certain forage plants (protein, carbohydrate, fiber, fat, and trace element contents); identification of individual organic molecules from plants utilized by traditional peoples for medicinal purposes (for example, digitalis) or of potential value as industrial raw materials (for example, jojoba or *Simmondsia chinensis*); or determination of genetic materials from wild cultivars of domesticated agricultural plants. Such analyses usually require laboratory techniques such as spectrometry, chromatography and even scanning electron microscopy. In general, such methods will be employed at the more detailed levels of plant resource assessment (levels IV and V). The increasing use of pesticides and herbicides worldwide will necessitate the use of laboratory analytical procedures for certain environmental monitoring programs.

The purpose of this section is to briefly review the array of techniques available, consider some special sample requirements, and describe the proper methods for handling and storing plant material samples.

6.1 Summary of Laboratory Techniques Available

It is beyond the scope of this report to provide specific details on the thousands of procedures and the great variety of instrumentation available for the analysis of plant materials for protein, carbohydrates, lipids, nucleic acids, terpenes, porphyrins, alkaloids, growth hormones, flavonoids, trace elements, pesticides, other chemical compounds. An excellent and essential source on this subject is the *Official Methods of*

Analysis of the Association of Official Analytical Chemists (13th ed. 1980). This book includes analytical methods for foods (cereal grains, sugars, cacao, coffee, and tea), flavorings (vanilla, various extracts, and spices), nuts and coconuts, livestock feeds, oils, fats, waxes, drugs, alcohols, hormones, vitamins, and other substances. Beyond this basic manual, a good library and recent scientific journals are necessary for selecting the most appropriate and up-to-date techniques for a particular analysis. Several of the techniques routinely used for analysis of plant materials to complement level V surveys are described in the following paragraphs.

Titration. In titration methods a reagent is added until a color change occurs or a precipitate forms, indicating that a balance point in a chemical reaction has been attained. These methods are simple but labor-intensive, and automated equipment is available for both reagent addition and data recording.

Spectrometry. Spectrometric techniques are used to determine the presence of many individual chemical elements (Thomas and Chamberlin, 1980). In addition to atomic absorption spectrometry, which is usually preferred, visible, ultraviolet, and infrared spectrophotometers are available.

Chromatography. Beginning with paper chromatography, this field has expanded enormously in recent years to include thin-layer chromatography, column chromatography, gas/liquid chromatography (Willard et al., 1981), and sophisticated and ultraquantitative high-performance liquid chromatography (Kuwana, 1980). Many pesticide laboratories are now highly automated for routine analytical procedures, including sample extraction, cleanup, injection, and final calculation (Zweig and Sherma, 1980).

Electrophoresis. This technique has greatly accelerated the elucidation and characterization of charged particles. It has been most effective in the analysis of proteins (Harborne and Van Sumere, 1975) and in the characterization of specific enzymes. A refinement of the technique is isoelectric focusing (Catsinpoolas and Drysdale, 1977), which has improved both the accuracy and quantification of these analyses. Electrophoresis and isoelectric focusing have been particularly helpful in establishing the phylogenetic relations of organisms. Plant breeders and molecular biologists find this methodology attractive as genetic engineering becomes a reality in the plant sciences.

Radioisotopes. Isotopic tracers have been invaluable in studies of metabolism and translocation. Techniques based on the use of isotopes include autoradiography, liquid scintillation counting, and radiochromatography (L'Annunziata, 1979). Like all sophisticated technology, these methods require considerable training and practice before experimentation begins. Environmental health and safety are of major concern when using these techniques.

Sometimes several methods are available for a particular type of analysis. In such cases, the final selection should be based on the level of sophistication required, speed of the analysis, and costs. Automated processes can allow quantification of many samples in a short period of time, and microprocessors have greatly speeded up both the procedures and the calculations.

Many laboratory techniques can be combined to design special analyses for answering particular questions. Problems related to quantification can be overcome if interfering substances (contaminants) can be removed from the sample by centrifugation, dialysis, bleaching or other standard procedures. Companies being considered to perform laboratory analysis of plant materials should be able to provide references demonstrating their competence in plant biochemistry (Bonner and Varner, 1976; Goodwin and Mercer, 1974; Morris and Morris, 1975). They should also be willing and able to work under the logistical and other constraints that are usually involved in foreign assistance programs.

6.2 Sampling for Laboratory Analysis

The basic considerations discussed in section 5.2 regarding an unbiased, representative sample of the population of interest apply to specimen collection for laboratory analyses. Once again, the objectives of the study are a critical determinant in the sampling design. For example, if the goal is to determine whether a plant species contains a particular chemical compound, a single specimen collected at random might be adequate. On the other hand, if the goal is to determine the variability of the amount of that chemical within the plant population in order to evaluate the economic feasibility of harvesting, transporting, and processing it as a commodity, then the sample size must be adequate to define the statistical variation.

Laboratory analyses can be expensive, and economics often dictate the intensity of sampling as well as the sophistication of the analysis. These two activities should be carefully matched in the overall program design. There is no justification for employing an advanced and precise analysis with an inadequate, nonrepresentative sample. Furthermore, if the variability within the population is so great that the data will be meaningless, one should reassess the desirability of doing the analysis at all.

Since most plant materials must be transported to laboratories for analysis, the total amount of material required should be held to a minimum. For example, if scanning electron microscopy is the intended analytical method, 10 square millimeters of material per sample is adequate. The critical decision concerns the number of samples required and experimental objectives.

6.3 Handling and Storage of Samples

Collection, handling and storage of plant material samples destined for laboratory analysis require advanced knowledge of the analyses to be performed. Fresh plant samples are exceedingly perishable; bacteria and fungi present at the time of harvesting continue to promote decomposition. Respiration of the material itself, or even volatilization with continued exposure to air, changes the nature of the original material. These processes can be accelerated by the high-humidity, high-temperature environment inside most sample containers.

In most cases, samples should be either dried or frozen as rapidly as possible, depending on the eventual analysis and the likelihood of enzymatic degradation and volatilization. Drying at 70°C is the standard procedure for determining dry weights. Elemental analyses can usually be performed adequately on heat- or freeze-dried samples mechanically ground to a specified particle size. Dried samples can usually be stored for long periods without further decomposition and shipped inexpensively in small, airtight glass, plastic, or metal containers.

When samples are being collected for the determination of metabolites subject to enzymatic degradation (for instance, carbohydrates), it is essential that the tissue be killed and all enzymatic activity cease as soon as possible so that the results of the quantitative analysis accurately reflect conditions at the time the sample was taken. Temperature is critical in storaging of such samples, and in some cases liquid nitrogen may be required for rapid freezing. Once the samples have been frozen, they may be stored on dry ice for extended periods. If a lyophilizer will be used for final drying of the sample, care must be taken to ensure that the sample never thaws during transport and storage.

6.4 Conclusion

Because of the complexities and specialized nature of laboratory analyses of plant materials, expert advice is needed in all phases of the program design. The recommended analytical methods can be explained and justified so that the appropriate methodologies are selected. The sampling program can be designed to ensure the proper degree of representation for statistically valid results and to match this with the sophistication of the technique selected. If other workers will be collecting the samples, clear instructions must be provided for proper handling, preparation, storage, and shipping of specimens. Finally, the services of a specialist should be available to assist in interpreting the results of the analysis.

Methods for Determining Human Uses of Plant Resources

Since one of the objectives of natural resource development programs in less developed countries is to improve the lives and futures of the indigenous peoples, it is reasonable that agricultural, forestry, and their other current use of plant resources be considered during the design and implementation of the program. On a more practical note, an understanding of the existing cultural framework and social organization provides at least three direct benefits to the resource development process:

1. Advantage can be taken of local knowledge of the natural resources.

2. Inadvertent destruction of resources relied on by the indigenous peoples can be reduced.

3. Development interventions that complement local subsistence strategies and are designed with and understood by the local people have a far better chance of ultimate success.

Thus, it is important to consider how the local inhabitants perceive, classify, value, and utilize their plant resources and how rights to plant resources are acquired, maintained, and transmitted.

In assessing the current use of plant resources by a human population, the social scientist typically would be concerned with two aspects, which are approached in different ways. The approaches are described in section 7.1.

The way in which people view, identify, and classify plant resources is the subject of ethnobotany, a branch of cultural anthropology. Ethnobotany involves the exhaustive use of native informants in the field as well as careful botanical identification. It requires a thorough knowledge of botany, linguistic techniques, and the native language of the people under study. It is time consuming and generally beyond the bounds of project or preproject development. In some cases, ethnobotanical studies will already have been carried out in the targeted region. Studies in ethnobotany show that not all people classify plants and other domains of nature in the same way.

Ethnobotanical knowledge is useful for mapping out the botanical world from the point of view of natives who have intimate contact with it. It can guide the investigator to the existence of unrecognized resources or prevent the unintentional destruction of resources valued by the indigenous population. For example, Brent Berlin of the University of California at Berkeley (personal communication), in the course of ethnobotanical research among the Aguaruna of Peru, discovered the existence of an abundant but previously undescribed tree species which is highly valued by the Indians as the principal source of glaze for their pottery. Even where a full-scale ethnobotanical study cannot be carried out, field investigators should concern themselves with some of the basic questions addressed by the discipline.

Human ecologists (whose training may have been in anthropology, geography, rangeland management, or another discipline) are concerned with the actual use and management of plant resources. They are highly field-oriented and make use of techniques from anthropology, ecology, and economics, such as interviewing, social surveys, time budgets, and soil samples. They observe how people utilize, transform, and exchange resources, and what additional features of culture are related to environmental adaptation. In one celebrated study, Rappaport (1968) suggested that the Maring people of highland New Guinea regulate their environment through their cycle of ritual pig slaughter.

Section 7.2 describes methods for analyzing human social response to changes in their environment. Social arrangements are highly sensitive to environmental perturbations as changes occur in access to resources, skills appropriate to resource capture, technology required, and so on. Social scientists have developed analytic categories for gauging the sociocultural response to environmental changes. In addition to these qualitative measures, there are many qualitative approaches to the assessment of change, including economic, demographic, nutritional, and other measures.

Both quantitative and qualitative approaches are important because the status of the social system may provide important indicators to the ecosystem of which it is a part. For studying the use of plants and for

IV. PLANTS

assessing change, one can utilize the three analytical methods outlined in chapter 5: comparative analysis (for instance, of homologous adaptive situations in different regions), cross-sectional analysis (of different communities exposed to different influences within the same general region), and longitudinal analysis (of changes in situ).

7.1 Investigating Plant Utilization Strategies

There are several methods used in field work. Interviews are very important.

7.1.1 General guidelines for conducting interviews.

The interview is a fundamental technique for data acquisition in the social sciences. Virtually any conversation with a knowledgeable person can be considered an interview and can yield invaluable information about human use of plant resources, provided a few precautions are taken. One important point to keep in mind is to choose informants carefully. Not all "natives" are equally well informed about their environments. For example, Brent Berlin (personal communication) found that Aguaruna men were systematically better informants than women when asked about wild food resources, while women had more detailed knowledge of domesticated plants. Interviewers must beware of respondents who are too anxious to please, especially those who would prefer to portray their compatriots in a favorable light to an outsider. The following general guidelines are recommended:

1) Specific time should be allotted for interviews; that is, they should not be combined with other study activities.

2) Interviews should be conducted near the informant's home and/or activity area and preferably in the local language.

3) Use of recording devices and constant note-taking should be avoided as they might inhibit responses.

4) Interviews should not be rushed; for the best results, rapport, trust, and interest must be established.

5) The first several hours of an interview should be unstructured to allow the informant time to develop his or her own frame of reference and indicate which plant resources are significant to him or her.

6) After the frame of reference has been established, the interviewer can carefully probe inconsistencies and hesitant answers, inquire about specific topics not covered, and follow up on subsidiary issues.

7) Direct questions should be posed in the frame of reference established by the informant, using similar terminology if possible.

8) To verify answers on key subjects, duplicate questions can be posed with subtle rephrasing. For example: "How many weeks do you spend clearing gardens each year?", and later, "How long did you spend clearing gardens last year?"

9) Repetitive interviews may follow a predetermined schedule of questions, provided the format is flexible enough to allow elaboration by the informant or follow-up questions by the interviewer.

Interviewers may often wonder whether the informants' beliefs about the efficacy of plants for certain purposes are true. This is a legitimate concern and should be pursued, but it should not obtrude in interviews. At this stage of investigation it is important to know primarily what the indigenous people value and what they value it for. It is also appropriate to explore the depth of their feelings (which may range from shallow to profound) and whether they believe there are alternatives to the way they do things. It is not uncommon for investigators to place too much stress on the weight of tradition and to overvalue the commitment members of a society have given to a lifeway. Some native people may believe in efficacy of certain plants as medicines but may also be willing to accept Western pharmaceuticals as equally effective.

7.1.2 Kinds of information required for plant resource use.

The initial objective of a survey to determine indigenous plant resource utilization practices is to compile an annotated list of those resources based on interviews and observations. People use plants for a tremendous variety of purposes. Ethnobotanical surveys should consider all food plants (cultivated and wild) as well as plants used for medicinal and magical purposes, building materials, fibers, clothing, tools, fuel, and animal feed. Care should be taken not to overlook infrequently used resources, such as foods gathered only during droughts or other emergencies. Some plant species may not be harvested but may be important as indicators of environmental conditions, such as the readiness of the soil for planting, or of the likelihood of finding certain animal species nearby. To the extent feasible, the investigator should find out the classes or categories to which different species belong. The classification may be very different from that used in Western botany, but may reveal important functional and adaptive aspects of plant use.

Not only individual species but also entire types may be significant to native groups as rangeland for livestock, hunting areas, escape or buffer zones from enemies, burial grounds, ritually or socially significant meeting places, potential agricultural plots, and so on.

For each species identified by the informant as useful, the following information should be compiled:

1) Local name of the plant (when possible, a specimen should be collected to determine the scientific name; flowers and/or fruits are usually required for proper identification).

2) What it is used for (multiple uses are common).

3) Where it is found (both physical location and distance traveled to obtain it; the latter can be an indicator of the intensity or potential intensity of use).

4) Frequency of use (daily, weekly, monthly, seasonal).

5) Quantities used (per person, family, animal, and so on).

6) Techniques used for collecting, propagating, cultivating, harvesting, processing, storing, and preparation. Food systems, including their reliability, nutritional quality, and potential productivity, are influenced by all these treatments.

Once the basic inventory of plant resource utilization has been completed, preferably by a combination of interviews and observation for verification, this information must be expressed in terms more useful to development planners. Two methods for this are summarized below. The first uses time expenditures to indicate relative importance; the second uses economic analysis to provide values applicable in cost-benefit equations. In some situations, quantitative methods for measuring human responses to environmental change (section 7.2.2) might also be useful.

7.1.3 Time allocation method.

Perhaps the single most accurate indicator of the importance of a plant resource is the amount of time people spend foraging for, cultivating, harvesting or processing it. When constructing time budgets, the level of detail depends on the resources at the disposal of the investigator.

The quickest and least expensive way to compile information on time allocation is to construct a *calendar of annual events* based on interviews. The investigator categorizes plant-related behavior into activities recognized by the informants (for example, forest clearing for gardens), then inquires how much time is spent on them. Some effort should be made to distinguish between the time actually spent at a task and the elapsed time over which the task is performed; because people usually do other things interspersed with a given task. Time allocation information from interviews should be verified insofar as possible by on-site observation. When time and budget constraints limit the investigator, key activities or particularly time-consuming tasks (such as weeding gardens or preparing manioc flour) should be emphasized.

A more accurate way to measure time allocation is by direct observation at selected periods over the annual cycle. These periods should be representative of the major seasonal activities such as planting, cultivating, watering, and harvesting. Records should be kept on the time spent on various activities by the "typical" people being observed. Altered behavior during the observation period and small sample size may limit the effectiveness of this technique.

Recently, a number of field workers have used random spot check sampling to study time allocation in entire communities (Johnson, 1975; Gross et al., 1979; Hames, 1978). Ideally, sampling is done at regular intervals for a year to obtain information on the annual cycle. This greatly increases the validity of the survey although it is costly in observer time. Random sampling has the advantage of yielding information on the full spectrum of activities, including those which are not the specific subject of the investigation and those which informants omit in their descriptions. Such data are potentially important if some change is to be introduced in the area because they will provide some idea of the trade-offs that will be involved. For example, if a plan is formulated to introduce high-yielding crop varieties requiring higher labor inputs, the increased labor time must be compensated by reducing labor time at other activities, such as hunting or trade.

No matter how they are obtained, time allocation measurements are extremely useful. They reflect the importance of a resource and provide a way of measuring the efficiency of production. By combining time allocation and yield data, the yield per unit effort can be calculated for such diverse activities as agricultural production (such as threshing), preparation (cooking), and hunting-gathering.

A very precise technique for selecting households for interviews and other field data collection is area frame sampling. In this approach, environmental characteristics are the basis for stratification; within strata, count units are identified and selected randomly as the basis for field work. This permits statistical analysis of the data. The approach was used by Fanale (1982) in analyzing herding patterns and pasture histories of Navaho Indians. Area frames have been built in a number of developing countries, and these would be useful for a baseline study of plant use and of property rights.

7.1.4 Economic analysis methods.

Standard methods for economic analysis include models of decision making which anthropologists have found useful in examining how people arrive at solutions to problems. These are applicable to virtually any situation where choices are made concerning resource allocation, regardless of whether the economy is monetized (Barlett, 1980; Chibnik, 1980; Ortiz, 1973).

When making decisions regarding development programs, planners have sometimes ignored the nonmonetized sectors of the economy, presuming them to be economic zeros. However, many people in underdeveloped areas hunt, gather, plant, and harvest food and other goods for themselves, and their families and for exchange without any currency being involved. Such production is usually not included in such measures as gross national product (GNP) or per-capita income, even though its cash equivalent may have been substantial. It is therefore important to consider the nonmonetized sector of a regional economy. Some programs designed to raise incomes in a given region have actually be detrimental to the natives' ability to feed and otherwise provide for themselves (Bodley, 1975).

This same concept applies to the land. If land is not contributing to recognized commodities and products such as agricultural crops or timber, it is often assumed in project planning that any development will be beneficial. Even if we ignore the "ecological services" of this vacant land for soil stability, biological diversity and nutrient exchange, we must consider its possible utilization by indigenous or even transient peoples. Economic analyses in these situations can be facilitated by the cross-sectional analysis described in section 5.3.2.

7.2 Methods for Determining Humans' Responses to Changes in Their Environment

The foregoing review of cultural anthropological methods was focused on determining the status quo: how people are currently living and utilizing the plant and other resources of their environment.

Determining the responses of humans to change in their environment, especially those induced by development programs, is also germane. Such determinations can be used to evaluate the results of project interventions or, preferably, to predict beforehand what the human responses will be. This is particularly important because of the central role of plant resources in subsistence strategies.

If the impact of development will be quite localized, acquiring adequate data for evaluation poses no great problems. If, as is more commonly the case, the action will have regional consequences, a stratified sampling design must be employed.

Sampling designs for determining human response to environmental changes are complicated by the fact that variation in human communities is not a simple function of environmental variation. Human activities and response patterns vary in a relatively homogeneous environment. Conversely, there may be a degree of uniformity across relatively sharp environmental gradients. Since the resource exploitations are culturally based, sampling must consider: (i) how many different ethnic groups occupy or exploit the area of concern and (ii) the relations of these groups to each other and to outside groups and institutions. Ethnic groups characteristically differ in their relations to the environment, and the investigator must be alert to such differences in order to include all socially significant resource utilizations in the sample. Interactions between ethnic groups can be quite varied. The adaptation of a particular group to an environmental zone may depend on its relations not only to resources but also to other groups in the area. For example, the transhumant Gujars in Swat State, Pakistan, have a mutualistic relation with the settled Kohistanis and frequently utilize summer pastures on the high slopes of the western part of the region. In the east, however, Kohistani herders require all of the summer pasture available and there the relationship between the same two ethnic groups is different, that is, exclusivistic (Barth, 1956).

Obviously, there are many possibilities for bias in sample site selection (such as easy accessability) and data summaries should include statements about these biases.

7.2.1 Qualitative methods for measuring human responses to change.

Qualitative measures of responses have the advantages of being relatively inexpensive and useful if conducted by a skilled and experienced individual. Such a person can discover in a few months what survey scientists might require a year to learn. The analytic categories are primarily part of the data language of social anthropology. Changes in the organizational profile of societies are diagnostic of changes in important processes, for instance, in responses to innovation. Any account of social response must begin with an account of the kinds of social units or groups in a society and how they function. For example, agricultural decisions may be made by a village headman in some societies, by household heads in others, and by mothers with young children in others. A development planner will profit from such knowledge in designing a program of intervention.

In virtually all societies, there are multiple channels of information and decision-making, ranging from the household up to the national government. In some situations, state controls are preponderant, as when agricultural credit is extended by a central bank only when farmers submit soil samples for tests. The existence of rival factions will strongly affect how information, techniques, or services will be distributed. Factions may be based on families, lineages, clans, villages, or political parties. Other important concepts of social organization are those of ownership and exchange. A common mistake made by Westerners is to assume that Western standards and definitions of ownership obtain elsewhere. In some societies, private ownership of grazing land, water supply, building materials, and so on does not exist. Moreover, exchange takes many forms other than monetized market exchange.

7.2.2 Quantitative methods for measuring changes.

There are numerous well-established methods for quantifying measurements of change (Blalock, 1960; Pelto and Pelto, 1978; Johnson, 1975). Some of them are particularly appropriate for measuring adaptive response to environmental changes. These are briefly discussed in the following paragraphs in relation to cash, nutrition, demography, and census.

Cash indicators. Cash-based measures of response to ecological change require significant cash flow within the population being studied, although it need not have a totally monetized economy. Agricultural modernization often leads to increased cash flow in a region. The amount and distribution of this cash flow can be quite revealing of the magnitude of changes.

Input-output analysis can be performed to determine the value of goods and services entering or leaving a delimited area or a particular industry (for instance, commercial timbering). Wherever cash income is the result of a substitution of one activity or mode of exchange for another, it is important to determine the equivalent value of the goods or work that was displaced and did not enter the money economy. For example, if subsistence farmers are recruited as salaried workers in a logging operation, the value of the lost food production and/or the increased activity costs must be deducted from the gross cash influx in order to properly assess the net effect of the change. This can be done on a regional, community, or household basis.

Another useful device for measuring impacts is the household budget. In monetized economies, income allocation (for food, shelter, clothing, education, and so on) is highly sensitive to fluctuations in employment, commodity prices, and other variables. A skilled interviewer can compile a substantial amount of data with only a modest expenditure of time, provided the necessary rapport to inquire into household budgets can be established.

Changes in income distribution relative to the total population can reveal the types of people who benefit from development programs. However, if no assessment was made of income distribution, it may be difficult to detect change over time with a single survey unless appropriate cross-sectional controls can be applied.

Nutritional indicators. Use of nutritional measures of environmental response is growing because environmental changes often have direct feedback on diet and nutrition and indirect feedback as a result of income redistribution. One major area of change involves the availability and distribution of nutrients. In many cases, destruction of forests has reduced the availability of wild plant and animal species used as food. Such differences can be detected in dietary surveys. The simplest kind of dietary survey involves asking a number of informants about the availability of certain foods in their locale. A more reliable method, "inventory recall," uses a checklist to determine whether a particular food was consumed within a specified recall period (typically 24 hours). Analysis of such data can yield quantified information about the presence or absence of certain nutrients in the diet, but it is difficult to express the results in terms of calories, grams of protein, or similar measures. The ultimate in dietary surveys involves direct observation of food consumption and meals in the homes of informants. This approach is time-consuming and intrusive and can influence the results.

The nutritional status of infants and growing children is a particularly sensitive indicator of overall health status and is highly reflective of environmental changes that affect nutrient availability, parasite vectors, activity levels, and so on. Gross and Underwood (1971) showed a correlation between income distribution, growth rates of stunted children, and the introduction of a new cash crop in northeast Brazil. Dewey (1981) used a similar approach in a study of a government-sponsored colonization program in Tabasco, Mexico.

Demographic indicators. If the investigator has access to raw data from the local census tract and can distinguish between responses to environmental alterations and those resulting from commodity price fluctuations, taxation policies, and so on, demographic data can be useful. Such measures as migration rates, migrant origins and destinations, mortality, fertility, fecundity, and morbidity can all be revealing. Loss of land due to a development intervention will result in migration to other areas, often cities. Fertility rates may be sensitive to the availability of resources, the demand for labor, mortality, and the desire for security in old age (White, 1973).

Morbidity may also be a sensitive indicator of response to environmental changes. It may reflect the introduction or removal of a particular disease vector caused by the modification of plant cover or some other variable. Morbidity is also sensitive to changes in nutritional status (Dewey, 1981).

Other census-based measures. Included in this category are size and construction of housing; availability of sanitary facilities such as flush toilets and running water, availability of educational, recreational and public health facilities, and level of employment. Changes in these and related variables may be indirect measures of alterations in the relations between human communities and their environment.

Classification of Plant Resources

8.1 The Importance of Classifications

Resource managers and development officers continually make decisions concerning the allocation and use, management, development, and/or restoration of natural resources. It is therefore imperative that they be able to communicate effectively with natural resource scientists and ecologists, who often translate their knowledge into maps, descriptive legends, and statistics. The mechanism for recognizing, describing, labeling, and comparing these resources is a classification system. Classifications allow us to put discrete and unique "handles" on the often rather indistinct components of natural resource systems. This is particularly important for resource inventories, which typically encompass great numbers of different resources and their variations. Even tabulating the number of hectares of forest types or agricultural crops is initally dependent on some classification system, which provides labels for the different categories of interest and specific criteria for recognizing them. This section concerns the concepts and utility of plant resource classification systems.

Everyone with some biological curiosity has observed that kinds of plants vary from place to place. Keen observers, even if they are not botanists, have noted that similar kinds of plants often seem to be repeated. These plant assemblages are frequently associated with some unique and discernible kind of environment such as rocky hills or river floodplains. Thus, the relations between plants and their environments are sometimes very obvious. In other instances, these relations are subtle and require an experienced plant ecologist to unravel them and define the differentiating characteristics of each assemblage.

The fact that vegetation occurs across the landscape in repeated units, or homologs, with very similar species composition and appearance, indicates that the effective environment is also quite similar. All species have certain physiological tolerances to environmental conditions such as temperature, available water, certain mineral elements, physical soil properties, and solar radiation. Although the ranges of these tolerances are quite variable between species, when considering assemblages of plants, the ranges of the environmental parameters in areas of similarity can be ascertained. This has important management implications. Management practices developed for a particular ecological site are directly applicable to other areas with similar effective environments. Thus, the real importance of classification is that it allows workers in different places to recognize areas of ecological similarity and communicate information on suitabilities, limitations, or management potential in a consistent manner.

Another important application of classification systems is in stratifying ecological types for sampling and monitoring. Vegetation changes over time, and it is necessary to monitor and document these changes in plant communities, land use, and landscape features in order to recognize trends so that planned action programs can be modified when necessary to achieve their objectives and avoid environmental deterioration. In documenting changes, it is useful to determine which are caused by humans (or by factors over which humans can exert some control) and which are natural. A good ecological stratification based on a sound classification system is needed in order to separate statistical errors due to sample site comparability from those due to time-related phenomena. Otherwise, important changes may go undetected.

In many international assistance projects, concern has been expressed about incompatibility of resource inventories where different projects join or overlap. This problem reduces the utility of the data for amalgamation into a broader regional picture. Inconsistencies may flag basic weaknesses in a particular inventory. They may be due to observation of different features or to differences in classification philosophy. One solution is to adopt a consistent concept of classification and some degree of standardization in a hierarchical legend system. However, excessive standardization will limit the flexibility necessary for addressing special information needs and improvements in inventory procedures.

Providing for the development of a suitable classification is frequently overlooked in project planning and budgeting. It must be done if a resource inventory is to be comprehensible, particularly in developing countries, where the vegetation is often not well known or has not been classified beyond the broadest levels. Sometime the existing classifications consist only of plant species listed in non-ecological "family" groups. These are not appropriate for detailed resource inventories or even for most reconnaissance surveys.

8.2 Vegetation Classification Terminology

In plant science, different terminology is used when referring to individuals or populations of individuals of the same kind than when referring to a "natural" grouping of different kinds of individuals. When the reference is general and one is speaking of individuals collectively without thinking of how they may be classified ecologically, morphologically, genetically, or in terms of evolutionary relations, the term used is *flora* or merely *plants*. When plant life is considered in the context of genetic and morphological similarity or in some phylogenetic or evolutionary context (but without respect to its environment), *varieties* (*cultivars* among domesticated plants), *species*, *genus*, and *family* are used in moving up the taxonomic hierarchy. When referring to plant life in the ecological context, terms such as *ecosystems*, *vegetation* or *plant communities* refer to assemblages of plants that have boundaries in space and time.

Plant communities. The concept "plant community" is perhaps the single most important basis for vegetation classification. A plant community is a recognizable assemblage of plants that grow together on a specific landscape because of generally similar tolerances for the habitat conditions that prevail there. For example, a woodland community dominated by a particular species of tree may also have a uniform understory of typically associated shrub species, a ground layer dominated by specific grasses, plus various forbs. Further, these species typically occur in the same relative proportions. Plant communities exist because the evolutionary interaction of biotic and abiotic factors resulted in segregation of species with similar tolerances in recognizable aggregations of plants. Wherever the effective environment (ecological summation of the habitat factors) falls within the tolerance range of each species, the entire group will tend to occur. This phenomenon leads to four corollaries that are of great significance to the analyst and manager of renewable natural resources:

1) Each kind of community is distinguished by a group of species showing coordinated patterns of relative abundance over the landscape.

2) Any large unit of vegetation is a mosaic of plant communities, whose distribution is governed by a corresponding mosaic of habitats.

3) The more heterogeneous the environment of an area, the more numerous are the kinds of plant communities it sustains.

4) Vegetation is a valuable criterion of degrees of similarity and difference among habitats (Daubenmire, 1968).

The term plant community can be used either in the concrete sense, referring to a single stand or assemblage of plants, or in the abstract sense, referring to a classification category in the array of plant community types that occurs across particular landscapes.

Ecotones. An ecotone is the zone gradation or transition from one plant community to another. Ecotones may be sharp in response to abrupt changes in the environment (radical changes in topography or soil type) or gradual in response to gradual changes in the environment.

As one considers an individual ecotone or transition between two communities across a gradual environmental gradient, the changing sequence in the vegetation from the presence to the absence of certain species represents a gradient in species composition across the ecotone from community A to community B. The ecotone is between concrete entities or stands of vegetation.

When one takes this concept to the abstract world of classification, numerous philosophical and conceptual problems may arise. "Stands" are concrete and real, whereas classification "entities" are abstract and unreal; to some degree they are even a function of one's approach to classification.

In mapping and landscape analysis the abrupt ecotone presents no problems, but gradual ecotones can present substantial challenges. One approach is to somewhat arbitrarily (but using the best possible judgment) draw a solid line somewhere in the transition zone. A second approach is to devise some map legend symbol, such as a dotted line, to indicate the ecotonal situation. A third approach is to consider stands that are obviously intermediate in their characteristics between two classification entities to be *intergrades*, if it is indeed necessary to separately map certain of them. Normally, intergrades will have more characteristics in common with one of their associated classes and, in practical mapping, can be included in that class. A fourth approach is to incorporate ecotonal types into the classification system, with attendant recognition criteria, then map corresponding areas as separate types.

Experience in Mali shows how a classification can be designed to accommodate ecotonal situations. In Mali there is often a very broad transition zone (over 100 km) between the Sahel and Sudan biogeographic provinces. In this area, there are many distinct plant communities which could be mapped, whose dominant species include indicator plants from both provinces. This was accommodated in the classification by creating a third class at the province level for Sudano-Sahelian transition communities. The principle is applicable to any scale of investigation, but requires a very experienced plant ecologist and usually the agglomerative approach to classification.

Ecotones are biologically significant areas. By definition, they include species from both adjacent "pure" plant communities and thus often exhibit very high species diversity. They are often considered prime wildlife habitat because they provide the "services" of two or more habitat types in a single area. Thus, it is a mistake to adopt an arbitrary approach to dealing with ecotones. Their significance to the ultimate objectives of the project should be considered in due course.

8.3 Types of Classifications

One may take a utilitarian or an ecological approach to classification development. The utilitarian approach is often easier, but it has the disadvantage that it is strongly oriented to a particular land use or management interest and rarely has much value except for this purpose. For example, if the concern is forestry or fuel wood, a simple classification could be developed that describes plant species and tree densities and ignores everything else. An ecological approach has a broader spectrum of uses because it strives to depict the natural characteristics of the landscape and to address the entire ecosystem rather than a single component.

In the ecological approach to vegetation classification there are two points of departure. One may work from the general (or broad classes) to the specific or from the specific to the general. The former approach is called *divisive* and the latter *agglomerative* approach. In most cases a divisive approach will work better and be much more efficient in developing nations, because too little is ordinarily known about the vegetation to use the agglomerative method. A divisive approach has been developed and used for about 15 years in resource inventory work in the United States, Canada, and four developing nations (Poulton, 1972).

The most appropriate classification format for natural resource inventories is a hierarchy. Hierarchical classifications are designed to provide information at different levels of detail, depending on whether it is needed for a site-specific management action, regional inquiries pertaining to site selection, or national planning designs. Information is disaggregated as one moves down the hierarchy from generalized levels to levels of increased detail (divisive approach). Conversely, detailed information is progressively aggregated and generalized as one moves up the hierarchy (agglomerative approach). Each class at any level is exclusive of all others. Hierarchies are also open-ended, readily accommodating new information. Although the divisive approach is better suited for constructing a classification in developing countries, once this task is accomplished, the flexibility of the hierarchical format provides for either divisive or agglomerative information retrieval at any level, depending on user need.

8.4 Sampling for Classification Development

Sampling is done during classification development because the classification units must be characterized in a descriptive legend. For reconnaissance surveys this can usually be done in narrative form with little or no quantitative data; the usual procedure is to rank the observed parameters. As map scale and examination intensity increase, however, measurement data become progressively more critical and sampling design more important.

In vegetation classification, the major variation is from area to area or among stands. Therefore, it is essential to develop an adequate sample size in terms of number of stands examined. If this is done, even at the expense of measurement accuracy within a stand, classification results will be much more defensible. Objective measures of individual parameters with high standards of accuracy and precision for each stand are not necessary for classification. It is much more important to examine enough separate stands on a multifactor basis to detect sufficiently strong clustering around loci in multidimensional space--in other words, to define the class. It is this clustering of the features of the ecosystem that establishes valid and useful classifications.

Stratification of the sampling location is also important for vegetation classification development. Samples are stratified to ensure that each individual data set represent a unique and homogeneous stand. For each parameter examined, this approach gives one valid data point in the distribution of each of the multidimensional features that characterize the class. Data analysis enables identification of discriminating features which can be used to recognize each ecosystem or classification category. Stratification of sample locations is based on species prominence, composition, and community structure within the stand.

Ecological stratification is difficult for many people because they perceive plants only as individuals and not within a community framework. For successful classification development this difference in scientific capability must be recognized, and only experienced plant community ecologists should make decisions about sampling stratification.

Stratification can also be difficult because some species tend to have a clustered distribution. The question arises of whether this represents a stratum difference or is the normal variation in that particular community parameter. This can be answered by considering correlations with other ecosystem parameters. If both landform and soils are relatively uniform across the patterns of plant clustering, then only a single stratum is involved. Otherwise, more than one ecosystem is represented and a separate sample is called for in each stratum as defined by the relative homogeneity of landform, soil, and vegetation.

8.5 Key Considerations in the Development of a Vegetation Classification System

The first consideration is whether the classification will be ecologically based (that is, as natural as possible) or purely utilitarian. These terms were defined in section 8.3. For resource inventory purposes, and especially for baseline studies designed to provide a data base for analysis and management of multiple natural resources, the ecological approach is superior.

Second, since resource inventories encompass geographic areas where nonvegetative subjects are also found across the landscape, it is beneficial to have a classification system suitable to any aspect of the earth's landscape. Thus, barren lands and water resources should be included, as well as such cultural features as cities, extractive industries, and agriculture. Of particular concern for vegetation inventory projects in many arid and semiarid developing countries is the distinction between sparsely vegetated and barren lands. Commonly encountered barren lands are rocklands, sand dunes, and salt flats. A practical threshold for this distinction is 5 percent vegetative cover. Below this amount, one must be careful in ascribing economic value to the plant resources.

Third, since abiotic features of the landscape figure prominently in natural resource identification and mapping, an ecological classification should incorporate pertinent aspects of thes features.

Fourth, and a very important consideration, is that criteria for distinguishing between different categories and levels must be logical and parallel. Only when this is the case can a classification system be consistently applied and summarized at a particular level of detail.

The fifth and sixth considerations are related to the advantages of hierarchical formats, as described in section 8.3. Besides making possible the agglomerative and divisive approaches, such systems are open-ended and can readily accommodate new information. If a digital legend is developed, the classification is also amenable to automated data processing.

The seventh and last consideration is the compatibility of the classification with remotely sensed information. We saw in chapter 3 (levels of assessment) and chapter 4 (remote sensing) that this technology is well

8.6 Example of a Hierarchical Resource Classification System

The example presented in Table 8.1 was originally developed by Poulton and his graduate students at Oregon State University under the auspices of the NASA Earth Resources Technology Program. It was based on intensive field evaluation in Oregon and Arizona, with an emphasis on practical usage (Poulton, 1972). Over the past 10 years, it has been refined and successfully applied to resource inventories in such diverse areas as the western United States, Alaska, Canada, Iran, Oman, Tanzania, and Mali. The practicality and usability of the system has been demonstrated for vegetation communities ranging from tropical rain forest to a variety of arid and semiarid lands and even tundra.

Table 8.1 is an abridged version of this classification. Furthermore, it represents only the symbolic portion of the entire classification, which also includes descriptive and interpretive components. However, it provides an overview of the organization of the system. It is hoped that this example will save potential users time and effort when they are deciding what kind of classification system and format might be most appropriate for a resource inventory. The example has the following primary classes.

```
100 - Barren lands
200 - Water resources
300 - Natural vegetation
400 - Cultural vegetation
500 - Agricultural production
600 - Urban, industrial, transportation
700 - Extractive industry, natural disasters
800 - Open
900 - Obscured lands
```

Table 8.1. Earth surface feature and land use classification: symbolic and technical legend classes [Abridged from Poulton (1972)]

```
Primary classes
    Secondary Classes
        Tertiary classes
            Quaternary classes

100 - Barren lands
    110 - Dry or intermittent lake basins
    120 - Aeolian barrens (other than beaches or beach sands).
    130 - Rocklands
        131 - Bedrock outcrops (intrusive and erosion-bared strata)
        132 - Extrusive igneous (lava flows, cinder, and ash)
        133 - Gravels, stones, cobbles, and boulders (usually transported)
        134 - Scarps, talus, and/or colluvium
        135 - Patterned rockland (nets or stripes)
```

140 - Shorelines, beaches, tide flats, river banks
150 - Badlands (barren silts and clays, related metamorphic rocks and erosional wastes)
160 - Slicks (saline, alkali, soil structural barren)
170 - Mass movement
190 - Undifferentiated complexes of barren lands

200 - Water resources
    210 - Ponds, lakes, and reservoirs
        211 - Natural lakes and ponds
        212 - Man-made reservoirs and ponds
    220 - Water courses
    230 - Seeps, springs, and wells
    240 - Lagoons and bayous
    250 - Estuaries
    260 - Bays and coves
    270 - Oceans, seas, and gulfs
    280 - Snow and ice
    290 - Undifferentiated water resources

300 - Natural vegetation
    310 - Herbaceous types
        311 - Grassland, steppe, and prarie
        312 - Meadows
        313 - Marshes
        319 - Undifferentiated herbaceous types
    320 - Shrub/scrub types
        321 - Microphyllous, nonthorny scrub
        322 - Microphyllous thorn scrub
            322.1 Acacia
        323 - Succulent and cactus scrub
        324 - Halophytic shrub
        325 - Shrub steppe
        326 - Sclerophyllous shrub
        327 - Macrophyllous shrub
        328 - Microphyllous dwarf shrub
            328.1 Spruce-fir
            328.2 Mountain heath
    330 - Savanna-like types
        331 - Tall shrub/scrub over herbaceous layer
        332 - Broad-leafed tree over herbaceous layer
        333 - Coniferous tree over herbaceous layer
        334 - Mixed tree over herbeaceous layer
        335 - Broad-leafed tree over low shrub layer
        337 - Mixed tree over low shrub layer
        339 - Undifferentiated savanna-like types
    340 - Forest and woodland types
        341 - Conifer forests
        342 - Broad-leafed forests
        343 - Conifer/broad-leafed mixed forests and woodlands
        344 - Broad-leafed/conifer mixed forests and woodlands
        349 - Undifferentiated forests and woodland types

400 - Cultural vegetation
    410 - Cultural herbaceous types
    420 - Cultural shrub/scrub types

                430 - Cultural forest and woodlands
                490 - Undifferentiated cultural vegetation types

500 - Agricultural production
        510 - Rain-fed agriculture
                511 - Cereal grain
                        511.1 Sorghum
                        511.2 Wheat
                512 - Tree, shrub, and vine crops
                513 - Mixed cropping
        520 - Irrigated agriculture
                521 - Field crops
                522 - Vegetable and truck crops
                523 - Pasture
        530 - Agriculture
        580 - Agricultural production facilities
        590 - Undifferentiated agriculture

600 - Urban, industrial, and transportion
        Developed regions   *Less developed regions*
        610 - Residential   610 - Cities
        620 - Commercial/services   620 - Towns/villages
        630 - Industrial   630 - Scattered villages
        690 - Undifferentiated   690 - Undifferentiated

700 - Extractable industry and natural disasters
        710 - Nonrenewable resource extraction
                711 - Sand and gravel
                712 - Rock quarry
                715 - Metals
                719 - Undifferentiated
        730 - Natural disasters
                731 - Earthslides and earthquakes
                732 - Fires (burn areas)
                733 - Water (floods)
                739 - Undifferentiated

900 - Obscured land
        910 - Clouds and fog
        920 - Shadows
                921 - Cloud
                922 - Relief
        930 - Smoke, haze, and smog
        940 - Duststorms and sandstorms
        990 - Undifferentiated obscured lands

Sources of Information

9.1 Local Sources of Information

A variety of information sources in host countries can be relevant to foreign assistance projects. These include field journals, maps, photos, and reports, including final reports from previous assistance projects. Most of this information is found in universities, government agencies concerned with natural resource management, nongovernmental conservation organizations, and certain military offices (for instance, cartographic units). Identification of and access to this informaiton are usually through personal contacts and personal examination of the available materials. Although it may seem paradoxical, there is often more information of this type in the least developed countries due to the disproportionately large number of foreign assistance projects there. Acquisition of the information requires hard work of an investigative nature, but can be well worth the effort, considering the alternative means for obtaining the same information. Although the data and information are not always presented in standard units or accepted formats, they are usually valid and generally understood by the host country principals.

In almost every country there are intelligent, well-educated private individuals who are knowledgeable about the extent and availability of in-country information sources. These persons can be very helpful in identifying specific information or data and their exact location.

9.2 Archives

Another valuable source of data and information is archival repositories dating back to the times of colonial occupation. These include old government records, legal documents, and official proclamations and correspondence. Much of this historical information can be found in national capitals, but it is often incomplete and usually not cataloged for direct access. The best sources of colonial documents are the capitals of the former colonial powers, such as London, Paris, Brussels, Amsterdam, and Rome. Within host countries, a valuable source of information is the old government publishing houses and "official" newspaper offices. For example, for former British colonies, an excellent information source is found by examining the historical gazettes that reported all government actions. One almost has to see first-hand examples of the kinds of information extractable from these sources to fully appreciate their value for natural resource development and management.

9.3 Computer Libraries

In the United States, there are now library computer retrieval services that access specific documents by using key words. The cost of these services is usually only a few cents per citation. The retrieval product is typically a full literature citation, which may or may not be accompanied by an abstract or summary. If a complete copy of the text is desired, it can be obtained from libraries or from commercial reprint services.

Computer libraries are usually restricted to citations from scientific journals, books, symposia, and reports on certain government-sponsored research, most of which are also cataloged by the National Technical Information Service (NTIS). As a result, the information tends to represent the United States or Europe, it emphasizes peer-reviewed literature, and it omits the so-called gray or soft literature which can frequently be of greater value. In addition, access to the full range of pertinent material is through key words, and information that is not indexed is not retrievable.

9.4 Bibliographies

Topical bibliographies (indexed and/or annotated) are available on specialized subjects in nearly all areas of natural resources. Most have been published on a restricted basis and are limited to major university libraries. Frequently, one must consult a specialist in the discipline to learn of the existence of a bibliography on a particular topic.

9.5 Herbaria

In most countries there are universities, government agencies, museums, or individuals that maintain herbaria of the local flora. These are the most valid local sources for taxonomic identification and/or confirmation and should be used for these purposes. Botanical protocols suggest that visiting botanists contribute at least two sheets of material to the local herbarium when removing herbarium specimens from the country. Usually, herbaria and herbarium personnel are also valuable sources of information on the flora and vegetation of their country.

References Cited

Aldrich, R.C. 1981. Limits of aerial photography for multi-resource inventories. In *Arid lands resource inventories, an internatl. workshop.* USDA Gen. Tech. Rep. WO-28, pp. 221-227. Washington, DC: U.S. Department of Agriculture.

Anderson, W.H. and F.X. Wallner. 1978. *Small format aerial photography: A selected bibliography.* Sioux Falls, S.D.: USDI EROS Data Center.

Avery, T.E. 1977. *Interpretation of aerial photography.* 3rd ed. Minneapolis, Minn.: Burgess Publishing Co.

Barth, F. 1956. Ecologic relationships of ethnic groups in Swat, North Pakistan. *Amer. Anthro.* 58:1079-1089.

Barlett, P.F. 1980. *Agricultural decision making: Anthropological contributions to rural development.* New York: Academic Press.

Bement, Robert E. 1969. A stocking rate guide for beef cattle production on blue grama range. *Jour. of Range Management.* 22:83-86.

Berlin, B. 1982. Personal communication.

Blalock, H.M. 1960. *Social statistics.* New York: McGraw Hill.

Bodley, J.H. 1975. *Victims of progress.* Menlo Park, Calif.: Cummings Publishers.

Bonham, C.D., L.L. Larsen, and A. Morrison. 1981. *A summary of techniques for measurement of herbaceous and shrub production, cover and diversity on coal lands in the West.* Office of Surface Mining Region 5. Denver, Co. (Available from Range Science Dept., Colorodo State Univ., Ft. Collins, Co.)

Bonner, W.J. and J. Morgart. 1981. Landsat: a sampling frame for arid land inventories. In *Arid lands resource inventories, an internatl. workshop.* USDA Gen. Tech. Rep. WO-28. pp. 230-238. Washington, D.C.: U.S. Department of Agriculture.

Bonner, J. and J.E. Varner, eds. 1976. *Plant biochemistry.* 3rd ed. New York: Academic Press.

Boserup, E. 1965. *The conditions of agricultural growth.* Chicago: Aldine Pub. Co.

Brokensha, D.W., D.M. Warren and O. Werner. 1980. *Indigenous knowledge systems and development.* Lanham, Md.: Univ. Press of America.

Brockway, L. 1979. *Science and colonial expansion: The role of the Royal Botanical Gardens.* New York: Academic Press.

Catsinpoolas, N. and J. Drysdale. 1977. *Biological and biomedical applications of isoelectric focusing.* New York: Plenum Press.

Chibnik, M. 1980. Working out and working in: the choice between wage labor and cash cropping in rural Belize. *Amer. Ethnol.* 7(1): 86-105.

Colwell, R.N. 1983. *Manual or remote sensing.* 2nd ed. Falls Church, Va.: Amer. Soc. Photogrammetry.

Corbett, Q. 1978. Short duration grazing with steers - Texas style. *Rangeman's Jour.* 5:201-203.

Curtis, H. and N.S. Barnes. 1981. *Invitation to biology.* 3rd ed. New York: Worth Publishers.

Daubenmire, R. 1968. *Plant communities: A textbook of plant synecology.* New York: Harper and Row.

Dewey, K. 1981. Nutritional consequences of the transformation from subsistence to commercial agriculture in Tabasco, Mexico. *Human Ecology* 9(2): 151-188.

Estes, J.E. and L.W. Senger. 1974. *Remote sensing: techniques for environmental analysis.* Santa Barbara, Calif.: Hamilton Publishing Co.

Fanale, R. 1982. Navajo land and land management: A century of change. Unpub. PhD. Dissertation. Washington, D.C., Catholic University of America.

Furley, P. 1980. Development planning in Rondonia based on naturally renewable surveys. In *Land, people and planning in contemporary*

Amazonia, ed. F. Barbira-Scazzocchio. Occasional Pub. #3. Cambridge: Cambridge Univ. Ctn. Lat. Amer. Stud.

Geertz, C. 1963. *Agricultural involutions.* Berkeley: Univ. of Calif. Press.

Gladwin, C. 1976. A view of the Plan Puebla: an application of hierarchical decision models. *Amer. Jour. Agric. Econ.* 58(5): 881-887.

Goodland, R.J.A. and H.S. Irwin. 1975. *Amazon jungle: Green hell to red desert.* Amsterdam: Elsevier Scient. Pub. Co.

Goodwin, T.W. and E.I. Mercer. 1974. *Introduction to plant biochemistry.* Oxford. Pergamon Press.

Gross, D. 1970. Sisal and social structure in northeastern Brazil. Unpub. Doctoral Dissertation. New York, Columbia Univ.

Gross, D. and B.A. Underwood. 1971. Technological change and caloric costs: sisal agriculture in northeastern Brazil. *Amer. Anthro.* 73(3): 725-740.

Gross, D., G. Eiten, N.M. Flowers, M. Francisca L., M. Ritter and D.W. Werner. 1979. Ecology and acculturation among native peoples of central Brazil. *Science* 206:1043-1050.

Hames, R.B. 1978. A behavioral account of the division of labor among the Yekwana indians of southern Venezuela. Ph.D. Dissertation. Santa Barbara, Univ. of Calif.

Harborne, J.B. and C.F. Van Sumere, eds. 1975. *The chemistry and biochemistry of plant proteins.* New York: Academic Press.

Heady, H.F. 1975. *Rangeland management.* New York: McGraw Hill.

Honigman, J., ed. 1973. *Handbook of social and cultural anthropology.* Chicago: Rand McNally.

Johnson, A. 1975. Time allocation in a Machiguenga community. *Ethnology* 14(3): 301-310.

Krebs, C.J. 1972. *Ecology: The experimental analysis of distribution and abundance.* New York: Harper and Row.

Kudat, A., D. Bates and F.P. Conant. 1981. *Rural roads in Kenya.* Nairobi: Kenyan Ministry of Trans.

Kuwana, T. ed. 1980. *Physical methods in modern chemical analysis.* Vol 2. New York: Academic Press.

Langley, P.G. 1978. Remote sensing in multi-stage, multi-resource inventories. In *Proc. intergrated inventories of renewable natural resources workshop.* USDA Gen. Tech. Rep. RM-55, pp. 192-195. Washington, D.C.: U.S. Department of Agriculture.

L'Annunziata, M.F. 1979. *Radiotracers in agricultural chemistry.* New York: Academic Press.

Lee, R.B. 1969. !Kung bushman subsistence: an input-output analysis. In *Environment and cultural behavior*, ed. A.P. Vayda. Garden City, N.Y: The Natural History Press.

Lillesand, T.M. and R.W. Kiefer. 1979. *Remote sensing and image interpretation.* New York: John Wiley and Sons.

Love, R. M. 1961. The range: Natural plant communities or modified ecosystems? *J. British Grassland Soc.* 16(2): 89-99.

Margolis, M. 1973. *The moving frontier: social and economic change in a southern Brazil community.* Gainesville: Univ. of Florida Press.

McKell, C.M. and B.E. Norton. 1981. Management of arid land resources for domestic livestock forage. In *Arid land ecosystems.* Vol. II, eds. Goodall, D.W. and R.A. Perry. Cambridge: Cambridge Univ. Press.

McKell, C.M., J.P. Blaisdell, and J.R. Goodin, eds. 1972. Wildland shrubs - Their biology and utilization. USDA For. Serv. Gen. Tech. Rep. INT-1. Washington, D.C.: U.S. Department of Agriculture.

Meyer, M.P. and P.D. Grumstrup. 1978. *Operating manual for the Montana 35mm aerial photography system.* 2nd ed. St. Paul, Minn.: Univ. of Minn. Remote Sensing Laboratory.

Morris, C.J.O.R. and P. Morris, eds. 1975. *Separation methods in biochemistry.* 2nd ed. New York: John Wiley and Sons.

Moran, E. 1980. Mobility and resource use in Amazonia. In *Land people and planning in contemporary Amazonia*, ed. F. Barbira-Scazzocchio. Occas. Pub. #3. Cambridge: Cambridge Univ. Ctn. Lat. Amer. Stud.

Mouat, D.A., J.B. Bale, K.E. Foster and B.D. Treadwell. 1981. The use of remote sensing for an integrated inventory of a semi-arid area. *Jour. Arid Environ.* 4:169-179.

Nash, A. 1979. Practical problems of remote sensing activities in international assistance programs. In: *Proc. Remote sensing for natural resources symposium*, pp.303-312. Moscow, Id.: Univ. Idaho.

National Academy of Sciences. 1977. *Resource sensing from space: prospects for developing countries.* Washington, DC: NAS.

Northington, D.K. and J.R. Goodin. in press. *The botanical world.* St. Louis: C.V. Mosby Co.

Odum, E.P. 1971. *Fundamentals of ecology.* 3rd. ed. Philadelphia: W.B. Saunders Co.

Official methods of analysis of the assn. of official analytical chemists. 13th ed. 1980. Arlington, Va.: Assn. of Official Analytical Chemists.

Ortiz, S. 1973. *Uncertainties in peasant farming: A Colombian case.* N.Y.: Humanities Press.

Paine, D.P. 1981. *Aerial photography and image interpretation for resource management.* New York: John Wiley and Sons.

Panzer, K.F. and B. Rhody. 1981. Applicability of large scale aerial photography to the inventory of natural resources in the Sahel of Upper Volta. In *Arid lands resources inventories, an internatl. workshop.* USDA Gen. Tech. Rep. WO-28, pp. 287-299. Washington, DC: U.S. Department of Agriculture.

Pelto, P.J. and G.H. Pelto. 1978. *Anthropological research: the structure of inquiry.* 2nd ed. Cambridge: Harvard Univ. Press.

Plucknett, D.L. and N.J.H. Smith. 1982. Agricultural research and third world food production. *Science* 217:215-220.

Poulton, C.E. 1972. A comprehensive remote sensing legend system for the ecological characterization and annotation of natural and altered landscapes. In *Proc. 8th Internatl. Symp. on Remote Sensing of Environ.* Ann Arbor, Mich.: Environmental Research Institute of Michigan.

Poulton, C.E. 1979. Uses of landsat imagery in a resource analysis of the Masai steppe region in northern Tanzania. In *Proc. remote sensing for natural resources symposium*, pp. 361-392. Moscow, Id. Univ. of Idaho.

Pratt, D.J. and M.D. Gwynne. 1977. *Rangeland management and ecology in East Africa.* London: Hodder and Sloughton.

Prescott-Allen, R. and C. Prescott-Allen. 1982. The case for in site conservation of crop genetic resources. *Nat. Resource Jour.* 18:15-20.

Rappaport, R.A. 1968. *Pigs for the ancestors. Ritual in the ecology of a New Guinea people.* New Haven, Conn.: Yale Univ. Press.

Reeves, R.G., ed. 1975. *Manual of remote sensing.* Falls Church, Va.: Amer. Soc. Photogrammetry.

Rhody, B. 1981. Combined forest inventory using 35 and 70 mm aerial photography with an extensive ground survey in Pinus caribaea plantations in the Savannah Plain, Venezuela. In *Arid lands resources inventories, an internatl. workshop.* USDA Gen. Tech. Rep. WO-28. pp. 471-479. Washington, D.C.: U.S. Department of Agriculture.

Rhode, W.G. 1978. Potential applications of satellite imagery in some types of natural resource inventories. In *Proc. workshop on integrated inventories of renewable natural resources.* USDA Gen. Tech. Rep. RM-55. Washington, D.C.: USDA.

Sabins, F.F., Jr. 1978. *Remote sensing: principles and interpretation.* San Francisco: W.H. Freeman.

Sayn-Wittgenstein, L. 1979. The role of landsat in operational resource inventories. In *Proc. remote sensing for natural resources symposium*, pp. 419-423. Moscow, Id.: Univ. of Idaho.

Stein, S. 1957. *Vassouras: a Brazilian coffee county (1850 - 1900).* Cambridge: Harvard Univ. Press.

Stoddart, L.A., A.D. Smith, and T. Box. 1975. *Range management.* 3rd ed. New York: McGraw Hill.

Thomas, L.C. and G.J. Chamberlin. 1980. *Colorimetric chemical analytical methods.* 9th ed. New York: John Wiley and Sons.

Treadwell, B.D. and J. Buursink. 1981. The Mali land use project: a multiple resource inventory in west Africa. In *Arid lands resources inventories, an internatl. workshop.* USDA Gen. Tech. Rep. WO-28. pp. 159-164. Washington, D.C.: USDA.

White, B. 1973. Demand for labor and population growth in colonial Java. *Human Ecology* 1:217-236.

Williams, J.T. 1982. Genetic conservation of wild plants. *Nat. Resource Jour.* 18:14-15.

Willard, H.H., L.L. Merritt, Jr., J.A. Dean and F.A. Settle, Jr. 1981. *Instrument methods of analysis.* New York: Van Nostrand Reinhold.

Zweig, G. and J. Sherma, eds. 1980. *Analytical methods for pesticides and plant growth regulators.* Vol. 11. *Updated general techniques and additional pesticides.* New York: Academic Press.

Suggested Readings

HARVESTING METHODS (PRODUCTION)

Herbaceous

Box, T.W. 1960. Herbage production in four range plant communities in south Texas. *J. Range Manage.* 13:72-76.

Boyer, W.D. 1959. Harvesting and weighing vegetation. In *Techniques and methods of measuring understory vegetation*, pp. 11-16. USDA For. Serv. Southern For. Exp. Sta. and Southeastern For. Exp. Sta.

Campbell, R.S. and J.T. Cassady. 1949. Determining forage weight on southeastern forest ranges. *J. Range Manage.* 2:30-32.

Dasmann, W.P. 1948. A critical review of range survey methods and their application to deer range management. *Calif. Fish and Game* 34:189-207.

Francis, R.C., C.V. Baker, G.M. Van Dyne and J.D. Gustafson. 1971. A study of the weight estimator method of botanical analysis. IBP Tech. Report No. 117. Fort Collins, Co: Natural Resource Ecology Laboratory

Frischknecht, N.C. and A.P. Plummer. 1949. A simplified technique for determining herbage production on range and pasture land. *Agron. J.* 41:63-65.

Getz, L.L. 1960. Standing crop of herbaceous vegetation in southern Michigan. *Ecology* 41:393-395.

Hilman, J.B. 1959. Determination of herbage weight by double sampling: Weight estimate and actual weight. In *Techniques and methods of measuring understory vegetation*, pp. 20-25. USDA For. Serv. Southern For. Exp. Sta. and Southeastern For. Exp. Sta.

Hughes, M.K. 1971. Ground vegetation biocontent and net production in a deciduous woodland. *Oecologia* 7:127-135.

Hughes, R. H. 1959. The weight-estimate method in herbage production determination. In *Techniques and methods of measuring understory vegetation*, pp. 17-19. USDA For. Serv. Southern For. Exp. Sta. and Southeastern For. Exp. Sta.

Hutchings, S.S. and J.E. Schmantz. 1969. A field test of the relative weight estimate method for determining herbage production. *J. Range Manage.* 22:408-411.

Joint Publication of American Society of Agronomy, American Dairy Science Association, American Society of Animal Production, American Society of Range Management. 1952. Pasture and range research techniques. *Agron. J.* 44:39-50.

Jones, M. and J.O. Thomas. 1933. IX. The estimated productivity method. *Agric. Prog.* 10:241-244.

Ovington, J.D., D. Heithamp and D.B. Lawrence. 1963. Plant biomass and productivity of prairie, savanna, oakwood and maize field ecosystems. *Ecology* 44:52-63.

Pase, C.P. 1958. Herbage production and composition under immature ponderosa pine stands in the Black Hills. *J. Range Manage.* 11:238-243.

Pechanec, J.F. and G.D. Pickford. 1937. A weight estimate method for determination of range or pasture production. *J. Amer. Soc. Agron.* 29:894-904.

Roberts, R. Alun. 1933. VIII. The percentage productivity method. *Agric. Prog.* 10:246-249.

Wilm, H.G., D.F. Costello and G.E. Kipple. 1944. Estimating forage yield by the double sampling method. *J. Amer. Soc. Agron.* 36:194-203.

Shrubs

Barrett, J.P. and W.A. Guthrie. 1969. Optimum plot sampling in estimating browse. *J. Wildl. Manage.* 33:399-403.

Wight, J.R. 1967. The sampling unit and its effect on saltbush yield estimates. *J. Range Mange.* 20:323-325.

INDIRECT METHODS (PRODUCTION)

Herbaceous

Blankenship, J. and D. Smith. 1966. Indirect estimation of standing crop. *J. Range Manage.* 19:74-77.

Crafts, E.C. 1938. Height-volume distribution in range grasses. *J. For.* 36:1182-1185.

Goebel, C.L., L. DeBano and R.D. Lloyd. 1958. A new method of determining forage cover and production on desert shrub vegetation. *J. Range Manage.* 11:244-246.

Heady, H.F. 1957. The measurement and value of plant height in the study of herbaceous vegetation. *Ecology* 38:313-320.

Heady, H.F. and G.M. Van Dyne. 1965. Prediction of weight compostion from point samples on clipped herbage. *J. Range Mange.* 18:144.

Hickey, W.C., Jr. 1961. Relation of selected measurements to weight of crested wheatgrass plants. *J. Range Manage.* 14:143-146.

Kelley, A.F. 1958. A comparison between two methods of measuring seasonal growth of two strains of *Dactylis glomerata* when grown as spaced plants and in swards. *Br. Grassl. Soc. J.* 13:99-105.

Kittredge, J. 1945. Some quantitative relations of foliage in the chaparral. *Ecology* 26:70-73.

Neal, Donald L. and Lee R. Neal. 1965. A new electronic meter for measuring herbage yield. USDA For. Serv. Res. Note PSW-56. Washington, D.C.: U.S. Department of Agriculture.

Payne, G.F. 1974. Cover-weight relationships. *J. Range Manage.* 27:403-404.

Wright, H.A. 1970. Predicting yields of two bunchgrass species. *Crop Sci.* 10:232-235.

Shrubs

Basile, J.V. and S.S. Hutchings. 1966. Twig diameter-length-weight relations of bitterbrush. *J. Range Manage.* 19:34-38.

Brown, J.K. 1976. Estimating shrub biomass from basal stem diameters. *Can. J. For. Res.* 6:153-158.

Harniss, R.O. and R.B. Murray. 1976. Reducing bias in dry leaf weight estimates of big sagebrush. *Range Manage.* 29:430-432.

Kinsinger, F.E. and G.S. Strickler. 1961. Correlation of production with growth and ground cover of whitesage. *J. Range Manage.* 14:274-278.

Lyon, L.J. 1968. Estimating twig production of serviceberry from crown volumes. *J. Wildl. Manage.* 32:115-119.

Morris, M.J., D.L. Neal, and K.L. Johnson. 1970. Estimating shrub component production with an electronic capacitance instrument. *Amer. Soc. Range Manage. Proc.* 23:9. (Abstract).

Rittenhouse, L.R. and F. Sneva. 1977. A technique for estimating big sagebrush production. *J. Range Manage.* 30:68-70.

Whittaker, R.H. 1965. Branch dimensions and estimation of branch production. *Ecology* 46:365-370.

OCULAR ESTIMATION   (COVER)

Herbaceous

Daubenmire, R. 1959. A canopy coverage method of vegetational analysis. *N.W. Sci.* 33:43-64.

Horton, J.S. 1941. The sample plot as a method of quantitative analysis of chaparral vegetation in southern California. *Ecology* 22:457-468.

Morris, M.J. 1973. Estimating understory plant cover with related microplots. USDA For. Serv. Res. Paper RM-104. Washington, D.C.: U.S. Department of Agriculture.

Pearson, H.A. and H.S. Sternitzke. 1974. Forest-range inventory: A multiple-use survey. *J. Range Manage.* 27:404-407.

Poulton, C.E. and E.W. Tisdale. 1961. A quantitative method for the description and classification of range vegetation. *J. Range Manage.* 14:13-22.

Smith, A. D. 1944. A study of reliability of range vegetation estimates. *Ecology* 25:441-448.

Shrubs

Coetsee, G. and D.P. LeRoux. 1971. Polygonal plot method for estimating the canopy-spread cover of shrubs. *Proc. Grassl. Soc. S. Afr.* 6:176-180.

LINE INTERCEPT (COVER)

Herbaceous

Burkhardt, J.W. and E.W. Tisdale. 1976. Causes of juniper invasion in southwestern Idaho. *Ecology* 57:472-484.

Canfield, R.H. 1941. Application of the line interception method in sampling range vegetation. *J. For.* 39:388-394.

Chew, R.M. and A.E. Chew. 1965. The primary productivity of a desert shrub (*Larrea tridentata*) community. *Ecol. Monogr.* 35:355-375.

Dasmann, W.P. 1949. Deer-livestock forage studies on the interstate winter deer range in California. *J. Range Manage.* 2:206-212.

Hormay, A.L. 1949. Getting better records of vegetative changes with the line interception method. *J. Range Manage.* 2:67-69.

Larson, R.W. 1959. Use of transects to measure low vegetative cover. In *Techniques and methods of measuring understory vegetation*, pp. 48-54. USDA For. Serv. Southern For. Exp. Sta. and Southeastern For. Exp. Sta.

Parker, K.W. and D.A. Savage. 1944. Reliability of the line interception method in measuring vegetation on the southern Great Plains. *Agron. J.* 36:97-110.

Pond, F.W. 1961. Basal cover and production of weeping lovegrass under varying amounts of shrub live oak crown cover. *J. Range Manage.* 14:335-337.

Van Dyne, G.M. and W.G. Vogel. 1967. Relation of *Selaginellal densa* to site, grazing and climate. *Ecology* 48:438-444.

Walker, B.H. 1970. An evaluation of eight methods of botanical analysis on grasslands in Rhodesia. *J. Appl. Ecol.* 7:403-416.

Wilde, S.A. 1954. Floristic analysis of ground cover vegetation by a rapid chain method. *J. For.* 52:499-502.

Shrubs

Stephenson, S.N. and M.F. Buell. 1965. The reproducibility of shrub cover sampling. *Ecology* 46:379-380.

POINT METHOD (COVER)

Herbaceous

Burzlaff, D.F. 1966. The focal-point technique of vegetation inventory. *J. Range Manage.* 19:222-223.

Crocker, R.L. and N.S. Tiver. 1948. Survey methods in grassland ecology. *Br. Grassl. Soc.* 3:1-26.

Cwik, M.J. and J.D. Dodd. 1975. Vegetation changes on an old rice field following herbicide treatment. *Southwest Nat.* 20:379-390.

Evans, R.A. and R.M. Love. 1957. The step-point method of sampling. A practical tool in range research. *J. Range Manage.* 10:208-212.

Fenton, E.W. 1933. IV. The percentage area method. *Agric. Prog.* 10:234-238.

Fisser, H.G. 1961. Variable plot, square foot plot and visual estimate for shrub crown cover measurements. *J. Range Manage.* 14:202-207.

Goodall, D.W. 1943. Point quadrat methods for the analysis of vegetation. The treatment of data for tussock grasses. *Aust. J. Bot.* 1:457-461.

Goodall, D.W. 1952a. Some considerations in the use of point quadrats for the analysis of vegetation. *Aust. J. Sci. Res.* 5:1-41.

Heady, H. F. 1957. The measurement and value of plant height in the study of herbaceous vegetation. *Ecology* 38:313-320.

Henson, P.R. and M.A. Hein. 1941. A botanical and yield study of pasture mixtures at Beltsville, Maryland. *J. Amer. Soc. Agron.* 33:700-708.

Holscher, C. E. 1959. General review of methodology on use of plant cover and composition for describing forest and range vegetation. In *Techniques and methods of measuring understory vegetation*, pp. 39-44. USDA For. Serv. Southern For. Exp. Sta. And Southeastern For. Exp. Sta.

Ibrahim, K.M. 1971. Ocular point quadrat method. *J. Range Manage.* 24:312.

Klapp, E. 1935. Methods of studying grassland stands. *Herb. Rev.* 3:1-8.

Levy, E.B. and E.A. Madden. 1933. The point method of pasture analysis. *N. A. J. Agron.* 46:267-279.

Long, G.A., P.S. Poissonet, Jr., A. Poissonet, P.M. Daget and M.P. Gordon. 1972. Improved needle point frames for exact line transects. *J. Range Manage.* 25:228-229.

Matern, B. 1972. The precision of basal area estimates. *For. Sci.* 18:123-125.

Nerney, N.J. 1960. A modification for the point-frame method of sampling range vegetation. *J. Range Manage.* 13:261-262.

Owensby, C.E. 1973. Modified step-point system for botanical composition and basal cover estimates. *J. Range Manage.* 26:302-303.

Poissonet, P.S., P.M. Daget, J.A. Poissonet and G.A. Long. 1972. Rapid point survey by bayonet blade. *J. Range Manage.* 25:313.

Radcliffe, J.E. and N.S. Mountier. 1964a. Problems in measuring pasture composition in the field. Part 1. Discussion of general problems and some considerations of the point method. *N. Z. J. Bot.* 2:90-97.

Radcliffe, J.E. and N.S. Mountier. 1964b. Problems of measuring pasture composition in the field. Part 2. The effect of vegetation height using the point method. *N. Z. J. Bot.* 2:98-105.

Rothery, P. 1974. The number of pins in a point quadrat frame. *J. Appl. Ecol.* 11:745-754.

Spedding, C.R.W. and R.W. Large. 1957. A point-quadrat method for the description of pasture in terms of height and density. *J. Br. Grassl. Soc.* 12:229-234.

Stanton, F.W. 1960. Ocular point frame. *J. Range Manage.* 13:153.

Warren-Wilson, J. 1959. Analysis of the distribution of foliage area in grassland. In *The measurement of grassland productivity*, ed. J.D. Ivins, pp. 51-61. London: Butterworth & Co.

Warren-Wilson, J. 1963. Estimation of foliage denseness and foliage angle by inclined point quadrats. *Aust. J. Bot.* 11:95-105.

Winkworth, R.E. 1955. The use of point quadrates for the analysis of heathland. *Aust. J. Bot.* 3:68-81.

BELT TRANSECT (COVER)

Herbaceous and Shrub

Box, T.W. 1960. Herbage production in four range plant communities in south Texas. *J. Range Manage.* 13:72-76.

Larson, R.W. 1959. Use of transects to measure low vegetative cover. pp. 48-54. In *Techniques and methods of measuring understory vegetation*, pp. 48-54. USDA For. Serv. Southern For. Exp. Sta. and Southeastern For. Exp. Sta.

Lindsey, A.A. 1955. Testing the line-strip method against full tallies in diverse forest types. *Ecology* 36:485-494.

Lindsey, A.A. 1956. Sampling methods and community attributes in forest ecology. *For. Sci.* 2:287-296.

Lindsey, A.A., J.D. Barton, Jr. and S.R. Miles. 1958. Field efficiencies of forest sampling methods. *Ecology* 39:428-444.

Woodin, H.E. and A.A. Lindsey. 1954. Juniper-pinyon east of the Continental Divide, as analyzed by the linestrip method. *Ecology* 35:473-489.

VARIABLE PLOT (COVER)

Herbaceous

Bitterlich, W. 1952. The angle-counting sample. *Forstwissens Chaftl. Zentralbl.* 71:215-225.

Cooper, C.F. 1963. An evaluation of variable plot sampling in shrubs and herbaceous vegetation. *Ecology* 44:565-569.

Hyder, D.N. and F. Sneva. 1960. Bitterlich's plotless methods for sampling basal ground cover of bunchgrass. *J. Range Manage* 13:6-9.

Laycock, W.A. 1965. Adaptation of distance measurements for range sampling. *Range Manage.* 18:205-211.

Shrubs

Cooper, C.F. 1957. The variable plot method of estimating shrub density. *J. Range Manage.* 10:111-115.

Fisser, H.G. 1961. Variable plot, square foot plot and visual estimate for shrub crown measurements. *J. Range Manage.* 14:202-207.

SUBQUADRATS (COVER)

Clapham, A.H. 1936. Area-dispersion in grassland communities and the use of statistical methods in plant ecology. *J. Ecol.* 24:232-251.

Fenton, E. W. 1933. V. The point quadrat method. *Agric. Prog.* 10:238-242.

Smartt, P.T.M., S.E. Meacock and J.M. Lamert. 1974. Investigations into the properties of quantitative vegetational data. I. Pilot study. *J. Ecol.* 62:735-759.

PHOTOGRAPHIC METHODS (COVER)

Claveran, A.R. 1966. Two modifications to the vegetation photographic charting methods. *J. Range Manage.* 19:371-373.

Pierce, W. and L. Eddleman. 1970. A field stereo photographic technique for range vegetation analysis. *J. Range Manage.* 23:218-219.

Pierce, W.R. and L.E. Eddleman. 1973. A test of stereo photographic sampling in grasslands. *J. Range Manage.* 25:313.

Ratliff, R.D. and S.E. Westfall. 1973. A simple stereophotographic technique for analyzing small plots. *J. Range Manage.* 26:147-148.

Wein, R.W., A.N. Rency and E.E. Wein. 1974. Sampling procedures for determining plant cover and standing crop over large areas of Canada's High Arctic. Report to Can. Wildl. Serv. Ottawa: Can Wildl. Serv. Mimeo.

Wimbush, D.J., M.D. Barrow and A.B. Costin. 1967. Color stereophotography for the measurement of vegetation. *Ecology* 48:150-152.

STATISTICS AND SAMPLING

General Statistics

Steel, R.G.D. and J.H. Torrie. 1960. *Principles and procedures of statistics.* New York: McGraw-Hill.

Snedaker, G.W. and W.G. Cochran. 1967. *Statistical methods.* Ames, Iowa: Iowa State Univ. Press.

Cochran, W.G. and G.M. Cox. 1960. *Experimental designs.* New York: John Wiley and Sons.

Cochran, W.G. 1953. *Sampling techniques.* New York: John Wiley and Sons.

Quantitative Ecology

Greg-Smith, P. 1964. *Quantitative plant ecology.* Washington, DC: Butterworths.

Kershaw, K.A. 1964. *Quantitative and dynamic ecology*. London: Edward Arnold.

Pielou, E.C. 1975. *Ecological diversity*. New York: John Wiley and Sons.

Miscellaneous

Bonham, C.D. and C.W. Cook. 1977. *Techniques for vegetation measurements and analysis for a pre- and post mining inventory*. Science Series No. 28. Colorado State University. Range Science Department.

U.S. Forest Service. 1963. *Range research methods*. U.S.D.A. Misc. Publ. No. 940. Washington, D.C.: USFS.

Little, T.M. and F.J. Hills. 1972. *Statistical methods in agricultural research*. AXT-377. Davis, Calif.: Agri. Ext., University of Calif.

National Academy of Sciences-National Res. Council. 1962. *Basic problems and techniques in range research*. Pub. 890. Washington, D.C.: NAS.

Part V -- Wildlife

Raymond Dasmann, Chair
George Petrides
Carleton Ray
Gary Klee
Thomas Lovejoy

V. WILDLIFE

Introduction

1.1 Scope

This part is directed toward people who are concerned with evaluation and management of animal populations in the context of economic development. It is a guide to the methods and techniques for wild animal population studies that will produce information useful in the design of development projects. Emphasis is placed on those species that are commonly termed wildlife, meaning mammals, birds, fish, reptiles, amphibians, and some invertebrates of recognized economic, ecologic, or cultural importance. Included also is some discussion of domesticated species that occupy wild lands and, in consequence, interact with wildlife.

1.2 Why Study Wildlife

In areas where economic development is planned or anticipated, wild animals are usually present. Their presence can contribute to development in either a positive or negative fashion, but cannot safely be ignored, lest benefits be lost or losses incurred. Before any change in land use is contemplated, an evaluation of the wildlife potential may indicate the presence of economically valuable wildlife resources, or, alternatively, the possibility of encountering wild-animal-related obstacles to economic growth in the region.

The presence of wild animals in an area can be of great importance to local communities, since aquatic and terrestrial wildlife are frequently important sources of protein and other useful substances. Economic development may either include protection of these wildlife resources to encourage their continued use, or it may be detrimental. Intensive agricultural development, for example, may lead to direct conflicts with wild species, followed often by the loss of the larger terrestrial species and the destruction of fisheries through pesticide and soil erosion contamination or drainage. This can result in lowered nutrition for local people even if the development produces gains in production of cash crops. A careful evaluation of wildlife potential could lead to economic development which could enhance the contribution of wildlife to human welfare. Evaluations could at least lead to modification of land use plans so they would not significantly decrease wildlife values.

Wild areas with abundant and diversified wildlife populations, or the potential to support such populations, may yield more to local communities if they are developed and managed as protected wildlife areas rather than intensively cultivated or grazed. Economic benefits in such areas may derive from scientific interest in the genetic diversity of wild animals; from the attraction of wildlife for tourism and recreation; from the managed, sustainable yield of wild animal crops; or simply from the ecological contributions of protected areas to soil stability, watershed protection, and overall biological productivity. Development of wildlife as a crop to be harvested through recreational or commercial fishing or hunting can sometimes be the most sustainable and economically valuable form of land use.

The cultural values of wild areas and wildlife to local populations should be considered in development planning. Not only may wild species be significant as sources of food, medicine, and other products of direct use, but they often have important traditional roles in local belief systems. Wildlife can also be a source of pests, predators, diseases, or parasites incompatible with economic development. An obvious example is the presence of African sleeping sickness, or trypanosomiasis, in areas that might otherwise be occupied by people and their domestic animals. The likelihood of such problems arising should be considered before investment is made. Certainly, it is neither safe nor sensible to proceed with development activities that are not ecologically or culturally sustainable, knowledge of the functions of wildlife in both natural and cultural systems is essential if long-term benefits are to be achieved.

> The term *natural area* or *natural system* is used in this report to indicate areas or systems in which wild, undomesticated species of plants and animals are predominant. These areas or systems are subject to low-intensity management. They are distinguished from cultural areas or systems in which human settlements or domesticated plants and animals provide the dominant aspect of the landscape.

1.2.1 Wildlife studies: Some background concepts.

Wild, domestic, and feral animals. Domestic animals have been derived from their wild progenitors through human care and manipulation of their genetic characteristics. They should be distinguished from tamed or gentle wild species such as Indian elephants which may live in association with people but still retain their wild characteristics. *Feral* animals are domestic species which have reverted to living in a wild state. Some domestic animals, including various breeds of dogs, poultry, sheep, and cattle differ so greatly from their wild ancestors that they cannot survive apart from human care. By contrast, other breeds of cattle and poultry and goats, some cats, and dogs can revert to living in a feral state and can successfully compete with their wild relatives. Most domestications of wild animals took place during Neolithic times or earlier. There has been little recent effort to domesticate new species, although attention is now being given to the domestication of eland, oryx, and some other grazing animals in Africa.

Domestic and wild grazing animals have coexisted in wild lands in many parts of the world over long periods since the pastoral way of human life with its strong dependence on domestic animals developed in areas already occupied by wildlife. However, the intensification of the production of domestic species in an attempt to gain higher economic yields often leads to conflict with wild species which may act as predators, competitors, or the hosts of diseases which infect domestic animals.

In general, domestic animals are more easily studied than wild species since they are under human ownership and control. However, many of the techniques discussed in this book can be applied equally well to either domestic or wild species.

Ecosystems and habitats. Techniques for studying wildlife can produce information about the distribution, abundance, productivity, behavior, and other characteristics of animal populations. To properly assess the role of wildlife, however, such techniques should be integrated with studies of water (part II) and soils (part III), which in the presence of solar radiation provides the nutrient support base for vegetation (part IV), which in turn provides the food supply for animals and is modified by animal action. In order to develop information on interactions within the total ecosystem, all these subsystems must be examined (see part I). This usually calls for a team approach combining the skills and expertise of hydrologists, soil scientists, plant scientists, and others along with wildlife specialists. Fortunately, in most areas information is already available on many ecosystem characteristics, which can be useful in assessment of animal interactions.

> The term *ecosystem* identifies an interacting web of populations of species that each perform a function in the capture, transfer, and use of solar energy and in the cycling of nutrients from soil, water or air. Each species plays a role in governing the flow of energy and materials; each feeds back on other parts of the system to help regulate such flows. No species exists in isolation from the total system. None can be removed, decreased, increased, or otherwise affected without effects on the total system and its components. Protection of ecosystems provides protection for all the component species within those systems and is the most effective means of species conservation.

Within each ecosystem, various animal species find their particular *habitat*--an area that meets their requirements for food, water, shelter from weather and enemies, and that provides suitable sites for reproduction and raising young. Further, animals exist only in suitable habitats. The continued presence of a species in an area indicates that the habitat meets at least the minimum needs of a species, since no wild animal survives for long if its environment becomes unsuitable. An abundance of that species indicates a highly suitable habitat.

The absence of a species from a particular locality, however, does not necessarily indicate that the area is an unsuitable habitat. The species could have been extirpated by hunting, by an introduced disease, or by some other condition such as depletion or overuse of waterholes by livestock. Thus, with restocking and adequate protection, or alleviation of other problems, the area may once again support the species. The usefulness of examining historic records for evidence of wildlife presence, and the importance of identifying the limiting factors during habitat assessment are elaborated later in this report.

Although the importance of ecosystem management requires emphasis in any discussion of wildlife studies, it must be borne in mind that ecosystems as such are not readily apparent. They are recognized and distinguished most often by their biological components (meadow, forest, woodland) or their physical makeup (pond, estuary, sandy desert). Even the biotic *community* as such, the forest, for example, may have less significance to people than its component individual trees. Similarly, in planning for the conservation or management of wildlife, the wildlife *population* is the unit of direct concern, since this is what increases, diminishes, or remains stable while individual animals are born and die. Nevertheless, to local people and to many nature lovers, the individual animal is the object of interest and concern, and the population is an abstraction. Many pastoral people focus their regard on individual cows, bulls, and calves, whereas to the economically oriented animal-husbandry expert, the productivity of the herd is the object of interest. This difference in perspective can lead to misunderstanding and conflict at the local level unless the development or management expert keeps these distinctions in mind.

1.2.2 Integrating development and conservation.

Potential conflicts between the development of an area for human economic benefit and the protection of the environment, including its wild animal life, often become apparent when changes in use of land or water are contemplated. Areas of highly productive soils where water can be made available are logical sites for intensive agricultural development. Most kinds of wildlife will be to some extent incompatible with such development. Other areas within the same region, because of soil, water, topography, or other factors, will have only marginal value for agriculture, but a high potential for some combination of uses that includes wildlife management. Thus, conflicts which are unavoidable in any one area being considered for development can usually be reconciled within a regional or national context. Or, put another way, it is not realistic to maintain wild bears in Paris or elephants in Accra, but it is realistic to plan for the survival of bears in Europe, and elephants in Africa, since they have long-term importance to humanity, in addition to whatever innate rights they may have to continue in existence.

The reasons for seeking reconciliation between development and conservation are stressed in the *World Conservation Strategy* (IUCN, 1980), which lists the following conservation objectives:

1) To maintain essential ecological processes and life-support systems (such as soil regeneration and protection, the recycling of nutri-

ents, and the cleansing of waters) on which human survival and development depend.

2) To preserve genetic diversity (the range of genetic material found in the world's organisms), on which depends the functioning of many of the above processes and life-support systems, the breeding programs necessary for the protection and improvement of cultivated plants, domesticated animals, and microorganisms, as well as scientific and medical advances, technical innovations, and the security of many industries that use living resources.

3) To ensure the sustainable utilization of species and ecosystems (notably fish and other wildlife, forests and grazing lands), which support millions of rural communities as well as major industries.

To accomplish these objectives and at the same time move forward toward economic goals, priorities must be set and compromises accepted (see Soils, part III). Obviously, a unique wild area, previously little disturbed, supporting a rich and diverse fauna, should be protected for the scientific, ecological and educational benefits which will accrue over time. If such areas are developed as national parks, they will also produce economic returns from tourism and recreational use. Any other uses, except perhaps the carefully controlled and sustainable harvest of key resources, can be expected to degrade the area and sacrifice long-term, almost immeasurable values for lesser benefits. Other areas usually exist which have only moderate or low potential for wildlife and can be developed with little conflict with wildlife values. However, unique wild areas may also be the site of mineral deposits upon which the economy of a country may depend, while many areas with low potential for wildlife may also have low potential for any other form of productive use. A strategic approach to development and conservation is therefore needed, in which potential losses and gains can be balanced in order to achieve sustainable and profitable development gains. Some attention to such strategies is given in section 2 of this report. To maintain the widest range of genetic diversity, it is important to plan to protect examples of the full variety of natural ecosystems and their associated animal life. Even where these include areas with high economic potential for other forms of development, representative and viable samples of the original ecosystems should be maintained. Table 1.1, prepared by the authors of this report, shows some comparisons between wildlife values and compatible economic development activities.

Scientific, ecological, or economic objectives must, however, always be examined from the viewpoint of the people who are engaged in or will be affected by a development plan. Regardless of how worthy it may appear from an international viewpoint to have protected areas or wildlife utilization schemes, these will not succeed without the support of local people, whose current attitudes and practices may not be the same as those expressed in the national capital, or by international experts.

1.3 Objectives

Methods and techniques for the study of wildlife will vary depending on the goals and objectives to be achieved. Suiting methods to goals is

V. WILDLIFE

TABLE 1.1 Conservation values and compatible development.

Wildlife Value	Development Compatibility
1. Unique natural area including undisturbed remnants of formerly more extensive natural communities—last of its kind or only one of its kind with species not found elsewhere.	Develop for scientific, educational or touristic values to maintain unique species or communities and protect indigenous flora and fauna. Limited nondestructive uses may be permissible such as controlled hunting-gathering, or selective removal of limited numbers of certain species.
2. Critical habitat on which the biological productivity of a much broader area depends: e.g., some rainforests, mangroves, coastal marshes, rookeries, sea bird colonies, etc.	Develop as above. Other uses may be compatible if carefully planned and controlled to protect wildlife values.
3. Highly productive wildlife habitat, but without exceptional characteristics of 1 or 2, e.g., extensive savanna or steppe supporting natural communities.	Develop as above or manage for commercial or recreational wildlife harvest. Other forms of development, such as extensive livestock grazing or forestry use may be possible if measures are taken to protect primary wildlife values.
4. Natural areas of a type common or widespread within the country and moderately productive wildlife habitat areas.	Develop to achieve primary economic goals while providing for wildlife protection as an important secondary resource.
5. Heavily disturbed areas and nonproductive sites with few wild species and low wildlife potential.	Develop under sound land and resource management principles.

important in minimizing costs. Long-term, relatively costly, intensive studies are not called for when a quick survey will yield the information necessary to achieve a particular objective. Collating and analyzing existing information may sometimes rule out the need for any further study. There are four categories of objectives which have been the most frequent justifications for wildlife studies: (i) establishment of protected natural areas including reserves or national parks; (ii) development of sustained-yield wildlife utilization, including fisheries development schemes; (iii) environmental impact statements for a proposed project; and (iv) pest control activities. More recently, multi purpose national or regional resource inventories or profiles have come into use. Each calls for a different intensity of effort.

1.3.1 Protected natural areas.

Relatively undisturbed areas supporting a rich and diverse fauna and flora are often obvious candidates for protective status, i.e., nature reserves, national parks, or some similar category. Protected areas can also be established to restore disturbed or depleted species and biotic communities.

Existing information may be sufficient for determining the general location and characteristics of areas suitable for protected area status. However, before the boundaries of such areas can be decided, more detailed information is usually needed concerning the distribution and abundance of the species involved, and the extent to which these species travel, either seasonally or during periods of extreme weather conditions. Once protected natural areas are established, their long-term values can be maintained or increased through baseline and monitoring studies described in section 3.4. In general, therefore, this objective calls for intensive, long-range, and sophisticated techniques--although the initial objective of establishing the reserve can be achieved with less intensive investigations.

1.3.2 Wildlife utilization.

Areas well suited to wildlife may often yield highest economic returns through the development and utilization of wildlife populations. This may involve planning for stream, lake, lagoon, or estuarine fisheries or for the development of game farming or game ranching activities that will make use of the natural productivity of wild species. Such activities depend on determining the sustainable yield from wild populations and regulating the utilization at this level. Initial studies need not be intensive and can take advantage of existing information. However, continued utilization and any increase in sustainable yield will depend on increasingly sophisticated measurements of wildlife populations, their characteristics, and their relationships to their total ecosystem. The situation is comparable to the establishment of more totally protected areas--the commitment of funds and resources may be initially small, but a long-term commitment is made to increasingly intensive and rigorous investigations. With commercial utilization schemes, income from development can be expected to keep pace with the expenses of further investigations.

Numerous examples are available from Zimbabwe (Dasmann, 1964), Kenya, Ghana (Asibey, 1974 and 1976), and elsewhere (Prescott-Allen and Prescott-Allen, 1982) in which the feasibility of managing wildlife for the yield of meat, hides, or other animal products has been established. Income from such operations has been shown to exceed that obtainable from domestic cattle where both are utilizing areas of marginal land unsuited for intensive domestic livestock management. Such income can be further increased if fees are charged to recreational hunters--safaris and the like--and the wildlife is managed for its trophy value as well as its meat production.

Moreover, use of an area for wildlife production does not necessarily exclude domestic animals. Many species with different food habits or foraging behavior will utilize areas of mixed vegetation more thoroughly and produce greater yields than single species or small numbers of species. Mixes of wild and domestic stock in various combinations should be considered to achieve maximum sustainable yields.

1.3.3 Environmental impact statements.

Many wildlife investigations are intended to measure the loss of wildlife values, or the enhancement of such values, as a result of development activities concerned primarily with the use of other resources. Depending upon the value of the area for wildlife, and the species that occupy that area, environmental impact surveys will vary in intensity. An area that supports little wildlife and has low potential can often be evaluated from previous studies with little need for more than cursory field investigations. Other areas with high wildlife potential can fall into the situation of sections 1.3.1 and 1.3.2, with the need for more extensive studies. Areas that support rare, threatened, or endangered species may require the most intensive level of study before any development that would further threaten those species can be considered.

1.3.4 Problem species.

Areas being considered for development that are likely to fall into any of the three categories above may also contain problem species of wildlife likely to interfere with the success of the planned activity. Such species may include exotic animals that have become established within a planned nature reserve--e.g. mongooses in Fiji or Cuba; red deer in New Zealand; minah birds in Hawaii. They may include predators dangerous to human populations or domestic livestock--lions, tigers, jackals. They may include bird or mammal pests to agriculture, settlement, or pasture, such as *Quelea* finches in Africa, rabbits on Pacific islands, rats in urban areas. Surveys should include identification of potential problem species, that are likely to become problems, their habitat requirements, and estimates of activities likely to decrease or accentuate pest problems. Particular attention should be paid to habitat changes likely to reduce pest problems. Removal of habitat is usually less expensive and more long-lasting than is direct population control.

1.4 Where Are You?

1.4.1 Ecological regions.

The usefulness of specific techniques and methods varies with the ecological region under consideration. This volume is concerned primarily with the tropics and subtropics, which encompass a wide variety of major natural communities. Included are near-shore marine environments, coastal swamps and marshes, estuaries, lagoons, freshwater lakes, streams, humid tropical forests, dry forests, savannas, steppes, deserts, and high mountain habitats.

1.4.2 Classification.

These natural areas are described under a number of different systems of ecological classification. Three of the most widely used international systems are biomes, biogeographic provinces, and life zones.

Introduction

Biomes are regions of relatively uniform climate characterized throughout by similar vegetation and animal life. Examples are the tropical rainforest, the savanna, or the desert biomes (Fig. 1.1).

Biogeographic or *biotic provinces* are areas throughout which plant and animal *species* are relatively similar and distinct from those of adjacent provinces (Figs. 1.2, 1.3). They may be subdivisions of a biome or may include the entire biome of a single continent. Examples are the grasslands of central North America, the pampas of South America, and the steppes of Eurasia, all of which are separate biogeographic provinces, although all are part of an overall grassland biome.

The *life-zone system* now widely used is that of Holdridge (1947), based on bioclimatic criteria. Precipitation, temperature and potential evapotranspiration are used to delineate a more detailed system of classification than either biomes or biogeographic provinces. Thus, Holdridge subdivides humid tropical forests into *rainforest*, *wet forest*, and *moist forest*, and these are further subdivided according to altitude and temperature into *lowland*, *lower montane*, *montane*, and *subalpine categories* (Fig. 1.4).

There are many ways in which natural areas can be classified, each useful for a particular purpose. It is usually advisable to continue whatever system is most widely used in the country where development is taking place, since this makes it possible to relate new research findings to earlier work. If another scheme must be used, an effort should be made to cross-reference the new system with the earlier work.

Techniques for studying wildlife will differ according to the ecological area under consideration. Aerial surveys may be the easiest and most direct method for determining the presence of larger wild animals in deserts, steppes, and open savannas, but not for surveying closed forests. Methods for study of aquatic populations (see part II) generally differ from those for terrestrial species, although the same cautions concerning sampling bias and data evaluation are applicable.

Wildlife abundance and diversity varies greatly with biome. High diversity is characteristic of humid tropical forests and marine coral reefs, but species abundance may be limited. Great abundance of large ground dwelling herbivores is characteristic of subhumid tropical savanna regions, but is less typical of closed forests, steppes and deserts. The difficulties of determining species distributions, densities, and productivity are greater in the areas of highest species diversity. Thus, in tropical rain forests, despite decades of study, many of the smaller species may yet remain to be discovered and described, whereas the wildlife of the savanna is relatively well known.

In Table 1.2, created by the panel, abundance and diversity are considered in relation to taxonomy and biomes, to map scales and methods and to the extent of human use and the urgency for different types of surveys. Taxonomic classes of fauna do not "scale" with geographic areas; they do with "order" levels in soils, i.e., distribution of species, genera, families and orders and their taxonomic rank bear little relation to areal distribution, per se.

420 V. WILDLIFE

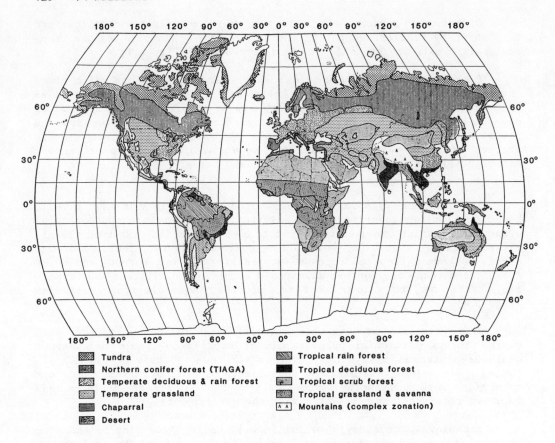

Figure 1.1 Schematic map of the major biomes of the world. [Source: Odum (1966)]

Species are the focus of interest in wildlife studies at all levels of scale with studies of populations or demes, the breeding units within species, being studied in the most intensive types of surveys. The larger taxonomic units--the genera, families, and orders--show phylogenetic relationships but do not correspond to larger and larger geographical areas in any consistent manner across groups of animals. There is a wide range of variation in the number of species or genera in any of the larger taxonomic groups. For example, there are two genera of elephants or giraffes in their respective families but 49 in the highly diversified family, Bovidae, which includes the antelopes and cattle of Africa, Asia, Europe, and North America.

Biogeographic generalizations tend to be on the scale of biomes, continents, and zoogeographic provinces. Thus, a comparison of the number of mammalian families and species per 100,000 square miles in the New World tropics and in Africa (considering those continental areas having an annual rainfall over 120mm) ranges between 0.6 and 0.7 families and 9.4 and 11.3 species, respectively (Keast, 1969). An examination of the eleven families of hoofed animals or ungulates that occur in the lowland rain

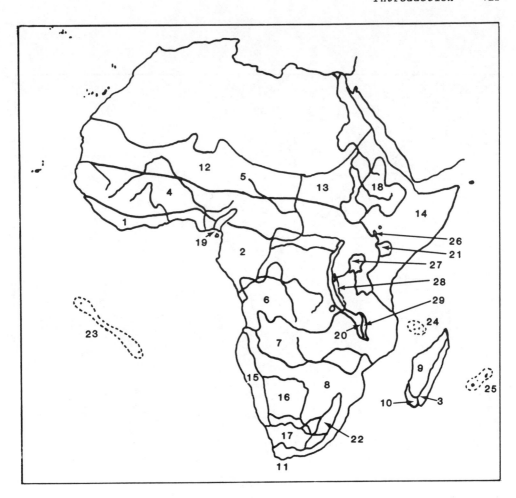

1. Guinean Rain Forest
2. Congo Rain Forest
3. Malagasy Rain Forest
4. West African Woodland/savanna
5. East African Woodland/savanna
6. Congo Woodland/savanna
7. Miombo Woodland/savanna
8. South African Woodland/savanna
9. Malagasy Woodland/savanna
10. Malagasy Thorn Forest
11. Cape Sclerophyll
12. Western Sahel
13. Eastern Sahel
14. Somalian
15. Namib
16. Kalahari
17. Karroo
18. Ethiopian Highlands
19. Guinean Highlands
20. Central African Highlands
21. East African Highlands
22. South African Highlands
23. Ascension & St. Helena Islands
24. Comores Islands & Aldabra
25. Mascarene Islands
26. Lake Rudolf [Turkana]
27. Lake Ukerewe [Victoria]
28. Lake Tanganyika
29. Lake Malawi [Nyasa]

Figure 1.2 Biogeographic provinces of Africa. [Source: Udvardy (1975)]

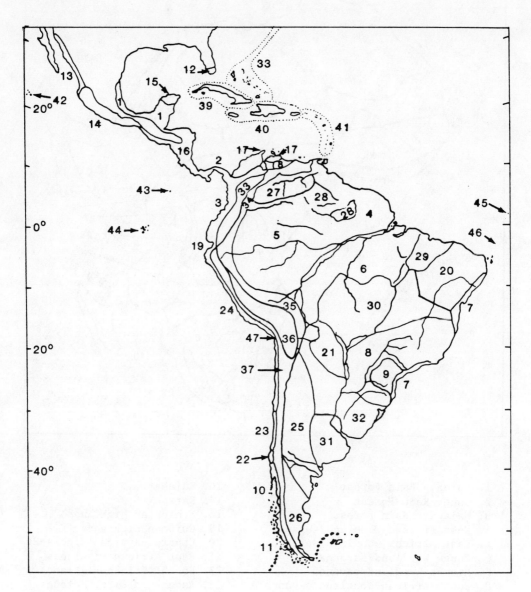

Figure 1.3 Biogeographic provinces of Latin America. [Source: Udvardy (1975)]

Legend for Figure 1.3

1. Campechean
2. Panamanian
3. Colombian Coastal
4. Guyanan
5. Amazonian
6. Madeiran
7. Serra do mar
8. Brazilian Rain Forest
9. Brazilian Planalto
10. Vildivan Forest
11. Chilean Nothofagus
12. Everglades
13. Sinoloan
14. Guerreran
15. Yucatecan
16. Central American
17. Venezuelan Dry Forest
18. Venezuelan Deciduous Forest
19. Ecuadorian Dry Forest
20. Caatinga
21. Gran Chaco
22. Chilean Araucarian Forest
23. Chilean Sclerophyll
24. Pacific Desert
25. Monte
26. Patagonian
27. Llanos
28. Campos Limpos
29. Babacu
30. Campos Cerrados
31. Argentinian Pampas
32. Uruguayan Pampas
33. Northern Andean
34. Colombian Montane
35. Yungas
36. Puna
37. Southern Andean
38. Bahamas-Bermudan
39. Cuban
40. Greater Antillean
41. Lesser Antillean
42. Revilla Gigedo Island
43. Cocos Island
44. Galapagos Islands
45. Fernando de Noronja Island
46. South Trinidade Island
47. Lake Titicaca

forest and tropical grassland biomes of Latin America and Africa illustrates this point further. The following numbers of families are represented by species in these biomes: two in the African grasslands, one in the African rain forests, one in the Latin American rain forests, two in both biomes of Latin America and five in both biomes of Africa (Keast, 1969). The two other families of ungulates are not represented in these four biomes at all. Thus, it may be possible to conserve representatives of the gene pool of a species and representatives of larger taxonomic groupings in an area of a given size, but it is not possible to generalize that one could protect a genus or family by subsequent increases such as doublings in the size of a protected area occupied by a species.

Species may be habitat specialists or generalists and both types of species may be found within a given higher taxonomic group. Widespread species can be "generalists" found in several habitats and even more than one biome. Wild pigs and peccaries, for example, have distributions that extend from lowland to highland conditions and savanna to tropical rainforests. Some widespread species, e.g., caribou and swordfish, are "specialists." More commonly, species are associated with a particular habitat within a biome and such relationships are indicated and verified by site specific surveys. Table 1.2 should be compared with similar tables in Soils (Table 4.1) and Plants (Table 2.1).

1.4.3 Use intensity gradient.

Wildlife investigations will be more or less demanding and time consuming depending on both the biomes to be considered and the existing or

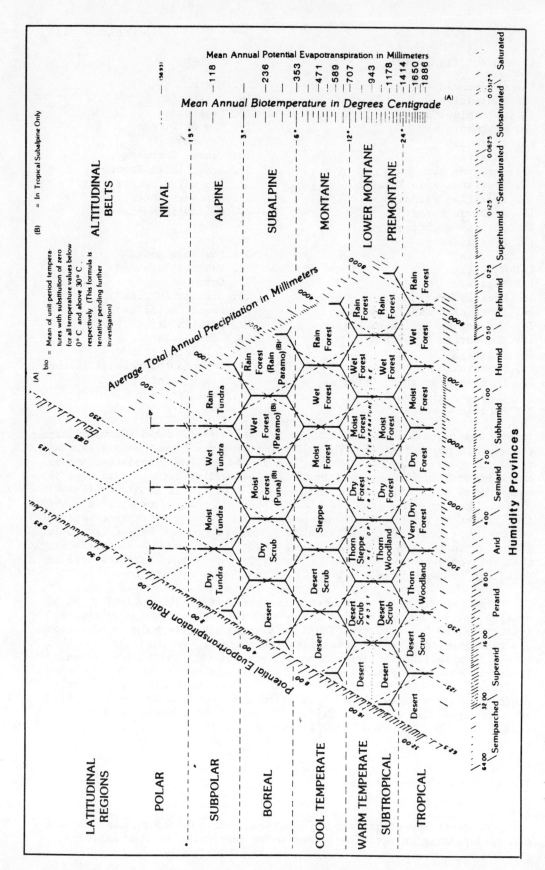

Figure 1.4 Life zones and classification systems of Holdridge. [Source: Holdridge (1947)]

TABLE 1.2 Taxonomy, data sources and methods for evaluating wildlife.

Type of survey	Strategic assessment	Reconnaissance survey	General survey	Intensive site-specific surveys
Data Sources and map scales	Landsat imagery; Existing and sketch maps; 1:200,000 to 1:1,000,000; Small scale →	Landsat imagery; 1:50,000 to 1:120,000; Increasing scales →	Landsat imagery; Aerial photography; 1:50,000; variable →	1:10,000 to 1:20,000 or larger; Large scale →
Data needs and methods	Existing regional data relevant to biomes, biogeographic provinces and life zones; "Surveys" of written and verbal accounts	Generally qualitative data on presence or absence of target species and estimates of habitat condition; Site visits to "ground truth" interpretations	Intermediate detail on animal numbers and habitats; Ground methods: surveys, censuses, inventories	Intensive analysis of wildlife populations, communities, and habitats; Site-specific and animal specific methods: visual, trap, radio telemetry, etc., Labor intensive
Human land use and intensification gradient	Overview of all uses	Wilderness areas and areas used extensively for hunting, gathering, subsistence livestock and crop agriculture	Extensive uses resulting in mosaic of protected areas for wildlife and areas for commercial crops, forests, livestock management and multiple uses; or game	Intensive human uses. Requires specialized surveys for pest or endangered species and critical habitats; wildlife areas and natural ecosystems highly fragmented
Taxonomy	Families, genera, species	Species	Species	Species, populations, demes
	Does not scale with size of survey area; species of interest at all levels, especially indicator species, species groups, communities, guilds		Species, populations	Species, populations
Relation to project	Type and intensity of survey depends on compatibility of wildlife and project use of same or adjacent areas for project sites			

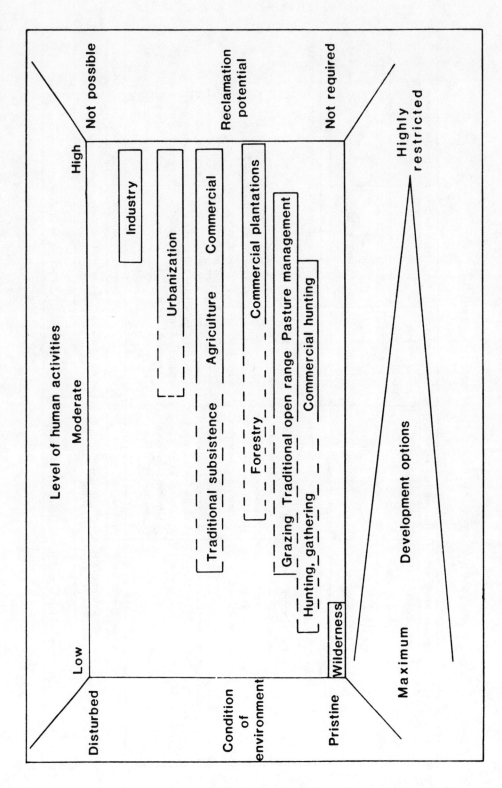

Figure 1.5 The use intensity and impact gradient for terrestrial ecosystems.

contemplated use to be made of the area. A use intensification gradient is illustrated in Fig. 1.5 which applies to terrestrial environments. In fresh water and marine environments a similar gradient may be defined, ranging from relatively undisturbed to highly disturbed aquatic environments; that is

uninhabited estuarine system;

    low-intensity subsistence fishing;

        small-scale human settlement with some resultant pollution;

            high-intensity commerical fishing and transportation;

                urban industrial seaport.

Such a gradient helps clarify development options. Wilderness can be left alone or converted to any other use-intensity category. It provides the widest range of development options for the future. Farther along the intensity gradient the option to move back toward wilderness diminishes and with intensive agricultural use virtually disappears. Although reclamation of such areas is possible, costs increase and likelihood of restoring the original character of the area decreases along the intensity gradient. Wild areas can thus be regarded as development capital, to be invested only with caution and maximum knowledge about possible impacts. Countries without such areas must live entirely on the interest from existing investments.

The extent to which wildlife investigations are needed also varies along the use intensity gradient. A decision to develop wildlands should be taken only after careful examination of the wildlife values. The decision to move from less intensive to more intensive agriculture or to urbanization could be taken with relatively little study of wildlife if remaining wildlife values are marginal. The potential effect of development activities on wildlife should always be considered, however, since effects may be felt on adjacent or even distant areas where wildlife is important. For example, pollution often has great effects far from its point of origin--acid rain of industrial origin, for example, can be devastating to wildlife in remote areas. Similarly, transportation networks may have little direct effect on wildlife through actual construction or use of the road, railroad, or port, but they can have enormous secondary effects through opening up remote areas to settlement or providing access for increased hunting, trapping, and fishing.

As will be discussed at length in chapter 6, outside experts and planners must remember that the people living in a development area or using it seasonally are usually well informed about conditions of life within that area. Local knowledge of wildlife may be detailed and exact, although seldom expressed in Western scientific terminology. Indigenous practices may be of great importance for both conservation and utilization of wildlife. Local attitudes toward wildlife can literally determine whether a development project will succeed or fail. It is vital, therefore, to involve local inhabitants from the beginning, to build from their knowledge, and work with them toward development or conservation goals they both appreciate and understand.

1.5 Approaches to Wildlife Investigations

1.5.1 Strategy and tactics.

Development of a strategy for evaluation and conservation of wildlife within a country, region, biome, or ecosystem should precede detailed investigations. Elements to be considered in preparing such a strategy include the relative urgency of conservation or development and the relative feasibility of accomplishing the conservation or development objective in each area under consideration.

The strategy should include identification of critical areas of great importance to wildlife (such as the location of rare or endangered species or ecosystems); determination of the probability of conflict between wildlife and other existing or proposed uses of the area and plans to mediate or alleviate such conflicts; and determination of personnel, equipment, finances, and need for education and training to prepare personnel for wildlife work. Another important strategic activity is a review of the current institutional framework, including those governmental organizations that draft and enforce wildlife laws. Also essential is a review of existing indigenous and scientific information, to determine need for further investigation. The outcome of the strategic overview is the establishment of priorities for wildlife investigations or management based on the relative values of wildlife and other resources and the feasibility of carrying out the proposed activity.

In summary, the strategy: (i) determines the priority requirements for achieving the objectives; (ii) identifies the obstacles; and (iii) proposes the most cost-effective ways of overcoming those obstacles (*World Conservation Strategy*, IUCN, 1980). Further details on preparing a strategy are discussed in chapter 2.

Most actual field work will be tactical rather than strategic. Within the broad strategy, the tactics are the specific techniques to be employed and the deployment of finances, equipment, and personnel required to carry them out. Tactics and techniques are the subject of chapter 5.

1.5.2 Surveys and inventories.

Before wildlife studies are undertaken, it is essential to decide on the degree of accuracy needed to achieve specific objectives. Far too often, detailed information, acquired at high cost was not needed and not utilized--a truth obvious to readers of many an environmental impact report.

Wildlife *surveys* are intended to determine the distribution of wildlife and location of habitats. Surveys may be simple examinations of existing maps and aerial photographs, or they may use remote sensing techniques or detailed on-the-ground examination of an area.

Wildlife *inventories* or *censuses* are intended to reveal the numbers of different species and the structure of the populations in terms of sex and age classes. Rarely is it feasible, or initially important, to determine absolute numbers, except for domestic animals or rare or endangered species. Usually population estimates which indicate relative abundance of species within an order of magnitude are sufficient to begin management.

1.5.3 Baseline studies and monitoring.

Baseline studies of wildlife are intended to describe the existing status of wild animal populations and habitats, with the expectation that changes in populations and habitats will be monitored at periodic intervals in the future. These are important whenever a change in land use is expected either to benefit or adversely affect wildlife.

Baseline studies must be more intensive and detailed than surveys or inventories, since it is important to determine not only what changes take place in wildlife populations but *why* they are occurring. This involves detailed investigations of *ecosystem dynamics*, with attention to changes in weather and climate, soils and water, vegetation and plant species, as well as changes within the wildlife populations.

Stations where intensive baseline and monitoring activities are to be carried out should be coordinated nationally and within such international systems as the UNESCO Man and the Biosphere (MAB) program, as they are important in maintaining a global overview of changes within the biosphere.

1.5.4 Management options.

Management of wildlife ranges from complete protection, as in strict nature reserves, to attempts to control or eliminate wild species from local areas. Between these extremes, management may include development of an area for tourism and recreation (though primarily for landscape and nature protection), as in national parks; controlled utilization in the form of fisheries development, game ranching or farming, or sport hunting; or the introduction of wildlife for these purposes.

Wildlife populations are dynamic, changing over time in response to their reproductive capacity, the distribution and abundance of predators and disease organisms, or to other changing characteristics of their habitats. Thus simple management practices may not always have the desired effect.

Total protection may seem to be an obvious policy for increasing the abundance of species that were previously depleted by excessive human use; however, such protection alone may fail unless other factors in the environment are also considered. For example, habitat change through normal ecosystem development can be unfavorable to species which require disturbed habitat conditions. Thus, many vegetation types and associated wildlife are dependent on fire for their continued existence. When fire is excluded through complete protection of an area, these species may disappear. Controlled use of fire to change vegetation is favorable to those forms of wildlife that prefer the early successional growth that springs up after burning, but will be unfavorable to species that prefer mature, stable vegetation. In consequence, the use of controlled burning as a management method must be evaluated for each area on the basis of the ecological requirements of species involved.

Cropping of wildlife through recreational or commercial hunting and fishing stimulates reproduction of some species, leading to high, sustainable yields. Other species may show little reproductive response to a reduction in their numbers, or may show a negative response. Large predators or herbivores which normally occupy stable habitats--elephants, eagles, whales--are less likely to show a positive reproductive response to a reduction in their numbers.

It is clear that management plans must take into account not only the distribution and abundance of wild animals, but also the interactions among species in the ecosystem and the role that each species plays within the system. Safeguards can be identified and implemented if baseline and monitoring studies are initiated before any management change takes place.

Strategic Assessment

2.1 Description

Strategic Assessment (SA) is a survey method designed to provide a systematic overview of existing ecological information as a guide for evaluating interactions between species and their habitats in a resource development context at a regional or countrywide level. The goals of the SA are to (i) predict the consequences of development actions and (ii) establish priorities for conservation. While a strategic assessment is useful for short-term decisions, may be essential for long term planning, where conservation and development are to accompany one another. The basic method can be employed in local evaluations by using more detailed information, although that is not the emphasis of this presentation.

Strategic assessment uses information acquired from previous surveys, and it identifies gaps in the data base to be filled by future inventories or baseline studies. More importantly, it is a technique that organizes a wide array of information into a geographic reference system. It also serves as an analytical method for identifying certain relationships between animals and their environments; for identifying potential conflicts between animals and development actions; and for establishing priorities and alternatives for decision making.

The SA approach has several distinctions. First, the method is *selective* in the choice of data for compilation and analysis. For instance, there is an emphasis on the use of "indicator" species, which are representative of whole groups of animals or specific environmental conditions. Second, it is an *integrative* technique which compares data in terms of time (such as seasonal needs of the animal) and space (animal distributions and geographically overlapping activities). Third, SA is an *open-ended data file*, and can readily accommodate new information. Fourth, strategic assessment can be very *cost-effective* because it reveals ways to avoid costly duplication, and it can predict where future problems are likely to occur. Fifth, it is goal specific; that is, different data tactics would be employed for an ecologic preserve where endangered species protection has the highest priority than for a coastal management plan where fish harvest management has the priority.

Unfortunately, SA is not undertaken during predevelopment planning as often as it should be. This is because some view it as a delaying tactic, which merely adds expense, while others prefer to view each development project as a separate unrelated activity. The error of this latter reasoning is aptly demonstrated by the definition of the word "strategy": What general would deploy his troops for a battle before studying the war strategy? "Tactics", or specific interventions, may win battles, but only strategy wins wars.

The explanation to follow was synthesized by the Wildlife panel and their consultant from their own knowledge of the field; it is a distillation of various approaches to SA. The methods and examples set forth reflect current best practice in regional ecological survey. The SA method provides a standardized pattern and organizing theme for regional surveys as a substitute for a succession of extemporaneous efforts. (Note: "Wildlife" and "animal" are used interchangeably herein).

A generalized Strategic Assessment process is summarized in the flowchart in Fig. 2.1. The key tasks are: (i) develop a data base using a series of compendia (compilations of types of data) and sketch maps; (ii) analyze the data base using the concepts of critical habitat and ecological support systems to identify animal-environment interactions in both time and space; (iii) rank the importance of the critical habitat and ecological support areas; (iv) identify existing and potential human-natural resource interaction areas; and (v) evaluate general management and development options to minimize or avoid conflicts.

2.2 Development and Organization of a Data Base

It is quite apparent that the evaluation of compatible uses and the resolution of conflicts among resource uses is determined by national goals and policies at the highest levels, and often involves value judgments of great complexity and controversy. If these judgments are to have any objective basis, they must originally derive from a data base which summarizes information in consistent and compatible formats, and presents this information in a clear and understandable manner. How data may be acquired is treated in the other sections of this report. Here, we speak of cataloguing the data already available in order to take maximum advantage of what is presently known for purposes of strategic planning.

Recalling that strategic assessment is a selective procedure, the first task is to generate a list of subjects to be considered. We select those data that are most instructive and reliable for planning purposes.

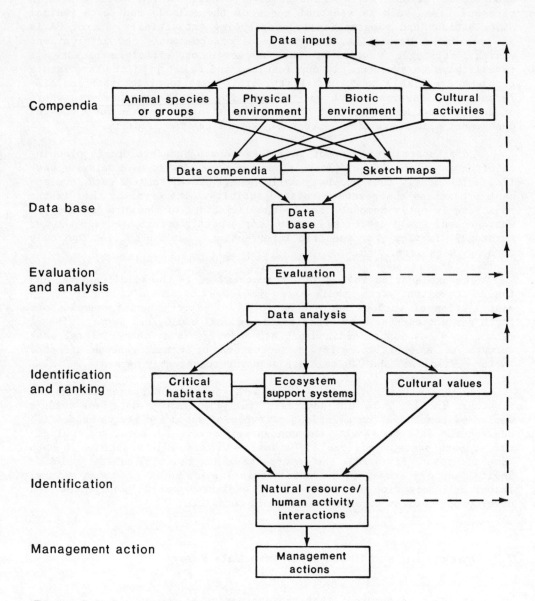

Figure 2.1 A generalized strategic assessment process.

Strategic Assessment

How much information is needed for a strategy consistent with sustained natural resource use and the maintenance of ecological processes? For a strategic assessment of animal-environment-development interactions, the subjects of interest can be divided into four general categories: (i) animal species; (ii) physical environment; (iii) biotic environments; and (iv) cultural activities and institutions. Comparisons within and among these categories are facilitated by using checklists and sketch maps that increase the consistency and comparability of the data.

The second task is to fill in sample checklists, or data compendia, for each of the four categories shown in Fig. 2.1 as relevant to wildlife species. An example compendium is given in Table 2.1 for guidance only--topics can be added or deleted to suit specific needs.

Table 2.1 An example "animal species" compendium for the Northern Pintail in Alaska (U.S.A.). [Source: G.C. Ray (unpublished)]

---

1. CATEGORY

    Bird

2. CLASSIFICATION

Class	Order	Family	Subfamily
Aves	Anseriformes	Anatidae	Anatinae

3. NAME

    Northern pintail, *Anas acuta acuta*.

4. LEGAL STATUS

    Protected under Migratory Bird Treaty Act of 3 July 1919, 40 Stat. 755, amended; and Title 5, Alaska Administrative Code, Article 3, AAC 81.115, and Article 9, 5AAC 81.349.

5. RANGE

    a. Worldwide (A.O.U., 1957; Godfrey, 1966)
    Holarctic, from northern North America and Eurasia to southern Eurasia, Africa, Hawaii, northern South America, and the West Indies.

    b. Region of Concern (Gabrielson and Lincoln, 1959; Portenko, 1972; Kischinskii, 1980)
    Throughout Alaska, northern Yukon, and Northwest Territories (south of Banks and Victoria Islands), and in eastern Siberia.

6. DISTRIBUTION

a. Discrete Populations
None known.

b. Concentrations (Bailey, 1948; Gabrielson and Lincoln, 1959; Irving, 1960; Pitelka, 1974; Bellrose, 1976; King and Dau, 1981)
 i. Natural
 Principally on the great tundra marshes of the Yukon-Kuskokwim delta, Kotzebue Sound, and the arctic coast of Alaska.

 ii. Commercial
 Not taken commercially.

7. HABITAT (Irving, 1960; Bergman, 1974; Johnsgard, 1975; Johnson et al., 1975; Palmer, 1976a; Bergman et al., 1977; Derksen et al., 1977; King and Dau, 1981)

a. Type
Summer: coastal tundra wetlands and along interior river valleys.
Winter: not present in region of concern.

b. Physical/Chemical
In region of concern, snow- and ice-free, open, marshy tundra among shallow lakes, lagoons, and streams; fresh to marine waters.

8. LIFE HISTORY (Bent, 1925; Gabrielson and Lincoln, 1959; Johnsgard, 1975; Bellrose, 1976; Palmer, 1976a)

a. Social Behavior
Non-colonial; male displays but is not territorial. Flock in groups of similar age and sex during fall migration; some mixed flocks occur.

b. Biological Associations
May nest near gulls and terns. In non-breeding season, frequently associate with other species of dabbling ducks, e.g., mallards, wigeon, and teal.

c. Nutrition (Johnsgard, 1975; Bellrose, 1976; Palmer, 1976a)
 i. Feeding type
 Grazer, herbivore; ducklings omnivorous.

 ii. Food
 Mainly seeds and tubers of submerged or floating aquatic vegetation including bulrushes (Scirpus), smartweed (Polygonum), pondweed (Potamogeton), and wigeon grass (Ruppia). Aquatic invertebrates important in diet during early growth of ducklings.

 iii. Feeding behavior
 Dabble and tip up to feed; also dive to shallow depths.

 iv. Feeding location
 Surface and bottom of shallow lakes and streams in tundra, principally near shore.

d. Reproduction (Bent, 1925; Bailey, 1948; Gabrielson and Lincoln, 1959; Godfrey, 1966; Bergman, 1974; Pitelka, 1974; Johnsgard, 1975; Bellrose, 1976; Palmer, 1976a; Bergman et al., 1977; Derksen et al., 1977; King and Dau, 1981)

   i. Mode
   Sexual; internal fertilization; dioecious.

   ii. Location
   Nest throughout region of concern on marshy tundra of coastal plain and interior valleys, near lakes and streams; also along brackish estuaries, occasionally on open coast. Nest is shallow depression lined with down, usually in dense vegetation.

   iii. Behavior
   Pair bond developed on wintering grounds; pairs arrive together on nesting grounds in spring; copulate in water; "rape flights" common. Male usually departs a few days after incubation begins, but occasionally may assist female in care of young.

   iv. Biology
   Eggs laid early June; clutch size 6-12 eggs; incubation 21-22 days. Renesting by unsuccessful females fairly common.

e. Development (Bellrose, 1976; Palmer, 1976a)
Eggs hatch late June; female moves brood to water within 24 hours. Ducklings very mobile; persistently defended by female, generally resulting in low mortality of young; fledged late July. Reach sexual maturity and breed at one year.

f. Growth (Johnsgard, 1975; Bellrose, 1976; Palmer, 1976a)
Males average 1,000 g (maximum 1,500 g); females average 820 g (maximum 1,090 g). Maximal life duration 8 years.

g. Movements (Bent, 1925; Gabrielson and Lincoln, 1959; Portenko, 1972; Johnsgard, 1975; Bellrose, 1976; Palmer, 1976a; Kischinskii, 1980; Arrive mid-May in region of concern; depart August-September. Almost 85% of pintails from Alaska migrate directly or indirectly to California. Birds on arctic slope of Alaska migrate to Mackenzie delta and south along Mackenzie River to Alberta. Birds from Chukchi and Bering move (a) through Yukon-Kuskokwim delta area, interior Alaska, and British Columbia, and to Alberta and southward; (b) across base of the Alaska Peninsula and along Pacific coast to Puget Sound; or (c) to Izembek Lagoon and across North Pacific to Klamath Basin in northern California. Some Siberian birds also move overland to Koryak coast. Spring migration essentially reverse of fall migration.

9. FACTORS INFLUENCING POPULATIONS

   a. Natural (Bent, 1925; Gabrielson and Lincoln, 1959; Godfrey, 1966; Palmer, 1976a; King and Dau, 1981)
   Heavy predation by ravens, gulls, and foxes on eggs and young. Flooding of nests a factor in low, wet areas.

b. Man-related
Heavily hunted, with extensive but well-managed harvest.

c. Potential (King and Sanger, 1979; King and Dau, 1981)
Ranked 103rd of 176 species in northeast Pacific in oil vulnerability index. Moderately susceptible to oiling; results in death. staging areas (e.g., Izembeck Lagoon) potentially subject to oil spills.

10. POPULATION SIZE (Palmer, 1976a; King and Dau, 1981)

Number using Bering Sea habitats 1.2 million in 1981; about one-fifth of North American population. One of the most abundant North American ducks, second only to mallard.

11. MANAGEMENT

May be taken legally in season in North America. Population monitored by U.S. Fish and Wildlife Service.

12. PERSONS CONSULTED

None.

Mapping, the third task, is developed from the data base. It focuses attention on the areal extent and other spatial relationships between environments that support wildlife and the areas proposed for intensified human uses. Maps for all data must be at the same scale. Small regional scales are required for Strategic Assessment because the analysis includes very large areas, such as entire countries, and even adjacent resource systems. Scales of 1:1,000,000 or 1:500,000 are appropriate. Many of the available topographic map series utilize the Universal Transverse Mercator projection. Two other useful map series are the World Aeronautical Charts (WAC) and the Operational Navigation Charts (ONC) which are scaled at 1:1,000,000 and are based on the Lambert Conformal Conic Projection. Both of these projections are compatible with Landsat satellite imagery because of its orthogonal nature. A useful feature of the WAC/ONC maps is their degree and minute longitude/latitude grid system, which can be used to reference specific features and areas. These maps are available for the entire world, and can be readily purchased through the U.S. National Ocean Survey and elsewhere. Scales, sampling, use of remote sensing and geographic information systems are discussed in chapter 3 of part I.

It should be emphasized that this regional scale of mapping is quite different from the large-scale maps required for local "tactical" decisions necessary for site specific developments. Such detailed maps *can not* be used for countrywide "strategic" planning.

The next task is to duplicate as many copies of the base map as are necessary to include all relevant data from the compendia. This base map can be reproduced rapidly and inexpensively as architectural "black line" prints. An example of a hypothetical base map with a reference grid is

Strategic Assessment   437

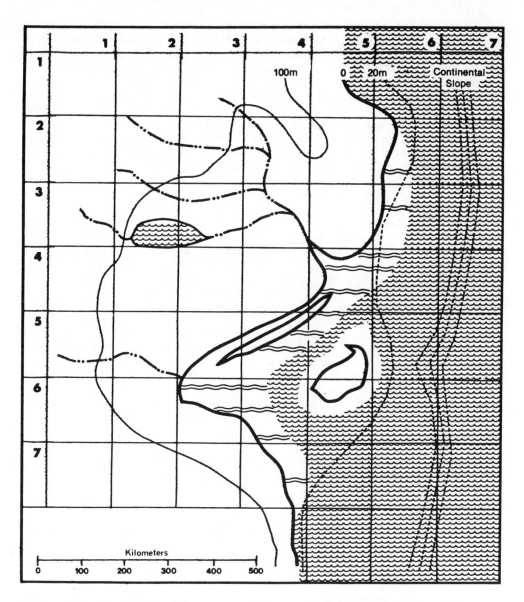

Figure 2.2 Hypothetical base map with superimposed accounting grid.

presented in Fig. 2.2. The use of this grid accounting system is described in the section 2.3, Analysis of the Data Base.

Extreme detail is not required at these small map scales; rather, the objective is to delineate areas so that regional interactions can be observed. Areas exhibiting apparent conflicts are "noted" for a closer look if necessary. Such techniques as the pantograph, overhead projection (camera lucida) and copy grids are quite suitable and inexpensive for transferring information from various maps to the SA Map Base.

Although the gathering and presentation of data into compendia and maps may sound like a grandiose scheme which requires economic assistance in itself to be implemented, it is worth remembering that this data base is within a countrywide framework. It neither requires an extraordinarily large number of parameters as implied by a complete inventory of all species nor a refined level of detail.

The final task related to development of the data base is evaluation of the reliability and comparability of the data compiled. Data for certain topics, even within the same categories, may be inconsistent, gathered by very different means, incomplete, or difficult to compare for a variety of reasons. Ecological and economic data are particularly difficult to assess together. Water quality data may be gathered at a very different scale than is the case for the species which live in the water body; water quality for drinking is different from that for survival of aquatic species. Thus, comparisons may be problematical due to the intensity of the research, mapping, or the purpose for which the data were gathered. The following is an example checklist used with map evaluation:

1. What Map Shows: single source, compiled, new? Clarify and rate reliability

2. Evaluation

   a. Justification

      1. Value: of subject matter, of itself and for analyses.

      2. Choice: of data mapped--why presented as it is ("county" vs. site, etc.)

   b. Completeness and Quality of Data

      1. Mapped Data: result of comprehensive analysis, partial, etc.

      2. Scale: adequate for element, too generalized, etc.

      3. Confidence: poor to excellent, obvious data needs?

      4. Problems encountered: data availability, form, predigested or not.

   c. Need for Additional Study

      1. As a part of this project

                a. Specific sites and/or problems

                b. Specific data gaps

          2. As a consequence of this project

     d. Relationships to other categories

          1. For analysis and/or synthesis

                a. possible conflict and/or compatibility

                b. cumulative and/or synergetic

          2. Seasonal needs or problems

          3. Other.

Data compendia are summary documents of data compiled from literature searches, interviews with well-informed individuals, reconnaissance surveys and more detailed scientific studies. It is important to record the sources of the information, judgments concerning its reliability, and the presence of data gaps, and conflicting information and attitudes. It is especially important to note the appropriateness of map scales for particular data and whether scales have been changed, for example, when detailed maps are summarized at smaller scales.

2.2.1 Selection of species.

A complete inventory or species list is not the desired objective here, and in fact may not even be possible because of the dynamic nature of wildlife populations. A more useful approach is to select certain species as "indicators," i.e., representative of particular ecological conditions, sensitive to some human perturbation, or closely associated with other species having similar requirements. Additional species may be selected because they are unique, rare, or important in an economic or social context.

The following criteria are suggested for selecting species for inclusion in the data base:

1) Economic value: Wildlife are economically valuable not only as tourist attractions and export products (e.g., food, research animals, pets), but also have a home-based economic value for meat, hides, and other products. Most species of economic importance will be readily identified in various government reports although the extent of subsistence utilization has rarely been evaluated (Prescott-Allen and Prescott-Allen, 1982). Local residents will be familiar with economically important animals, as will regional authorities within areas targeted for development.

2) Ecological importance: Included in this category are those species which play significant roles in their ecosystems, are good indicators of specific environmental conditions, or are known to be particularly sensitive to changes in some ecological factor. These animals are important

in the data base because they can represent the ecological systems and processes intrinsic to ecosystem maintenance. An experienced ecologist should be consulted for advice.

3) Social value: Many species are subjects of beliefs and practices of local human populations: These may have aesthetic, artistic and educational values; values related to status and rites-of-passage (such as killing and wearing skins of the large cats); and valued totems; or for spiritual and magical beliefs. Although these values tend to be more important in certain cultures, all societies ascribe particular values to some animals (e.g., the bald eagle in the United States). Information on socially valuable animals can usually be acquired from sociological and anthropological literature, or from national cultural centers. Local peoples should be consulted. The importance of determining socially significant animals can not be overstated, as infringement on certain of these values can result in delays and ultimately may require costly modification of development programs.

4) Rare, threatened, or endangered species: Listings of species in this category are usually available from both world and national conservation agencies. The importance of these animals ranges from scientific and ecological values to social and economic values.

5) Availability of data: The data base will require specific data on each of the species selected, so the availability of those data must also be a criterion for species selection. Scientific or economic studies completed on certain animals often provide data which can be readily incorporated into the data base.

Sketch maps showing desired types of information are given in Figs. 2.3, 2.4, and 2.5. The maps are designed as sketches of life cycle phases and required habitat.

2.2.2 Physical environment.

The major factors of the physical environment of importance to the SA process are climate, physiographic features, and geological/edaphic characteristics. Physical factors not only set the limits on animal distribution, but are critical determinants in regional development planning. It is within this context that we begin to look beyond the animal species themselves, and attempt to place them within the overall environmental setting. Precipitation and temperature regimes impose definite boundaries on certain agricultural interventions, with areas beyond these agricultural regions relegated to livestock husbandry and wildlife utilization. Physiographic features, particularly mountainous landforms, can even impose limitations to livestock use, thereby relegating the area to wildlife habitat and certain resource extraction activities such as mining or timber cutting (assuming access and transportation are economic). Many lowland topographic features, such as floodplains and riverine settings with their attendant alluvial soils, are primary areas for agricultural development, which often results in conflicts with wildlife and livestock use. Although geologic and edaphic (soil) factors may be less obvious than such environmental factors as lack of precipitation and rough

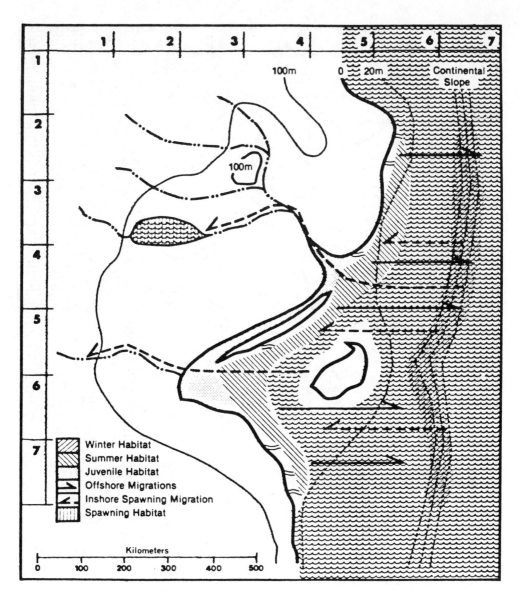

Figure 2.3 Example seasonal distribution map--a coastal fish of commercial value

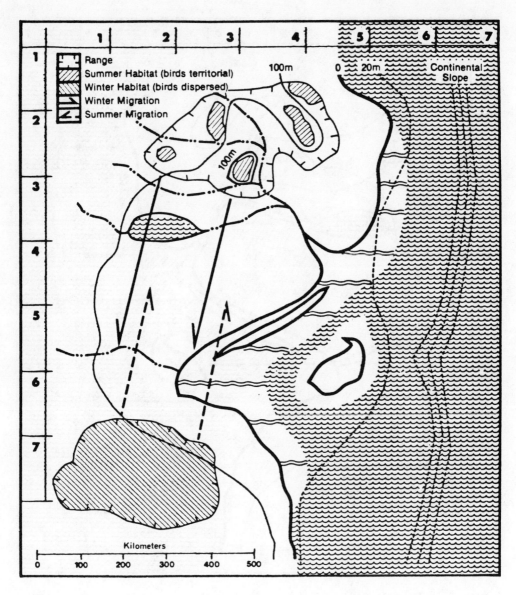

Figure 2.4 Example seasonal distribution map--a migratory bird species.

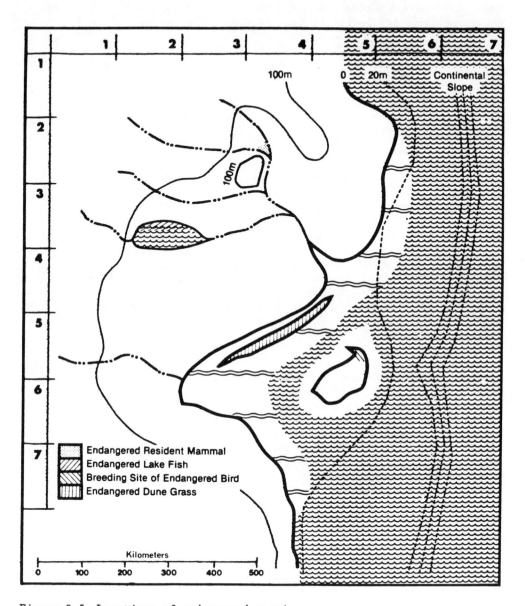

Figure 2.5 Locations of endangered species.

terrain, they are equally important in establishing the ultimate land uses and developments. Ground water supplies, soil mineral composition and soil characteristics as drainage and infiltration rates, effective root depth, salinity and fertility are crucial to agriculture.

2.2.3 Biotic environment.

Section 1.2.1 described animal habitat as the arrangement of food, cover, water, and other special requirements that provide the basic necessities for a species. A composite habitat map is shown in Fig. 2.6. This category addresses the "living habitats," and in particular, those attributes related to food and cover. Vegetation has long been recognized as the expression of these requirements for terrestrial and some aquatic animals (for further discussion, see part IV).

For many strategic assessments, the broad ecological classifications mentioned in section 1.4.2 (biomes, biogeographic provinces, or life zones) will provide an adequate guide to habitat types. Depending on the data already available for the country, general descriptions such as "clayey uplands with desert scrub", "loamy bottomlands with broadleaf deciduous trees", or "muddy benthic deltaic coasts" may suffice. (For additional description of vegetation classification and inventories, see part IV, plants.)

Since habitats are generally dynamic, one should also consider habitat "condition" and "trends." Condition is defined as the present state of the habitat relative to the potential climax state for the site. Trend is defined as the direction of change in habitat condition. These concepts are useful when dealing with habitats which have been disturbed by fire, overgrazing, agriculture, dredging or other development and which are regenerating through natural successional stages towards their potential climax type.

2.2.4 Cultural activities and institutions.

This category includes *economic* activities, such as agriculture, commercial development of plant and animal resources, and exploitation of mineral and fossil fuel deposits. It also includes the *social characteristics* of population size, and ethnic composition, health status, and land tenure systems, as well as *political* and *legal* factors such as ordinances, regulations, and administrative areas, including reserves and national parks. Even such factors as transportation networks, former economic assistance projects, and necessary import commodities are included. Inclusion of import commodities should remind program planners to take advantage of opportunities to increase self-sufficiency in various products during project planning. For instance, the harvest of trees should be considered if tree cutting is needed to clear additional agricultural lands, or as a salvage operation before damming and flooding an area, especially when this reduces the need for expensive importations. Simplified examples are given in Figs. 2.7 and 2.8.

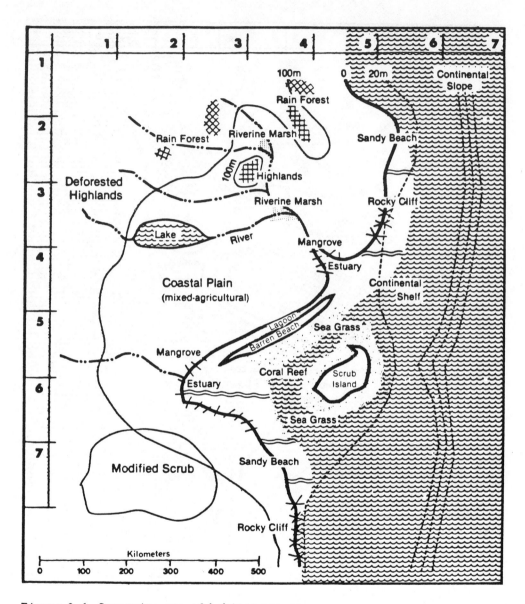

Figure 2.6 Composite map of habitat types.

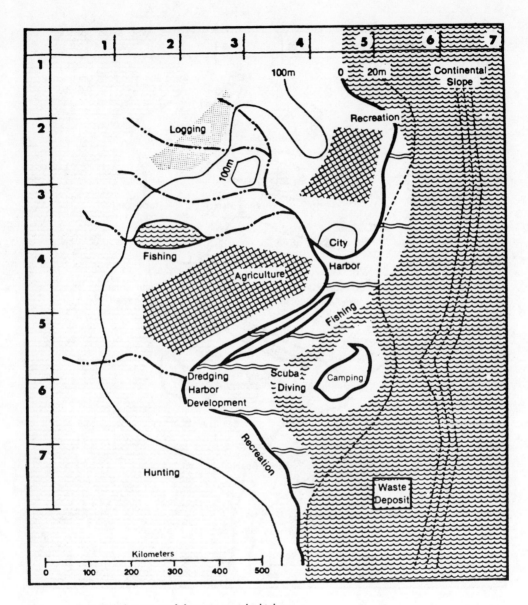

Figure 2.7 Land use and human activities.

Strategic Assessment 447

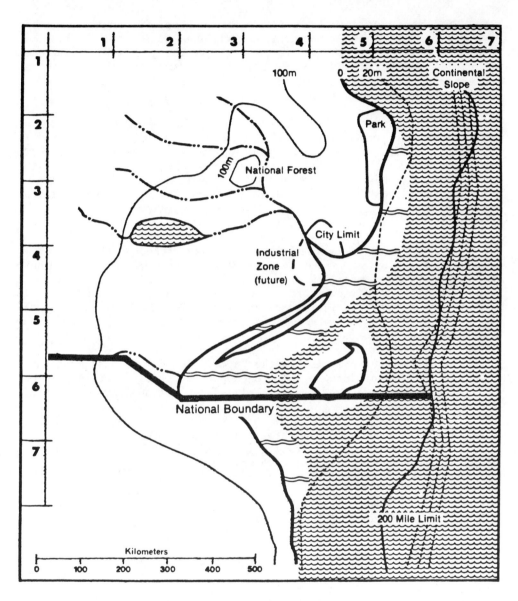

Figure 2.8 Political and institutional jurisdictions.

2.3 Analysis of the Data Base

The basic analytical procedure to be applied to the data base involves the use of simple map overlay techniques to identify spatial correlations between animal species, environmental parameters, and human activities. Selected information from the data base is drawn on transparencies and these are overlaid in various combinations, depending on the particular comparison to be made.

It is important to recognize that a simple spatial correlation does not necessarily indicate functional interdependency; i.e., two species living in the same area may or may not be competing, or even interacting strongly. However, overlay analytical techniques suggest possible relationships between animals and their environments and human actions that are of concern at the regional or strategic level of decision-making. Further, they suggest the kinds of questions that must be answered in advance of tactical, site-specific actions.

One must also recognize that an overlay technique should be by an experienced ecologist, aided by persons well-acquainted with the data. Overlays lack the capability of expressing the dynamics of processes and quantification is difficult; for such extended analysis, computer analysis or multivariate statistics are necessary. Nevertheless, at the strategic stage, overlays can provide a "gestalt" for guidance into future analyses, and at low costs.

To facilitate analysis, whether accomplished by manual means or by computer, it is useful to be able to compare successive data inputs on a standardized per unit area basis. This is accomplished by superimposing the identical coordinate *grid* over each of the individual maps. An example of a grid system is shown on the hypothetical base map (Fig. 2.2). Data on such diverse topics as animal species present, average annual precipitation, plant community, soil permeability, and human land use may be compared by tabulating them for each grid cell.

2.3.1 Species-environment relationships in critical habitat areas.

The objective of this SA activity is to understand the relevant animal-environment interactions that are commonly included when the term "natural systems" is used. In this analysis, one attempts to place areas of special biological importance to populations, i.e., "critical habitats," within an ecological context. This approach goes beyond an analysis of limiting factors, in which one seeks to identify the most critical habitat components under current conditions. One attempts to include all of the areas important to the animals; thus, breeding grounds as well as feeding areas are included, even if food supplies appear to be the more important current limiting factor.

When making this analysis, it is useful to think of animals in terms of the dynamics of their life history. Animals' lives, expressed as the species "niche," are essentially a collection of activities (e.g., eating, resting, breeding) which are carried out at different locations or habitats (feeding in the surf, resting in the estuary, nesting on the cliffs). A niche describes a species' place in an ecosystem in terms of energetics and social structure. Animal distributions recorded without consideration of their behavior and ecology do not tell us whether the animal is a good indicator of certain environmental conditions, or, more importantly, whether it is particularly responsive to changes in its habitat. Adaptive strategies of animals may be divided along a spectrum between those characterized by large body size, slow growth and reproduction, long life spans and stable environments; and those characterized by small body size, high reproductive and migration rates and population turnover, and disturbed or unpredictable environments. Generally, species in the first category and those with the most specialized diets and niches are the most sensitive to habitat change.

In practice, this analysis is performed by taking the maps for each species or animal group considered in the Data Base and overlaying these on maps of physical and biotic environments. The associations express the relationships among animals and environments (Fig. 2.9).

The procedure may also be subject to cell by cell analysis resulting in a "presence/absence" or "use/non-use" assessment for each grid cell. At this point, it is important to understand that there is no implied value ascribed to any of the cells. Any areas known to be of particular significance (i.e., related to limiting factors) can be indicated. Value judgments, whether from the ecological or human perspective, will be considered later in the Strategic Assessment process. The analysis thus far is only intended to identify areas which are important in providing biological necessities at some time in an animal's life.

The majority of data analyzed will be qualitative, i.e., it is either based on the professional judgment of an experienced biologist, or derived from the best available knowledge cited in the published or other archival sources. When quantitative data are available, they can be entered into the grid-accounting system. While the addition of quantitative data increases the complexity of the analysis, it also increases its usefulness over a purely qualitative presence-absence analysis.

That the determination of critical habitats requires an intimate knowledge of animals should be apparent. Since detailed, site specific information is often not available for many species, it becomes particularly important to select the most appropriate species according to the criteria presented in the previous section on Development of the Data Base. If the species are well chosen, they should be representative of the local environmental conditions, and thus circumvent the necessity of dealing with large numbers of species.

After the habitat needs of individual species are considered, it is important to take a broader view and consider communities of species and their ecological support systems. An ecosystem approach helps to identify fragile environments and areas of high natural productivity or narrow tolerances to perturbations. Overlay comparisons can only suggest critical areas, the question should be repeatedly asked: "What are the ecosystem services and processes (such as nutrient availability or succession), which cause these areas to be as they are?" A few subjects to be considered in answering this question are rainfall distribution, frequency, and intensity; soil fertility and erosion potential; vegetation types and productivity; and distribution and duration of water sources. The actual list is dependent on the particular ecosystem or life zone. If, for exam-

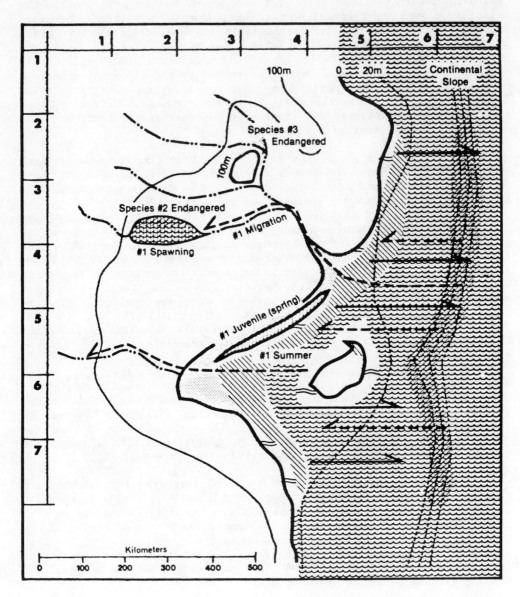

Figure 2.9 Simulated overlay of animal species and habitats (from Figs. 2.3 and 2.5).

ple, the goal of the analysis is to examine factors that influence coastal zone productivity, then one would concentrate on such variables as tides, salinity, rainfall and river flow characteristics, winds, and coastal landforms. On the other hand, if the goal is to identify areas with a high probability for desertification, then one might examine areas subjected to concentrated rainfall, long droughts, and high winds, and various forms of human use.

The ecosystem comparison focuses on ecological processes and services and may include support areas that are spatially separated, distant from critical habitat areas. Watersheds are an excellent example of this relationship: ground surface runoff characteristics and in-stream flow characteristics control the nature of the marsh or estuary at the river's mouth and hence its associated fauna. Thus, critical habitat analysis identifies "core" areas for the species, and the ecosystem process analysis identifies complementary "support" areas. The ecosystem approach provides less well defined "boundaries" of the more easily conceived biological habitats. Whyte (1977) appropriately distinguishes "core" and "buffer" areas.

Ecosystems should be thought of as functional units. "Ecosystem" describes an intrinsic dynamic organization of great importance to resource managers. Selective pressures on ecosystems influence the amount of biomass, the species composition, and specific turnover rates for nutrient recycling, regeneration, and other ecological processes, and management must be determined by the nature of ecosystems if utilization is to be sustainable.

2.3.2 Species-environment-human resource use interactions.

The preceding simplistic analysis of species "core" areas and ecosystem "support" areas identified geographic areas of possible ecological importance (further analysis may be required-see above). The next step is to determine where current and anticipated human resource uses may conflict spatially with existing species needs and ecosystem stability.

The same overlay procedure is applicable. The maps prepared for the data base for such activities as distribution and intensity of agriculture; subsistence or other activities; preserves, national parks or educational sites; and industrial sites will show human-natural resource spatial relationships (see Fig. 2.10). Spatial overlap alone does not indicate incompatability, but it does suggest areas for further scrutiny. For instance, subsistence hunting in many cases may have been within sustained yield limits. When market hunting results from economic incentives or when indigenous populations increase, however, a potential resource use conflict is established. Accordingly, once the spatial relationships between human activities and natural resources are identified, it is necessary to return to the data compendia or other supplemental information sources to determine the degree of utilization and whether a conflict exists or is likely to develop. More detailed analyses will usually be required. Nevertheless, one of the major advantages of the grid accounting system is to provide a tabulation of the multiple resources and activities that might be occurring in the same area.

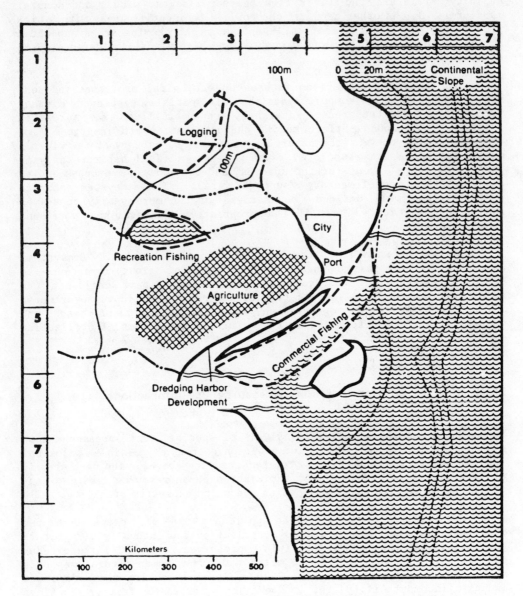

Figure 2.10 Human activity and natural resource spatial conflicts.

2.4 Management Options

Not all human social and economic activities are deleterious nor all "traditions" beneficial, and all natural resources are not of equal value in the perceptions of various segments of society. Therefore, it is especially important to ask, *before development begins*: What are alternative uses of resources and how many could be implemented before the resources begin to suffer? Such questions almost certainly will evoke conflicting answers. For example, disrupted ecosystems, containing "pioneer" or "weedy" habitats, are particularly difficult to assess. Are these to be restored and thus allowed to reach a "climax" state? Or are they more productive in their present condition? The latter is often true. Thus for what purposes may such habitats be used, how much management effort must be expended to maintain them, and what might the return be in either monetary or social terms? These are strategic-level questions, requiring an efficient, objective, and systematic approach towards decision-making, with as many elements of society involved as possible. Conflicts and incompatibilities of resource uses are bound to arise in planning for *conservation with development*. While there are no easy solutions at least there can be a systematic approach to decisions.

2.4.1 Assigning relative values.

The preceding analysis ascribed no particular values to the biological or ecological importance of each grid cell. Prior to decision-making, however, it is necessary to determine how important a particular resource might be. Values attributed to natural goods and services of an ecosystem, and the beneficial or adverse impact of human activities on these natural resources are, for the most part, relative, not absolute values. Further, economic approaches to "cost benefits" will rarely suffice, except for a limited segment of society. For a more complete discussion of methods, refer to Whyte (1977).

There are at least three criteria that can be considered from an ecological standpoint. It is often suggested that areas of high *species diversity* have a special ecological value. The SA analysis, however, simply counts the number of species in each cell without any indication of whether the species may be in the peripheral reaches of their range or not. Nevertheless, areas of *high diversity* in either species or habitats deserve "another look" to see to what extent such grid cells have special characteristics and support species concentrations. High diversity cells are hints--nothing more--that the areas may be of high value. Great care must be exercised, however, in equating diversity with productivity; in most areas, highest productivity occurs in rather simple ecosystems (e.g., estuaries).

The second criterion of ecological value is *uniqueness*. Unique areas are literally "one of a kind" within a given biophysical region, and have a special value for preserving not only the genetic integrity of organisms restricted to that area, but unique ecological processes as well. A correlate to uniqueness is the third criterion: *representativeness*. For example, the presence of a series of forest types with very similar species composition found throughout a given biophysical region, provides the opportunity to manage one or more of these areas as "test cases" of the impacts of certain activities, while at the same time preserving some other of these representative areas as they presently exist. Such areas may be prevalent within a region or nation, and therefore offer unique opportunities for conservation as well as development. Such criteria and representativeness and uniqueness must be referable to a classification system. Udvardy (1975) proposed a terrestrial system. Ray et al. (1983) propose a coastal and marine system. Compatible systems should also be developed at a country level.

Other criteria concern humans more directly. There are a host of ways to rank the importance of areas by economic, social, and educational values. Firsthand knowledge of local and national institutions will help to elucidate this subject. One should realize, however, that the norms, traditions, and values which characterize societies are not necessarily the same as the preferences of individuals. The crucial factor in assigning importance to these values is to define the criteria explicitly and then apply them consistently.

2.4.2 Management considerations.

"Management", includes both conservation and development--e.g. protection of species and unique habitats, restoration of disturbed ecosystems and increasing the productivity of a wide variety of habitats. There are a number of examples of how productivity has been enhanced by management (e.g., forestry and aquacultural developments, open system aquaculture, lake restoration with increased fish production, modern pastoral practices), but the unfortunate fact is that such management often is not systematic or consistent in its results.

One of the major reasons for the management failures so prevalent in both conservation and development is that ecological processes, which sustain ecosystems, are usually not understood or considered. Ecosystems can only be sustainable over time by protecting the particular processes that drive them. For example, forests are different from estuaries in the way energy is used to recycle materials. Forest recycling times are generally very long; aquatic ecosystems are "fast" in that their material is passed through the trophic structure rapidly. In addition, routes of nutrition may be through herbivores or through microbial, detrital, or planktonic organisms. An understanding of these differences is essential for management, but they are often poorly known. It is also true that one cannot await the outcome of research on "ecosystem process ecology" to take initial management action. This is the precise reason for the strategic assessment which provides a preliminary strategy for management, and for the monitoring that is required to improve management practices over the long term.

Thus far, in a preliminary way, we have identified the locations of candidate areas for management--areas of intrinsic biological or ecological significance (section 2.3.1)--and have considered utilization by humans (section 2.3.2). SA has also made it possible to rank conflict-potential areas according to some defined criteria of importance (section 2.4.1). These are all strategic actions. Management is also a strategic activity initially because it is concerned with a number of options for utilization. Examples of strategic management options include evaluations of whether complete protection of especially rare habitats or of animal species is warranted, whether traditional uses are compatible with development plans (and vice versa), and whether development plans should be modified because they adversely affect either the biologically critical habitats or ecological support systems for particular species.

Areas selected as being of special importance, and especially those where human activities are threatening the sustainability of resources or ecological processes, are candidates for special management status. It must be understood that there are a range of special management categories which allow a number of management options. The International Union for the Conservation of Nature has taken the lead in developing a classification system of natural areas. A matrix for protected areas and the activities that may be carried out within them is shown in Table 2.2.

It is suggested that this matrix be used as an initial guide for management of areas that deserve a range of management from strict protection to multiple-use areas. The possibility of multiple use should constantly be considered during any development design or management decision process. Careful multiple use programs can often alleviate certain conflicting activities and augment utilization potential. For example, a breeding area may be utilized outside the breeding season for certain economic activities and the form of utilization may actually enhance the breeding habitat. Fishing may be acceptable in certain coastal, aquatic reserves. It will be noted in the matrix that both the contained resources and the objectives for management determine the type of management area or protected area.

The final step, from the strategic point of view, involves assigning areas to appropriate use categories. An endangered species within a unique area belongs within a "sanctuary"; an area of high productivity, containing species of high economic value, might be assigned to a multiple use area. Of course, the matter is not so simple; often, several uses and even endangered species will occur together. In such cases, subsets of areas, or adjusting uses according to the timing of critical biological and ecological events, may be considered.

Further considerations are:

1) Under which jurisdictions do the areas identified and selected come? Crosscutting political boundaries may complicate the application of legal codes to habitats. Compiling a data compendium of political jurisdictions (local, regional, national, and international) would result in a more complete understanding of the areas in question.

2) The importance of ecological support areas cannot be overstated. These areas will usually comprise over 90 percent of both land and inshore sea regions. Without their conservation, the core areas will not be sustainable over time. Unfortunately, the concepts, much less the management tools, for conservation and development of ecological support areas are only in the early stages of development. Watershed management is a case in point. It was only recognized in recent decades that lumbering and damming activities alter nutrient supplies and hence the sustainable productivity of entire watersheds (Obeng, 1981). Obviously, entire watersheds cannot be called "core areas" or this would include the major

TABLE 2.2 Alternative categories for managing lands for conservation and development. [Source: Miller (1978)]

	Alternative Management Categories				
Objectives for Conservation and Development	National Park	Natural Monument	Scientific or Biological Reserve	Wildlife Sanctuary	Resource Reserve
Maintain sample ecosystems in natural state.	(1)	(1)	2	(1)	—
Maintain ecological diversity & environmental regulation.	(1)	(1)	(3)	(1)	(1)
Conserve genetic resources.	(1)	(1)	3	(1)	—
Provide education, research & environmental monitoring.	(2)	(2)	(1)	(2)	—
Conserve watershed production.	3	3	3	3	—
Control erosion, sediment & protect downstream investments.	3	3	3	3	—
Produce protein from wildlife; support hunting and fishing.	—	—	—	—	—
Provide for recreation and tourism.	(2)	4	—	4	—
Produce timber on sustained yield basis.	—	—	—	—	—
Protect sites and objects of cultural, historical, archaeological heritage.	(1)	4	—	—	—
Protect scenic beauty and green areas.	(1)	(1)	3	3	—
Maintain open options through multipurpose management.	—	—	—	—	(1)
Support rural development through rational use of marginal lands and provision of stable employment opportunities.	(3)	(3)	(3)	(3)	(4)

1 Objective dominates management of entire area.
2 Objective dominates management of portions of area through "zoning."
3 Objective is accomplished throughout portions or all of area in association with other management objectives.
4 Objective may or may not be applicable depending upon treatment of other management objectives, and upon characteristics of the resources.

TABLE 2.2 *(cont.)*

	Alternative Management Categories					
National Forest	Game Reserves, Farms, & Ranches	Protection Zones	Recreation Areas	Scenic Easements & Rights-of-Way	Cultural Monuments	Watershed Programs, River Valley Corps*
2	4	4	4	4	4	4
(1)	(3)	(3)	(3)	(3)	(3)	(3)
3	3	3	3	3	3	2
2	4	4	2	4	2	2
(2)	3	(1)	3	3	4	(1)
(1)	3	(1)	3	3	4	(1)
(2)	(1)	—	—	—	—	2
(2)	2	—	(1)	3	4	2
(2)	—	4	—	—	—	2
4	—	—	4	—	(1)	2
3	3	3	(1)	(1)	4	3
(1)	—	3	3	3	—	(1)
(1)	(1)	(3)	(1)	(3)	(3)	(1)

( ) Major purposes for employing management systems.
—Not applicable.
*In the case of the Watershed Programs or River Valley Corporations, the areas normally include towns, agriculture, and other land uses.

portion of all land and inshore coastal zones. However, viewed from the ecosystem or ecological process level, the management of watersheds and the protection of core areas within them becomes clearer. Species are not the focus nor are specific cores. Rather, management focuses on maintaining processes such as nutrient turnover rates, bacterial decomposition, or nutrient flows.

The major challenge of the future is the development of practices that ensure sustainable resource use, and these practices must be conservative because the margins for error are much more restricted than previously. In this respect, the coastal zone--that broad interface, or ecotone, of land and sea which incorporates the coastal plain and

continental shelves--deserves special attention. This is the area where the majority of the world's people lives. Harbors for marine transportation--the major present use of the sea--are usually sited in exactly the areas of highest inshore productivity: i.e., naturally protected estuaries, bays, and lagoons. There is presently a lack of attention to coastal zone area designation and management, although many agencies are concerned with marine resources and pollution. Traditions developed for the coasts and seas are quite different from those for the land, and there are overlapping and confusing jurisdictional responsibilities and a general lack of application of designation of areas for special management. This is most apparent for the multiple use category. It is fortunate that the concept of "marine parks" and "marine sanctuaries" seems to be well-founded and is becoming more widely applied, but other strategies for development with conservation of those vital resources are also essential.

2.5 Summary of Strategic Assessment

We have proposed the Strategic Assessment method to assist in making decisions pertaining to *development with conservation* in a regional context. Detailed lists of "data inputs" have been discussed, and an analytical procedure has been described which incorporates specific wildlife species and their habitats, the ecological support systems which produce and sustain those habitats, and the human interactions with these "natural systems." Finally, we have suggested criteria to assigning "values" to certain of these parameters, and a conceptual approach to considering different options for management decisions.

Finally, we wish to point out that there is a special reason for including the topic of strategic assessment methodology in the Wildlife part of this report, although it may seem to be a more general synthesis. The reason is that wildlife invariably ends up on the lower rungs of the development ladder, and while there may be much discussion about conservation and animal preservation, it is rare that such preservation becomes the primary determinant behind particular development decisions. The burden has definitely been on the animal ecologists to demonstrate "compatible alternatives" to proposed development scenarios. Innovative approaches are a necessity, and we believe this analytical process can assist in recognizing and demonstrating viable options.

The need for Strategic Assessment is a consequence of the imposing problem it is intended to address. We have attempted to explain both the "why" and the "how" for each step of the Process. We hope the reader will evaluate the merits of the SA not just by thinking about all the steps involved, but by *doing* them and by extending and improving the process while applying it to particular problems and areas.

Methods for Determining Existing Status of Wildlife

This section presents a brief overview of the wide variety of approaches and techniques used to acquire information on wildlife species and their environments for inventory and baseline studies. It is concerned mainly with the techniques and tactics of field studies. For additional information on many of these methods, the reader is referred to basic manuals on wildlife (e.g., the *Wildlife Management Techniques Manual*, Schemnitz, 1980) and various vertebrates (Australian NPWS, 1979; Burnham et al., 1980; Caughley, 1977; NRC, 1981; Norton-Griffiths, 1978; Ralph and Scott, 1981; and Seber, 1973).

3.1 Decisions in Method Selections

3.1.1 Choice of species.

The choice of species for a given study can be critical to developing an appropriate program to assess the status of wild animals and their associated habitat. Significant factors in choosing species are ease of study, relative abundance, particular roles in the ecosystems, and relevance to particular management objectives. Birds, for example, are easier to study than nocturnal, arboreal mammals. More abundant species are easier to study than rarer ones. Bird or mammal species involved in pollination or dispersal of seeds play more significant roles in their ecosystems and reflect changes in ecosystem status better or more rapidly than others. Certain species may be more relevant than others to a particular development or management objective. Examples are: seed eating mammals where forestry activities are planned, primates where their harvest for scientific research is involved, or sooty terns where the status of a local tropical marine fishery is concerned. When endangered or nuisance animals are at issue, there is little choice in species selection. Nonetheless, when options do exist and indicator species can be identified, careful attention should be given to the selection of the species for study.

When the identification of a species is questionable, voucher specimens should be collected and sent to a systematist or museum for identification. While this is not necessary for most of the large wildlife species, it is appropriate for many of the smaller, relatively poorly known species. Guidelines are available for the collection of voucher specimens (Lee et al., 1982).

3.1.2 Choice of methods.

Selection of a study method depends not only on the species involved (netting and banding birds, tranquilizing and radio-collaring tigers) but on several other considerations as well. If population decline and habitat change are of concern, more benign methods are required than if control of a pest (e.g., coyotes, rabbits) is the goal. Every method has inherent limitations: visual observation of unmarked animals may preclude the identification of individuals if there is no distinguishing variation. Conversely, catching and tagging animals may alter their behavior. Some techniques are more efficient than others, but more elaborate ones may reveal information unavailable through quick methods. Costs, labor, and time are all important.

3.1.3 Matching methods and habitats.

Different habitats impose certain limitations on methods for wildlife evaluation and study. The extremes of these limitations can be illustrated by comparing a grassland and tropical rain forest. In open grassland environments, visual studies are far more practical than in the closed shaded environment of the tropical forest, where an observer is lucky to just glimpse an animal before it disappears into the tangled vegetation. Accordingly, techniques of netting, trapping, and baiting are more appropriate in forests.

There is also a tremendous contrast between the two habitat types in number of species (diversity). Tropical grasslands often have few large grazing species, which typically occur in large numbers (sometimes in vast herds), and an even smaller number of predatory species. In many cases, a great deal can be ascertained about the grassland community as a whole by studying or monitoring, perhaps in considerable detail, the status of two or three of these species. The tropical rain forest, with its great species diversity, presents a different situation. Most species will be comparatively rare, and will occur at very low densities of individuals per hectare. One suitable approach is to sample periodically a particular part of the forest community (such as by mist-netting birds in the lower portion of the forest), to monitor whether the diversity (equal to number of species in the simplest of formulations) remains at characteristic levels. When evaluating this information, it is critical to have an adequate sample to detect true variation in diversity since neither random sampling variation *nor* natural variation can be expected to show annual changes in the community.

Another approach useful in forest environments is to note the presence or absence of particular species sensitive to environmental change (i.e., indicator species). In an Amazonian forest, for example, such species include macaws, large predatory birds (e.g., the harpy eagle), and large mammals such as jaguar, tapir, anteaters, peccary and certain species of monkey. Each region and habitat type will have certain species which are sensitive indicators. These can be used to provide a rough index to the condition of a biological community, the diversity and complexity of which defy simple measurement and monitoring.

3.1.4 Other considerations.

Measuring Habitats vs. Animal Populations. It cannot be stressed too strongly that every animal's distribution and abundance is determined by its habitat. If the habitat is absent from a particular area or is deficient in some important aspect, the animal population will be absent or markedly diminished. Often, therefore, measurements of the extent of suitable habitat will be an effective index to the distribution and abundance of its associated fauna.

Relative vs. Absolute Abundance. Usually, it is more important to know whether an animal population is, say, twice, half or one-tenth as abundant as it used to be, than it is to know exactly how many individuals are involved. This is equally true when the abundance of an animal species in one area is compared with that of another. The substantial costs required to determine absolute numbers usually preclude this method in all but unusually favorable circumstances or in the study of rare and unique species.

Population Characteristics. The sex and age (or body-size) proportions within animal populations may be important to determine whether reproduction and survival are normal (see section 3.4.3). Such information assists in defining the *trend* of the population.

Variables Affecting Results. The time of day and the season of the year can have profound effects on the numbers of animals detected. Some species are nocturnal; many diurnal animals are most active near dawn and dusk; migratory species may be absent during certain seasons; some animals are most obvious at times of mating, others may be reclusive then; and so on. Unless a measurement of animal abundance is required at a specific time of day to resolve a particular problem, it is generally best to undertake animal counts during the times that they are most active and evident. In any case, it is important to adopt standardized procedures and always to record the environmental conditions under which a population measurement is made. Ecologists reviewing the report later can then interpret the degree to which the work was done under suitable conditions. The date, hour, weather conditions, exact location, methods used and personnel involved always should be recorded. It is never harmful to note additional information so long as the above essentials are included.

Repetitive Observations. It is essential that the status of important plant and animal species and communities be assessed on repeated occasions to detect changes which may occur. This is the essence of baseline and monitoring activities, which are elaborated later in this section.

Mathematical Considerations. Measurements of the size and characteristics of wildlife populations are subject to many difficulties. Animals usually are not easy to see or to count. Their sex is not always obvious and age is usually difficult to determine. Variations in observed conditions which are due to chance alone are considerable. Replicated series of measurements often tend to yield large variances. Confidence limits within less than 20 percent of mean values generally are not attainable.

Accuracy is more important than precision. Wrong answers, even if highly precise and "repeatable," are still wrong answers. The emphasis here is on census and other procedures which will improve the accuracy of results. Index counts and other surveys meant to be repeated annually or seasonally should be made under carefully recorded and standardized conditions so that statistical analyses can be used to assess significant differences.

3.2 Resource Inventories and Habitat Surveys

We have seen that both biotic and abiotic features of the natural environment, as well as human actions, combine to form living ecosystems, which, in turn, provide animal habitats. There is extensive literature on methods for resource inventories. Much of this literature emphasizes soils, water, and vegetation characteristics of land. More germane to this book are the techniques which can be used to define and evaluate how these resources and ecosystems interact to provide the habitat for various animal species.

When considering methods for animal habitat studies, the concept of the component parts of a habitat can be useful. The three fundamental components of animal (both wild and domestic) habitats are *food*, *cover*, and *water*. A fourth component is *special requirements*, such as snag trees for certain birds, salt for ungulates, specific denning/breeding sites for a variety of animals, and perhaps such intangible items as isolation for animals like the grizzly bear. Two other attributes of habitat are *spatial arrangement* and *temporal relationships*. Spatial arrangement is the juxtaposition of the various features of the habitat, the three dimensional relationships of all features of a habitat, and is basic to defining the shape and extent of an animal's home range. (The home range of an animal must include all of the basic necessities of life for that animal.) Temporal relationships include such dynamic ecological parameters as seasonal variations in climate, successional trends, or stability--stability is defined as the ability of the habitat to mitigate impacts, whether natural (floods and some fires), or imposed by humans (Treadwell, 1979).

Various study programs for describing habitats can be evaluated by asking the following questions: What habitat attributes does a given technique describe? Are there methods which provide data on several attributes? Have any of the five attributes been overlooked?

Although vegetation types or plant communities are not synonomous with habitat types, there is such a close parallel between the two for many wildlife species of interest that vegetation is often used as a first approximation of terrestrial habitats. The fact that plant communities are "integrators" of many of the physical factors present in a particular locality (precipitation, soil texture, slope and exposure) lends further credence to this practice. Vegetation types are also closely identified with two of the fundamental components of habitat, food and cover, with plants providing food directly for the herbivorous majority of animals and indirectly for many carnivorous species via predator-prey relationships. Cover types for virtually all terrestrial animals can be related to vegetation types. These animal-vegetation associations can be demonstrated both at the generic level (i.e., antelope and grasslands) and at the specific level (collared peccary, *Pecari tajacu*, and prickly pear cactus, *Opuntia phaeacantha*). If a particular plant species is usually present within a plant community, observation of the vegetation type may be used to infer the presence of both a particular plant species and the related animal. However, the actual presence of the animal must also be directly determined for each specific area.

A last consideration when selecting methods for habitat analysis, is to remember to evaluate the habitat from the animal's perspective. Forest cover mapping might suggest which general areas might provide suitable habitat but, considering the example of the black bear in the United States, only certain trees provide any food or cover. The shrub understory (where the bear actually lives), however, provides most of the food sources as well as resting, breeding and escape cover.

3.2.1 Habitat mapping.

Remote sensing technology offers a variety of tools for delineating the spatial attributes of habitats (location, areal extent, juxtaposition, etc.) and for recognizing characteristic vegetation types and occasionally certain plant species as well as numerous other habitat features such as topography, water sources and substrate types. The infrared capabilities of some photographic films and scanning devices can also be useful for detecting insect and fire damage, plant diseases, pollution, suspended sediments in water, and even plant vigor.

For regional (large area) studies, *satellite imagery* is generally the most appropriate format and is used routinely to produce resource maps at scales of 1:200,000. Landsat satellite images are available for most of the world and can be purchased from the Earth Resources Observation Systems (EROS) Data Center, Sioux Falls, South Dakota, U.S.A. or seven other regional distribution centers worldwide (NRC, 1981).

When the study requires increased detail, or larger scale maps covering a smaller area than the regional perspective, *aerial photography* is recommended. Coverage can be acquired in black and white, color and/or color infrared films, and at scales ranging from 1:120,000 to 1:1,000 and even larger. Aerial photos are available virtually worldwide, although obtaining this existing coverage can be rather involved due to the large number of governmental and private/commercial organizations engaged in this activity. Typical photo-scales in developing countries range from 1:10,000-20,000 (for local, intensive development sites) to 1:50,000-80,000 (countrywide). Black and white is still the most common type of film.

One of the most cost-effective capabilities of remote sensing which is particularly well suited to habitat mapping is *multistage sampling*. This technique employs *multiscale coverage* of the study area. The *small-scale* satellite imagery is used to provide an overview of the entire region and to identify or select localities for more detailed studies (these areas may include potential development sites, or may be critical habitat for a particular species). Depending on the type of intervention anticipated, *medium-scale* aerial photography (1:25,000-1:50,000) is next acquired for the actual habitat mapping. The last stage, if warranted, can utilize *large-scale* aerial photos (1:1,000-1:10,000) to provide detailed information on such features as vegetation density and species composition, or surface erosion conditions. See the remote sensing discussion in part I and Table 1.2 in this part.

Multidate coverage is often available for the same area; comparisons between periodic overflights can often yield valuable information on trends in vegetation succession, land use, and habitat changes.

A last vital aspect of remote-sensing-based studies is *ground data collection*. "Ground truth" serves not only to verify delineations and identifications, it also fills in the blanks where interpretation alone was inadequate. A major advantage of remote-sensing-based surveys, compared to those based only on traditional groundwork, is the possibility of extrapolating information from sample sites to unsampled areas. This is of course risky where wildlife is concerned and requires verification on each site. The aerial perspective enables the interpretor to map with accurate boundaries, but the quality of the final map is often directly proportional to the intensity of the supporting ground data collection activities.

Although remote sensing is a proven and useful tool for habitat mapping, there will be situations in developing countries where aerial photography or satellite imagery is either unavailable or inappropriate. In such circumstances, topographic or other existing base maps may be helpful.

3.2.2 Food and water: Habitat analysis methods.

Evaluation of the food and water resources available for terrestrial animal species is a very broad subject. A multitude of techniques, ranging from statistically valid research on the effects of grazing to the nutrient content of selected grass species, to management-level estimates of relative abundance of key forage plants, to rapid subjective surveys of habitat "suitability," have been devised and applied by wildlife biologists, forestry technicians, range managers and a host of other resource management disciplines.

From the standpoint of evaluating the food resources of the habitat, *carnivores* are perhaps the easiest animal group to deal with although often the most difficult to see. First, being at the top of food webs, they are fewer in both number and species. Second, most predator-prey relationships (although not necessarily the dynamics), are rather well documented. In those cases where data are inadequate, such methods as direct observation, inspection of prey carcasses, and analysis of predator stomach contents and feces should provide the required information. One must not forget that humans also act as predators; therefore hunter sociology should be considered.

For *herbivorous* species whose food habits are well known, it is merely a matter of going afield to locate their food plants, determine their amounts, and consider the effects of seasonal and annual fluctuations in the supply. Unfortunately, food requirements for the majority of animals are not so well known, or perhaps the investigation is concerned with many different animal species. Either case necessitates a more comprehensive methodology. Since animals eat species of plants, not vegetation *types*, studies must be at the plant species level. Thus, the first task is to compile a *floral list* of the plant species found in the area. Then the *amount* of each of these species present can be measured in a variety of ways, the more common being *frequency*, *density*, *abundance*, *cover* (crown or basal) and *biomass* (weight). Of course, the degree of effort (time and cost) expended on acquiring data on plant species and amounts is clearly dependent on the objectives of the study (see Plants-part IV).

After determining what food sources are available (the floral list and corresponding amounts), it is usually necessary to determine the *utilization* of these resources. Utilization studies assess the animal's food habits or its impact on the plant resources. There are three general approaches: (i) *direct observation of animals* (usually through binoculars) during feeding activities (the feeding-minutes method records the actual time spent eating each plant species as an index of importance); (ii) *evidence of animal use on plants*, such as the tops of grasses removed by grazing, or the tips of branches removed by browsing (exclosures can be useful to protect areas from animal use to enable a direct comparison to be made of both the individual species and the amounts of each used); and (iii) *analysis of stomach contents or feces* by laboratory procedures which can yield percentages of each species consumed. Holechek et al. (1982) provide an excellent summary and literature review of these methods for determining range herbivore diets. Additional information on vegetation study techniques are included in part IV, in general plant ecology texts (Daubenmire, 1968) and in documents produced by the Range Inventory Standardization Committee (RISC) of the Society for Range Managment.

It is important to recognize that the principal foods of a consumer species, whether herbivore or carnivore, are not necessarily the preferred foods. The *principal foods* are those which a species population eats in greatest quantities and which form the largest percentages of food items in the animals' diet. *Preferred foods*, in contrast, are those which are sought out by the animal and consumed to a greater degree than would be indicated by their abundance.

To determine food preference values (Petrides, 1975), measurements are made of the quantities of all plant food materials available and consumed (Table 3.1). Based on the total vegetation available, the percentage availability (a) of each plant species can be determined. Considering only the vegetation eaten, the percentage of each species in the diet (d) also can be ascertained. Preference values $p = d/a$ then indicate whether a food species is *preferred* (above 1.00), *neutral* (exactly 1.00) or *neglected* (under 1.00). Forages avoided by herbivores are assigned zero values. Loehle and Rittenhouse (1982) and Hobbs (1982) describe certain restrictions on the use of this simple ratio technique. The first authors reviewed five forage preference indices finding no clear statistical advantage of one over the other, but concluded that sampling problems combined with inadequacies of the indices themselves might prevent an actual representation of true dietary preferences. In particular, they question the validity of using standing crop as the sole independent variable. Hobbs (1982) states that it is essential to calculate error associated with the estimated index, or it may be misleading and not statistically valid. He provides a method for determining the confidence intervals

TABLE 3.1 Calculation of food preference ratings and the dietary importance of forage species using hypothetical data. [Source: Petrides (1975)]

Forage "Species"[1]	Quantities[2]		Percentages			(p) Preference ratings[5]
	(A) Available	(R) Removed	(a) Available	(d) Diet[3]	(r) Removed[4]	
B	1000	900	10	30	90.0	3.00
C	1500	750	15	25	50.0	1.67
E	2000	600	20	20	30.0	1.00
F	2400	300	24	10	12.5	0.42
G	2500	210	25	7	8.4	0.28
H	400	150	4	5	37.5	1.25
I	100	90	1	3	90.0	3.00
J	100	0	1	0	0.0	0.00
Totals	SA = 10,000	SR = 3,000	100	100	—	—

[1]Species are designated by letters which are without meaning: A and D are not used for species to avoid confusion with symbols for availability and diet.
[2]Numbers or volumes of food items available and removed (eaten).
[3]Percentage of each species removed as related to all food removed and comsumed: $d = 100R/SR$.
[4]Percentage of available forage (A) which is consumed: $r = 100 R/A$.
[5]Note that the relative values of precentages of foods removed (r) directly parallel the preference rating values: $p = d/a$.

around preference indices. Wide confidence limits often may indicate that preference values are subject to considerable chance variation. Nevertheless, mean preference ratings (Table 3.1) can be useful indicators of forage values and of *range trend and condition* (see section 3.4.4).

In tropical rain forests, the large number of available forage species may make it impractical to determine the percentage composition of an entire local flora (Williams and Petrides, 1980). Food preference studies are possible even where forage species cannot be identified in the field. In such situations, unknown species can be designated species A, B, C, etc., until proper identifications can be made. Specimens of such unknown plants should be collected in duplicate and labelled. One specimen then can be submitted for herbarium identification while the second can be kept for proper labelling and reference after the botanical report is received.

Carnivores may display prey preferences and these values may be calculated in the same way as forage preferences (Pienaar, 1969).

When contemplating food habit studies as an aspect of habitat evaluations, one should keep in mind the spatial and temporal attributes of the habitat. Feeding areas must be within the home range of the animal to be of use, *and* within daily range of water for many species. Thus, all forage areas might not contribute to the actual habitat being used, and some areas appearing as particularly good feeding grounds might not be accessible to the animal. Furthermore, animal distribution is often patchy and not uniform even when the habitat appears homogeneous. The spatial attribute includes the vertical as well as the horizontal dimension. Many species subsist on browse during all or part of the year, but the browse must be within reach. Severe hedging and the cutting of fuelwood thus imposes spatial limitations on habitat use. Temporal aspects of habitat use are also important; many annual forage ranges are underutilized because seasonal water supplies dry up too early. Most wildlife species and domestic stock are not too dependent on water quality (although there are reports that bighorn sheep in the American Southwest will not use waterholes fouled by wild burros). Thus it is often the distribution and availability (spatial and temporal attributes) of water supplies with respect to other habitat elements (food, cover), which often function as a limiting factor. This is especially true in the more arid environments.

3.2.3 Factors limiting species abundance.

It can be assumed that each species strives instinctively to reproduce and increase its numbers. If it were otherwise, the species would soon become extinct. The factors which limit population expansion often are environmental. The identification of limiting factors is a practical objective of the wildlife biologist in evaluating the food, water, shelter, disease protection, and breeding-place requirements of wildlife.

The factors most often blamed by the public are rarely the real ones. Hunting can reduce population size, but it does not do so where properly regulated, i.e., restricted to the harvest of the annual population surplus. Predation is seldom responsible--most species have long become adapted to living with their enemies. Most predators, tend, in the longrun, to be limited by prey abundance rather than the reverse.

The effects of introduced (non-native) plants, animals, microbes, and industrial chemicals must be evaluated. Competition for forage by domestic livestock may be likely. Even a previous period of overuse by animals or people may have depleted or altered habitat productivity. There is no certain shortcut to success in solving ecological problems.

3.2.4 Habitat fragmentation.

While habitat destruction has been the greatest cause of decline and loss of plant and animal species, scientists have become increasingly aware in recent years that habitat *fragmentation* presents an additional problem. It has become apparent that the ecological dynamics of an isolated tract of habitat is very different from those of the very same tract when it was part of continuous habitat. Simply put, isolated pieces of wildlands lose or shed species after isolation until they ultimately become a much less diverse biological community. It is in a sense an ecosystem decay process.

Habitat fragmentation has become a dominant feature of the landscape. With it, ecosystem decay has been set into motion all over the globe and has become a fairly dominant process itself. Relatively little is known about it, yet the management of all such isolated fragments has to be dealt with.

An associated problem that can occur with habitat fragmentation stems from the removal of habitat barriers. The fragmentation of forest can remove a forest barrier between two grassland species which may then hybridize. In certain instances, one species will genetically swamp the other. This is a particularly difficult problem to solve once a barrier has been removed and the hybridization gets underway. In such instances, it may be important to try and protect some genetically pure stock of the species being swamped and to establish it in an isolated area or on an island.

Clearly, consideration of size is a critical part of the planning and establishment of national parks and biological reserves; otherwise species which are intended to be protected by such areas might well be lost.

3.3 Animal Inventories, Censuses, and Population Indices

3.3.1 Aerial censuses.

For large animals in open environments, aerial counts usually are the most efficient and least expensive. Such censuses must be carried out, however, by experienced personnel and under favorable conditions.

Both pilots and observers must be properly trained. Most such flights are at altitudes of 75-125 meters above the ground. Precautions must be taken to use a small but safe aircraft, to follow a prescribed flight pattern and to record all animals seen. The use of radar altimeters by pilots and tape recorders by observers usually are essential to maintain an accurate altitude of the plane and undistracted counts of animals seen on each individual transect.

Aerial censusing is not simple. Animals easily can be missed. Even elephants may be overlooked when they are standing under trees or in dense bush. Discussions on aerial counts in Norton-Griffiths (1978) and Australian NPWS (1979) should be read prior to planning such censuses.

3.3.2 Aerial inventories with photographic records.

A logical extension of visual censusing from aircraft is the addition of cameras to provide a permanent record of the ground scene or animal groups for analysis at a later date. This technique has been used successfully for many species of wildlife and livestock. A variety of camera systems (including economical 70mm and 35mm formats), camera mounts, and film types can be used (Anderson et al. 1980).

Heyland (1973) provides a comprehensive introduction to this subject which discusses films, filters, processing, mission planning, and interpretation. Photo-interpretation is a key factor. Cost-efficient use of aerial photography for censusing usually precludes acquisition of very large scale photos in which the animals appear as they would if viewed through binoculars. Accordingly, characteristics other than resolution, such as gross morphological features, groupings, habitat preferences, associated species, and changes in appearance from season to season, are used to identify animals.

3.3.3 Ground observations.

Animals may be counted from vehicles, by persons mounted on riding stock, or by people on foot. The procedure is best done when driving cross-country, usually by four-wheel drive vehicle. This approach is recommended over aerial counts where vegetation hides at least some animals from airborne observers, but does not prevent traverses by vehicle.

Persons mounted on horses, camels, elephants or other riding animals may be able to traverse terrain and penetrate vegetation which is inaccessible to motorized vehicles. Some wild animals are not alarmed by riding animals and in those cases such transportation may be superior to the motor car for making censuses. Note the cover illustrating radiotracking with the use of elephant. In other cases, wildlife may be as startled by mounted observers as by those on foot. Body motions by observers often are the cause.

People on foot tend to scare animals. Startled animals which are not detected will prevent a complete count. If they can properly be assumed to comprise a constant proportion of the population, they may not distort an index count, although this assumption is often questioned. The number of deer seen per mile by hunters while walking (sitting produced fewer deer) was found however, to be a reliable index to deer abundance in Georgia, U.S.A. (Downing et al., 1965).

Transect routes either follow randomly selected straight lines through an animal's habitat or proceed along established roads or paths. This procedure is called a strip "census" if the area in which animals are seen can be computed. The method requires that all individual animals which occur within the visible strip be counted and that the area be adequately sampled.

Several formulas have been developed for measuring strip width. Some are based on the diagonal distance between the observer at the time of sighting and the spot from which the animal is first startled. Others involve the perpendicular distance between the line of travel and the point of animal sighting. Kelker's formula involving perpendicular distances divided into belts of equal width generally may be the most appropriate procedure (Robinette et al., 1974). A detailed monograph on estimating density of biological populations using line transect sampling can be found in Burnham et al., 1980.

Even where strip width is not calculated, the average number of animals seen along an established and repeatedly counted route nevertheless may provide a useful index to species abundance. This index is of value both in judging changes in abundance for a species from one year to another in the same area, and to detect differences in the abundance of a species between areas. This method has also been used to survey fishes on coral reefs.

For large mammals, a more useful procedure is the "visibility profile." This type of census normally is made using a vehicle on a route where animal distances from the line of travel can be determined. The distances at which adult animals can be seen from a vehicle depend on both the density of the vegetation and the topography. Such distances have been measured in several ways. In the United States, one method was developed (Hahn, 1949) to count white-tailed deer. It requires that an assistant walk directly away from the car until hidden waist-deep (deer-height) by vegetation. The individual in the vehicle then calls out and the person on foot measures his return distance, usually by pacing. Such measurements are taken from random points along the census route. The average distance then is taken to represent the width of the visible road-strip and doubled to represent the total average strip width on each side of the route. The doubled strip width multiplied by the length of the route measures the area involved. The areas of those portions of the strip which are not visible from the line of travel must be deducted from the total. The average number of deer seen in this area during a number of evening counts reveals their average density.

In Africa, plywood outlines of animals of various sizes have been used (Lavieren and Bosch, 1977, Hirst, 1969) in a similar manner to determine the visibility profiles for several species all of which would be counted along established routes. The visible strip widths for each of the several size categories of species will vary, of course, depending on the size class of the species involved.

Another method of determining the visibility profile along a census route (Hirst, 1969) is to measure or estimate distances from the census route at which startled animals of each species disappear from view. Such "disappearing distances" then are averaged as a measure of the half-width of the strip in which all animals being counted are presumed to be visible.

Animal censuses often are attempted on irregularly shaped plots rather than transects. The simplest census may be made merely by averaging the number of resident animals seen from one or more observation points on a number of comparable and otherwise suitable occasions. The sizes of the areas under observation must be measured, and the extent to which they are representative of wider zones appraised. Aerial photos can be used for this purpose.

In some places, the animals being counted may not readily be seen from fixed points and yet the terrain is accessible. Often, then, it will be feasible to cover the study tract completely by vehicle or riding animal and to map the locations of each form of wildlife. Care must be taken during the count to disturb the animals as little as possible and to keep track of moving individuals so that they are tallied only once.

The method of drives, or beats, by lines of men either on foot or mounted on animals often offers excitement and dramatic action. Accurate animal censuses seldom are achieved in this way, however, whether an area is first surrounded by stationary observers or only lines of beaters are employed. This is because individual animals may remain motionless and hidden while others may pass invisibly back through the line of beaters. On a relatively small area surrounded by game-proof fences and driven by a very large number of men, however, deer were accurately counted in Georgia, U.S.A. (Downing et al., 1965). Under some conditions, as for gallinaceous birds (or even tigers in South Asia guided between long lines of cloth "fencing"), this method may indicate the presence or abundance of a particular species. The results of drives under less than ideal conditions probably are better accepted as indices to abundance than as complete tallies of the animals present.

3.3.4 Using habitat parameters to indicate animal abundance.

Certain habitat parameters can be used to infer animal abundance. This is especially true where grazing and/or browsing herbivores leave evidences of their feeding activities.

Range *condition*, *trend*, and *utilization* (see Plants, part IV) can reveal whether the area is over or understocked if food preferences are known. If preferred food species are scarce or observed to be decreasing in abundance, and certain less preferred species are increasing, the area is probably overstocked. (The range management terms applied to these plant types are *decreasers* and *increasers*, respectively. *Invader* species are plants not normally found in the plant community. They are usually unpalatable to the consumer animal, but take hold due to the poor condition of the site. This sequence of events describes a *downward trend* or degradation of the site.) Conversely, if the preferred foods are abundant, then the area is either at its carrying capacity, or possibly even understocked. (Recall the difference between preferred and principal foods, in section 3.2.2).

The *carrying capacity* concept can also be used to infer animal numbers under certain circumstances. Carrying capacity can be defined as the maximum number of animals that a specified area can maintain in a healthy and vigorous condition within a stated time interval without damage to the habitat. Thus, if the *stocking rate* (number of animals) is known either from previous studies, or from reliable, long-term harvest records, and the trend is stable (i.e., neither overused nor underused), one can infer that the stocking rate approximates the actual population. The carrying capacity concept is elaborated in section 3.4.6.

3.3.5 Animal signs.

Nests, burrows, "houses," dens and other home and breeding places of birds and mammals are among the evidences that wildlife occurs in an area. "Salt licks," browsed twigs, grass cuttings, gnawed bark, destroyed trees, scent posts, droppings and tracks are other signs of animal presence.

Frequently these signs may serve as *indices* to annual or seasonal changes in population density. The average number of colonies of nesting birds per square kilometer and/or the number of nests per colony, for example, may be useful data to plot changes in a species' relative abundance. If the average numbers of adults and young per nest also can be determined, then an actual population census can be ascertained for the time and area studied.

Under some conditions, animal tracks can aid in censuses. In Uganda, after favorable showers the previous afternoon, early morning surveys (Petrides and Swank, 1965) along river banks revealed the tracks of individual hippopotamuses directed away from the water for the night's grazing and toward the channel upon their return. Survey of such track pairs could be useful to confirm "head counts" of hippopotamuses which, in large groups, almost always have some members submerged.

Track counts can also be made on dusty or snow-covered roadways. Old tracks can be erased by dragging brush piles behind a jeep or, conceivably, behind draft-animal or human assistants. On flat lands, track counts in Florida, U.S.A., were found to be correlated with deer abundance on nearby areas (Tyson, 1959). In the western United States and possibly other mountainous areas where animals migrate up and down the slopes annually, the average number of animals whose tracks cross the cleared path per day (night, hour, etc.) can be counted. Tracks heading opposite to the major migration direction are presumed to be a result of wandering and are subtracted from the total. Considerable variance occurred in track counts observed on cleared lowland plots in Georgia, U.S.A. (Downing et al., 1965), yet such counts commonly are accepted as indexes to animal abundance.

Where the topography is such that the animals being tracked are certain to have entered, say, a high valley from which the snows of the previous winter had driven them, the track count may constitute a census.

Counts of the fecal *pellet groups* of hoofed animals are regarded with amusement by lay people unacquainted with the difficulties of animal censusing and with sampling theory. In some areas and for some ungulate species, however, this may be the best census method available. The results obtained differ from those of most other methods, however, in that the average number of animals seasonally present per day is calculated rather than a total number during some brief time period.

The pellet-group census method has been employed principally with North American deer; it has also been used to census tropical ungulates (Eisenberg and Lockhart, 1972). It is applicable, though, to any species which defecates in a consistent manner and at a definite rate under environmental conditions which preserve the feces for a week or so. Though Downing et al. (1965) found that dung beetles very quickly destroy fresh deer pellets in Georgia, U.S.A., the disintegration of feces is slower in drier areas and in dry seasons.

Most and perhaps all deer species tend to defecate numerous small pellets several times per day. Although Downing and his coworkers (1965) found greater variation in defecation rates during their study of white-tailed deer, most deer biologists in the United States and Canada have accepted results indicating an average of 12.6-13.0 dung groups defecated per day for white-tailed deer (*Odocoileus virginianus*) and mule deer (*O. hemionus*). Recently Dinerstein and Dublin (1982) found an average defecation rate of 28.0 times per day for captive chital (*Axis axis*) living on high quality forage at the Seattle Zoo, Washington, U.S.A. This value corresponded more closely with their census data obtained by several methods during field studies in Nepal than did the lower value. These results on an Asian deer show that it is necessary to determine experimentally species-specific and diet-specific defecation rates, rather than just extrapolate old values to new areas such has been done with the well-established rate for North American deer.

In contrast to counts of groups of fecal pellets, average defecation rates of individual pellets have been determined for several species of hares and rabbits. Based on these standards, counts on cleared areas have been used to estimate population densities (Hartman, 1960). Similarly, the defecation rate of fecal boluses has been determined for the Asiatic elephant (Benedict, 1936). Jachmann and Bell (1979) used such data combined with a determined rate of decompositon of elephant feces during the period of their study in Malawi.

All of the above methods utilized animal signs left by the animal during its normal activities and no attempt was made to influence the animal's movements. Linhart and Knowlton (1975), however, used "scent stations" to attract and determine the relative abundance of coyotes. Rather than just driving the roadways looking for tracks, they established scent stations at regular intervals around the study area which attracted coyotes in the *immediate* vicinity. Visitation was determined by tracks in a dusted circle around the station. Stations were checked periodically, and relative abundance (not density of individuals) was measured by the frequency of visits. This technique should be successful with all canids, and perhaps with modifications, might be applicable with other types of animals (Roughton and Sweeny, 1982 and Roughton, 1982).

3.3.6 Automatic remote cameras for recording wildlife activities.

Remote, automatic-triggered cameras can be used to acquire data on animal presence, their characteristics, behavior and some activities (Gysel and Davis, 1956). Cameras are set up at locations where animals are likely to be encountered (e.g., nesting sites, waterholes, established trails) and in situations when limited manpower precludes continual observations for several days or perhaps weeks. Camera systems can include inexpensive instant-print types, 35 mm, and movie cameras. Triggering mechanisms can range from simple mechanical devices to photocells sensitive to emitted body heat or movement.

3.3.7 Mark and recapture methods.

1) *Lincoln-Petersen mark ratio method.* Generally considered to be the most basic animal census procedure, this method is called the Lincoln Index by wildlife biologists and the Petersen Index by fisheries biologists. It is less confusing to have the two personal names involved than it is for the method to be called an index. The procedure is not an index to relative abundance; it is an estimator of the number of animals in the population--a census method.

The mark ratio method is based on a simple concept but its validity depends on assumptions that:

a) Marked animals do not lose their tags or marks.

b) Marked and unmarked animals are equally subject to emigration (movements away from the study area) and mortality.

c) There is no reproduction or immigration (movements into the study area).

d) Either the marked or recovered sample of animals is an unbiased cross section of the population.

While each assumption is important in achieving an unbiased recovered population sample, there is increasing evidence that individual animals are not equally subject to capture or even to visual observation. Within a population, some individuals invariably are less susceptible than others to being trapped, netted or sighted.

It is especially important in applying the mark ratio method to recover a sample of the population by a different method than was used to capture the marked specimens. The marking method must be selected with this in mind. Rabbits marked by trapping might be "recovered" through visual observation of color-dyed tails (Geis, 1956); hoofed animals immobilized by dart-gun perhaps could be neck-banded and viewed from aircraft (Rice and Harder, 1977); birds or bats banded in mist nets might be recovered by shooting (Whitcomb, 1962). If not recovered by a different method, the resulting population estimate may be biased toward the minimum, applying only to those individuals willing to be trapped or unable to avoid capture by the original device or method. A new design for capture-recapture indices has been proposed by Pollock (1982). His method is intended for long-term studies, and is claimed to be "robust" to population heterogeneity (age, sex, social status, etc.) and varying trap response. It allows an analysis that uses methodology from both closed and open population models.

The mark ratio method has been used in other ways in wildlife studies. Robinette et al. (1956) tagged a known number of dead animals in a starvation area. Another team of workers ascertained the tagged-untagged ratio, and the total number of dead animals in the area was estimated. Geis (1956) wanted to know how many cottontail rabbits were shot by hunters but not recovered. He secured some dead animals from highway kills, placed metal tags in their ears and scattered a known number of such carcasses over the study area. Cooperating hunters reported the numbers of tagged and untagged dead rabbits found and thus enabled total "crippling" losses to be ascertained.

2) *Continuous marking-recapture method.* The basic concepts of this procedure were developed by Schnabel (1938), Schumacher and Eschmeyer (1943) and Hayne (1949) to apply where a population is live-trapped and marked continually over a period of time. The assumptions under which the method operates are the same as for the mark ratio method.

In a population where each animal is equally subject to capture, the fraction of unmarked individuals captured in successive samples should gradually decline. When graphed against the total number of animals previously marked, the plot of such declining fractions will tend to form a straight-line slope (Fig. 3.1). The extension of that regression line to the horizontal axis (by either graphic or algebraic methods) will then indicate the population size, i.e., the number of animals that would have to be marked to achieve a zero-proportion of unmarked specimens.

Unfortunately, there are virtually no vertebrate populations whose members are equally subject to capture in traplines as they are usually operated. Almost invariably, the wilder individuals in a population tend to be more wary of the traps and excluded from the census. Sometimes this will be revealed by a curvilinear set of graph-points. On other occasions, curvilinearity is obscured by sampling variation and the consequent scatter of graph-points. Hartman (1960) encountered a rare and possibly unique situation in a snowshoe hare population in Ontario. The hares did display an equal probability of capture, and he concluded that the homogeneous nature of trap response in his study was due to the fact that he moved his traps daily to different locations.

Where censuses are based only on data derived from continuous use of the same traps and trap locations, the results are probably underestimates of the population size. If the same traps or trapping method is to be used both to mark and recapture animals, then it is essential to move the traps as Hartman did. Perhaps a better technique would be to tag the animals with visible marks (e.g., dyes, bleaches, plastic collars, tag-streamers, or ear-discs) then to "recapture" them, when possible, by visual observations. Radio-transmitter collars offer another, but more expensive, way to distinguish marked and unmarked animals.

Probably individuals of all species live within more or less established boundaries. Home ranges can be mapped for animals which are caught, marked, and recaptured repeatedly in a grid of live-traps. Home range areas also can be plotted through the study of radio-collared specimens. Though the work is not easy, population densities of resident animals can be determined in this way.

3.3.8 Census methods based on changes in relative abundance.

1) *Change-in-ratio method.* A census by this method is possible wherever two population segments are readily distinguishable in the field (by sex, age, size, color, etc.) either at a distance or upon capture. The method requires that the original ratio between the two categories be established and then altered by removing a measured number of specimens from one or both segments (Paulik and Robson, 1969).

As with all census procedures, this method assumes that certain conditions prevail. In this method, the assumptions are:

a) The population ratios are true and representative of the population.

b) The removal (kill) figures are accurate.

c) Immigration, emigration, reproductive and survival rates remain stable during the study or affect the observed population segments proportionately.

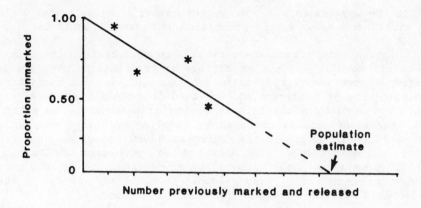

Figure 3.1 Estimation of the number of animals in a population using a continuous mark-ratio method.

Variation in wariness is an often-overlooked factor that can affect animal observations. Such variation can be expected to occur in most, if not all, populations. If pre- and postseason ratios are equally biased, then population estimates should not be affected. Although it has long been suspected (Allen, 1938), it now seems evident (Petrides, unpublished) that male ringneck pheasants in Michigan are more susceptible to observation than females so that observed hunting-season sex ratios are distorted. It is believed that corrected pre- and postseason values may be assessed only from regression slopes of successive rates of sightings but the collection of suitable data probably would be feasible only where intensive hunting is practiced over relatively short periods.

The change-in-ratio method can be adapted to a number of situations, i.e., wherever a ratio occurs which is subject to measurable change. Changes in ratios between two age groups, for instance, or even between two species, could be used in place of the two sexes usually considered.

A more recent variation of the change-in-ratio method is presented by Eberhardt (1982). He describes the use of an index calibrated by removal data for estimating population size, and calculates confidence limits. He also uses an estimated variance technique to delineate when his "index-removal" method may be appropriate, and compares his approach with the use of marking methods described previously.

2) *Declining yield method.* Also known as the DeLury method, after a Canadian biologist, this procedure is based on the idea that, other conditions being equal, the time required to capture an animal changes inversely with its abundance. Where one quail per hour might be the kill rate at the beginning of a hunting season, a quail every two hours would be the average rate of return when the population had been reduced by half. The method has application mainly on managed and intensively hunted areas, where the numbers of hunters and of animals killed, the hours or days hunted (and the time that each animal is killed all can be determined. In such situations, too, corrected sex ratios may be ascertainable from separate declining-yield estimates of the two sexes.

3.3.9 Local knowledge.

Though lacking some elements of quantification and precision, the opinions and recollections of local residents should not be overlooked as a source of information. The older people especially may recall whether habitat changes, hunting practices, and animal abundances have changed over the years. Whether local water sources, firewood, and/or forage are more or less abundant than in earlier times may indicate the causes and nature of changes in habitats. Increases in human population densities also may be related to wildlife distribution and abundance. However, anecdotal information must be used with great caution.

3.4 Baseline Studies and Monitoring

Baseline studies are conducted to gather more precise and detailed information about animal populations and their habitats where their significance or importance justifies intensive and costly effort. If knowledge of dynamic ecosystem characteristics as population fluctuations and changes in habitat productivity are the objectives, a monitoring program to measure these trends is included. Although the baseline study can be as simple as recording the location and abundance of a species so subsequent monitoring can observe the changes in these parameters, usually it is a more comprehensive effort which entails recording various ecosystem characteristics in the attempt to learn *why* the changes in the distribution and numbers of animals had occurred. Some of the obvious ecosystem parameters to be measured include weather and climatic factors; vegetation information such as plant species, distributions, abundance and utilization by animals; soil characteristics such as compaction, fertility loss or increased erosion potential; and changes in the distribution, quantity, and quality of water. A more complete listing of variables which might be considered can be found in the data compendia presented in section 2 of this part. The remainder of this section will deal primarily with techniques and measurements used in baseline studies of animal populations.

Information of importance in these baseline studies include not only numbers, density, and distribution, but also sex and age structure of populations, natality and survival of young, mortality and its causes, and mobility of the species (both migratory travels and daily and seasonal movements) within its residential area (home range). Behavioral studies include particularly the tendencies of individuals to maintain or defend home ranges or portions thereof (territoriality) and other tolerance to crowding; and relations to other species including those it feeds on (plants or animal prey), predators that feed on it, or others that interact with it through food competition or in other ways. Complete information on all these points is rarely obtainable; however, the more one knows, the easier it is to evaluate the place of a species in its ecosystem and the ways it will be affected by changes. Monitoring of any or all these points will lead to increased ability to manage and protect the species. Nevertheless, the investment of time and money in detailed baseline and monitoring studies is justified only where the objectives clearly demand such intensive studies, which usually require highly trained scientists and technicians for long periods of time.

3.4.1 Population structure and dynamics.

Population characteristics vary with time. While some variation is normal, extreme changes in sex and age proportions may signal that important changes in the status of the population are underway. The proportions of a population according to sex, age and social grouping can serve as baseline information against which to compare future data in order to detect trends in population growth or decline, and to account for their environmental causes (see Papageorgiou, 1979).

Sex and age proportions can be determined from such population characteristics as reproductive and survival rates, rate of population increase, average and maximum longevity, expectations of further life at various ages, generation and population turnover time, and minimum and maximum ages of fertility (Dittus, 1981).

Unfortunately, the considerable effort required will prevent the collection of detailed population data in many areas. For species which are important in the local economy or cultural heritage, however, it is desirable at least in major habitats, to measure the sex ratio of adults and the number of young per adult female. If the birth season is not spread out over many months, these counts are best done when the young have nearly achieved full growth. Then they are distinguishable as young and the heavy mortality of early life will have occurred. Appraisals made then will indicate whether population losses will likely be balanced by reproduction.

Sex structure of populations can be determined among live free-living animals for species in which sexual dimorphism is apparent; these include larger mammals, some birds, reptiles, amphibians, and fish. Among others (small mammals and many birds, for example) sex ratios can only be determined from captured or dead animals and the information may be biased in the same way that censuses derived from trapping or shooting animals are biased. Age structure is most easily determined among species which take a long time to mature and which display differing body configurations, or fur or feather colorings when immature. Otherwise weighing, measuring or skeletal examination of captured or dead animals is required, with the same danger of bias. Dentition, including rate of replacement of juvenile teeth and wear on adult teeth can be used to determine age classes. Growth rings on tooth sections or earbones, rings on horns, or the size and configuration of antlers provide information on aging. There are many techniques, but all are time consuming and involve population sampling techniques which avoid bias. Traditional knowledge is sometimes helpful, but inclusion of scientific expertise is required.

Where it is important to determine natality rates and the productivity of populations, one must distinguish between actual fertility, which is influenced by the nutrition and health of the animals, and the actual survival of the young produced. Fertility studies often require examination of pregnant (usually dead) females, although approximations can be made by egg counts--for birds, reptiles, amphibia, or fish--or counts of newly hatched young among the same groups or newly born young among mammals. Mortality, usually highest among the youngest age classes, declines with approaching maturity. Replacement or turnover rates of populations are determined by counts of young nearing maturity. These in turn determine the productivity and potential yield where a species is to be cropped or harvested. Although time consuming, such studies are essential if sustained-yield cropping of wildlife is to be carried out, just as one must know the production of young to maturity to determine marketing rates from domestic herds.

Where detailed information on sex and age classes, natality, and survival is available, this information may be presented and evaluated in life tables or fecundity tables.

3.4.2 Life tables: Survival-fecundity tables.

For any population in which the numbers of individuals alive at each age can be determined, there is a basis for calculations of the average expectation of further life at each age. Hence, average and maximum longevity can be estimated and rates of survival at various ages determined.

Assembling such data in actuarial life tables requires the assessment of age in many individuals. This may be done by following the fate of individuals ringed, tagged, or otherwise marked as a cohort at an early and known age. Or it may be accomplished by examining a cross section of animals either from (i) the living population or (ii) a collection of useful (usually skeletal) remains. In either of the latter cases, it is essential that procedures be available for determining age in the species concerned. Where the sample is derived from skeletal remains (the classic case among mammals is that of the Dall sheep in Alaska by Murie, 1944), it may not be true that the sample is representative of the living population. It may offer a reasonable correlation, however, and will provide useful comparative information.

Fecundity tables similarly compile the rates at which young females are born to females of each age. This information may be combined with life table data to yield the net reproductive rate, generation time and other population characteristics. Normally, the fecundity characteristics for males are assumed to be the same as for females.

The mathematical modeling of populations to predict future growth potentials currently is a popular area of research. To date, however, these resemble econometric models in having considerable theoretical benefits but poor forecasting values.

3.4.3 Body growth and population productivity.

Though of more direct value in scientific studies than in the immediate establishment of management policies, the collection of growth data and charting of growth curves may assist in the determination of rates of gain, in estimating age in young specimens, and combined with life table data, in the evaluation of population productivity (Petrides and Swank, 1966).

Often local wildlife species are immune or resistant to local diseases (e.g., most wild hoofed animals in Africa are not affected by the tsetse-fly-borne trypanosomiasis or sleeping sickness of domestic stock). If disease-resistance or some other characteristic of the wild species is important in meat-production considerations, then studies of rates of gain in body growth might be important in deciding whether husbandry of the species under wild (game ranching), semiwild or domesticated (game farming) conditions is warranted. Such values of wildlife should not be overlooked, especially in protein-deficient regions. Those values may be very important, but it must be understood that maintaining once wild animals under captive conditions is a by-product of wildlife management. It does not accomplish the objective of preserving wild species under natural conditions.

3.4.4 Behavior and travels of animal populations.

Wariness, dominance, territoriality, home range, and migration characteristics are animal characteristics which may affect management policies and population size. The size, shape, and character of nature reserves must take into consideration the behavioral requirements of their faunas.

Studies of behavior usually require detailed studies of recognizable individual animals. Among sedentary species with distinctive markings such studies can be carried out without capturing or handling individuals. Usually, however, species must be marked for long-term recognition. Markings used can be rings on bird legs, colored ear-tags or collars on mammals, or other devices. Where short-term studies are involved, dyes and bleaches on fur or feathers are useful. For studies of travels and migrations, radiotelemetry is now widely used. Radio transmitters are attached by collars, clips, or implantations. The animals broadcast on individual frequencies and their locations are followed by using receivers in airplanes or vehicles, carried by observers on foot, or mounted on fixed towers. Detailed information on home range, habitat selection, food preferences, territorial defense, and migratory travels can be obtained in this way. Needless to say, it is expensive.

3.4.5 What to monitor.

The essence of monitoring is that it is a continuing process, which means it can be expensive and labor intensive. Consequently, it is important to be highly selective in determining *when* monitoring is appropriate and in choosing what to monitor. Monitoring is an activity which figures prominently in the recurrent cost of projects.

The measurements involved in monitoring activities can be considerably fewer than those involved in establishing the baseline. Inherent in the baseline study concept is the need to collect initial data on a wide variety of parameters, but monitoring focuses on continued data collection only of those parameters identified as important to meet specific research or management information needs and is a much more selective process.

During baseline data collection, a much increased understanding of the ecosystem is achieved. This often enables the recognition of attributes which can function as indicators of ecosystem changes. Such attributes are prime subjects for monitoring. One example of a reliable indicator situation--i.e., the trends in preferred food species--was discussed in section 3.2.2. As a second example, raptors have been recognized as indicator species in situations where pesticide accumulation is suspected. Of course, all species are indicators of one sort or another. Their increases and decreases reflect some sort of environmental change, once random fluctuation is taken into account. As a consequence, gross changes in relative abundance of species can often merit recording and subsequent endeavors to interpret their significance.

Another useful approach to habitat monitoring is the examination of limiting factors (section 3.2.3). There are other situations when it might be prudent to monitor a variety of habitat parameters to augment the state of knowledge of a particular ecosystem's dynamics. Many of the methods used for these studies are discussed in the parts II, III, IV. One example of this approach is the use of exclosures to deny animals access to an area, thus allowing a comparison of grazed and ungrazed habitat. Such areas can provide an early indication of rangeland degradation (changes in plant species composition, abundance or vigor, acceleration of surface soil erosion, etc.), before the situation becomes apparent in the condition of the grazer. Exclosures can also be used for separating the effects of grazing between wild and domestic species.

When monitoring animal population dynamics, there are three essential aspects that need to be assessed on a continuing basis: the birth rate, the actual conditions of the animals (healthy and well-fed versus sickly and underweight), and whether there are any marked changes in abundance. These can often be done on a somewhat qualitative basis without a need for painstaking precision. Such data should provide a reasonable indication of whether habitat quality is declining or whether productivity is increasing or declining because of hunting or some other influence.

3.4.6 Carrying capacity.

Questions that are frequently asked concerning animal populations, wild and domestic, are whether or not a particular area is: (i) overstocked and consequently in danger of a serious deterioration in habitat followed by a die-off or population collapse; (ii) understocked and with the likelihood of a marked increase in numbers; or (iii) properly stocked; that is, in a dynamic balance between animal numbers and available food, water, and resting, breeding, or escape cover. Related to these questions are those concerning the causes of the particular condition. All of these questions relate to the *carrying capacity* of a particular habitat. This can be defined as the maximum number of animals that a specified area can maintain in a healthy and vigorous condition, within a stated time interval without damage to the habitat. There are several aspects to this definition which are important to understand. First, carrying capacity is *not* a static concept. Climatic conditions have been known to change over time, and these changes definitely affect the carrying capacity. Second, the maintenance of these animals is *not* at the expense (detriment) of the habitat, i.e., it is achieved with the surplus productivity of the food source. Third, this definition implies that the true carrying capacity is the number of animals which can be supported during the *least* favorable years. This is a crucial point for management considerations, because climatic factors also fluctuate over the short-term, often with profound effects on the environmental productivity. See chapter 5, Techniques for Management.

Some species are essentially self-limiting in their populations, which is to say they do not tolerate crowding above a level at which their needs can be met, a level which may change with increases or decreases in food or other resources in short supply, such as the number of watering points in the dry season. These species are generally *territorial* during all or part of the year, meaning they repel intruders of their own species from areas which they maintain as exclusive territories. Many predators and some herbivores are territorial in behavior.

Other species may be limited in their numbers primarily by predation, and these as a rule become more vulnerable to predators as their numbers increase relative to the availability of secure places in their environment, and less vulnerable as their numbers decrease. Predators seldom eliminate a prey species, but frequently hold its population at a lower density than the food supply would otherwise support.

The final category, which includes many kinds of grazing herbivores, is limited primarily by food supply and its distribution in relation to other necessities of life. These species include much of the spectacular big game of Africa and their equivalents in Asia and Latin America. They are subject to periodic buildups in population during years of favorable rainfall and die-offs in subsequent years of drought and diminished forage production. They have the capacity to overgraze and damage their environment and thus reduce its carrying capacity.

Unfortunately, for economic motives or through long-standing cultural practice, many people allow their domestic livestock to follow population patterns similar to wild herbivores, increasing to excessive numbers during favorable years, overgrazing and damaging their habitat (range or pasture), and then frequently dying off during drought years, or in mountain areas in periods of cold and heavy snowfall. Repetition of this pattern over a period of years causes vegetation destruction, soil erosion, and formation of erosion pavements, soil compaction, and diminished soil water-holding capacity, leading in turn to more rapid runoff and diminished water supply during dry seasons--in other words, to diminished carrying capacity.

Similar patterns of over use can develop in humid areas where water is not in short supply, if the numbers of livestock and wild grazing animals are not controlled. The danger in such areas is not from desertification, but from the removal through grazing of all nutritious forage and invasion of the area by other species of low or no nutritional value.

Techniques for analyzing animal numbers in relation to carrying capacity fall into two categories, those based on examination of the animals, and those based on habitat examination.

Inspection of animals should be carried out during the season of greatest environmental stress, usually the end of the dry season in the tropics. If animals at this time are obviously in good health, reasonably well fleshed, not weak or scrawny in appearance, it will be apparent that they have received adequate nutrition. If there is also a good survival of young animals relative to breeding females, this is further evidence that carrying capacity has not been exceeded. If the reverse is true and survival of young is poor while most adults are bony, scrawny, sick or weak, overpopulation may be suspected. Veterinarians should be consulted if the animals are in poor condition, although there is no obvious habitat deterioration, since a variety of diseases not directly related to malnutrition could also be involved.

It should be realized, however, that animals can remain in reasonably good condition during the earlier (and most easily reparable) stages of habitat deterioration. Attention should, therefore, be given to the condition of the vegetation, especially during the season of greatest climatic stress. Look for: (i) an overabundance of or an increase in bare areas including an excessive number of animal trails; (ii) evidence of excessive erosion, such as exposed plant root crowns, lichen lines on rocks, erosion pavement (usually with a layer of small pebbles on soil surfaces), soil drifting to form dust piles behind vegetation, rills, or gullies (for further information, see part III); (iii) heavy use of those species on which grazing animals prefer to feed; (iv) reproduction and spreading of species on which animals do not appear to feed. (Techniques for habitat analysis are discussed in greater detail in part IV.) Bear in mind that excessive use will be present in certain areas where animals prefer to gather and usually around dry-season water holes. It is the relative extent of such areas that must be considered in relation to total forage availability.

The techniques for treating the problems of overpopulation are simple to state but often most difficult to put into effect. They involve control over animal numbers, meaning in the long run, if not immediately, removal of excess numbers. Immediate solutions to the problem can involve making new pastures available (e.g., through water development), but as noted earlier, this can simply spread and accentuate the problem unless numbers can be kept within the new carrying capacity. But encouraging pastoral people to reduce the size of their herds usually involves providing them with some other guaranteed means of economic security.

Reduction of wild animal populations is most economically done by encouraging hunting, trapping or other forms of removal. With species that are locally abundant, but scarce or absent in other areas of suitable habitat, trapping and transplanting them to new areas may be justified. The encouragement of increased hunting may have political consequences that are forbidding, for example, when the excess numbers are within a national park. The great increase and subsequent die-off of elephants in Tsavo National Park, Kenya, was particularly the result of a political dilemma, accentuated by disagreement among experts whether to allow hunting in the park and how to compensate farmers adjacent to the park for wildlife damages.

It remains true, however, that where populations are not self-limiting, and not controlled by predation, control of numbers through human agency provides the only alternative to the "boom and bust," buildup and die-off, cycle with all of its potential consequences for habitat change or deterioration.

3.5 Aquatic Habitat and Fisheries

"Wildlife," as defined in this volume, includes fishes, which sometimes present quite different problems for research and management than terrestrial wildlife. An exhaustive treatment would not be appropriate here; rather, this section attempts to contrast methods for aquatic wildlife with what has been said above for terrestrial wildlife.

3.5.1 Population status.

Our knowledge of aquatic species and their habitats is not nearly as extensive as for terrestrial ones. An exception to this is the fact that commercial and subsistance species have been caught in large quantities for a very long period of time. The numbers of individuals caught are generally much larger than for terrestrial wildlife, and because many species "school" in single-age-class groups, the quantitative data available for population dynamics studies are often very extensive. In such cases, historical commercial fisheries records can provide valuable insights into original status.

The status of aquatic, as well as terrestrial, species is subject to change in two quite different ways: (i) individual species depletion or enhancement, and (ii) community structure alteration. For fresh water systems, the former is predominant; for coastal and marine systems, the latter is predominant. Fresh water lakes, streams, and rivers present exceedingly disjunct habitats, which allows a high degree of speciation; i.e., every tributary of a river or habitat of a lake may support a different species of a single genus, each of which may be subject to depletion to the point of extinction. By contrast, coastal and marine systems are much more continuous. Their species are generally much more widely distributed and less subject to extinction, but overfishing and habitat alteration have resulted in large-scale community alteration. Whether such changes result in ecosystem instability or lesser fisheries yield must be evaluated on a case-specific basis and can be very difficult to evaluate.

3.5.2 Ecosystem characteristics.

Terrestrial and aquatic ecosystems differ in many ways, but for present purposes, one is essential, i.e., the differing emphasis on *biomass* versus *productivity*. Both terrestrial and aquatic ecosystems are based upon plants, but terrestrial plants accumulate very large biomass. By contrast, aquatic plant (phytoplankton) biomass is usually smaller than the herbivores that eat it; they make up for the lower biomass (= standing crop) by high productivity rates. The consequence is high turnover rates measured in hours, days or weeks, whereas the turnover rates of the carbon and other material incorporated in trees, for example, is measured in decades or even centuries. This leads to very different objectives for monitoring of fish populations and their habitats.

There are exceptions: some aquatic plants are rooted (e.g., pond weeds and sea grasses) and others are attached to the substrate (e.g., algae). These are somewhat equivalent to grasslands in that they are "grazed" and in that the rooted species derive many of their nutrients from the sediment, which is roughly equivalent to terrestrial soil.

3.5.3 Censusing and monitoring.

An exceedingly large body of literature and a diverse array of methods exists for fisheries censusing. As stated above, gathering a large data base for fisheries is relatively simpler than for terrestrial animals; a single cast of a net may yield hundreds to thousands of individuals. By contrast, habitat assessment presents formidable difficulties. No photographic remote sensing method is able to penetrate more than a few meters, even into waters of great clarity. Therefore, one is forced into often difficult and expensive underwater techniques. There is no need to emphasize the giant advance in aquatic research since the development of scuba diving equipment and other man-in-the-sea techniques. These have added immeasurably to remote sampling by grabs, dredges, acoustics, photography, and the like, but one is always faced with the fact that assessment of aquatic habitats is more expensive and time consuming than assessment of terrestrial ones.

With these caveats in mind, the methods outlined above for terrestrial species and habitats do apply to aquatic ones. Transects, marking, and sampling must still be done. Habitat indicators and food preferences must be determined, as must "signs" of species presence or activity. Local knowledge must still be incorporated, and life tables--which for fish usually take the form of population dynamics models--must be constructed. And, perhaps most essential, the behavior of populations and their ecology relative to the factors limiting abundance must be determined.

It is obvious that, due to relatively poor knowledge of aquatic habitats and fish populations, the initial emphasis may fall upon survey or research, then later turn to monitoring when some of the important unknowns are clarified.

3.5.4 Habitat management.

Some aquatic areas are subject to very specific habitat modifications and management, directed toward specific and predictable purposes, for example, ponds for aquacultural purposes. For others, one must rely upon the natural productivity and processes of ecosystems for which modification toward predictable purposes remains a far future possibility; the coasts and oceans have been modified, but it is as yet far from certain what the near-term impact on fisheries has been or what the future outcome will be.

In this respect, aquaculture presents an interesting alternative for future food production. However, much aquaculture, particularly of marine and estuarine species, depends upon wild populations for eggs or larvae and upon unpolluted waters for raising healthy fish or invertebrates. It often also depends upon a harvest of natural, wild products for food for the species being cultured. A concomitant is, of course, that aquaculture itself can be a source of pollution, resulting in eutrophication of receiving waters. All these matters must be carefully borne in mind in order to realize the full potential of aquaculture.

Methods for Determining Past Status of Wildlife

One cannot assume that present wildlife status and habitat conditions are as they were in the past. In addition, the ability to predict future trends is to some extent dependent on knowing how the present situation differs from that in previous times. For these reasons there is particular value in gaining information about previous status.

Published faunal lists from the nineteenth century or earlier can often provide information of this sort but unfortunately sometimes only on a presence or absence basis. The reliability of the author is a critical element in determining the usefulness of such information. Some lists were based on actual museum collections, many of which are still intact. In such instances, it is possible to return to the original data and even, when necessary, to verify or correct the original identifications. The value of such collections varies a great deal, however, and in some cases, the associated data are meager, poor, or nonexistent. Sometimes, information on the locality or date of the collection can only be determined by reading the expedition account.

Such collections rarely if ever can be taken as an unbiased sample of the local populations. Abundant species tended to be "undercollected" and were sometimes even ignored as being uninteresting to collectors who pursued their quest for the rare and the new. Nonetheless, a great deal of information relative to population and habitat change resides as yet untapped in the collections of the natural history museums of the world.

Most of the larger museums are located in Europe and North America remote from many of the countries where the collections were made. Consequently, these data may not be easily accessible to scientists in developing nations. Yet the value of such information argues for making the effort to gather and interpret it.

Historical records can often be surprisingly valuable in providing clues to the former presence of species. Depending on the quality of the observer, historical records also sometimes provide an indication as to whether a particular species was absent in the past, although such "negative" data will always be subject to some degree of question. Historical records can also provide some indication of past environmental conditions--for example, the cedars of Lebanon, which are confined to a ten-acre walled plot today, were once much more widespread. Similarly, historical accounts of fishing in lakes in areas such as the Adirondacks of New York state are important in establishing the changes in those lakes that can be attributed to acid precipitation.

Historical accounts, including those of early expeditions, provide useful indications of the abundance of particular species, for example, the American bison, the great bustard of Europe, or the quagga (extinct) of Southern Africa. It is equally useful when such records indicate a species to have always been relatively rare, because it may well indicate that present rarity is at least partly due to the ecological requirements of the species as opposed to being caused solely by human activities.

The accuracy and usefulness of such historical information vary widely. Some regions and types of habitats were far better explored than others. Tropical rain forests remain the least known of all terrestrial biological formations, for example, but those of Southeast Asia are far better known than those of South America.

Potentially useful information can be gleaned from less conventional sources such as oral history handed down by indigenous people. Useful glimpses of past animal life, and attitudes and practices regarding wildlife can often be found in petroglyphs (rock paintings), in wall paintings of antiquity such as Egyptian tombs, and in more recent forms of art.

In addition to the review and evaluation of historical records and collections suggested above, considerable information can sometimes be acquired from the analysis of faunal remains from archeological sites, a study known as zooarcheology (Cornwall, 1956; Chaplin, 1971). The analysis of archeological sites occupied within the last few thousand years provides an opportunity to identify animals from the more pristine environments because these sites were usually occupied under climatic conditions similar to those still extant today.

There is a wide range of sources, tools, and techniques for gathering, recording, and manipulating traditional resource data--from literally using children's "Tinker Toys" to the use of elaborate computers (Chambers, 1980). The following are merely a few of the many possibilities:

Libraries are a good place to find records of indigenous knowledge. There are very few places (including the most remote island villages) that have not been studied at one time or another by early travelers, historians, anthropologists, or geographers. Since traditional villages are generally connected to district centers that have small libraries and museums, chances are good that some historical documents can be found locally. If not, regional libraries (particularly university libraries) have very extensive collections. For instance, the University of Guam (Agana, Guam) houses information on the island districts of Micronesia and the University of Hawaii (Honolulu, Hawaii) is a major center of information for the whole Pacific Basin. Libraries, however, will generally only provide information on past systems of traditional resource management, and at that, the information will be rather meager and anecdotal (i.e., very few books, periodicals, survey reports, academic papers, maps, atlases, or government statistics deal directly with traditional forms of conservation).

Government documents are a good source of information regarding when or why traditional conservation techniques died within a region. It is often political pressures (e.g., colonial governmental attitudes toward pastoral nomads) that instigate the breakdown of cultural traditions. And, in some cases, information from village documents can be obtained regarding existing systems of conservation, such as a Council of Elders (a traditional governing body) that keeps records on villagers fined for having hunted or fished illegally.

Missionary records often contain some information that can be useful, such as a church's attitude and policy toward the indigenous religion. In addition to better understanding the missionary's role in cultural breakdown, these descriptions sometimes contain tidbits of information on the role of animals in traditional religions and society.

Interviews are a must if traditional resource systems are to be recorded and understood. They can be in the form of an oral history (recorded recollections of a village elder), an informal discussion, or a structured and standardized questionnaire. In all cases, however, key informants must first be identified. As with any society, there are simply "those that know" and "those that don't." Specifically, the people that need to be identified (which isn't always easy) and interviewed are the traditional "game wardens" and "fisheries ecologists" that oversee the region. These people have the expertise and technical knowledge that must be recorded; the other locals will only have, at best, a rudimentary understanding of the problem. However, the information obtained from the ecologically aware informants can be later double checked through group interviews with the other less-knowledgeable villagers. This step helps the key informants recall more detail.

Interviews can be used to discover all the following and more: (i) the whereabouts of past and present game preserves; (ii) game laws (e.g., hunting seasons and bag limits); (iii) attitudes towards keeping pets; (iv) systems of time reckoning that relate to resource utilization (e.g., fishing seasons based on the lunar cycle); (v) fishing rights to specific geographic regions; (vi) knowledge of animal behavior (e.g., life histories, feeding habits, mating seasons, and habitat requirements of terrestrial wildlife and marine species; and (vii) specific regulation according to traditional laws.

Observation is another means of gathering resource information. This can be in the form of direct observation (merely watching from a distance) or participant observation (joining in and learning by doing). The benefit of observation as a research technique is that researchers can learn for themselves. The drawback to observation, however, is that it requires enormous amounts of time, is often impractical or awkward (especially true for participant observation), and requires professional training if the results are to be useful.

Nevertheless, on terrestrial hunting parties, the participant observor can identify (i) wild animals that are sought for food, hides, or antlers; (ii) wild animals that are avoided and/or protected; and (iii) attitudes toward animal habitats (e.g., treatment of nests, bird eggs, young, or pregnant animals). On fishing expeditions, the participant observor can record (i) marine species sought, avoided, and/or protected; (ii) construction and use of fishing hooks, lines, traps, and ponds; and (iii) locations fished vs. locations avoided.

Diaries kept by village members are important, since the researcher cannot be at all places at the same time. When individual household members cannot be recruited to keep a diary, local university or high school students that have a familiarity with the study site can sometimes substitute. Diaries can be used to record (i) types and duration of work effort in the household as it relates to fishing or hunting (including who is doing the hunting or fishing, the frequency, and the period); (ii) kind and amount of food consumed (e.g., percentage of food from wild animals vs. domesticated animals and agricultural produce, which might entail the daily weighing of butchered meat); (iii) food preferences and avoidances; (iv) relationships between human sexual activity, i.e. taboos, and frequency of hunting (although accurate record keeping in this instance may be questionable); and (v) changes in dietary habits (including wild animals consumed) with the spread of western civilization.

Photography, in its multitude of forms, can be used to record all kinds of important data regarding resource management. For example, conventional still photography can be used to record indigenous trees and their associated wildlife, or terrestrial and marine species in the field for later identification and discussion with village resource managers.

Whereas early ethnographic films can be viewed to discover some traditional conservation techniques (hunting techniques, fishing methods, local food habits), new ethnographic films can be created to record remaining conservation systems for the purpose of "reeducating" those locals that have forgotten the old way (and often the better way) of doing things.

More sophisticated forms of photography, such as aerial photography and remote sensing, can be used to identify (i) habitat manipulation and maintenance (e.g., fire traditionally used for game control and clearing); (ii) indigenous trees on which various forms of wildlife depend; (iii) introduced grazing animal ranges vs. traditional game reserves; and (iv) succession of vegetation types and associated wildlife important to the particular cultural group. Aerial surveys, however, can only tell so much and therefore must be "ground-checked" (confirmed) by using indigenous technical knowledge (see remote sensing section in part I).

Maps of all kinds can be used to record existing or recent resource information. For example, vegetation maps can be used to plot flora/fauna relationships; topographic maps can serve as base maps to plot ancient archeological sites, forest groves, and wildlife preserves; and marine maps can be used to locate and plot existing shoals, coral reefs, and shoreline areas that are traditionally fished.

The scope of wildlife and fisheries mapping using indigenous technical knowledge is practically unexplored. With cartographic skills, the researcher can map (i) migration patterns in those societies that vary their pattern of occupation according to the environmental limitations of the seasons, thereby lessening the intensity of wildlife exploitation; (ii) migration patterns in those societies that move to new locations only after the old areas become depleted; (iii) human settlement patterns in association with water holes for wild animals to see which ethnic groups may avoid such areas; (iv) hunting ranges (the extent or territory covered by a tribe in pursuit of game); (v) regions, such as natural forests, that were believed to be inhabited by ancestral spirits and thus become protected areas; (vi) game preserves; (vii) past tribal war zones and associated buffer zones (the latter being beneficial to wildlife); and (viii) those areas left "fallow" because of past or present human population shifts or migration trends.

Statistics and computers are useful tools for the manipulation of resource data. Statistics, for example, can be used to calculate: (i) the nutritional values of individual intakes of wild animal foodstuffs; (ii) the changing acres devoted to game preserves; (iii) tribal population pressures on wildlife reserves; (iv) the effects of modern organized poaching of animals within traditional wildlife preserves; (v) the effects of new technologies (e.g., high-powered, radar equipped fishing boats, rifles, snowmobiles) within traditional wildlife preserves; (vi) frequency and severity of drought and its effect on local flora and fauna (which impacts directly on the risks and advantages of hunting and herding in a particular region). Regarding studies in traditional systems of resource management, computers are particularly useful in both time allocation analysis and the creation of energy flow models that compare traditional with modern systems of managing resources.

In most cases, the gathering of indigenous technical knowledge will require not one but several of the above techniques in combination, from the simplest to the most complex. For example, to discover Tasbapuani turtle grounds, Professor Bernard Nietschmann used techniques ranging from children's Tinker Toys in the field to computers back at his home university (Johnson 1978). Miskito Indians of Eastern Nicaragua know where every bank and reef is located and can find them in almost any weather condition. Coral reefs, Nietschmann was told, could be "felt" on the surface even though they were practically invisible from above--their location being revealed in the way a canoe behaves in the ebb and flow of cresting waves. But how was he to gather and map this information?

His first step was to identify the most knowledgeable fisherman (the key informant) within the village. Since the Miskito fishing grounds were very large, a method was needed to represent the underwater features on a small and manipulable scale. Consequently, he needed a mapping "tool" that was easy to handle, consisted of several components, and most importantly, was available in the field. What met all of these requirements was a box of children's Tinker Toys.

Nietschmann's key informant took the Tinker Toys and positioned and repositioned the individual pieces (using the traditionally used practice of backsighting and triangulation) until the toy pieces resembled an aerial view of the turtle banks, islands, and shoals. Once his informant had the pieces arranged in proper order, Nietschmann sketched the pattern and labeled each symbol with its respective Miskito name. He then photographed the exact location of the Tinker Toys on the floor. From this sketch, a base map was prepared that he took on fishing expeditions (participant observation) for the purpose of double checking the data as well as adding new information that was accidentally overlooked during the land-based interview.

Finishing touches on the map were done back at Nietschmann's home university, which involved redrawing and reducing the map to proper scale, and photographing the final paste-up. The point of all this is that although expensive equipment may be used and sometimes required for the final preparation of a document, the gathering of resource data in the field may only require the very simplest of materials. It does, however, require professional training.

Techniques for Management

5.1 Passive vs. Manipulative Procedures

There is a dichotomy in approaches to management of wildlife, often with a profound underlying philosophical difference. One school prefers the passive, hands-off approach, essentially taking the view that if just left to itself nature will take care of its own. The other argues that humans have already affected the face of the planet to such an extent that only active manipulative management can possibly be appropriate. The reality is that management encompasses both manipulative and passive approaches: the only problem that arises is if the passive approach is allowed to be a fatalistic one which fails to recognize when the passive approach itself is failing. Certainly the habitat fragmentation problem would indicate that there are many instances where active management is necessary to alter changes triggered by humans. Yet clearly there are situations where a pristine waterbody or watershed is managed best by leaving it alone.

5.2 Core vs. Buffer Areas

The two approaches to management are recognized in the reserves design approach exemplified by the Man and Biosphere (MAB) reserve program of UNESCO. Such reserves contain a central core area of sufficient size to maintain their own integrity and diversity, along with an outer buffer area. The latter serves two purposes, one being that expressed in its name, i.e., to provide a protective zone around the pristine core, and a second to provide a relatively natural (but *not* undisturbed) area where research relating to various uses and manipulations of the particular ecosystem can be undertaken. In coastal marine environments buffer areas should include sources of nutrient supply for species in the protected core of a marine park or reserve.

5.3 Controlling Habitat vs. Controlling Species

An interest in controlling species stems from a desire to improve their lot (endangered forms) or reduce their effects (pest species). This is always a very complicated situation and one meriting a cautious approach because the relationships between a species and its biological community are always multiple and can be affected in not easily foreseen ways. For example, the use of poison to reduce pest species often also affects detrimentally valued species in its food chain. Introductions of predators and diseases affecting the pest do not necessarily confine their effects to the targeted species. The essential point is that *affecting one species in an ecosystem will inevitably affect others*.

Sometimes manipulating the habitat itself rather than the species directly is an easier and more effective approach. If an endangered bird requires old rotting trees for nestholes, then altering forestry practices to leave old trees is one of the most important management policies to encourage. If a particular successional stage of vegetation is important for an animal, periodic burning is the most effective procedure. If migratory shorebirds can benefit from extensive mudflats, temporary lowering of reservoir levels is an important technique to use. Similar habitat management can diminish the populations of pest species, but it is always important to determine in advance whether any valued species might also be adversely affected.

Propagation of endangered or game species is an important technique in aiding the recovery or enhancement of a wild animal species. It is not to be undertaken lightly nor is it ever proper to view it as a substitute for conservation of the species in nature. In some instances it is a critical boost to efforts to protect a wild population or an important insurance step; in yet others it may hold the only hope for the species.

In most cases (i.e., when there is natural habitat remaining which is reasonably secure) reintroduction should be the ultimate goal of the exercise. This is not a simple matter and requires study and monitoring to increase the possibility of success. Failure rates have been high in the history of reintroduction.

5.4 Exotic Introductions

Introduction of animals for decorative, recreational or other purposes has been a far more frequent activity than is usually recognized. In the majority of successful cases the results have been detrimental to local wildlife. This has been true whatever the purpose, whether stocking of trout or bass for fishing or introducing mongoose for pest control or animals for hunting. *Introduction should only be undertaken in rare instances and only with considerable advance study* and with the possibility of reversal if adverse consequences become apparent.

5.5 Animal Rescues from Hydroelectric Flooding

There is something instantly appealing about rescuing animals from areas about to be or in the process of being flooded. It has a humane appeal and also it seems sensible to avoid the waste of such animal resources. Yet it is important to recognize that such undertakings can be expensive and labor intensive. More importantly, the reintroduction of the rescued animals is a considerably more complex undertaking than releasing the animals into some remaining wild habitat. Such habitats may already be filled to carrying capacity with other individuals of the same species. If not, they may not be suitable for some subtle and inapparent reason; i.e., there may be an important reason why the species is not already there.

5.6 Maintenance and Restoration

As noted earlier (section 5.3), maintaining habitat in a particular successional stage is a management goal which can require active manipulation (e.g., fire, grazing). Similarly, restoration of a habitat can require active interventions such as restocking or the introduction of critical or key plant species. The important corollary is that even if the remaining available habitat for a reserve is clearly too little to protect its diversity of plant and animal life, there is no reason not to correct for this by including adjacent land areas where the return of natural vegetation can be encouraged.

5.7 Other Essential Management Policies

It is always critical that plans to protect natural areas include consideration of economic and cultural as well as ecological benefits. Such benefits should occur not only to the nation as a whole but also to the local people. The more the local populace is integrated into the management and activities of protected areas the more likely they are to develop and maintain an interest in their future and security.

Few protected areas can achieve their goals without real, active protection. Simple designation of a protected area, without resources and commitment for protection and management, is almost inevitably doomed to failure.

Careful study and testing should precede any major new management technique or procedure. Funds spent on fish ladders to permit migratory fishes to pass by new dams have sometimes been wasted because the design proved inadequate, or no provision was made to protect downstream migrants. Aquaculture efforts have been defeated by failure to consider the effects of pesticides borne by waters drained from sprayed areas. Game farming or game ranching usually fails, not for biological or ecological difficulties, but through failure to consider the effects of cultural and political attitudes on the cropping and marketing of wildlife products. Always the total system involved, including the human cultural component, must be the unit of development or management.

Cultural Ecological Assessment

In areas long occupied by people who pursue hunting and gathering, subsistence agriculture, or pastoralism as their way of life, a long-term relationship and familiarity is developed between human beings and the wild species of animals and plants that surround them. The ability to discover medicinal or food values in wild plants, and the skill to make use of wild animals as food sources can lead to enhanced nutrition and health. Conversely, failure to recognize poisonous or other dangerous species can be fatal. People who overuse or destroy useful wild species and their habitats will be less likely to survive and thrive than those who learn to protect their wild resources. It follows that a reservoir of knowledge concerning wild species will be built up among indigenous peoples, and often will be maintained even though changes in life styles result from assimilation into a national or global socioeconomic system. Understanding traditional beliefs and customs and making use of this traditional knowledge should be a basic prerequisite to any development program intended to benefit local economies. Indeed, this knowledge can make the difference between success and failure of development projects, and it can save the cost of rediscovering what is already well known.

6.1 Value Systems and Use

What are the major types of values that tropical peoples have attached to wild animals? How might they be classified? These questions must first be answered in order to place the human-animal relationship in proper perspective. As it turns out, most human values associated with wild animals can be listed as either *Economic/Practical* (food source, wealth object, recreation or scientific benefits), or *Aesthetic/Intellectual* (religious-spiritual, aesthetic, artistic, or educational values).

6.1.1 Economic/practical values.

Our human ancestors hunted animals and collected plants until the process of domestication began ten to twenty thousand years ago. Animals and animal products have provided *wealth* objects for various cultures. One example is the Maasai, a pastoral people of East Africa, whose herds of long horned cattle form the basis of their wealth (Layall-Watson, 1965). Many cultures find great prestige in wearing or displaying the skins of ferocious animals. Foreign markets also influence the value of animals, for example, rhinoceros are hunted for their horns so that Chinese pharmacists can prepare and sell aphrodisiacs (Bennett, 1975).

In some countries, such as Kenya, *recreation* has replaced wealth as the most important value associated with wild species. Historically, the use of wild species for recreation, whether for game viewing or sport hunting, has been primarily a European or Western value.

6.1.2 Aesthetic/intellectual values.

Animals have long been associated with the *religious/spiritual* needs of traditional societies. Some examples include India (the concept of grateful animals assisting their benefactors); Thailand (the white elephant--symbol of good fortune); and Oriental cultures (the belief that Buddha is reincarnated and lives within the white elephant) (Tocher and Milne, 1974). Human fulfillment, psychic need, compassion, concern, and sensitivity are all notions that have been historically associated with wild animals.

Many traditional cultures also looked at animals as a source of beauty, i.e., for satisfaction of *aesthetic* needs, as illustrated by the following African saying:

> To look at the Zebra, look at the leg;
> to look at the body and colour, you will
> be charmed (Chavunduka, 1978).

Wild species were everywhere more than "meat and manure." Traditional Oriental cultures, for example, are renowned for their love of nature and wild animals for their *artistic* value, as displayed in their literature, painting, and gardening (Schafer, 1962).

Traditional cultures have also tended to personify animals in ways which provide lessons for their children, i.e., assigning an *educational* value to wild animals. Navajo Indians, for example, would not kill chipmunks since these animals could show travelers where to find food and water (Tocher and Milne, 1974).

The most recent value to be associated with wild animals and pristine landscapes is the appreciation of *scientific* benefits such as the maintenance of genetic diversity and ecological stability.

The economic/practical vs. the aesthetic/intellectual value sets, are often at odds with each other. Striking a balance between these two cultural value systems is frequently at the heart of wildlife management conflicts, not to mention the entire global conservation/development issue.

6.2 Traditional Systems of Wildlife and Fisheries Resource Management

6.2.1 What are traditional systems?

By *traditional* is meant natural resource exploitation systems developed by the peoples of indigenous cultures living within a single, or perhaps two or three closely related ecosystems. Some traditional practices may be complex, or involve ingenious devices, but they are often characterized by a low level of technological achievement.

Several conservation practices that recur in traditional societies in different regions are associated with the subsistence resource base of these societies. For traditional cultures that are dependent on local ecosystems for their survival, persistent overexploitation of the resource base can result in a scaling down or disappearance of the culture. This applies equally to hunting, fishing, and subsistence agricultural societies that may overhunt, overfish or neglect to maintain the soil fertility of the land they occupy. Thus hunting-gathering societies--such as the Bushmen of the Kalahari Desert; a fishing society dependent on the productivity of nearby reefs, such as the Yapese of Micronesia; and an agricultural society, such as the Tsembaga of New Guinea--have resource management systems with traditional practices for the regulation of resource use from local environments, whether desert, or coral reefs, or humid forestlands.

Traditional cultures, presumably aware of the finite resources of their environment, have *usually* treated their ecosystems gently, trying not to violate the inherent ecological equilibria. Local constraints in the form of taboos, hunting bans, marine preserves, and so on, preserve the delicate balances between humans and nature. This is not to say that all traditional societies lived within the bounds of their environment; there are too many cases of resource degradation to ignore. One must, therefore, be cautious not to overstate or romanticize successful traditional societies that have been able to adjust to the limitations and potentialities of their bounded ecosystems.

6.2.2 The need for studying traditional systems.

One can glean from the literature various comments about taboos, hunting and fishing restrictions, marine preserves, wildlife sanctuaries, or sacred forest groves, some of which undoubtedly conserved resources. Many of these customs have been studied systematically, but resource managers who have been trained in developed countries have paid little attention to what traditional cultures have had to teach. This has led to strategies for protecting wildlife in tropical and subtropical countries that were developed by Europeans and other expatriates, based on their experiences and traditions in temperate regions. But Africa, Asia, and Latin America have unique wildlife and human needs that require unique regional and cultural solutions. For example, plans for protecting African wildlife call for setting aside large reserves for its exclusive use. This concept is totally alien to African tradition (Beyer, 1980; Lusigi, 1981), and Africans neither appreciate nor support a conservation effort so foreign to their traditional patterns of thought and livelihood (Lusigi, 1977). If established parks and protected areas in Africa are to survive, accommodations must be made that recognize cultural needs of local peoples.

The primary purpose of this section on cultural ecological assessment is to introduce the range of traditional wildlife resource management techniques that can be found in the tropics and to stimulate planners to investigate, record, and incorporate into future development plans those appropriate traditional conservation practices that exist or existed within their local regions.

6.2.3 Traditional resource-using systems: A regional overview.

Oceania. Traditional Pacific island cultures had a variety of methods that undoubtedly had an effect on restoring, maintaining, and increasing wildlife populations. Habitat maintenance, game preserves, game laws (such as hunting seasons and bag limits), and prohibitions against certain sacred plants and animals all played a role in conserving wildlife resources (Table 6.1). It is interesting to note that some island cul-

TABLE 6.1 Traditional wildlife resource management.

Conservation Practice	Examples	A	PD	R	S	References
Habitat Maintenance	*Oceania*					
	Australia	*		R		Powell 1975:18
	Africa					
	Widespread	*		R		Chavunduka 1978:61
	Cameroon	*		R		Beyer 1980:12
	Kenya	*		R		Isack 1976:90–94
	Botswana	*		R		Silberbauer 1972:296–298
	South Asia					
	India	*		R		Sinha 1972:384–385, 388–389, 396–397
	Latin America					
	Amazon Basin		**	R		Carneiro 1961
						Moran 1979:276
						Linares 1976
	East Asia					
	China	*		R		Schafer 1962:288–289
Game Preserves	*Oceania*					
	Pukapuka	*		R		Beaglehole & Beaglehole 1938:73
	Pukapuka		**	R		Allen 1976:8–9
	Marshalls	*		R		Tobin 1952:12
	New Guinea		**	R		Clarke 1971:66–67
	South Asia					
	India	*		R		Murton 1980:89
	Latin America					
	Peru	*		R		Guggisberg 1970:111
	East Asia					
	China	*		R		Schafer 1962:286–288
Game Laws	*Oceania*					
Hunting Seasons	Tokelau		**	R		Allen 1976:8
	Africa					
	Widespread	*		R		Chavunduka 1978:61
	South Asia					
	India	*		R		Murton 1980:89
	East Asia					
	China	*		R		Schafer 1962:304
Bag limits	*Oceania*					
	Widespread	*		R		Allen 1976:217
	Micronesia	*		R		Owen 1969:303
						Elliott 1973:217
	Africa					
	Tanzania	*		R		Pratt 1977:35
						Cloudsley-Thompson 1967:187
	Latin America					
	Central Andes	*		R		Bennett 1975:234
	East Asia					
	China	*		R		Schafer 1962:286–288

A Ancient
PD Present Day
R Recognized Conservation Practice
S Secondary Conservation Practice

498 V. WILDLIFE

TABLE 6.1 (cont'd.)

Conservation Practice	Examples	A	PD	R	S	References
Prohibition of nonsa-cred animals	*Africa* Kenya		**	R		Beyer 1980:24 Isack 1976:90
	South Asia India	*		R		Gardner 1972:414
	Latin America Nicaragua		**		S	Nietschmann 1973:111
	Brazil		**	R		Ross 1978
Prohibition of sacred animals	*Oceania* Tonga	*			S	Gifford 1929:325
	Samoa & Society Islands	*			S	Handy 1927:130
	Hawaii & Niue	*			S	Loeb 1926:171
	Africa Widespread	*			S	Chavunduka 1978:71
	Latin America Paraguay		**		S	Clastres 1972:154
Prohibition of sacred plants	*Oceania* Niue	*			S	Loeb 1926:171
	Marshalls	*			S	Tobin 1952:23
Regulations against keeping pets	*Oceania* Pukapuka	*			S	Beaglehole & Beaglehole 1938:73
Magico-religious taboos	*Oceania* Australia	*		R		Powell 1975:18
	Pukapuka	*		R		Beaglehole & Beaglehole 1938:73
	Tonga	*		R		Gifford 1929:325
	Niue	*		R		Loeb 1926:171
	Micronesia	*		R		Tobin 1952:23–28 Owen 1969:303 Elliott 1973:217 Allen 1976:8
	Africa Kenya	*		R		Lusigi 1981:88 Isack 1976:90–93
	Widespread	*			S	Chavunduka 1978:61
	Botswana	*		R		Silberbauer 1972:321
	South Asia India	*		R		Murton 1980:88 Bennett 1975:110
	Latin America Nicaragua	*			S	Nietschmann 1973:110–112
	Colombia		**	R		Reichel-Dolmatoff (1971, 1976)
Fines and punishment	*Oceania* Niue	*		R		Loeb 1926:171
	Marshalls	*		R		Tobin 1952:27
	Latin America Peru	*		R		Guggisberg 1970:11

A Ancient
PD Present Day
R Recognized Conservation Practice
S Secondary Conservation Practice

tures even specifically forbade the keeping of animals as pets (Beaglehole and Beaglehole, 1938).

Restrictions on wildlife utilization were enforced by a series of taboos and penalties (Owen, 1969; Beaglehole and Beaglehole, 1938; Gifford, 1929). Tobin's field study in the Marshall Islands aptly illustrates the interrelatedness of game reserves, religious ritual, avifauna and certain marine resources. He states that certain islands have been used since time immemorial as game reserves. These islands provided habitat for numerous species of sea birds and nesting fowl as well as turtles, whose flesh and eggs were a valuable protein source for the indigenous atoll peoples. Divine sanction was requested prior to collecting these foods. A gathering party proceeded in single file behind their high chief, each individual carefully stepping in the footprints of the person in front of him so it would appear as if only one person had been there (Tobin, 1952).

Many traditional wildlife management practices are still extant in the Pacific. For example, the Council of Elders still regulates bird collection in the Tokelau Islands, and traditional game reserves are still kept in the Pukapukas (Allen, 1976). Clarke observed that the sacred groves (*Komung*) of the Maring serve not only as a source of seed for the recolonization of rain forest trees, but also as hunting preserves (Clarke, 1971). Some plots of rain forest called the "false komung" were not charged with spiritual danger but each was left uncut as a kind of private hunting preserve for its claimant.

Certainly, the conservation value of taboos or prohibitions on the capture of certain sacred animals, the picking or harvesting of certain sacred plants, and the keeping of animals as pets was not always recognized. Most practices were so recognized, however, and these included habitat maintenance (Australia), game preserves (Pukapuka), hunting seasons (Tokelau), bag limits (widespread), taboos (Tonga), and strict fines and punishment (Niue).

The effectiveness of Pacific island traditional wildlife conservation techniques has often been cited in the literature (Allen, 1976; Elliott, 1973; and Owen, 1969), although these island cultures did not always do what was best in terms of conservation management. Nor is there always hard evidence to support the effectiveness of traditional systems of resource management. Two contrasting evaluations of the Maori in New Zealand exemplify the problem of analysis. Cumberland and Whitelaw (1970) write, "he (the Maori) hunted, gathered and fished but always within the limits imposed by the *taboo* and an inherent reverence for nature. His keen sense of conservation meant that he made little long-term impression on the land's surface." Another New Zealand specialist, R. Gerald Ward, Professor of Human Geography at the Australian National University, has this to say about the Maori as a conservationist (Ward, pers. comm.):

> I am a little skeptical of some studies of traditional conservation practices as they have not always distinguished between the traditional idea and the reality of what happened. For example, the New Zealand Maori has been acclaimed as a great conservationist and in theory it may be true, but in fact they succeeded in de-foresting a large part of the country with fire, often wantonly lit, while according to tradition showing great reverence for the forest.

TABLE 6.2 Selected lapsed or dying-out attitudes and practices that might be revived, reinforced, or modified in Oceania. [Source: Modified from Klee (1980, p. 275)]

Attitudes and Practices	Revived and/or reinforced	Modified	Not revived, not reinforced, not discouraged
Terrestrial animals			
Habitat maintenance	x		
Game reserves	x		
Game Laws			
Hunting seasons	x		
Bag limits	x		
Prohibition of sacred animals	x	x	
Prohibition of sacred plants	x	x	
Regulations regarding pets	x		
Magico-religious taboos	x	x	
Fines and punishment	x	x	
Marine fisheries			
Environmental awareness	x	x	
Master fishermen	x	x	
Marine tenure systems			
Fishing rights to specific areas	x	x	
Fishing seasons	x		
Specific species regulation	x		
Food avoidances			x
Traps and fish ponds	x		
Methods of food preservation	x		
Magico-religious taboos	x	x	
Fines and punishment	x	x	

Africa. Habitat manipulation and maintenance was one of a variety of techniques used throughout tropical Africa to conserve and preserve wildlife resources. In Cameroon fire was used for game control and clearing (Beyer, 1980). Certain trees, on which various edible insects fed, were also preserved by many African cultures (Chavunduka, 1978). In Botswana, the G/wi Bushmen stayed within the carrying capacity of their territory by deliberately varying their pattern of occupation according to the environmental limitations of the seasons, thereby lessening the intensity of resource exploitation (Silberbauer, 1972). Traditional Kenyans prohibited settlements too close to water sources that were used by gerenuks, impalas, and oryx to keep children and other individuals from disturbing the water and its aquatic creatures (Isack, 1976).

Restricting hunting to certain seasons was another technique used by traditional Africans to reduce hunting pressure during the breeding season or when animals were accompanied by nursing young (Chavunduka, 1978). Although the Maasai of Tanzania occasionally killed lions or other aggressive animals, they traditionally adopted an attitude of "live and let live"--an informal "bag limit" or restriction on the number of animals killed (Pratt and Gwynne, 1977).

In Kenya, the indigenous religions refer specifically to the preservation of natural things, and made it taboo to kill more than what was needed for survival (Lusigi, 1981). Likewise, Silberbauer notes that the G/wi Bushmen of Botswana would not kill more than they actually needed for fear of angering N'adima, their Supreme Being (Silberbauer, 1972).

The killing of certain nonsacred species was also prohibited by numerous African cultures. Honey bees were protected throughout Africa (Beyer, 1980). Crows and woodpeckers were protected for their alarm purposes in Kenya, and traditional Kenyans also avoided certain animals like warthogs, zebras, ant bears, porcupines, elephants, hippos, and all carnivores which were classified as unclean (Beyer, 1980; Isack, 1976).

Some animals were classified as "sacred" and therefore off limits to hunters. *Totemism* was a widespread practice in traditional Africa, and this custom tended to operate in favour of game preservation (Chavunduka, 1978). Each tribe or clan adopted the name of an animal as its totem or "identification tag"; a practice which helped separate one clan from another. Since the great ancestor of the tribe or clan was often portrayed in the likeness of the totem animal, no member of the tribe or clan could harm or eat the meat of that animal. Eland, buffalo, elephant, and fish were often made totem animals, thus receiving protection.

Many traditional African land conservation practices were an integral part of religious or belief systems. Africans believed, for example, that certain natural forests were inhabited by ancestral spirits and were therefore sacrosanct, so none of the trees and wildlife habitats could be cut down or disturbed in any way. Entire habitats are protected by the Kenyan belief that certain areas of land are inhabited by evil spirits that had been removed from a possessed person. Animals and plants remain undisturbed in these areas since people avoid visiting or walking through them. Superstition also played a direct role in animal preservation. For example, Africans would not come near hyenas, since many believed them to be agents of sorcerers. Bushmen usually avoided hunting lions since they believed lions caused the eclipses of the moon in order that they might more easily steal into the Bushman's hut (Tocher and Milne, 1974). Other African cultures also believed lions to be vehicles for tribal spirits and so did not hunt them. Chavunduka records that the striped grass-snake was not harmed for it was believed to be the medium for the spirit connected with rainmaking (1978).

Many snakes (sign of a safe journey), wasps (a general good luck omen), spiders (killing one was believed to invite bites by others), and insects like safari ants (a sign of rain) are rarely if ever killed (Isack, 1976).

Isack (1976) reports that at least one tribe, the Boran people that live in northern Kenya still practice most of their ancient customs that protected many species of wildlife from being harassed or killed. Certainly other African cultures exist that continue their ancient traditions of wildlife resource management.

The concentration of plains herbivores which Western societies now seek to conserve probably exists only because of a tolerance for wildlife in African society which is sometimes lacking in other cultures. The recorded contribution of African cultures to the survival of wildlife on the continent include the Maasai (Pratt and Gwynne, 1977; Cloudsley-Thompson, 1967); the Bushmen (Lee, 1972); and the Borane (Isack, 1976). Other cultural groups, such as the Hadza of Tanzania, have paid little attention to the conservation of resources (Simmons, 1979).

There are also examples of inadvertant conservation. For example, large shade trees, important for tribal meeting places, are rarely if ever cut down. Isack notes that even traditional tribal rivalries have a tendency to preserve wildlife habitats (1976). For example, the Boran and Samburu tribes each fears the other, so they have retreated from their tribal borders by as much as fifty miles. This leaves a stretch of shrubland about one hundred miles wide devoid of human settlement and activity.

Asia. For at least two and a half millennia, the Birhors of northeastern India conserved forest habitats by periodically moving their *tandas* (isolated settlements) to new forest locations as the resources in the old areas became depleted. This was a conscious attempt to maintain the long-term yield of the forest and its resources (Sinha, 1972). Although not directly for wildlife protection, this was certainly a form of habitat maintenance, since the forests, jungles, and marshes of South Asia formerly were inhabited by an enormous variety of mammals, reptiles, and birds (Murton, 1980).

In pre-British India, Murton notes that South Asians achieved a relationship between humans and nature which protected human survival (1980). He does note, however, that the strong religious sanctions against killing were often offset by just as powerful hunting practices, especially against predators by peasant farmers as well as princely elites.

Game preserves were also used to protect wildlife. Small preserves were almost everywhere in South Asia, but they were established primarily for ruling princes. Hunting seasons were also established across the country for birdlife and major game animals (Murton, 1980). The Paliyans, hunters and gatherers in the hills of peninsular India, had a policy of not exploiting numerous nonsacred animal species within their environment--a form of specific species protection (Gardner, 1972).

Hinduism played a major role in wildlife protection in India. In general, "...Hindus regard all animals with respect and protect animals to a degree that might astonish even the more avid animal lovers in Western society" (Bennett, 1975). The belief that there is no separation of man, nature and the supernatural was widespread throughout traditional India (Murton, 1980). Although the Hindu religion protected animals to a degree, Bennett also reminds us that not all persons in the region were Hindus (1975). Therefore, we must look for additional reasons to explain how so much of India's wildlife has survived to the present. Examples of other practices that played roles in wildlife protection were the fact that hunting was not commonly practiced by the masses in South Asia, and that ancient South Asian folklore instilled the fear of large carnivores (and thus their avoidance) amongst the citizenry.

In ancient China, habitat maintenance for wildlife was a common practice (Schafer, 1962). Old Chinese almanacs warned against gathering eggs, destroying nests, and hunting young or pregnant animals. Game preserves, park reserves, and natural gardens were also protected and often were stocked with all sorts of wild animals and birds for the pleasures of royalty. Although the huge imperial parks of Han were established for multiple uses (agriculture, grazing, hunting, and fishing), these parks were reserved for the ruler and those whom he wished to honor. Schafer also notes that hunting seasons and bag limits were resource management techniques used to assure controlled harvesting of plants and animals to the advantage of the state. There were resource wardens who designated what kinds of creatures were entitled to protection and the degree of protection afforded (Schafer, 1962).

Latin America. Game preserves were used by royalty and nobility to preserve wild animals for sport in South America. In Peru, for example, the Incas strictly controlled meat eating and bag limits with heavy fines and punishment for poachers (Bennett, 1975). The privilege of hunting in established wildlife preserves was reserved for the Inca caste and the regional tribal rulers.

Religious beliefs also played a part in relationships between humans and animals in Latin America. For instance, numerous spirits, beliefs, and cultural constraints controlled or limited the daily harvest of many animals available to the Miskito culture of Nicaragua. Since it was believed that many spirits and supernatural animals existed in the forests, rivers, lagoons, and seas, traditional Miskito Indians rarely hunted at night when some animals were more easily taken. Miskito Indians also had taboos against killing certain nonsacred species such as the white-tailed deer (Nietschmann, 1973). Similar taboos against killing and eating deer were found amongst the Bayano Cuna of Panama and the Kuikuru of Brazil.

Many of these prohibitions on nonsacred species are still extant today in Central America. In Nicaragua, for example, Nietschmann sees "food preferences" as an adaptive mechanism for maintaining acceptable levels of exploitation (1973). Miskito Indians still do not eat certain animals (e.g., opossum, howler monkey, squirrel), they rarely eat beef, pork, or brush rabbit, and restrict the eating of other animals (e.g., green turtle heads or Hicatee turtles by pregnant women). As a further example, river-adapted groups show little interest in wild game and surround its exploitation with taboos. The Kalapalo and the Kuikuru of Central Brazil have cultural prohibitions that emphasize the poor return of labor invested in hunting (vs. fishing) in a riverine environment (Ross, 1978).

Taboos against the harvesting of certain so-called sacred animals still persist in Paraguay, where the Guayaki Indians do not kill or eat certain birds that serve important mythological functions (Clastres, 1972).

The Tukano of the Vaupes River in the Columbian northwest Amazon continue to see human society and the local fauna as sharing the same pool of reproductive energy (Reichel-Dolmatoff, 1971 and 1976). Since it is believed that the fertility of both men and animals have a fixed limit, it is important for the Tukano to seek an equilibrium in human sexual activity so that the animals of the forest can reproduce and, in turn, serve as nourishment for the human population. This cultural mechanism controls depletion of wild game by limiting both the sexual activity of the hunter and the frequency of hunting.

Habitat maintenance is still practiced by many South American tribes (Moran, 1979). Carneiro (1961) claims that tribal villages rarely reach carrying capacity, but they split up well before overshooting their resources. This practice no doubt helps preserve forest habitats, including their animal populations. Linares (1976) has noted that the practice of "garden hunting" (hunting in the vicinity of planted gardens) maintains habitats by shifting the hunting activity from the forest interior to the edge of the forest. According to Linares, this practice eliminates seasonality and scheduling problems and increases the biomass of selected animals that live at the forest edge (for example, armadillos, rodents, and small deer that are caught while raiding gardens). Still being practiced by inland South American groups, such as the Guaymi and the Cuna, garden hunting may be considered to be a substitute for animal domestication.

Other Areas. No doubt other tropical regions of the world have traditional wildlife conservation techniques that still persist in the 20th century. As Tocher and Milne have noted, these ancient ideas and practices still persist among indigenous peoples on every continent (1974).

6.3 The Disruption and Breakdown of Traditional Practices

Despite the apparent values in some of the traditional resource conservation systems, many of the underlying rules and regulations of these practices have been breaking down for decades. The major reasons for these disruptions are: (i) increased contact with outside cultures; (ii) human population increases and the associated expansion of agriculture; and (iii) the failure of colonial education, administration, and enforcement systems. Other minor factors involved in this breakdown are probably in the hundreds, they include such events as the advent of political independence; organized poaching; new technologies and weaponry; and dietary changes. Bodley (1975) believes the breakdown of traditional conservation systems has generally resulted from the invasion of *tribes* (subsistence peoples) by *states*, although this oversimplifies the changes.

6.3.1 Foreign cultural contact.

Oceania. Pacific Islanders began to lose their traditional conservation ethics and practices with the first foreign cultural contact. Ship captains introduced domesticated animals that out-competed many indigenous animals for their habitats. Missionaries substituted religions that had neither a conservation ethic nor any tolerance for traditional customs or practices that regulated and protected wildlife sanctuaries and marine preserves. World War II increased the pace and degree of contact with alien cultures. As military activity swept through the Pacific region, cultures that synchronized their activities to the cycles of nature (e.g., the blossoming of flowers, trees, and shrubs; habits and movements of terrestrial animals; the migration and spawning cycles of fish) began to have their cultural institutions shattered. In recent history, the search for natural resources; tourism; volunteer groups (American Peace Corps); strategic military desires; and industrial adventures (Palau supertanker controversy) have all increased island contact with outside cultures (Bodley, 1975; Klee, 1980).

Africa. The ivory trade marked the beginnings of an alienation from nature and wildlife that was new to Africans, not to mention the origin of the slave trade which resulted from the need for porters to carry the ivory elephant tusks from inland areas to the sea (Lusigi, 1981). Large-scale slaughter of wildlife was expanded by colonization, settlement, and two world wars.

More recently, in an effort to save wild animals as an international heritage, outside governments have applied political pressure on African governments to establish national parks and reserves and enact strict game laws. These conservation policies pushed many Africans from their traditional homelands and ways of life and ignored the needs of the people for subsistence hunting.

Other forms of direct governmental intervention have disrupted traditional African resource-using systems. Many governments have attempted to rid their territories of pastoral nomads. Although many pastoral nomads have evolved successful ecological adaptations to extremely arid conditions and have survived for thousands of years, central governments often maintain that the nomadic way of life is inefficient, irrational, a wasteful use of natural resources, and an infringement on the rights of sedentary "civilized" farmers. Perhaps more to the truth, the mobility of nomads makes it difficult for governments to impose central control and demand allegiance (Bodley, 1975). Also, the best pasture, kept by pastoralists as drought reserves, were often thought by outsiders to be unused and hence available for alienation to farmers.

Asia. Traditional attitudes toward nature conservation in Asia began to break down with the advent of colonization. Like African countries, recent negative attitudes toward wildlife conservation can be linked to the resentment of former colonial influences (UNESCO, 1974). After independence, many Indians temporarily rejected such vestiges of colonialism as British systems of wildlife reserves, sanctuaries, and forest and hunting regulations. As a result, India's wildlife and forest resources were reduced.

The spread of industrialization and its need for global resources has resulted in the acculturation of many tribal peoples who once effectively preserved their resources (e.g., the Chittagong Hill peoples of Bangladesh (Bodley, 1975). World political considerations within the past twenty years have forced regional governments and their allies to question the "loyalty" of tribal peoples that have occupied the interior hilly uplands of Southeast Asia. Peaceful as well as guerrilla activities have been used to win the support of these tribal peoples (Bodley, 1975).

Other areas. Historically, countries within Latin America and elsewhere have met with similar types of outside cultural contact that have disrupted their indigenous peoples and their customary wildlife conservation techniques.

6.3.2 Human population increases and associated agricultural development.

Oceania. Initially, island human populations were drastically reduced due to foreign contact and the ensuing changes in diet, diseases, acquisition of firearms, and so on (Klee, 1980). Once state intervention took hold, however, the population increases began. Missionaries halted traditional population control mechanisms (abortion, infanticide, intertribal warfare); Western medicine lowered birth mortality and generally improved health and longevity; and outside pressures created new incentives to increase family size. Rapid population growth was the inevitable consequence. Where swidden agriculture was once practiced at a rate which permitted vegetation to regenerate fully before recutting, human population pressures now force the abandonment of traditional conservation practices that protected forests and wildlife habitats.

Africa. In recent years, many African countries have undergone an explosive increase in population, and these areas are now faced with the difficulty of restoring a balance between population size and available natural resources (Chavundaka, 1978; Simmons, 1979). Population pressure has resulted in an increased need for arable land, and some families have now been allocated land for cultivation in areas previously used exclusively for grazing by wild animals and/or domestic stock. As traditional wildlife habitats and indigenous animals decreased, so did the many practices of traditional wildlife conservation.

Asia. The explosive growth in human numbers in South Asia resulted in deforestation, the spread of agriculture, and increased urban settlements. Many mammals and other vertebrates were decimated in this need to exploit the land (Bennett, 1975). As Murton (1980) has noted, the occupation of forest and once fallow lands has resulted in the weakening (and often elimination) of traditional conservation practices and "human habitats" have replaced what were once "animal habitats."

Other Areas. In the countries of Latin America (Denevan, 1980), and certainly elsewhere, human population expansion and its associated agricultural dispersion and intensification has also reduced wildlife populations.

6.3.3 Failure of education, administration and enforcement systems.

Educational systems that neglected to teach traditional environmental knowledge and conservation techniques further stimulated the breakdown of traditional cultural controls over resource managment in Oceania (Klee, 1980), Africa (Chavunduka, 1978), and no doubt other areas.

Traditional cultural controls over wildlife resources also deteriorated as a result of colonial administration systems that did not incorporate aspects of traditional resource management and enforcement into their own resource management policies. This notion is well stated by Allen (1976):

> Unfortunately, the British did not understand this remarkable institution, and so did not incorporate it into the new administrative structure they created when they established indirect rule. As a result, the office of *vaka vanua* (forest and garden crop ecologist) lost most of its authority and prestige, the number of forest and garden crops placed under *tapu* (taboo) declined . . . Each of the Lau Islands also enjoyed the services of a *ndau ni nggoli*, a master fisherman and authority on the island's fish lore

and fishing techniques . . . Again, unfortunately, the British did not understand this institution, and the *ndau ni nggoli* has lost status.

6.4 Reconciling Traditional Systems with the Modern World

Resource utilization conflicts between traditional subsistence tribes and modernized consumptive states result from the basic incompatibility of the two cultural systems, but there are signs that the best of the tribal old way can and in some cases is being incorporated into the newer state systems so that more sustainable social and ecological systems can be developed. However, before advocating a carte blanche program of revival and assimilation of traditional practices, several cautions should be considered.

First, traditional cultures had destructive as well as constructive and conserving practices, and the modern-day resource manager must examine these traditional cultures to see what if any conservation elements may be worthy of reviving and reinforcing.

Second, it is not always desirable, practical, or plausible for tropical peoples to revert to former ways. Many are no longer isolated, and those who have entered a cash economy may have a particularly difficult time accepting old ways of doing things.

Third, AID field officers and other conservation/development planners interested in preserving traditional systems of resource management are faced with an unfortunate paradox. It is often their (European and North American) predecessors who initially broke down the ecologically sound traditions that they now wish to preserve or revive. In one generation, Westerners frequently had demands or requests which broke down conservation practices; in subsequent ones, they ask that they be restored. Levels of responsibility and the "who are you supposed to believe" are issues.

Some social indicators do support the partial reconciliation of viable traditional practices and modern development. If planners can identify traditional practices and work within the framework of these concepts, remnants of traditional systems may continue to contribute to both social and ecological stability.

6.4.1 Indicators of acceptance of extant traditional practices.

Those social indicators that support the possibility of at least partial retention of traditional systems of resource management are as follows: the presence of traditional social structures and resource controls under modern conditions; the increased recognition of the value of traditional forms of resource management; the resurgence of ethnic pride and the rebirth of conservation mindedness; the "fallowing" of rural resource areas; and an increased interest in wildlife.

Presence of traditional conservation techniques. The very fact that traditional technologies and conservation practices can still be found under modern conditions is clear evidence that traditional ecological awareness, the sheer force of custom, and the ability of people to weigh the pros and cons of new methods against known ways and to "selectively adapt" are factors at work in maintaining these traditional practices. Examples of these adaptations can be illustrated by various forms of traditional marine tenure systems still found in the Pacific, such as: (i) fishing rights to specific areas (Pukapuka; Hawaii); (ii) specific species regulation (Yap); (iii) optimum fishing seasons according to traditional time reckoning (Belau); and (iv) the use of closed seasons (Yap) (Klee, 1980).

Increased recognition of traditional systems. Whereas indigenous cultures have long known the value of their traditional conservation techniques, international conservation organizations are only now beginning to take a second look at these systems of resource management. For example, the International Union for the Conservation of Nature and Natural Resources recently stated that "South Pacific islanders possess a rich store of knowledge of their environment; and that the traditional conservation practices of many South Pacific cultures were once highly effective and, if supported or adapted to modern conditions, could continue to be so." (IUCN, 1976). The conference concluded by urging governments and other international organizations to record, preserve, and possibly incorporate this environmental knowledge into future conservation management plans.

Africa can serve as an example of increased awareness by ecologists of the role of cultures and their relationship with animals and wildlife preservation. Lusigi, for instance, reminds us that in African savanna systems, humans have traditionally played a very significant role in maintaining the ecosystem--by grazing livestock, by predation, or by regulating bush through the use of wood for cooking and building (1981). According to Lusigi, excluding human predation and other activities from the national parks resulted in the starvation of thousands of elephants in Tsavo Park and the death of others in Nairobi National Park, Kenya.

Resurgence of ethnic pride and rebirth of conservation mindedness. A third major social indicator that traditional conservation techniques "might work" under modern conditions is the fact that several regions within the tropics are experiencing a resurgence in ethnic pride and related rebirth in conservation mindedness. Many school systems, for example, are now encouraging the teaching of traditional values, attitudes, and skills (South Pacific Regional Environment Programme, 1982).

Furthermore, many cultures are beginning to say *no* to "progress for progress sake," such as the recent Palauan denial of a proposal for a supertanker oil facility on Kossoll Reef in the northern Palauan Islands. The islanders rejected the joint Japanese, Iranian, and American proposal on the grounds that it threatened their traditional cultural institutions and marine sanctuaries (Klee, 1980).

The "fallowing" of rural resource areas. Peoples are currently migrating from their traditional rural lands to the city in many tropical countries, as has been documented recently for Pacific islanders (Klee, 1980) and Africans (Chavunduka, 1978). This migration allows these rural areas (with their animals and fisheries) to be "fallow"--abandoned or partially unused for a period of recuperation. According to Yen, this fallowing provides an opportunity to conserve valuable natural resources as "banks" for future exploitation by peoples under a new economic and political order emanating from the towns (1975).

This trend towards urbanization also has another conservation value. As more and more people are drawn into an urban community, increased numbers of peoples begin to appreciate nature and the unspoiled open spaces from which they came, and urbanized communities tend to respond more readily to conservation measures than their rural counterparts. According to Chavundaka, there is evidence of this in Africa, where urbanized Africans are increasingly pursuing excursions, recreation, and holidays in rural surroundings (1978). In this example, one element of cultural change (rural fallowing) indirectly aids another supportive element (rebirth of conservation mindedness), and both are social indicators of the possibility of preserving traditional systems of resource management in the future.

Increased interest in wildlife: Not only are some tropical countries beginning to have a renaissance in general conservation, but some countries are experiencing an awakening interest in wildlife *per se*. This is the case in several African countries (Layall-Watson, 1965) as evidenced by the creation and expansion of a zoo at Entebbe (Uganda); the Kisoro Gorilla Sanctuary (Uganda); the Moremi Reserve (Bechuanaland); and the Amboseli Reserve (Kenya). Most importantly, many of these reserves remain under tribal control.

6.4.2 The reconciliation process: A Pacific Island example.

Within the Pacific, terrestrial animals were protected by a number of traditional conservation techniques, beliefs, and enforcement systems, many of which could be revived and/or reinforced. For example, such wildlife resource practices as habitat maintenance, game reserves, hunting seasons, bag limits, and regulations regarding the keeping of pets, might be revived without too much difficulty. Traditional restrictions on sacred plants and animals might be slightly modified to include "endangered" plants and animals.

Development planners can augment this process by assuring that local knowledge is incorporated into the development process. This knowledge will include life histories, the feeding habits, mating seasons, habitat requirements and other basic knowledge of various forms of local wildlife and fish. Much information can also be learned about types and migration cycles of avifauna; the breeding requirements and location of various insects; the location and traditional regulation of wildlife reserves; traditional hunting and fishing rights and practices; methods of food preservation; closed seasons; specific species regulation; and many other characteristics and practices.

Reestablishing the degree of environmental awareness, the concept of "master fishermen," fishing rights to specific geographic areas, taboos, and related fines and punishments, would often require a high degree of modification of present practices. For instance, the "conservation ethic" behind an indigenous religion might be revived and reinforced without doing the same for the magical aspects (i.e., many citizens support the Ten Commandments without actually believing and supporting the origin of those ideas). The severity of fines and punishment might be modified to meet internationally accepted standards. Food avoidance based on class or sexual differences probably could not (and should not) be revived or reinforced as the notion of equality, as well as ethnic pride, is sweeping through the islands. However, if food avoidances according to class or sex lines remain to some degree in a particular culture, the practice should not be discouraged for it does play a role in conserving resources.

Again, the development officer can aid this reconciliation by identifying local sources of information. For example, much can be learned regarding the lunar periodicity of spawning of reef fish; the location and traditional regulation of marine preserves; the fishing grounds used by a particular village; the effect of rainfall, winds, currents, and temperatures on fishing conditions and the habits of certain fish. A great deal of information can be gathered regarding the times, places, and seasons of optimum fishing; peculiarities of different islands and different parts of the coasts of larger islands in relation to fish habits and migration; the incidence of toxic plants that might render fish poisonous; the traditional fishing rights, closed seasons, specific species regulation, and food avoidances; the optimum days for particular fishing techniques; the construction and proper use of traps and ponds for fish conservation; and the various methods of preserving fish. According to Handy and others, "the experienced native fisherman is possessed of a store of precise knowledge that may be truly characterized as natural science" (Handy, 1932; also see Johannes, 1975 and 1981; and Klee, 1972 and 1980). Development planners have much to gain by incorporating this "natural science" into their future management policies.

6.5 Conclusion

There is a growing recognition that traditional cultures are carried by "... real people who have developed unique adaptations to unique environments" (Bodley, 1975), and that the preservation of cultural diversity (in addition to ecological diversity) will be critical for the long run survival of mankind (Dubos, 1965; Klee, 1980; Meggers, 1971; Rappaport, 1971; Watt, 1972). While there will no doubt be potential conflicts between modern conservation goals and traditional conservation techniques, they can probably be resolved by cooperation and understanding. What is important to remember is that devoting time to preserving traditional systems of resource management is a far more worthy task than learning how to remove traditional obstacles to modernization--the all-too-frequent practice of the past.

Organizations and Directories

7.1 Organizations

The following is a list of organizations with memberships and programs concerned with environmental education or with management and research on wildlife. The list is not exhaustive but rather offers a starting point for entry into an information network.

African Wildlife Foundation, Inc.
P.O. Box 48177, Nairobi, Kenya.
In U.S.: 1717 Massachusetts Ave., NW Washington, D.C. 20036

> Finances conservation projects and wildlife management training (scholarships) in Africa. The scientific and education staff provide technical assistance to national parks and conservation education programs in school.

Environmental Liaison Centre.
Nairobi, Kenya.

> A membership organization paritally supported by United Nations Environment Program (UNEP). It provides information on environmental topics ranging from wildlife to urban planning.

Food and Agricultural Organization of the United Nations (FAO).
Viale delle Terme di Caracalle, 00100, Rome, Italy.

> A specialized agency of the UN with 147 government members. It supports programs aimed at improving the production and distribution of food and agricultural products, including fish and forestry products to improve levels of nutrition and standards of living.

Friends of the Earth International.
Amsterdam, Netherlands.

> A conservation group with organizations in several countries that promotes the rational use of the earth and its resources.

International Foundation for the Conservation of Game.
Paris, France

International Union for Conservation of Nature and Natural Resources (IUCN).
Avenue du Mont Blanc. CH-1196 Gland, Switzerland

> A non-governmental organization with states, government agencies, and non-governmental organizations as members. It promotes scientifically based action for the conservation of wild living resources through publications and the activities of 6 commissions (Ecology; Education; Environmental Planning; Environmental Policy, Law and Administration; Natural Parks and Protected Areas; and the Survival Service Commission). Publications include Red Data Book (a list of threatened vertebrate and plant species); UN List of Natural Parks and Equivalent Reserves; and books on a variety of symposium topics relating to conservation and development.

512   V. WILDLIFE

Wildlife Society, Inc.
5410 Grosvenor Lane, Bethesda, Maryland 20814.

> A non-profit organization of professionals active in wildlife research, management, and education. It promotes sound management of both wildlife resources and the environmental resources upon which the fauna depend. Publications include Journal of Wildlife Management, and Wildlife Monographs.

United Nations Education, Scientific and Cultural Organization (UNESCO).
Paris, France.

> The Man and Biosphere (MAB) Global Program has developed a network for Biosphere Reserves worldwide. It has national programs in many countries and sponsors some individual research and training programs. Publications include reports of symposia, training manuals in environmental resource management and environmental profiles of many developing countries.

United Nations Environmental Program (UNEP).
P. O. Box 30552, Nairobi, Kenya.
New York Liaison office: P. O. Box 20, New York, New York 10017.

World Wildlife Fund.
Gland, Switzerland.
In U.S.: 1601 Connecticut Avenue, N.W. Washington, D.C.   20036

> A private group with affiliates in 27 countries. It finances scientific research, educational, and wildlife and habitat preservation projects in many countries. It assists in creating national parks and reserves and in conserving endangered species.

7.2   Directories of Organizations

Examples of annotated listings of organizations with natural resource use and management information include:

National Wildlife Federation's *Conservation Directory 1983.* 27th ed. National Wildlife Federation, 1412 Sixteenth St., N.W. Washington, D.C. 20036.

> Annual directory of organizations, agencies, and officials. It provides annotated listings for international groups and organizations in the United States and Canada, with country listings for other national governments.

World Directory of Environmental Organizations. Trozyna, T.C. and Coan, E.V., eds. 1976. Sierra Club Special Publications, International Series, No. 1, 2nd ed. Claremont, Calif.: Sequoia Institute.

> Annotated listing of intergovenmental organizations, international non-governmental organization ands national organizations.

References Cited

Allen, D.L. 1938. Ecological studies on the vertebrate fauna of a 500-acre farm in Kalamazoo, Michigan. *Ecol. Monograph* 8:347-436.

Allen, D.L., ed. 1956. *Pheasants in North America.* Harrisburg, Pa.: Stackpole Co.

Allen, R. 1976. Ecodevelopment and traditional natural resource management in the South Pacific. Paper presented at Second Regional Symposium on Conservation of Nature, 14-17 June 1976, Apia, Western Samoa. Mimeographed.

Anderson, W.H., W.A. Wentz and B.D. Treadwell. 1980. A guide to remote sensing information for wildlife biologists. In *Wildlife management techniques manual.* 4th ed. rev., S.D. Schemnitz, pp. 291-304. Washington, D.C.: The Wildlife Society.

Asibey, E.O.A. 1974. Wildlife as a source of protein in Africa south of the Sahara. *Biological Conserv.* 6:32-39.

Asibey, E.O.A. 1976. The effects of land use patterns on future supply of bushmeat in Africa south of the Sahara. Working party on wildlife management and national parks, Fifth Session. FAO: AFC/WL:76/6/4. Rome: FAO. Mimeographed.

Australian NPWS (National Parks and Wildlife Service). 1979. *Aerial surveys of fauna populations.* Special publication 1. Canberra, Australia: Australian Government Publishing Service.

Beaglehole, E. and P. Beaglehole. 1938. *Ethnology of Pukapuka.* Honolulu: B. P. Bishop Museum Press.

Benedict, F.G. 1936. *Physiology of the elephant.* Publ. 474. Washington, D.C.: Carnegie Inst.

Bennett, C.F. 1975. *Man and earth's ecosystems.* New York: John Wiley & Sons.

Beyer, J.L. 1980. Africa. In *World systems of traditional resource management*, ed. G.A. Klee, pp. 5-38. London: Edward Arnold.

Bodley, J.H. 1975. *Victims of progress.* Menlo Park, Calif.: Cummings Publ. Co.

Burnham, K.P., D.R. Anderson and J.L. Laake. 1980. Estimates of density from line transect sampling of biological populations. *Wildl. Monogr.*, No. 72, pp. 202. Washington, D.C.: The Wildlife Society.

Carneiro, R. 1961. Slash-and-burn agriculture: A closer look at its implications for settlement patterns. In *Men and cultures*, ed. A.F. Wallace, pp. 229-234. Fifth International Congress of Anthropological and Ethnological Sciences. New York: International Union of Anthropological and Ethnological Sciences.

Caughley, G. 1977. *Analysis of vertebrate populations.* New York: John Wiley and Sons.

Chaplin, R.E. 1971. *The study of animal bones from archeological sites.* New York: Seminar Press.

Chambers, R. 1980. *Shortcut methods in information gathering for rural development projects.* Paper for World Bank Agricultural Sector Symposia, Institute of Development Studies. Brighton: University of Sussex.

Chavunduka, D.M. 1978. African attitudes to conservation. *Rhodesia Agr. J.* 75:61-63.

Clarke, W.C. 1971. *Place and people.* Berkeley, Calif.: Univ. California Press.

Clastres, P. 1972. The Guayaki. In *Hunters and gatherers today: A socio-economic study of eleven such cultures in the twentieth century*, ed. M. G. Bicchieri, pp. 138-174. New York: Rinehart & Winston.

Cloudsley-Thompson, J.L. 1967. *Animal twilight: Man and game in Eastern Africa.* London: Foulis & Co.

Cornwall, I.W. 1956. *Bones for the archeologist.* New York: MacMillan.

Cumberland, K.B., and J.S. Whitelaw. 1970. *New Zealand.* Chicago: Aldine Press.

Dasmann, R.F. 1964. *African game ranching.* Oxford: Pergamon Press.

Daubenmire, R. 1968. *Plant communities.* New York: Harper and Row.

Denevan, W. 1980. Latin America. In *World systems of traditional resource management*, ed. G.A. Klee, pp. 217-244. London: Edward Arnold.

Dinerstein, E. and H.T. Dublin. 1982. Daily defecation rate of captive axis deer. *J. Wildl. Manage.* 46:833-835.

Dittus, W.P.J. 1981. Primate population analysis. In *Techniques for the study of primate population ecology*, pp. 135-197. Washington, D.C.: National Academy Press.

Downing, R.L., W.H. Moore, and J. Kight. 1965. Comparison of deer census techniques applied to a known population in a Georgia enclosure. *Proc. 19th Annual Conf. SE Assoc. Game Fish Comm.* 19:26-30.

Dubos, R. 1965. *Man adapting.* New Haven: Yale Univ. Press.

Eberhardt, L.L. 1982. Calibrating an index by using removal data. *J. Wildl. Manage.* 46:734-740.

Eisenberg, J.F. and M. Lockhart. 1972. An ecological reconnaissance of Wilpattu National Park, Ceylon. *Smithson. Contrib. Zool.* 101:1-118.

Elliott, E.F.I. 1973. Past, present and future conservation status of Pacific Islands. In *Nature conservation in the Pacific*, eds. A. B. Coston and R. H. Groves. Canberra: Australia National Univ. Press.

Gardner P.M. 1972. The Paliyans. In *Hunters and gatherers today: A socio-economic study of eleven such cultures in the twentieth century*, ed. M.G. Bicchieri, pp. 404-447. New York: Rinehart and Winston.

Geis, A.D. 1956. A population study of the cottontail rabbit in southern Michigan. Ph.D. Thesis, Michigan State University.

Gifford, E.W. 1929. *Tongan society*. Honolulu: B.P. Bishop Museum Press.

Guggisberg, C.A.W. 1970. *Man and wildlife*. London: Evans Brothers.

Gysel, L.W. and E.M. Davis, Jr. 1956. A simple automatic photographic unit for wildlife research. *J. Wildl. Manage.* 20(4): 857-858.

Gysel, L.W. and L.J. Lyon. 1980. Habitat analysis and evaluation. In *Wildlife management techniques manual*. 4th ed. rev., ed. S.D. Schemnitz, pp. 305-329. Washington, D.C.: The Wildlife Society.

Hahn, H.C. 1949. *A method for censusing deer and its application in the Edwards Plateau of Texas*. Austin, Tex.: Texas Game, Fish and Oyster Comm.

Handy, E.S.C. 1927. *Polynesian religion.* Honolulu: B. P. Bishop Museum Press.

Handy, E.S.C. 1932. *Houses, boats, and fishing in the Society Islands.* Honolulu: B. P. Bishop Museum Press.

Hartman, F.H. 1960. Census techniques for snowshoe hares. M.S. thesis, Michigan State University.

Hayne, D.W. 1949. Two methods for estimating population from trapping records. *J. Mammal.* 30:399-411.

Heyland, J.D. 1973. Increase the accuracy of your airborne censuses by means of vertical aerial photographs. *Trans. Northeast Sect. Wildl. Soc., Fish and Wildl. Conf.* 30:53-75.

Hirst, S.M. 1969. Road-strip census techniques for wild ungulates in African woodland. *J. Wildl. Manage.* 33:40-48.

Hobbs, N.T. 1982. Confidence intervals on food preference indices. *J. Wildl. Manage.* 46:505-507.

Holdridge, L. R. 1947. Determination of world plant formations from simple climatic data. *Science* 105:367-368.

Holdridge, L.S. 1967. *Life zone ecology.* San Jose, Costa Rica: Tropical Science Ctr.

Holechek, J.L., M. Vavra, and R.D. Pieper. 1982. Botanical composition determination of range herbivore diets: A review. *J. Range Manage.* 35:309-315.

Isack, H.A. 1976. An African ethic of conservation. *Natur. Hist.* 85:90-95.

IUCN (International Union for Conservation of Nature and Natural Resources)-UNEP-WWF. 1980. *World conservation strategy. Living resource conservation for sustainable development.* Gland, Switzerland: IUCN.

Jachmann, H. and R.H.V. Bell. 1979. The assessment of elephant numbers in occupance by means of droppings counts in the Kasungu National Park, Malawi. *Afr. J. Ecol.* 17:231-239.

Les jeune les atlas Afrique. 1975. *Atlas de la Haute-Volta.* Paris: Jeune Afrique Institute Geographique National.

Johannes, R.E. 1975. Exploitation and degradation of shallow marine food resources in Oceania. In *The impact of urban centers in the Pacific*, eds. R. W. Force and B. Bishop, pp. 1-60. Honolulu: Pacific Science Association.

Johannes, R.E. 1981. *Words of the lagoon: Fishing and marine lore in the Palau District of Micronesia.* Berkeley, Cal.: Univ. of California Press.

Johnson, K.L. 1978. Mapping fishing grounds. Paper presented in seminar class at San Jose State University on November 1978. (Paper based on interview with Bernard Nietschmann on November 3, 1978 at the University of California, Berkeley).

Keast, A. 1969. The evolution of mammals on southern continents. VII. Comparisons of the contemporary mammalian faunas of the southern continents. *Quart. Rev. Biol.* 44:121-167.

Klee, G.A. 1972. The cyclic realities of man and nature in a Palauan village. Ph.D. dissertation, Univ. of Oregon.

Klee, G.A., ed. 1980. *World systems of traditional resource management.* London: Edward Arnold Publishers.

Lavieren, L.P. van, and M.S. Bosch. 1977. Evaluation des densites de grans mammiferes dans le Parc National de Bouba Ndjida, Cameroun. *La Terre et la Vie* 31:3-32.

Layall-Watson, M. 1965. Men and animals: Their fruitful co-existence in Africa. *Optima* 15:85-94.

Lee, R.B. 1972. The !Kung Bushmen of Botswana. In *Hunters and gatherers today: A socio-economic study of eleven such cultures in the twentieth century*, ed. M. G. Bicchieri, pp. 327-368. New York: Rinehart and Winston.

Lee, W.L., B.M. Bell and J.F. Sutton. 1982. *Guidelines for acquisition and management of biological specimens.* Lawrence, Kans.: Association Systematic Collelctions, Musuem of Natural History, Univ. of Kansas.

Linares, O. 1976. Garden hunting in the American tropics. *Human Ecol.* 4:331-349.

Linhart, S.B. and F.F. Knowlton. 1975. A preliminary report on determining relative abundance of coyotes by scent station lines. *Wildl. Soc. Bull.* 3:119-124.

Loeb, E.H. 1926. *History and traditions of Niue.* Honolulu: B. P. Bishop Museum Press.

Loehle, C. and L.R. Rittenhouse. 1982. An analysis of forage preference indices. *J. Range Manage.* 35:316-319.

Lund, H.G., M. Caballero, R.H. Hamre, R.S. Driscoll and W. Bonner. 1981. *Arid land resource inventories: Developing cost-efficient methods.* Proc. Internatl. Workshop, La Paz, Mexico, 30 Nov.-6 Dec. 1980. USDA Forest Service General Technical Report WO-28. Washington, D.C.: U.S. Government Printing Office.

Lusigi, W.J. 1977. The conservation unit approach to the planning and management of national parks and reserves in Kenya. Friesing, West Germany: Institute for Landscape Ecology. Mimeographed.

Lusigi, W.J. 1981. New approaches to wildlife conservation in Kenya. *Ambio* 10(N2-3): 87-92.

Meggers, B.J. 1971. *Amazonia: Man and culture in a counterfeit paradise.* Chicago: Aldine.

Miller, K.R. 1978. *Planning national parks for ecodevelopment.* Vols. 1 and 2. Madrid: Fundacion para la Ecologia y Parala Proteccion del Medio Ambiente.

Moran, E.F. 1979. *Human adaptability: An introduction to ecological anthropology.* North Scituate, Mass.: Duxbury Press.

Murton, B. 1980. South Asia. In *World systems of traditional resource management*, ed. G.A. Klee, pp. 67-99. London: Edward Arnold.

Murie, A. 1944. *The wolves of Mt. McKinley.* Fauna of the National Parks, Fauna Ser. No. 5. Washington, D.C.: U.S. Park Service.

Nietschmann, B. 1973. *Between land and water.* New York: Seminar Press.

Norton-Griffiths, M. 1978. *Counting animals.* Handbook No. 1, Serengeti Ecological Monitoring Program. Nairobi, Kenya: African Wildlife Leadership Foundation.

NRC. 1981. *Techniques for the study of primate population ecology.* Washington, D.C.: National Academy Press.

V. WILDLIFE

Obeng, L.E. 1981. Man's impact on tropical rivers. In *Perspectives in running water ecology*, eds. M.A. Lock and D.D. Williams, pp. 265-288. New York: Plenum Press.

Owen, R.P. 1969. The status of conservation in the Trust Territory of the Pacific Islands. *Micronesica* 5:1-303.

Papageorgiou, N. 1979. Population energy relationships of the Agrimi (*Capra aegagrus cretica*) on Theodorou Island, Greece. *Mamm. Depicta* 11:1-56.

Paulik, G.J. and D.S. Robson. 1969. Statistical calculations for change-in-ratio estimators of population parameters. *J. Wildl. Manage.* 33:1-27.

Petrides, G.A. 1975. Principal food versus preferred foods and their relations to stocking rate and range condition. *Biol. Conserv.* 7:161-169.

Petrides, G.A. and W.G. Swank. 1966. Estimating the productivity and energy relations on an African elephant population. *Proceedings 9th International Grasslands Congress*, Sao Paulo, Brazil, pp. 831-842.

Pienaar, U. de V. 1969. Predator-prey relationships amongst the larger mammals of the Kruger National Park. *Koedoe*. 12:108-176.

Pollock, K.H. 1982. A capture-recapture design robust to unequal probability of capture. *J. Wildl. Manage.* 46(3): 752-761.

Powell, J.M. 1975. Conservation and resource management in Australia, 1788-1860. In *Australian space/Australian time*, eds. J.M. Powell and M. Williams, pp. 1-18. Melbourne: Oxford Univ. Press.

Prescott-Allen, R. anc C. Prescott-Allen. 1982. *Economic contributions of wild plants and animals to developing countries.* US MAB mimeographed report. Washington, D.C.: US Department of State, UNESCO MAB Programme.

Rappaport, R.A. 1971. The flow of energy in an agricultural society. *Scient. Am.* 224:117-132.

Ralph, C.J. and J.M. Scott, eds. 1981. Estimating numbers of terrestrial birds. Studies in avian biology No. 6. *Proceedings of Internat. Symp. held at Asilomar, Ca., 26-31 Oct. 1980*. Lawrence, Kans.: Cooper Ornithological Society and Allen Press.

Ray, G.C., B.T. Hayden and R. Dolan. 1983. Development of a biophysical coastaland marine classification system. *Proceedings of the World National Park Congress Bali, Indonesia, 11-22 Oct. 1982.* Gland, Switzerland: IUCN

Reichel-Dolmatoff, G. 1971. *Amazonian cosmos.* Chicago: Univ. of Chicago Press.

Reichel-Dolmatoff, G. 1976. Cosmology as ecological analysis: A view from the rain forest. *Man* 11:307-318.

Rice, W.R. and J.D. Harder. 1977. Application of multiple aerial sampling to a mark-recapture census of white-tailed deer. *J. Wildl. Manage.* 41:197-206.

Robinette, W.L., D.A. Jones, J.S. Gashwiler, and C.M. Aldous. 1956. Further analysis of methods for censusing winter-lost deer. *J. Wildl. Manage.* 20:75-78.

Robinette, W.L., C.M. Loveless, and D.A. Jones. 1974. Field tests of strip census methods. *J. Wildl. Manage.* 38:81-96.

Ross, E. 1978. Food taboos, diet and hunting strategy: The adaptation to animals in Amazonian cultural ecology. *Curr. Anthrop.* 19:1-36.

Roughton, R.D. 1982. A synthetic alternative to fermented egg as a canid attractant. *J. Wildl. Manage.* 46:230-234.

Roughton, R.D. and M.W. Sweeny. 1982. Refinements in scent-station methodology for assessing trends in carnivore populations. *J. Wildl. Manage.* 46:217-229.

Schafer, E.H. 1962. The conservation of nature under the T'ang Dynasty. *J. Econ. Soc. Hist. Orient* 5:282-308.

Schemitz, S.D., ed. 1980. *Wildlife management techniques manual.* 4th ed., rev. Washington, D.C.: The Wildlife Society.

Schnabel, Z.E. 1938. The estimation of the total fish population of a lake. *Am. Math. Mon.* 45:348-352.

Schumacher, F.X. and R.W. Eschmeyer. 1943. The estimate of fish populations in lakes or ponds. *J. Tenn. Acad. Sci.* 18:228-249.

Seber, G.A.F. 1973. *The estimation of animal abundance and related parameters.* New York: Hafner Press.

Silberbauer, G.B. 1972. The G/wi Bushmen. In *Hunters and gatherers today: A socio-economic study of eleven such cultures in the twentieth century*, ed. M.G. Bicchieri, pp. 271-326. New York: Rinehart and Winston.

Simmons, I.G. 1979. *Biogeography: Natural and cultural.* London: Edward Arnold Publishers.

Sinha, D.P. 1972. The Birhors. In *Hunters and gatherers today: A socio-economic study of eleven such cultures in the twentieth century*, ed. M. G. Bicchieri, pp. 371-403. New York: Rinehart and Winston.

South Pacific Regional Environment Programme. 1982. *Report of the Conference on the Human Environment in the South Pacific, Rarotonga, Cook Islands, 8-11 March 1982.* Noumea, New Caledonia: South Pacific Commission.

Tobin, J.E. 1952. Land tenure in the Marshall Islands. *Atoll Res. Bull.* pp. 12-32. Washington D.C.: Smithsonian Press.

Tocher, R. and R. Milne. 1974. A cross-cultural comparison of attitudes towards wildlife. *Trans. N. Am. Wildl. Nat. Resour. Conf.* 39:145-150.

Treadwell, B.D. 1979. A provisional framework for defining black bear habitat. In *Proc. First Western Black Bear Workshop*, ed. A. LeCount, pp. 319-329. Phoenix, Ariz.: Ariz. Game Fish Dept.

Tyson, E.L. 1959. A deer drive vs. track census. *Trans. N. Am. Wildl. Nat. Resour. Conf.* 24:457-464.

Udvardy, M.D.F. 1975. *A classification of the biogeographical provinces of the world.* IUCN Occasional Paper No. 18. Morges, Switzerland: IUCN.

UNESCO. 1974. *Natural resources of humid tropical Asia.* Natural Resources Research, 12. New York: UNESCO.

Urquhart, G.M. and J.G. Ross. 1962. Game conservation in East Africa. *Vet. Rec.* 74:231-235.

U.S. MAB (Man and Biosphere Programme). 1981. *Environmental profiles.* Series by country. Mimeographed reports. Washington, D.C.: U.S. Department of State, UNESCO MAB Programme.

Ward, R. G., personal communication. September 27, 1976. Professor of Human Geography, Australian National University

Whyte, A. 1977. *Guidelines for field studies in environmental perception.* MAB Technical Notes No. 5. Paris: UNESCO.

Watt, K.E.F. 1972. Man's efficient rush toward deadly dullness. *Natur. Hist.* 81:74-82.

Williams, K.D. and G.A. Petrides. 1980. Browse use, feeding behavior, and management of the Malayan tapir. *J. of Wildl. Manage.* 44(2): 489-494.

Yen, D. 1975. Effects of urbanization on village agriculture in Oceania. In *The impact of urban centers in the Pacific*, eds. R. W. Force and B. Bishop. Honolulu: Pacific Science Association.

Suggested Readings

A. *Major sources for methods*:

Australian NPWS (National Parks and Wildlife Service). 1979. *Aerial surveys of fauna populations.* Special publication 1. Canberra, Australia: Australian Government Publishing Service.

Burnham, K.P., D. R. Anderson and J. L. Laake. 1980. Estimates of density from line transect sampling of biological populations. *Wildl. Monogr.* No. 72. Washington, D.C.: The Wildlife Society.

Caughley, G. 1977. *Analysis of vertebrate populations.* New York: John Wiley and Sons.

NRC (National Research Council). 1981. *Techniques for the study of primate population ecology.* Washington, D.C.: National Academy Press.

Norton-Griffiths, M. 1978. *Counting animals.* Handbook No. 1, 2nd ed. Serengeti Ecological Monitoring Program. Nairobi, Kenya: African Wildlife Foundation.

Ralph, C. J. and J. M. Scott, eds. 1981. Estimating numbers of terrestrial birds. Studies in avian biology, No. 6. *Proceedings of Internat. Symp. held at Asilomar, Ca. 26-31 Oct. 1980.* Lawrence, Kans.: Cooper Ornithological Society and Allen Press.

Schemitz, S. D., ed. 1980. *Wildlife management techniques manual.* 4th ed., rev. Washington, D.C.: The Wildlife Society.

Seber, G. A. F. 1973. *The estimation of animal abundance and related parameters.* New York: Hafner Press.

B. *Background reading:*

Ajayi, S. S. and L. B. Halstead, eds. 1979. *Wildlife management in savannah woodland.* London: Taylor and Francis.

Bailey, J. A., W. Elder, and T. D. McKinney. 1974. *Readings in wildlife conservation.* Washington, D.C: The Wildlife Society.

Brokaw, H. P., ed. 1978. *Wildlife and America.* Washington, D.C.: Council of Environmental Quality.

Brown, L. 1981. *Building a sustainable society.* Washington, D.C.: World Watch Institute.

Dasmann, R. F. 1964. *African game ranching.* Oxford: Pergamon Press.

Dasmann, R. F., 1981. *Wildlife biology.* 2nd ed. New York: John Wiley & Sons.

Dasmann, R. F., J. Milton and P. Freeman. 1972. *Ecological principles for economic development.* London: John Wiley. (Later editions in Spanish, Portuguese, Indonesian and Polish.)

Frankel, O. H. and M. E. Soule. 1981. *Conservation and education.* Cambridge: Cambridge Univ. Press.

Franklin, J. F., and S. L. Brugman. 1979. Selection, management, and utilization of biosphere reserves. *Proceedings of the United States - Union of Soviet Socialist Republics Symposium on Biosphere Reserves, May 1976, Moscow USSR.* U.S. For. Serv. Gen. Tech. Rept. PNW-82. Washington, D.C.: U.S. Department of Agriculture.

Holdgate, M. W., and M. J. Woodman. 1978. *The breakdown and restoration of ecosystems.* New York: Plenum Publ. Corp.

Holling, C. S., ed. 1978. *Adaptive environmental assessment and management.* New York: Wiley & Sons.

Holt, S., J. and L. M. Talbot. 1978. New principles for the conservation of wild living resources. *Wildl. Monogr.* 59:1-38.

IUCN, no date. *Data atlas (preliminary): Planning a marine conservation strategy for the Caribbean region.* Gland, Switzerland: IUCN.

IUCN. 1980. *United Nations list of National Parks and Equivalent Reserves.* Gland, Switzerland: IUCN.

Lewis, J. K. 1969. Range ecosystems viewed in the ecosystem framework. In *The ecosystem concept in natural resource management*, ed. G. M. VanDyne, pp. 97-187. New York: Academic Press.

Lusigi, W.J. 1978. Planning human activities on protected natural ecosystems. *Dissertationes Botanicae* 48. Vaduz: J. Cramer.

Notes and Queries. 1951. *Notes and queries in anthropology.* 6th ed., rev. and rewritten. London: Routledge and Kegan Paul.

Ray, G. C., M. G. McCormick-Ray, J. A. Dobbin, C. N. Ehler, and D. J. Basta. 1980. *Eastern United States coastal and ocean zones: Data atlas.* Washington, D.C.: NOAA Office of Coastal Zone Management and the U.S. Council on Environmental Quality.

Soule, M. E., and B. A. Wilcox, eds. 1980. *Conservation biology: An evolutionary perspective.* Sunderland, Mass.: Sinauer Assoc., Inc.

Thomas, J. W. and D. E. Roweill, eds. 1982. *Elk of North America: Ecology and management.* Harrisburg, Penn.: Stackpole Book Co.

United States, Council on Environmental Quality. 1980. *The global 2000 report to the President.* 3 vols. Washington, D.C.: U. S. Government Printing Office.

van Dobben, W. H., and R. H. Lowe-McConnell, eds., 1975. *Unifying concepts in ecology.* Report of the Plenary Sessions of the First International Congress of Ecology, 8-14 Sept. 1974. The Hague, Netherlands: Junk.

Wildlife Review. Issued quarterly by the US Department of the Interior, Fish and Wildlife Service. Washington, D.C.: Government Printing Office.

INDEX

A

abiotic sector, ecosystem
    functioning and, 18
accuracy, statistical
    definition, 360
acrisols, 210
actinomycetes, 86
aerial photography, 470
    animal census, 468
    color infrared, 349
    habitat mapping, 463
    repetitive coverage, 351
    scales, 349
    small format, 351-352
    timing, 349
    Tri-Metrigon, 349
    vegetation analysis, 349
agricultural development, 201
    defined in ecosystem terms, 202
    ecological objectives, 202
    economic objectives, 201
    planning, 201
agricultural intensification, 205, 317
    colonization and, 205
agricultural productivity
    increases, 190
    prime land use, 191
agricultural systems, 193
    indigenous, 193
agriculture, 265
    fossil fuels and, 317
    industrialization of, 317
    intensive nonmechanized, 322
    soil suitability for, 265-267
agroecosystems, 198-199
    adaptation for human use, 202
    sustained yield, 199
alfisols, 224
algae, 144
analytical methods
    See also water analysis
    alkalinity, 121
    ammonia
        cadmium reduction, 121
        Kjeldahl, 120
        phenate method, 121
    chlorophyll, 129
    diazotization, 120
    for foods and food components, 373

andosols, 212
animal census
    See also census; visibility profile
    aerial photography for, 468
    ground observations, 469-471
animal domestication, 315
animal populations, 461, 468-479
    See also wildlife
    characteristics, 461
    food supply, 483
    habitat measurement, 461
    member counts
        replication, 461
        variables affecting, 461
    predation, 482
    self-limiting species, 482
    territorial, 480, 482
animal samples, preservation
    and sorting, 128
animal species, strategic assessment of, 433
animals
    See also animal populations; wildlife
    domestic, 412
    feral, 412
    sacred, 509
appropriate methodologies,
    basis for selection, 330
aquaculture, 485
aquatic ecosystems
    development project effects, 60
    eutrophication, 80
    geochemical cycles, 59
    human demands on, 59
    intervention in, 59
    monitoring, 485
    overview of, 8
    precipitation and, 78
    properties of
        chemical, 10
        definitions, 9
        functional, 10
        physical, 9
    surface water
        storage, 78
        stream flow, 78
arenosols, 212
aridosols, 224
assessment levels

intensive, 347
levels, 335
parameters for, 347
of plant resources, 334
assistance projects, information sources, 394-395
atmosphere, ecosystem relationship, 65
autotrophic ecosystem level, 17
autotrophic organisms, 312-313
  definition, 312

B

bacteria
  biomass analysis, 129
  in plankton, 133
bacterial denitrification, 120
bag limits, 509
basal area measurement, 358
baseline
  data collection, 478
  studies, 477
    soil, 189
Benchmark soil concept, 227
Benchmark Soils Project, 236
benthic biota, 88
benthic organisms analysis, 125
benthos, 132
  pollution damage, 132
  population characteristics
    gonad indices, 133
    growth, 133
    reproductive cycles, 133
biocides, 86
biogeochemical cycles
  carbon, 74
  hydrogen, 74
  macronutrients, 74
  micronutrients, 74
  nitrogen, 74
  oxygen, 74
biogeographic provinces, 418
biological activities, 147
biological degradation of soils, 272
biological processes in soil, 194
biological reserves, 468
biomass
  See also plant biomass
  estimation, 145
  habitat analysis, 465
biomes, definition, 419
biota, critical levels, 88
biotic climaxes, 325
biotic community
  human significance, 414
  introduced species, 467
biotic environment, in strategic assessment, 433
biotic provinces, 419
bog vegetation, 92
Bureau of Indian Affairs soil surveys, 284
Bureau of Land Management soil surveys, 284
Bureau of Reclamation
  evaluation system, 236
  irrigation suitability classification, 237

C

cambisols, 212
canopy (crown) cover, 358
carnivores. See secondary and tertiary consumers
carrying capacity, 45, 191, 199, 243, 326, 471, 482, 492
  demographic trends and, 45
  long-term production and, 327
    range ecologist, need for, 328
  methods to increase, 11
  monitoring, 327
  rainfall effect, 45
census, animal, 468, 472
  See also strip census
  methods, 472-476
  in aquatic ecosystems, 486
chemical composition as plant characteristic, 324
chemosynthetic primary production, 94
China, intensive nonmechanized agriculture in, 322
chlorophyll analysis, 129
Class I-VIII soils, 230-231
classification development
  See also soil classification; vegetation classification
  for crop suitability, 235
  for ecosystems, 335
climate
  as exploitable resource, 70
  in strategic assessment, 440
climax communities, 17, 325-326
  existing vegetation and, 326
coastal plain, in strategic assessment, 458
coenocline in sampling, 155

colonization, effect on agricultural intensification, 205
communal labor, examples, 21
communication technology for earth resource management, 34
communities, natural
See natural communities
community, remote settlements as part of, 24
community composition, determination by ordination techniques, 155
community sampling
 canonical correlation, 156
 clustering, 156
 discriminant function, 156
 principal components analysis, 156
 reciprocal averaging, 157
compaction of soils
See soil compaction
comparative analysis of homologous vegetation units, 363
competition
 effect on biotic community, 467
 in marine benthos, 149
composite mapping, 28
See also mapping
conductivity measurement, 115, 125
conservation, 214
 cost benefits, 453
 development and, 453, 458
 ethic, 509
 practices, benefits of, 4
 techniques, traditional, 508
consultants, 284, 333
coral reef surveys
 for pollution damage, 132
 of zooplankton, 135
coral reefs, 88, 133
country environmental profiles, 333
critical habitats
 analysis, 451
 development projects and, 5
 in strategic assessment, 448
critical maximum, biota, 88
critical minimum, biota, 88
critical population density, 244
crop production, 267
crop suitability, classifications for, 235
crop yield, 283
 erosion effect prediction model, 283
 factors influencing, 269
 field productivity experiments, 269
 technology effects on model predictions, 283
crops, high-yielding, 316
cross-sectional analysis of homologous vegetation units, 363
cultivation factors, 244
cultivation map, 32
cultural ecology study methods, 240
cultural activities, in strategic assessment, 433
cultural systems, environmental effects of, 19
cultural/societal factors in, ecosystem analysis, 335
current annual growth, definition, 359
current measuring techniques, 110
 Eulerian, 111
  current meters, 111
  electromagnetic meter, 112
  Savonius rotor, 112
  thermistor flow meter, 112
 Lagrangian, 110
  drift bottles, 110
  drogues, 110
  swallow floats, 111
 volume transport, 112
customary range usage, 244
cyanobacteria, 86

D

data collection
 ground, 355
 levels, 33
 quality, 361
 sources, 33
 unbiased, 360
decision-making
 parameters needed, 361
 sampling needed, 361
demersal plankton, 136
demographic trends,
 carrying capacity and, 45
density determination, plants, 367-368
density measurement, 115

development
    conservation and, 4, 454, 458
    division of labor and, 22
    effect on ecosystems, 16, 48
    sustainable ecosystems and, 454
development planning goals, 355
    species diversity and, 453
development problems, soils information and, 218
diazotization analysis, 120
disclimaxes, 325
diseases, 61-63, 88
    as limiting factor in species abundance, 467
dissolved gases (water), 84
    carbonate-bicarbonate equilibrium, 85, 118
    carbon dioxide, 85, 118
    hydrogen sulfide, 85
    oxygen, 84, 116
diversity, genetic, 15
    See also species diversity
division of labor, development and, 22
domain, definition, 19
domestic animals, definition, 412
domestication
    animal, 315
    plant, 313
double sampling, 362
    See also sampling
downstream systems, impact of development on, 332
drought assessment, 78

E

Earth Resources Observation System (EROS), 261
ECOCLASS system, 30
ecological classification, 335
ecological management, basic concepts, 311
ecological problem-solving, 467
ecological processes
    critical, 203
    essential, 414
ecological succession, 325
ecological value, species
    diversity, 453
    representativeness, 453
    uniqueness, 453
ecology, definition, 16
economic activities in strategic assessment, 444
economic objectives in agriculture, 201
ecosystem, 413
    See also ecosystems, aquatic
    change indicators, 481
    characteristics, 477
    classification in, 335
    components, 323
    cross-sectional analysis, 363
    decay, 467
    functional relations in, 324
    geographic boundaries, 324
    homologous areas, 329
    indices, 95, 151
    level of characterization, 322
    management, 414
    monitoring, 329
    stratification, 324
        of landscapes, 329
    use of term, 322
    vegetation boundaries, 324
ecosystem concept, use to characterize landscapes, 328
ecosystem processes in strategic assessment, 449

ecosystems
    energy transfer in, 17
    natural, 193
    strategic assessment, 449, 451
    sustainable, 415, 454
    technology in, 193
    terrestrial, abiotic components in, 323
ecosystems, aquatic
    biomass, 484
    characteristics, 484
    phytoplankton, 484
    productivity, 484
    turnover rates, 484
ecotones
    definition, 387
    intergrades, 388
edaphic climaxes, 325
endangered species, 455
    plants, 333
    sanctuaries for, 455
energy transfer, ecosystems and, 17
entisols, 224
environment
    conscious awareness of, 21
    institutions and, 16
environmental analysis, 333
environmental change

census-based measures, 385
indicators
  cash, 383
  demographic, 384
  nutritional, 384
  measurement of human response to, 382-385
environmental management, bad effects of, 309
epibenthic dredging, 128
epibenthos analysis, 125, 127
epifauna analysis, 125
erosion, 33, 282
  control, 272
  estimation model for southern Africa (SLEMSA), 277
  Landsat monitoring, 257
  productivity and, 282
    model (EPIC) for, 282
  sheet, 277
    estimation model for southern Africa (SLEMSA), 277
  sustained, effect on crop yields, 283
  water, 257, 275, 281
    equation for, 276
  wind, 257, 275, 278
    equation for, 278
  yield relation, 281
ethnobotanical surveys, 379
ethnobotany, 377
eutrophication, 95, 150, 151
evapotranspiration losses, 103
extensive management of resources, 319

F

FAO "integrated approach." See integrated surveys
FAO Agroecological Zones Project, 236
farming system information, 203
feral animals, 412
ferralsols, 210
  acric, 223
  humic, 223
  orthic, 222
  plinthic, 223
  rhodic, 223
  santhic, 223
field assistants, 24
field methods to analyze ecosystems, 16

field productivity experiments, 269
field records, species sampling, 139
field work, 23, 355
  participant observation in, 23
filariasis, 61
fire climax, 325
fish fauna sampling, 138
  bones, 140
  electric shock, 139
  otoliths, 140
  poisons, 139
  spines, 140
fisheries
  population status in, 485
  resources, 91
  traditional resource management, 495
  yield, 484
fisheries, inland
  practices, 142
  sampling, 141
fixed nitrogen analysis, 120
flood assessment, 78
floristics, 324, 345
fluvisols, 213
food
  as limiting factor in species abundance, 467
  preference studies, 466
  purchased, 22
  utilization, in habitat studies, 465
Food and Agriculture Organization (FAO), 30
  See also FAO
  soil surveys, 284
food webs, 95, 149
  diagrams, 95
  energy flow, 150
  trophic levels, 313
foods
  analytical methods for, 373
  neutral or neglected, 466
  preference values, 465
  principal, measurement for consumer species, 465
forage
  preference ratings, 466
  production, 327
fossil fuels, in agriculture, 317
fouling organisms, 132
fuelwood cutting, impact on animal food supply, 466
fungi, 144

G

game reserves, 509
game ranching and farming, 417
gases. See dissolved gases (water)
genetic diversity, 15
geochemical cycles, aquatic ecosystems and, 59
geographic information systems, 42
geological/edaphic characteristics in strategic assessment, 440
geosmin, 86
gleysols, 213
grazing, 32
green revolution, 316
  energy dependence, 317
ground cover
  percentage estimation, 358
  relative to single species, 358
ground data collection, 355
  methods, 356
groundwater development schemes, downstream impacts, 346
groundwater quality, 79
  See also water quality
group labor in agricultural cycle, 21
guinea worm disease, 63

H

habitat
  changes, local knowledge of, 477
  critical analysis of, 451
  definition, 413
  food habit attributes of, 466
  past conditions, methods for determining, 486-490
  unsuitable, 414
habitat fragmentation
  barriers, 467
  genetic swamping, 468
  hybridization, 468
habitat maintenance, 509
  by South American tribes, 503
habitat monitoring
  animal abundance, 471
  appropriate, 480-481
  carrying capacity and, 481
  limiting factors, 481
  quality evaluation, 481
habitats
  aerial photography, 463
  animal population measurement, 461
  aquatic, 484, 485
  baseline studies, 477
  component parts study
    animal, 462
    plants, 462
  food resources analysis, 464
  legal codes, 455
  mapping 463
  plant resources analysis, 464
  productivity, 477
  suitability analysis, 464
hard water components, 122
health problems
  See also diseases
  chlorinated hydrocarbons, 63
  fertilizer effects, 63
  water supply effects, 60
herbaceous vascular plants, 310
  See also plants
herbicides, 373
herbivores
  See also primary consumers
  in marine benthos, 149
herds, labor dependence, 22
heterotrophic ecosystem level, 17
heterotrophic organisms, definition, 312
hierarchical classifications for resources, 25
high-yielding crops, 316
histosols, 213, 224
holoplankton, 133, 135
homeostasis, ecosystems and, 16
homologous vegetation units, analysis of, 363-364
household, as unit of production, 21
human parasites, 88
human water use. See water, measures of human use
humans as foragers, 313
hunting, as factor in species abundance, 467
hydrogen sulfide analysis, 121
hydrologic cycle
  air chemistry, 160
  climate forecasting, 160
  ecosystem boundaries, 57
  ecosystem relationship, 65
  levels of analysis, 159
  mesocale meteorological modeling, 160

weather, 160
hypolimnion
    organic matter, 82
    overturn, 82

I

ichthyoplankton, 133
image elements. See remote
    sensing programs, image
    elements
immature communities, 17
inceptisols, 224
indicator species
    autotrophic, 93
    fish, 91
    heterotrophic, 93
    phytoplankton, 89
    zooplankton, 89
indicators, 140
indigenous knowledge, use in
    project planning, 310, 316
indigenous populations, 239, 245,
    316
industrial pollutants, in plants,
    analysis of, 373
indigenous systems, 239-246
industrialization of agriculture,
    317
infaunal analysis, 125, 127
informants
    stratified sampling, 23
    subsistance activities and, 15
information sources, 335
inorganic solids in water. See
    water analysis, inorganic
    solids; water analysis,
    inorganic particulates
integrated inventories
    examples, 25
    resource, 323
integrated surveys, 26-27
    FAO, 27
    OAS, 27
intensification, 218, 232
    agriculture, 205, 317
    land usage, 47
    management of resources and, 6
intensive management of
    resources, 320
intercropping, subsistence
    systems and, 15
International Institute for Land
    Reclamation and Improvement
    (ILRI), 30
International Soil Museum soil
    surveys, 284
intersystem dependencies, 332
intertidal elevation, 108, 109
intertidal zones, 92
interviewing, in field work, 23
inventories
    elements of, 25
    integrated, 25
irrigation, 158, 236, 279
    economic considerations, 236
    soil degradation and, 210
    water logging, 279
irrigation suitability
    classification, 209, 237

J

jojoba, as example of specific
    inventory item, 334

K

key industry species, in
    secondary productivity, 94
Kofyar land use. See land use,
    Kofyar

L

laboratory analysis of plant
    materials, 373, 374-375
land capability, 225, 228, 243
    evaluation, 228
    indigenous populations and,
        239
    soil classes (I-VIII) and,
        230-231
land classification approaches,
    225-226
    genetic, 226
    landscape, 226
    parametric, 226
land suitability, 225, 239
    classes, 233
    indigenous populations and,
        239
    levels, 233
    principles, 232
land use, 216, 241, 243
    appropriate, categories of,
        455
    intensification, 47
    Kofyar, 240
    types, 234
    urbanization, 318

land use data, aggregation problems, 27
landholding, by ethnic population, 245
Landsat, 255, 350
 area frame sampling, 244
 boundaries, 256
 false-color composite, 350
 microprocessors, computer-compatible data and, 350
 multispectral response lack, 350
 plants, level I and II inventories with, 350
 soil parent material and, 257
 vegetation categories and, 350
landscapes inventory, 328
Langmuir circulations, phytoplankton concentration effects and, 137
legal factors in strategic assessment, 444
levels of assessment. See assessment levels
life form. See vertical distribution of vegetation
life-zone system, 419
light, effect in aquatic ecosystems, 84
light measurement
 Secchi disk, 116
 submarine photometer, 116
littoral vegetation
 analysis, 142
 macrophyte, 91
  sampling, 143
 periphyton, 91
 scuba equipment use, 142
longitudinal analysis, homologous vegetation units, 364
lotic waters sampling, 126
Lugols iodine, for phytoplankton preservation, 137
luvisols, 211

M

macrophyte sampling, 143
macroplankton, 133, 135
malaria, 61
Man in the Biosphere Programme, 332
management
 of aquatic habitats, 485
 conservation, 454
 critical policies for, 493-494
 crop production and, 267
 development, 454
 of ecosystems, 311, 414, 451, 453
 environmental, 310
 extensive resource, 319
 georeferenced information systems and, 42
 index, 321
 intensity, effects of, 321
 intensive resource, 319
 natural science and, 511
 of plant resources, 319
 of poor land, 268
 of resources, 319
 of significant areas, 454
 of similar environments, 329
 of soil resources, 199
 for sustained resource yield, 326
 of watersheds, 215, 455
 of wildlife, 491
  traditional, 496-506
 of wildlife ranges, 429
map evaluation, example, 438
map units
 average, 336-337, 339, 342, 344
 minimum, 336, 337, 339, 342, 344
mapping, 28, 31
 in strategic assessment, 436
 scale choices, 436
mapping scales, 33, 335, 344, 436
map data sets
 aggregation and disaggregation, 44
 digitized, 43
 geometric corrections, 43
 regional base, 438
marine benthos
 herbivores in, 149
 study protocol, 149
marine organisms
 See also individual organisms
 circulation and, 80
 currents and, 80
 intertidal area, 80
 salinity and, 81
  density-dependent, 82
  effluent plume, 81
  estuarine, 81
 splash zone, 80
 water temperature effects on metabolic rates, 81

    on tolerance, 81
  zonation and, 81
marsh vegetation, 92
mature communities, 17
medicinal molecules,
  analysis of, 373
megaplankton, 133, 135
men, wage labor, 22
meromictic lakes, 84
meroplankton, 133, 135
mesozooplankton, 133, 134
meteorological modeling, 160
methods
  ECOCLASS, 30
  for environmental management, 310
  for plant cover estimation, 365
  for soil properties, 44, 45
    biotic, 45
    chemical, 45
    functional, 44
    physical, 45
    sampling, 45
microbial activity in aquatic habitats, 144
microbial communities, structure, 144
microbiology
  biochemical techniques, 162
  new methods in 162
microbiota, 93, 144, 145
  nitrogen fixing, 145
microplankton, 133
microzooplankton analysis, 135
  hemocytometer, 135
  Sedgwick-Rafter cell, 135
miracidium, 61
modernization
  commercialization and, 318
  impacts
    on plant resources, 317
    on population growth, 318
  traditional peoples and, 316
  transportation changes, 318
mollisols, 224
monoclimax hypothesis, 325
morphoedaphic index, 142
multistage sampling, 328

N

Nahuatl land use, 241
nanoplankton, 133, 137
natural area, 6, 310
  classification, 418

natural communities
  classification, 418
  major, listed, 418
natural resources, 6, 310
  See also resources
  United Nations monitoring projects for, 30
natural systems, in strategic assessment, 448
nature reserves, animal behavior and, 480
nitosols, 211
nitrogen analysis, 120-121
nonrandom sampling, 360
  See also sampling
nonvascular plants, 310
nutrient budgets, 94, 147
nutrient cycling, 94, 147, 197
nutrient quality analysis,
  plants, 373

O

ocular estimates. See plant cover estimation methods
omnivores. See secondary and tertiary consumers
onchocerciasis, 61
Operational Navigation Charts (ONC) for strategic assessment, 436
ordination techniques, polar, 155
organic material, analysis of
  detrital, 124
  nonvolatile filterable residues, 123
  particulates, 124
organic matter in water, 86-87, 123
  detritus, 87
  pH, 87
  plankton, 87
oxisols, 224

P

Palmer cells, 137
  species identification
    chlorophyll ratios, 137
    hemocytometer, 137
    membrane filters, 137
participant observation, 23
pedons. See soil profiles
Peru, ancient, intensive non-

mechanized agriculture in, 322
pH, 124, 266
　of soils, 266
phaeophytins, 137
phenology, 329
Philippines, intensive non-mechanized agriculture in, 322
phosphate, in water, 119
phosphorous budget, 119
photoperiodic response, 84
photosynthesis, as measure of primary production, 94
phototransects, 351
physical degradation of soils, 272
physical environment, in strategic assessment, 433
physiognomy, 336, 344
physiographic features, in strategic assessment, 440
phytoplankton, baseline study of, 137
planimeter, 127
plankton, 133, 135
　sampling, 134, 136
planosols, 211
plant adaptation, 309
plant biomass
　See also biomass
　production, 359
　valid data for, 358
plant communities, 324, 387, 462
　as "integrators," 462
　ecosystem boundaries and, 324
　levels, 326
　successional status, 345
plant cover estimation methods, 324
　See also plant cover measurement
　inaccuracies in, 365
　ocular estimates
　　plot shapes, 365
　　quadrat areas, 365
plant cover measurement
　line-intercept technique, 366
　percentage, 358
　point method, 367
　　potential errors in, 367
　transect lengths, 367
plant density measurements
　methods, 367-370
　sample size, 369
plant diversity
　sampling, 359
　species composition, 359
plant domestication, hominids and, 313
plant information
　goals, 343
　sources, 343
plant materials
　herbicides and pesticides, 373
　laboratory analysis of, 373, 374-375
　nutrients in forage, 373
　sample care, 375-376
　sampling program, 375, 376
　specific organic molecules, 373
plant monitoring, 310
plant productivity, 359
plant resource assessments
　cultural/societal perspectives, 346
　levels, 334
　statistical reliability, 346
plant resource management, 319
plant resource surveys
　intraregional, 338
　management unit, 340
　regional, 337
plant resources, 310
　adaptation, 315
　assessment levels, 334
　characteristics, 324
　classification, 343, 385-386
　importance to indigenous peoples, 380-382
　random sampling, 360
　traditional use, 377
　utilization, 315
　variability, 329
plants, 464
　analyses to monitor changes, 12
　classification of, 386
　habitat analysis, 464
　human needs and, 309
　indigenous value of, 379
　nutrient quality analysis, 373
　overview of, 11
　taxonomy, 345
　types, 310
　varieties, 315
podzols, 212
political factors, in strategic assessment, 444
polyclimax hypothesis, 325
poor land, management of, 268

population
    life tables, 479
    limiting factors, 482-483
    longevity, age determination,
        478-479
population density, high, in
    strategic assessment, 458
population growth, 190
    impact of modernization, 318
population, animal
    See also animal populations;
        productivity
    characteristics, 477
        fertility, 477-479
        other, 477-479
        sex-age structure, 477
    dynamics, 478
    indices, 468-469, 471-476
    structure, 478
    variations, 478
populations, local, subsis-
    tence activities and, 15
precision
    importance of, 32
    statistical definition, 360
predation, as limiting factor in
    species abundance, 466-467
predator-prey interactions
    key industry species, 149
    keystone predators, 149
    loop analysis, 149
preliminary national recon-
    naisance, 335
primary consumers, 312
primary production
    in aquatic ecosystems, 84
    measurement, 94
    oxygen method, 148
    radiocarbon method, 148
primary productivity, impor-
    tance of, 312
principal components analysis, 156
problem analysis, 330-331
productivity
    as goal of development
        programs, 345
    habitat monitoring, 481
    measurement, 351
    soil, 262, 267
    variability, 329
program design, local sub-
    sistence strategies
    and, 376
projects, small, synergistic
    interactions, 332
protozoa, 135, 137, 144

Q

quadrat sampling designs, 361,
    369
    costs, 362

R

rangeland productivity
    measurements, 370
reclamation needs assessment, 272
regression equations, use with
    double sampling methods,
    362
remote sensing, 34-35, 42, 160,
    357
    accuracy, 32
    computer-compatible tapes, 39
    false color, 35
    image quality, 38
    Landsat, 34, 41
    multispectral scanner, 37
    precision, 32
    radar systems, 37
    receiving stations, 38
    "scenes", 38
    side-looking airborne radar,
        (SLAR), 37
    thematic mapper (TM), 37
    thermal scanners, 37
remote-sensing programs
    image interpretation, 352-353
    multistage sampling, 354
        three-stage system 355
    for plant resource studies,
        348
    problems, 352
    for soils survey, 249,
        254-461
    terrain feature-vegetation
        correlates, 353
    weighted sample selection,
        31, 354
renewable resources, 4
resource classifications, 25
resource exploitation, by
    ethnic groups, 382
resource inventories, 63, 203
    approaches, 25
    incompatibilities in, 386
    integrated, 323
    soil, 189
resource management
    concern for traditional sys-
        tems, 496-499
    extensive vs. intensive tech-

niques, 319
resource sampling
    data categories for, 31
    weighted sample units, 31
resources, 17-20
    See also entries for individual resource categories
    decisions to change, 21
    estimation, 17
    homogeneous units, 31
    managed changes, 20
    resilience, 17
    sampling, 31
rice. See green revolution
rotifers, 137
rural poverty
    funding and, 5
    species extinctions and, 5

S

salinity, 113
    determination, 115
salinization, 279
    of soils, 275
sampling, 359
    bias, 361
    costs, 362
    double, 362
    errors, 31, 362
    interviews and, 23
    nonrandom, 360
    random, errors in, 31
    sites, 342, 360
    in surveys, 33
sampling designs, 23, 31-32, 227
    quadrat, 361
    vertical stratification, 361
    weighted units, 31, 354
saprophytes, as level of ecosystem, 17
satellite hydrology, 160
schistosomiasis, 61
seaweed, as wetlands vegetation, 92
secondary and tertiary consumers, 312
secondary productivity, 149
    keystone species, 94
    predators, 94
sedimentation
    studies, 151
    unanticipated, 332
selection categories for
        assessment, 335
seral stages, 325
settlements, 24
sheet erosion, 277
shelter, as limiting factor in species abundance, 467
site description, 218
site determination for productivity measurement, 351
site-specific investigations, 342
sleeping sickness, 61
social characteristics, in strategic assessment, 444
sodic soil formation, 279
soil, 22, 191
    microbial processes in, 195
    biological, chemical, and physical processes in, 194
    spectral differences in, 259
soil analysis, 273-275
    chemical, 273
    microbiological, 274
    mineralogical, 274
    physical, 273
soil classification
    field methods, 262
    genetic, 220
    indigenous, 239
        studies of, 242
    natural, 217
    suitability, 209
    taxonomic, 217
    technical, 217
    units, 249
soil classification systems
    Australian, 220
    Canadian, 221
    in England and Wales, 221
    French, 221
    Kubiena's (European), 221
    Russian, 220
soil compaction, 276, 280
soil conservation, 214
soil degradation, 209
    assessment, 275
    human causes, 275
    irrigation, 209
    wind and water erosion, 275
soil depth, 266
    relation to agriculture, 265
soil erosion, 275, 327
soil fertility, 213, 232
    assessment, 271
    fertilizer use, 213
    loss, 276, 279

soil horizons, 263
soil individual, 218
soil information
   for conservation, 204
   for intensification, 204
soil management, 199
   technological changes
      and, 190
soil map units, 249-251
   comparison with classi-
      fication units, 249
soil microflora, 310
soil pH, 266
soil productivity, 267
   assessment, 262
soil profiles, 218, 263
soil properties, 205, 216
   chemical, 217
   morphological, 217
   physical, 217, 220
soil salinity, 266
soil structure, 266
soil survey
   consociation, 248
   1st order, 247
   2nd order, 248
   3rd order, 248
   4th order, 248
   5th order, 249
   scales, 249
soil survey maps, 246
   legibility, 246
   minimum size delineation,
      247
soil surveys, 200, 203
   costs vs. benefits, 251
   information needed for
      planning, 201
   institutional sources, 284
   management of information, 200
   Resource Assessment Micro-
      processor (RAMP), 201
   types, 249
   world, 203
Soil Taxonomy, 210-213, 223
   indigenous, 240
   official U.S. system, 223
soil texture, 266
soils, 203
   complex, 248
   development planning, 189
   distribution, 215
   engineering suitability, 238
   microbial activity across,
      273-274
soils information, 202

baseline, 203
inventory, 202
Soils Management and Support
   Services (SMSS), 203
solonchaks, 213
species
   abundance, 467
   composition, 329, 358
   endangered. See endangered
      species
   indicators, 439
   sustainable utilization, 415
species diversity, 152, 453
   analysis methods, 152-153
   high, concern in development
      planning, 453
   indices, 140
   ordination techniques, 155
species-area relations, 332
spodosols, 224
statistical accuracy, 360
statistical precision, 360
statistical reliability
   of data, 335
   as sampling goal, 359
statistically valid estimate, 360
statistics, role in plant resource inventories, 359
steady-state communities, 17
stereoscopic photos, 351
stoniness, 267
strategic assessment
   biological needs analysis, 449
   of biotic environment, 444
   of coastal plains, 458
   compatible alternatives, 458
   cost effective, 431
   of critical habitats, 448
   of cultural activities, 444
   data for, 458
      availability, 440
      quality, 438
   data base development,
      431, 448
   data compendia, 439
   of economic activities, 444
   ecosystem support approach,
      451
   high-density population in,
      458
   indicator species, 439
   justification, 458
   natural systems, 458
   niche, 448
   overlay technique, 448, 451
   of physical environment, 440

regional base map preparation, 438
species study choice, 459
  voucher specimens, 459
subjects to be considered, 433
tasks, 431
values, 458
of wildlife types, 430
stratification
  for resource sampling, 23, 31
  map, 31
  measurement, 115
  of vegetation, 359
strip census, 469
  disappearing distances, 469
  visibility profile, 469
subsistence, 14
  in agricultural societies, 496
subtropics, agricultural cycle, 21
sulfate analysis, 121
sustainable development planning, 201
sustainable utilization of species, 415
sustainable yield, production efficiency and, 311
sustained resource yields, 326
swamp vegetation, 92
swidden agriculture, productivity and, 322
synecological investigation, 326
systems
  environments as, 19
  hierarchical, 24

T

technological change and soil management, 190
technology
  crop production and, 267
  environment and, 16
  positive effects on erosion problems, 283
temperature measurements, 115
  infrared scanner, 113
  sampling, 113
  satellites, 113
  thermistor, 113
terrestrial ecosystems, abiotic components, 323
tides
  classification, 108
  effect on marine organisms, 80
  entrainment and, 80
  gauges, 108
  measurements, 108
tolerance range, biota, 88
tools, importance, 22
topographic climaxes, 325
topography, relation to agriculture, 266
total phosphorus, 119
traditional conservation practices
  Africa, 500-502
  Asia, 502
  Latin America, 503
traditional systems, 495-496
  fishing, 495-496
  hunting, 495-496
  resource exploitation, 382
traditional wildlife practices
  Africa, 504-505
  Asia, 505, 507
  deterioration of, 506
  Latin America, 503
  Pacific island cultures, 500, 504, 505
transects, 127
  methods for plant cover estimation, 366
transitory communities. See seral stages
translation, in field work, 23
transparency, effect in aquatic ecosystems, 84
transportation, modernization and, 318
trophic state index, 95
trophic levels, illustrated, 313
tropical rain forests. See food preference studies, tropical rain forest
tropical soils, intensity of biological, chemical, and physical processes, 195
tropics, agricultural cycle in, 21

U

ultisols, 224
ultrananoplankton, 133
United Nations Environment Program (UNEP), 214
U.S. Department of Agriculture (USDA)
  International Division soil surveys, 284

Soil Conservation Service
(USDA/SCS) soil surveys,
284
U.S. Department of Interior soil
surveys, 284
U.S. National Ocean Survey, for
strategic assessment, 436
Universal Soil Loss Equation
(USLE), 276
  in Australia, 277
  in humid tropical regions,
277
  in West Africa, 277
Universal Transverse Mercator
Projection Maps, for strategic assessment, 436
urbanization, land use and, 318
usage gradients, 48
use intensity gradient, illustrated, 423

V

vegetation, 310, 327
  classification terminologies,
386
  horizontal distribution,
358-359
  littoral, 143
  in strategic assessment, 444
  vertical distribution, 359
  wetlands, 143
vegetation biomass, 370
  double sampling techniques,
372
  estimation
    via harvesting methods, 371
    via plot shapes, 371
  measurement errors, 371
  regression (indirect)
measurement, 372
  production parameters and
correlated plant characteristics, 372
vegetation classification
  agglomerative, 388
  categories, 350
  divisive, 388
  key considerations, 390-391
  sampling, 389
vegetation cover, 31
  map, 32
  sampling, 31
vegetation descriptions, 335,
344
vegetation record, permanent, 351

vegetation units, homologous,
363-364
vegetation-soil-terrain
relations, 324
vertisols, 211, 224
visibility profile, 470

W

wage labor
  men, 22
  women, 22
waste water treatment
  pathogenic organisms, 146
  self-purification, 146
  toxic chemicals, 146
water
  See also aquatic ecosystems
  demand, 158-159
  as limiting factor in species
abundance, 467
  measures of human use, 157
water analysis, 120-123
  alkalinity components, 118,
121
  dissolved gases, 84
    carbon dioxide, 85, 118
    oxygen, 84, 116
    polarographic oxygen, 117
    Winkler method, 117
  dissolved inorganic solids,
118
  dissolved organics, 86, 123
  herbicides, 123
  inorganic particulates, 123
  pesticides, 123
  sample contamination, 118
  sample storage, 119
water balance, 96, 157
  direct measurement, 157
  ecosystem relationship, 65
  meteorological formulas, 157
  moisture-water budget, 157
water density stratification
  epilimnion, 82
  hypolimnion, 82
  metalimnion, 82
  thermocline, 82
water logging of soils, 275
water losses, 103
water management, 103-104
water quality, 86-88
  bacterial products, 86
  biocides, 86
  conductivity, 88
  particulates, 86

potential free acidity, 87
water supply
  losses, 103
  model, 103
water supply development studies
  dynamic programming methods, 103
  flood studies frequency function, 103
  hydroelectric projects, 102
  linear programming methods, 103
  rainfall-runoff, 103
  simulation, 102
  unit-hydrograph, 103
waterlogging of irrigated land, 279
watershed development impact, 332, 335
watershed management, 215
watersheds
  alteration of basic characteristics, 346
  in plant resource systems, 346
wave action
  effect on marine organisms, 80
  erosion, 80
waves, 108, 109-110
  gauges, 109
  radar measurement, 109
  significant height, 109
weather modification, 160
wetlands vegetation
  artificial substrates, 144
  periphyton (Aufwuchs), 92
    sampling, 144
  protozoa, 92
  seaweeds, 91
wheat. See green revolution
wild species, traditional uses, 412
wildlife, 13, 411, 414
  See also animal populations
  abundance, 419
  baseline studies definition, 429
  comparisons
    by geographic unit, 419
    by taxonomic unit, 419
  conflict with other land uses, 428
  cropping, 430
  cultural values and, 412
  diversity loss, 419
  economic development and, 411-412
  as indicators, 13
  priorities, 415
  safeguards, 430
  species comparisons, 423
  strategic assessments, 431
  taxonomic units, 419
  traditional practices, 495, 500-506
  surveys, 428
  utilization, 417
wildlife conservation techniques
  in Maori New Zealand, 499
  in traditional Pacific island cultures, 499
wildlife management, 429, 491, 495-506
wildlife species
  ecological importance, 439
  economic value, 439
  rare, threatened, or endangered, 440
  social value, 440
wildlife status, past, methods for determining, 486-490
wildlife studies
  inventories, 429
  objectives, 415
    environmental impact statements, 418
    establishment of natural areas, 416
    pest control, 418
    sustained-yield utilization, 416-417
  strategies, 428
  use intensity gradient, 427
women
  as heads of households, 22
  reproductive histories, 24
woody vascular plants, 310
working ecosystem, 334
  level relationships, 334
World Aeronautical Charts (WAC), for strategic assessment, 436

X

xerosols, 212

Y

yellow fever, 61
yermosols, 213
yields, as management concept, 326

Z

zonation, 82, 88, 90
zooplankton, 133, 135
   entrainment effects, 136
   eutrophication, 89
   pollutant effects, 89
   power plant entrainment, 89
   sampling, 136
zootic climaxes, 325